Strongly Coupled Plasmas

NATO ADVANCED STUDY INSTITUTES SERIES

A series of edited volumes comprising multifaceted studies of contemporary scientific issues by some of the best scientific minds in the world, assembled in cooperation with NATO Scientific Affairs Division.

Series B: Physics

RECENT VOLUMES IN THIS SERIES

Volume 27 – Topics in Theoretical and Experimental Gravitation Physics
edited by V. De Sabbata and J. Weber

Volume 28 – Material Characterization Using Ion Beams
edited by J. P. Thomas and A. Cachard

Volume 29 – Electron–Phonon Interactions and Phase Transitions
edited by Tormod Riste

Volume 30 – Many Degrees of Freedom in Field Theory
edited by L. Streit

Volume 31 – Many Degrees of Freedom in Particle Theory
edited by H. Satz

Volume 32 – Electron and Ion Spectroscopy of Solids
edited by L. Fiermans, J. Vennik, and W. Dekeyser

Volume 33 – Microscopic Structure and Dynamics of Liquids
edited by J. Dupuy and A. J. Dianoux

Volume 34 – Path Integrals and Their Applications in Quantum, Statistical, and Solid State Physics
edited by G. Papadopoulos and J. T. Devreese

Volume 35 – Correlation Functions and Quasiparticle Interactions in Condensed Matter
edited by J. Woods Halley

Volume 36 – Strongly Coupled Plasmas
edited by Gabor Kalman

Volume 37 – Coherence in Spectroscopy and Modern Physics
edited by F. T. Arecchi, R. Bonifacio, and M. O. Scully

This series is published by an international board of publishers in conjunction with NATO Scientific Affairs Division

A Life Sciences
B Physics — Plenum Publishing Corporation, London and New York

C Mathematical and Physical Sciences — D. Reidel Publishing Company, Dordrecht and Boston

D Behavioral and Social Sciences — Sijthoff International Publishing Company, Leiden

E Applied Sciences — Noordhoff International Publishing, Leiden

Strongly Coupled Plasmas

Edited by

Gabor Kalman
Boston College
Chestnut Hill, Massachusetts

Assistant Editor

Paul Carini
Boston College
Chestnut Hill, Massachusetts

SPRINGER SCIENCE+BUSINESS MEDIA, LLC

Library of Congress Cataloging in Publication Data

Nato Advanced Study Institute on Strongly Coupled Plasmas, Université d'Orléans, 1977.

Strongly coupled plasmas.

(NATO advanced study institutes series: Series B, Physics; v. 36)

"Lectures presented at the NATO Advanced Study Institute on Strongly Coupled Plasmas, held in Orléans-la-Source, France, July 6–23, 1977."

Includes index.

1. Plasma (Ionized gases)–Congresses. 2. Plasma waves–Congresses. 3. Coulomb functions–Congresses. I. Kalman Gabor. II. Carini, P. III. Title. IV. Series.

QC717.6.N38 1977 530.4'4 78-15698

ISBN 978-1-4613-2870-4 ISBN 978-1-4613-2868-1 (eBook)

DOI 10.1007/978-1-4613-2868-1

Lectures presented at the NATO Advanced Study Institute on Strongly Coupled Plasmas held in Orléans-la-Source, France, July 6–23, 1977

Director: Gabor Kalman
Co-Director: Marc Feix

Originally published by Plenum Press, New York in 1978

LECTURERS

R. Balescu	Service de Chimie Physique II Université Libre de Bruxelles
M. Baus	Service de Chimie Physique II Université Libre de Bruxelles
Ph. Choquard	Laboratoire de Physique Théorique École Polytechnique Fédérale de Lausanne
C. Deutsch	Laboratoire de Physique des Plasmas Université Paris XI
H. E. DeWitt	University of California Lawrence Livermore Laboratory
M. Feix	CRPE/CNRS Université d'Orléans
K. I. Golden	Department of Electrical Engineering Northeastern University
R. Grandey	KMS Fusion, Inc.
E. P. Gross	Martin Fisher School of Physics Brandeis University
J. P. Hansen	Laboratoire de Physique Théorique des Liquides Université Paris VI
S. Ichimaru	Department of Physics University of Tokyo
G. Kalman	Department of Physics Boston College
E. Schatzman	Observatoire de Paris
K. S. Singwi	Department of Physics and Astronomy Northwestern University
A. Sjölander	Institute of Theoretical Physics Chalmers University of Technology

PREFACE

The Advanced Study Institute on Strongly Coupled Plasmas was held on the campus of the Université d'Orléans, Orléans-la-Source, France, from July 6th through July 23rd, 1977.

15 invited lecturers and 50 other participants attended the Institute.

The present Volume contains the texts of most of the lectures and of some of the numerous seminars presented at the Institute.

The topic of strongly coupled coulomb-systems has been an area of vigorous activities over the last few years. Such systems occur in a great variety of physical situations: stellar and planetary interiors, solid and liquid metals, semiconductors, laser compressed plasmas and gas discharges are some of the most important examples. All these systems have the common feature that for one or more of their constituent charged particle liquids the potential energy to kinetic energy ratio is not small, and therefore the application of the traditional plasma perturbation techniques is not feasible. Many ingenious theoretical schemes have been worked out in order to attack both the related equilibrium and nonequilibrium problems, and also various methods have been borrowed from areas where problems not dissimilar to the ones arising in coulomb-systems had already been tackled. At the same time, computer simulations have led to a probably unparalleled accumulation of data on the behavior of an ensemble of classical charged particles. For the first time, the Institute assembled workers from various disciplines who had been involved with diverse aspects of the strongly coupled plasma problem. The lectures and seminars presented in this Volume reflect the variety of approaches and points of view, ranging from formal statistical mechanics and kinetic theory to applied solid state and plasma physics.

The Institute was sponsored by the North Atlantic Treaty Organization, which provided the lion's share of the financial aid that made the Institute possible.

Additional co-sponsors were the Centre National de la Recherche Scientifique (France) and the Department of Physics, Boston College (U. S. A.) who helped us both by offering further financial assistance and by furnishing their clerical and technical services for the purpose of the organization and running of the Institute. The Université d'Orléans made its campus facilities available. The National Science Foundation (U. S. A.) helped participants with travel grants. Various organizations allowed and encouraged the lecturers to use research funds provided by them for the preparation of contributions to this Volume; amongst them, special thanks are due to the Air Force Office of Scientific Research on my own behalf and on behalf of other authors of this Volume.

The organizational tasks from the very inception of the Institute were shared by its Co-Director, Professor Marc Feix, who later on assumed the primary responsibility for the day-to-day functioning of the Institute. His contribution to bringing the Institute into existence was indispensable and invaluable.

Special thanks are due to many individuals whose assistance, cooperation and collaboration were essential at various stages of the organization of the Institute and the preparation of this Volume:

- to Dr. J. Hieblot of the C.N.R.S. for helping with all organizational matters, and for the support received from the C.N.R.S.;

- to Professor R. L. Carovillano of Boston College for making Boston College facilities available;

- to Mr. P. Carini of Boston College, Assistant Editor of this Volume, and Secretary of the Institute, for editing, proof-reading and correcting manuscripts and for performing the many arduous and thankless tasks that arose in the line of his responsibilities;

- to Dr. D. G. Samaras of the AFOSR for encouragement;

- to Mrs. Judy Bredin for the difficult and unending task of meticulously typing, re-typing and editing the manuscripts of this Volume;

- to Miss Diep Chau, to Miss Sharon Thompson, to Miss Joyce Vickery, and to Madame Dominique Lhuillier for clerical help;

- to Dr. E. Fijalkow for his help with all local arrangements;

- to my wife, Suzana Kalman, and to Mrs. Laura Kruskal for organizing social programs;

and to all the lecturers and participants for their contributions.

Gabor Kalman
Director of the Institute

CONTENTS

La faiblesse qui ramène à l'ordre vaut mieux que la force qui s'en éloigne.

JOUBERT

MICROSCOPIC KINETIC THEORY OF FLUIDS*

Eugene P. Gross

Martin Fisher School of Physics
Brandeis University
Waltham, Massachusetts 02154

*Work supported by the National Science Foundation.

TABLE OF CONTENTS

MICROSCOPIC KINETIC THEORY OF FLUIDS

Eugene P. Gross

Martin Fisher School of Physics
Brandeis University
Waltham, MA 02154

I. INTRODUCTION

We will be concerned with kinetic equations and time-dependent correlation functions for neutral fluids and one component plasmas. There has been a great deal of work in the subject in recent years. We confine the discussion to theories that use the BBGKY hierarchy in an essential way. Many, by no means trivial, restrictions will be made. First, we limit ourselves to classical dynamics. Second, we work in the linear response or small amplitude disturbance domain. Third, the equilibrium correlation functions are used as input, with no attempt made to compute them. Indeed, it is the conjecture and hope of this type of theory that for dense gases or strongly coupled plasmas the time dependent hierarchy can be truncated at a relatively early stage. It is the use of exact equilibrium distributions that is counted on to save us.

These lectures constitute a simplified account of our recent paper "Formal Structure of Kinetic Theory" [Gross, 1976]. That paper leans heavily on Boley's analysis [Boley, 1974, 1975] and extension of our earlier, more naive approach [Gross, 1972; Bergeron, Gross and Varley, 1974]. In the present version we shift emphasis to a modified cumulant approach, which clarifies the relation to older kinetic theories of gases and plasmas. We pay attention to the criteria for constructive approximations that lead to correct short time behavior (Chapter IV). Extensive use is made of the notion of a one body additive operator and of the associated dressed particle approximations [Gross, 1974]. We use the hard sphere pseudo-Liouville hierarchy to illuminate the structure of approximations to the smooth potential hierarchy.

The reader should consult the work of Resibois and Lebowitz [Resibois and Lebowitz, 1975; Resibois, 1975; Resibois, 1976] for the hard sphere case. It is very much along the lines of the present discussion, is more detailed and goes farther. We also recommend the study of the extensive work of Mazenko and his collaborators [Mazenko and Yip, 1977]. The general relationship to the present approach has been studied by Boley and by Lindenfeld [Lindenfeld, 1977]. However the connection at the higher approximation levels, where the "going gets rough" for all kinetic theories, is not clear.

It is appropriate to consider both the neutral fluid and plasma systems from a unified point of view. The strongly coupled plasma is characterized by major collisional contributions as well as by Debye screening and plasmon effects [Hansen, this volume; DeWitt, this volume]. It is more closely related to neutral liquids than is the weakly coupled plasma. We will not obtain a new theory of strongly coupled plasmas. Instead, using the modified cumulants, we will reorganize the BBGKY hierarchy, so that short time and distance features are automatically handled with any desired precision. The reorganization allows us to see where existing, semi-intuitive approximations fit in, and helps to see what must be done to validate or improve them.

II. LIOUVILLE'S EQUATIONS AND MICROSCOPIC PREPARATIONS

A. Microscopic Initial Conditions

We are interested in studying the solutions of Liouville's equation

$$\frac{\partial}{\partial t} F_N(\vec{p}_1 \,..\, \vec{q}_N;\, t) + L\, F_N = 0 \quad , \qquad \int F_N \, dp_1 \,..\, dq_N = 1 \tag{2.1}$$

where L is the Liouville operator for smooth two body interactions

$$L = \sum_{L=1}^{N} \left(\frac{\vec{p}_i}{m} \frac{\partial}{\partial \vec{q}_i} - \frac{\partial V}{\partial \vec{q}_i} \frac{\partial}{\partial \vec{p}_i} \right), \qquad V = \frac{1}{2} \sum_{i \neq j} v(|\vec{q}_i - \vec{q}_j|) \tag{2.2}$$

The standard initial value problem of mathematical physics prescribes F_N at t = 0 in the entire phase space (Γ space). We call this a microscopic preparation of the system. The point to be stressed is that many experimental situations involve just such a preparation. This point of view is in contrast to that of older approaches to kinetic theory. It has come to the fore in modern theories of linear response and light and neutron scattering from fluids and plasmas [Berne, 1971].

Denote the Gibbs equilibrium distribution for a classical system by

$$\Phi = \frac{e^{-\theta H}}{Z} \quad , \quad H = \sum_{i=1}^{N} \frac{p_i^2}{2m} + V \quad , \quad \theta = \frac{1}{k_B T}$$

$$Z = \int \Phi \, dp_1 \, .. \, dq_N \equiv \int \Phi \, d\Gamma \tag{2.3}$$

The Gibbs distribution is annulled by the Liouville operator, viz. $L\Phi = 0$, and is thus a constant of motion.

One type of preparation of the system is to imagine that the system was exposed in the past to a static external potential, i.e.

$$U = \sum_{i}^{N} U(\vec{q}_i)$$

for $t < 0$. At $t = 0$ the external potential is switched off. We are interested in the 'relaxation' of the system. Assuming that the system is in its most likely state, we take the initial state to be the Gibbs distribution in the presence of the potential.

$$F_N(t = 0) = e^{-\theta(H+U)}/Z_U = \Phi \, e^{-\theta \Sigma U(\vec{q}_i)}/Z_U \quad .$$

$$Z_U = \int \Phi \, e^{-\theta \Sigma U(\vec{q}_i)} \, d\Gamma \tag{2.4}$$

Linear response theory studies deviations from Φ to the first power of the magnitude of U, but allows arbitrarily rapid and temporal variations. In addition, since U can be quite irregular, disturbances can be everywhere in space.

To first order in U we have

$$\mathcal{F}_N(t = 0) = \Phi \, [1 + F_N(t = 0)]$$

$$F_N(t = 0) = -\theta \left\{ \sum_{i=1}^{N} U(\vec{q}_i) - \langle \Sigma U(\vec{q}_i) \rangle \right\} \quad . \tag{2.5}$$

where $\langle A \rangle$ denotes the thermodynamic average $\int \Phi \, A \, d\Gamma$.

The initial deviation is thus one body additive. We can imagine a more general, fictitious preparation where the Hamiltonian for $t < 0$ was also momentum dependent, viz.

$$H_U = H + \sum_{i=1}^{N} U(\vec{p}_i, \vec{q}_i) \quad .$$

This leads to the most general one body additive preparation in phase space as the linear response initial condition.

For free particles even if the initial preparation is one body additive in only the coordinates, it becomes one body additive in momentum as well in the course of time.

At this point we establish the correspondence between the solutions of the Liouville equation and the time correlation function used in the theory of neutron and light scattering. This is similar to the relation between the Heisenberg and Schrödinger pictures in quantum mechanics. Suppose we are asked to compute a time correlation function for a system in equilibrium. Define

$$\langle A(0)\, B(t)\rangle \equiv \int \Phi\, A(0)\, B(t)\, d\Gamma \tag{2.6}$$

Φ and $d\Gamma$ can be expressed in terms of phase coordinates at $t = 0$. All quantities can be equally well expressed in terms of phase coordinates at time t.

$$\langle A(0)\, B(t)\rangle = \int \Phi\, A(0)\, e^{Lt}\, B(0)\, d\Gamma_0 = \int \Phi\, A\, e^{Lt}\, B\, d\Gamma \tag{2.7}$$

On the other hand one can look at these time correlation functions from the point of view of solutions of the Liouville equation (Schrödinger picture). With $\mathcal{F}_N = \Phi(1 + F_N)$ the formal solution is

$$F_N(t) = e^{-Lt}\, F_N(t = 0) \tag{2.8}$$

The expectation value of B is given as

$$\tilde{B}(t) \equiv \int \Phi\, B\, e^{-Lt}\, F_N(t = 0)\, d\Gamma \tag{2.9}$$

Since $L\Phi = 0$, integration by parts yields

$$\tilde{B}(t) = \int \Phi\, F_N(t = 0)\, e^{Lt}\, B(0)\, d\Gamma \tag{2.10}$$

Thus if we take $F_N(t = 0) = A(0)$ we obtain $\langle A(0)\, B(t)\rangle$.

In sum, the equilibrium time dependent correlation functions may be computed by solving the Liouville equation with a suitable microscopic initial condition involving (usually) one body or two body additive functions. Then one takes an inner product with a function of the same type. The inner product has the Gibbs Φ as a weight function and involves integration over phase space.

Let us discuss some of the implications of these elementary considerations for the reduced distributions of hierarchy theory (singlet, doublet, etc.). We are interested in distribution functions that are symmetrical with respect to particle permutation.

Introduce the singular phase space function (distribution)

$$N(\vec{p}\vec{x}) = \sum_{\alpha=1}^{N} \delta(\vec{p}_\alpha - \vec{p})\ \delta(\vec{q}_\alpha - \vec{x}) \tag{2.11}$$

This is a one body additive quantity, labelled parametrically by $\vec{p}$ and $\vec{x}$. This is at some fixed time, and we will mainly use the Schrodinger picture. The equilibrium singlet distribution is defined as

$$\langle N(\vec{p}\ \vec{x})\rangle \equiv \int \Phi\ d\Gamma N(\vec{p}\vec{x}) = \rho_0\ \phi(p) \tag{2.12}$$

The doublet distribution is obtained by removing the self-interaction, i.e.

$$N(\vec{p}\vec{p}^1,\ \vec{x}\vec{x}^1) \equiv N(\vec{p}\vec{x})\ N(\vec{p}^1\vec{x}^1) - \delta(\vec{p} - \vec{p}^1)\ \delta(\vec{x} - \vec{x}^1)\ N(\vec{p}\vec{x}) \tag{2.13}$$

Then

$$\langle N(\vec{p}\vec{p}^1,\ \vec{x}\vec{x}^1)\rangle = \phi(p)\ \phi(p^1)\ \rho_2(\vec{x},\ \vec{x}^1) \tag{2.14}$$

where $\rho_2(\vec{x},\ \vec{x}^1)$ is the equilibrium pair distribution function.

Usually we will use the slightly ambiguous notation, $\vec{p}_\alpha$ for a phase momentum, $\vec{p}_1$ for a parameter.

$$N(12) = N(1)\ N(2) - \delta(1 - 2)\ N(1) \tag{2.15}$$

i.e. $\vec{p}$, $\vec{p}^1$ are replaced by $\vec{p}_1$, $\vec{p}_2$ and $\vec{x}$, $\vec{x}^1$ by $\vec{x}_1\vec{x}_2$.

The same quantities are used to define time dependent distribution functions in a Schrodinger picture. Thus

$$f_1(1;\ t) = \int N(1)\ F_N(t)\ d\Gamma$$

$$f_2(12;\ t) \equiv \int N(12)\ F_N(t)\ d\Gamma \quad . \tag{2.16}$$

The deviations from equilibrium may be written as

$$\tilde{f}_1(1;\ t) = \langle N(1)\ F_N(t)\rangle$$

$$\tilde{f}_2(12;\ t) = \langle N_2(12)\ F_N(t)\rangle \tag{2.17}$$

Suppose now that $F_N(t = 0)$ is one body additive.

$$F_N(t = 0) = \sum_{\alpha=1}^{N} \psi(\vec{p}_\alpha \, \vec{q}_\alpha) - \langle \sum^{N} \psi(\vec{p}_\alpha, \vec{q}_\alpha) \rangle$$

$$= \delta N(\overline{1}) \; \psi(\overline{1}) \quad , \quad \delta N(1) = N(1) - \langle N(1) \rangle \tag{2.18}$$

where the bar indicates the integration $dp_1 \; dx_1$.

The microscopic preparation at the level of F_N fixes all of the reduced distributions. We have

$$\tilde{f}_1(1; t = 0) = \rho_0 \; \phi(p_1) + \left\{ \rho_2(\vec{x}_1 \; \overline{\vec{x}_2}) - \rho_o^{\;2} \right\}$$

$$\cdot \; \phi(p_1) \; \phi(\overline{p}_2) \; \psi(\overline{\vec{p}_2} \; \overline{\vec{x}_2}) \tag{2.19}$$

We note that this already involves the equilibrium pair distribution. If one wants to use the usual self-contained singlet kinetic equations (Boltzmann, Vlasov, etc.), to compute time correlation functions, one must employ initial conditions of the preceding type. The doublet distribution satisfies

$$\tilde{f}_2(12; t = 0) = \left\{ \rho_3(\vec{x}_1 \vec{x}_2 \overline{\vec{x}_3}) - \rho_0 \rho_2(\vec{x}_1 \vec{x}_2) \right\} \phi(p_1) \; \phi(p_2) \; \phi(\overline{p}_3) \; \psi(\overline{3})$$

$$+ \; \rho_2(\vec{x}_1 \vec{x}_2) \; \phi(p_1) \; \phi(p_2) \left\{ \psi(1) + \psi(2) \right\} \tag{2.20}$$

and involves the triplet equilibrium distribution. This would be used as an initial condition in a theory which is self-contained at the level of the second BBGKY equation. The standard second cumulant, defined by

$$c_2(12; t) = f_2(12; t) - f_1(1; t) \; f_1(2; t) \tag{2.21}$$

leads to a linearized cumulant

$$\tilde{c}_2(12; t) = \tilde{f}_2(12; t) - \rho_0 \; \phi(p_1) \; \tilde{f}_1(2; t) - \rho_0 \; \phi(p_2) \; \tilde{f}_2(1; t) \tag{2.22}$$

For an interacting system $\tilde{c}_2(12; t = 0)$ is not zero. This fact is important for truncation schemes. The schemes fix a relation between a higher order and lower order distribution functions. This relation is maintained as an approximation at all times (in violation of the exact situation), including $t = 0$. In a systematic microscopic theory this defect of conventional truncation schemes is overcome.

Another way of studying linear response theory is to have $F_N = \Phi$ for $t < 0$, and to add the impulse function $\rho(k) \; \delta(t)$ to the

Hamiltonian [Martin, 1968]. Here

$$\rho(k) = \sum_i e^{i\vec{k}\cdot\vec{x}_i}$$

is the Fourier component of the density. The linearized Liouville equation is

$$\left(\frac{\partial}{\partial t} + L\right) F_N = F_N(0)\ \delta(t)$$

$$F_N(0) = -i\vec{k} \cdot \vec{j}(\vec{k})\ \theta$$

$$m\vec{j}(k) = \Sigma \vec{p}_i\ e^{i\vec{k}\cdot\vec{x}_i} \tag{2.23}$$

The initial condition is proportional to the one body additive longitudinal current. The quantity $\int \rho^*(k)\ \Phi\ F_N(t)\ d\Gamma$ is the density susceptibility.

B. Short Time Behavior of $F_N(t)$ and Restricted Function Spaces

For smooth potentials the short time behavior may be studied by solving the Liouville equation as a power series in time.

$$F_N(t) = F_N(0) - t\ L\ F_N(0) + \frac{t^2}{2!} L^2\ F_N(0) + \ldots \tag{2.24}$$

If the initial condition $F_N(0)$ is completely general, one can say nothing at all. Since we are dealing with a first order equation in time the Liouville operator and thus just the first term in t completely characterizes the system. There is no possibility of finding a 'model' Liouville operator to compute the general N point time dependent correlation function.

Going to the other extreme of very special initial distributions, there are drastic simplifications. There are many 'model' operators which can replace the Liouville operator with some defined accuracy if one is intersted only in computing particular inner products. Suppose

$$F_N(0) = \sum_{i=1}^{N} e^{i\vec{k}\cdot\vec{q}_i}$$

and to order t^2

$$F_N(t) \simeq \int N(1) \left\{ 1 - it\frac{\vec{k}\ \vec{p}_1}{m} - \frac{t^2}{2!}\left(\frac{\vec{k}\ \vec{p}_1}{m}\right)^2 \right\} e^{i\vec{k}\cdot\vec{x}_1}\ dp_1\ dx_1$$

$$- \frac{t^2}{2!}\frac{i\vec{k}}{m}\ N(\overline{12}) \left| \frac{\partial V(\overline{12})}{\partial \vec{x}_1}\ e^{ik\overline{x}_1} + \frac{\partial V(\overline{12})}{\partial \vec{x}_2}\ e^{i\vec{k}\cdot\vec{x}_2} \right| \tag{2.25}$$

It is clear that the interactions drive the initial distribution out of the one body additive space in the course of time. As one proceeds further new spaces are brought in, but the terms in the lower order spaces are also modified. One is led to ask whether one can develop approximation schemes that are based on arbitrarily constraining the functional form of the N body Liouville distribution for all times.

Let the approximate F_N be constrained to be a sum of a finite number of functions

$$F_N = \sum_{n=1}^{M} a_n(t)\, \psi_n(\vec{p}_1 \ldots \vec{q}_N) \qquad (2.26)$$

The ψ_n are orthonormal with respect to the Gibbs Φ as weight factor. One finds a set of linear first order equations for the $a_n(t)$. For example, for a given wave vector $\vec{k}$ one can choose the five hydrodynamic functions which are Fourier components of the density, current and total energy density. One can then fit the microscopic initial conditions for the computation density autocorrelation function. This leads to the Euler perfect fluid equations with the exact sound speed for the interacting system.

Nossal and Zwanzig [Nossal and Zwanzig, 1967; Nossal, 1968] extended this set to include momentum and energy fluxes, in a search for high frequency collective modes in liquids. Whenever there is only a finite number of functions there is a discrete set of purely imaginary eigenvalues for each wave vector $\vec{k}$.

The discrete set of eigenvalues becomes a quasi continuum when one includes the continuously indexed (by $\vec{p}$ and $\vec{k}$) of one body additive functions. This leads to the modified Vlasov equation with the direct correlation function replacing the bare potential. This way of obtaining the equation was introduced by Zwanzig [1968; 1967] and by Akcasu and Duderstadt [1969; 1970].

One can also augment the function set by modifying the one body additive kinetic energy density to a function that includes the potential energy density. This has been studied by Bergeron [Bergeron and Gross, 1975], Gross and Lindenfeld [Gross and Lindenfeld, to be published; Lindenfeld, 1975]. It is also at the root of the successful semi-phenomenological theory of Forster and Jhon [1975].

One also obtains a quasicontinuum by including mode-mode coupling functions of the type $\psi_\alpha(\vec{k} - \vec{k}_1)\, \psi_\beta(\vec{k}_1)$. Here the $\psi_\alpha(\vec{k})$ are hydrodynamic functions and the system is continuously indexed by $\vec{k}$ and $\vec{k}_1$ [Keyes and Oppenheim, 1973; 1973; Gotze and Lucke, 1975].

In the present lectures we proceed along the route of generalizing the space of one body additive functions by defining larger and larger orthogonal spaces using n body additive functions. This turns out to be close to the standard older treatments of the BBGKY hierarchy, except for improvements at short times and distances.

With any chosen set of functions, one can formally close the theory with the aid of a memory matrix. Following Zwanzig and Mori, one divides the function space into the given part and an orthogonal part. If the orthogonal part has a vanishing initial condition, it can be formally eliminated to yield a memory matrix. A self-contained, but non-Markoffian description is obtained for the chosen set. When the set is the one body additive functions, this becomes a theory of the operator that is a rigorous replacement for the Boltzmann operator for the singlet distribution.

C. Stationary Variational Principles

A natural and flexible way to implement the program of constraining the functional form of $F_N(t)$ is in terms of a stationary variational principle [Gross, 1973; Hopps, 1971; 1976; Phythian, 1972; Mostellor and Duderstadt, 1974]. Let us work with the Laplace transform

$$\tilde{F}(S) = \int_0^\infty e^{-St} F_N(t)\, dt$$

$$F_N(t) = \frac{1}{2\pi i} \int_{c-i\infty}^{c+i\infty} \tilde{F}(S)\, e^{St}\, dS \tag{2.27}$$

Liouville's equation becomes

$$(S + L)\, \tilde{F}(S) = F_0 \equiv F_N(t = 0) \tag{2.28}$$

Suppose we want to compute the time correlation function $\int \Phi\, d\Gamma\, G_0\, \tilde{F}(S)$. We treat the Liouville equation as a constraint with a Lagrange multiplier function $\tilde{G}(S)$. Thus, we consider the functional

$$J = \int \Phi\, d\Gamma\, G_0\, \tilde{F}(S) - \int \Phi\, d\Gamma\, \tilde{G}\, \{(S + L)\, \tilde{F} - F_0\} \tag{2.29}$$

Variation with respect to $\tilde{G}(S)$ yields the Liouville equation, while the independent variation with respect to $\tilde{F}$ yields

$$(S - L)\, \tilde{G}(S) = G_0 \tag{2.30}$$

The stationary value of J is

$$|J| = \int \Phi\, d\Gamma\, G_0\, \tilde{F}(S) \tag{2.31}$$

This quantity has only a second order error for small deviations from the exact solution. The trial functions $\tilde{F}$ and $\tilde{G}$ are to be chosen to satisfy the initial conditions. (It is possible to find a new form for J to bypass this necessity). In the systematic approximation schemes to be discussed, this means that we are to solve the problem twice, with different initial conditions for $\tilde{F}$ and for the adjoint $\tilde{G}$. This increases the accuracy of the estimate of the correlation function that is desired.

There is a Schwinger (norm independent) form of the variation principle that is particularly powerful. It uses the functional

$$J_S = \langle \tilde{G} F_0 \rangle \langle \tilde{F} G_0 \rangle / \langle \tilde{G}(S + L) \tilde{F} \rangle \tag{2.32}$$

For example, with $F_0 = \rho(k)$, $G_0 = \rho(-k)$, is treated as a real variable); we take the simple non-interacting trial functions

$$\tilde{F} = (S + L_0)^{-1} F_0 \quad , \quad \tilde{G} = (S - L_0)^{-1} G_0 \quad .$$

This leads to the same density autocorrelation function as is obtained with the full one body additive trial function in the first variational principle. Very little use has been made of these properties of variational principles.

III. CONVENTIONAL APPROACHES TO THE BBGKY HIERARCHY

A. Standard Cumulant Theory

In this section we emphasize some features of the conventional derivations of kinetic equations for systems with smooth potentials and for hard spheres. One aim is to pinpoint those features that must be present in a good microscopic theory. A second aim is to exhibit as close a connection as is possible between modern and older theories [Wu, 1966; Klimontovich, 1967].

For both the smooth potential and hard sphere systems, after writing the appropriate hierarchy of equations, one introduces cluster functions or cumulants by the definitions

$$f_2(12) = c_2(12) + f_1(1)\, f_1(2)$$

$$\begin{aligned} f_3(123) = c_3(123) &+ c_2(12)\, f_1(3) + c_2(23)\, f_1(1) \\ &+ c_2(31)\, f_1(2) + f_1(1)\, f_1(2)\, f_1(3) \ \ldots \end{aligned} \tag{3.1}$$

These definitions hold at each time t. In equilibrium

$$\langle f_1(1)\rangle = \rho_0\ \phi(p_1) \quad , \quad \langle f_2(12)\rangle = \phi(p_1)\ \phi(p_2)\ \rho_2(\vec{x}_1\vec{x}_2)$$

$$\langle f_3(123)\rangle = \phi(p_1)\ \phi(p_2)\ \phi(p_3)\ \rho_3(\vec{x}_1\vec{x}_2\vec{x}_3) \tag{3.2}$$

with Maxwellian and static distribution functions. The corresponding equilibrium cumulants are denoted by $c_{eq}(\vec{x}_1\vec{x}_2)$, $c_{eq}(\vec{x}_1\vec{x}_2\vec{x}_3)$, etc. There is then a straightforward pattern of analysis which we will sketch for the case of the one component plasma (OCP).

For the one component plasma, with a compensating background, it is convenient to scale with lengths measured in terms of the Debye length $\lambda_D = (k_BT/4\pi\ \rho_0\ e^2)^{1/2}$, times measured in terms of the inverse plasma frequency ω_p and velocities measured in terms of $(k_BT_0/m)^{1/2}$. To have a close correspondence with the usually employed reduced distributions, we define

$$F_s = \eta_0^{-s} \int N(1,2,\ldots s)\ \mathcal{F}_N\ d\Gamma \tag{3.3}$$

Here η_0 is the density in Debye units ($\eta_0 = \rho_0\ \lambda_D^3$). The Gibbs distribution is now

$$\Phi = \prod_{\alpha=1}^{N} \phi(p_\alpha)\ e^{-V}$$

$$\phi(p) = e^{-p^2/2}(2\pi)^{-3/2} \quad , \quad V = \frac{\gamma}{2} \sum_{i\neq j} \psi(|\vec{x}_i - \vec{x}_j|)$$

$$\psi(x) = 1/|\vec{x}| \quad . \tag{3.4}$$

Here

$$\gamma = \frac{e^2}{kT} / \lambda_D \quad , \quad \gamma\eta_0 = 1/4\pi \tag{3.5}$$

γ is the ratio of the Landau close collision length to the Debye length. The hierarchy, with the compensating background takes the form [Fisher, 1964]

$$\left(\frac{\partial}{\partial t} + \vec{p}_1 \frac{\partial}{\partial \vec{x}_1}\right) F_1(1) = \frac{1}{4\pi} \frac{\overline{\partial\psi(\vec{x}_1 - \vec{x}_2)}}{\partial \vec{x}_1} \frac{\partial}{\partial \vec{p}_1} \{F_2(1\bar{2}) - F_1(1)\}$$

$$\left\{\frac{\partial}{\partial t} + S_0(12)\right\} F_2(12) = \frac{1}{4\pi} \left[\frac{\overline{\partial\psi(\vec{x}_1 - \vec{x}_3)}}{\partial \vec{x}_1} \frac{\partial}{\partial \vec{p}_1} + 1 \rightleftarrows 2)\right]$$

$$\cdot\ [F_3(12\bar{3}) - F_2(12)]$$

$$S_0(12) = \vec{p}_1 \frac{\partial}{\partial \vec{x}_1} + \vec{p}_2 \frac{\partial}{\partial \vec{x}_2} - \frac{\partial \psi(\vec{x}_1 - \vec{x}_2)}{\partial \vec{x}_1} \left(\frac{\partial}{\partial \vec{p}_1} - \frac{\partial}{\partial \vec{p}_2} \right) \tag{3.6}$$

The equilibrium plasma hierarchy has been the subject of much study. To first order in γ, except at small distances, one finds the Debye-Huckel pair distribution. For small γ, with the neglect of the triplet cumulant, there is an elegant discussion by Frieman and Book [1963]. A study of the coupled equations for the doublet and triplet cumulants has been made by O'Neil and Rostoker [1965]. There is a large literature on obtaining the static correlation functions when γ is not small. This is one of the concerns of the present Volume. The use of truncations of the standard cumulants does not appear to be a productive technique.

Let us now examine the linearized time dependent hierarchy. The linearized cumulants are defined by

$$\tilde{F}(1) \equiv \tilde{c}(1) \quad , \quad \tilde{F}_2(12) = \tilde{c}_2(12) + \phi_1 \, \tilde{c}(2) + \phi_2 \, \tilde{c}(1)$$

$$\tilde{F}_3(123) = \tilde{c}_3(123) + [\phi_1 \, \tilde{c}_2(23) + \tilde{c}(1) \, \phi_2 \phi_3 \, \eta(\vec{x}_2 \vec{x}_3)$$
$$+ \text{ cyclic perm}]$$

$$\eta(\vec{x}_1 \vec{x}_2) \equiv 1 + c_{eq}(\vec{x}_1 \vec{x}_2) \tag{3.8}$$

The hierarchy is

$$\left(\frac{\partial}{\partial t} + \vec{p}_1 \frac{\partial}{\partial \vec{x}_1} \right) \tilde{c}(1) - V(1)\, \tilde{c}(1) = \frac{1}{4\pi} \frac{\partial \psi(\vec{x}_1 - \vec{\bar{x}}_2)}{\partial \vec{x}_1} \frac{\partial \tilde{c}_2(1\bar{2})}{\partial \vec{p}_1}$$

$$\left\{ \frac{\partial}{\partial t} - V(1) - V(2) + S_0(12) \right\} \tilde{c}_2(12) = H(12) \tag{3.9}$$

Here $V(1)$ is the Vlasov operator, defined by

$$V(1)\, \tilde{c}(1) \equiv \frac{1}{4\pi} \frac{\partial \psi(\vec{x}_1 - \vec{\bar{x}}_3)}{\partial \vec{x}_1} \frac{\partial \phi_1}{\partial \vec{p}_1} \tilde{c}(\bar{3}) \tag{3.10}$$

The right hand side of the doublet equation is

$$H_2(12) = \{1 + P(12)\} \left\{ \gamma \frac{\partial \psi(\vec{x}_1 \vec{x}_2)}{\partial \vec{x}_1} \left[\phi_2 \frac{\partial}{\partial \vec{p}_1} - \frac{\partial \phi_2}{\partial \vec{p}_2} \right] \tilde{c}(1) \right\}$$

$$+ \{1 + P(12)\} \left\{ \frac{1}{4\pi} \frac{\partial \psi(\vec{x}_1 - \vec{\bar{x}}_3)}{\partial \vec{x}_1} \frac{\partial}{\partial \vec{p}_1} \right.$$

$$\cdot\ [c_{eq}(\vec{x}_2\vec{x}_3)\ \phi_2\ \tilde{c}(1) + \tilde{c}_3(12\bar{3})]$$

$$+\ c_{eq}(\vec{x}_1\vec{x}_2)\ \phi_2 V(1)\ \tilde{c}(1)\Big\} \qquad (3.11)$$

There are several noteworthy points. All terms on the right hand side are of order of γ or higher. When one eliminates $\tilde{c}_2(12)$ the right hand side of the singlet is of order γ. The process of introducing cumulants rearranges the hierarchy so as to bring out medium terms, with each member of a pair acted upon independently by a Vlasov operator. The second member is unaffected In a γ expansion, the plasma scaling suggests that we give priority to the medium terms [Rosenbluth and Rostoker, 1960]. The direct binary interaction $S_0(12)$ is unaltered, an oversimplification arising from the $\tilde{c}_3 \approx 0$.

We now outline the standard solution when the triplet cumulant and the direct interaction are neglected in the equation for the second cumulant. One introduces spatial Fourier transforms by

$$\tilde{\psi}(\vec{k}) = \int e^{i\vec{k}\cdot\vec{x}}\,\psi(\vec{x})\ dx \qquad (3.12)$$

The Green's function for the Vlasov equation obeys (for $t \geq 0$)

$$\left(\frac{\partial}{\partial t} - i\vec{k}_1\cdot\vec{p}_1\right) < \vec{p}_1|G(\vec{k}_1;\ t)|\vec{p}_4) > \frac{i\vec{k}_1}{4\pi}\,\tilde{\psi}(\vec{k}_1)\,\frac{\partial\phi(p_1)}{\partial\vec{p}_1}$$

$$\cdot \int < \vec{p}_3|G(\vec{k}_1;\ t)|\vec{p}_4 > dp_3 = \delta(\vec{p}_4 - \vec{p}_1)\ \delta(t) \qquad (3.13)$$

With the approximations indicated, the doublet equation involves the sum of Vlasov operators acting independently on the members of the pair. Thus the solution of the doublet equation with $\tilde{c}_2(\vec{p}_1\vec{k}_1\vec{p}_2\vec{k}_2,\ t = 0) = 0$ is [Dupree, 1961]

$$\tilde{c}_2(\vec{p}_1\vec{k}_1\vec{p}_2\vec{k}_2;\ t) = \int_0^t <\vec{p}_1|G(\vec{k}_1;\ t - t')|\bar{\vec{p}}_3><\vec{p}_2|G(\vec{k}_1;\ t - t')\ \bar{\vec{p}}_4>$$

$$\cdot\ \tilde{H}_2(\bar{\vec{p}}_3\vec{k}_1\bar{\vec{p}}_4\vec{k}_2;\ t')\ dt' \qquad (3.14)$$

One must add a solution of the homogeneous equation to find a solution satisfying the microscopic initial condition to compute time correlation functions. The analytic structure involves the product, at the same time, of two Vlasov Green's functions. This feature reappears in all 'dressed particle' generalizations.

$\tilde{H}_2$ involves only the singlet distribution function. Thus when we insert the solution for $\tilde{c}_2$ into the right hand side of the first BBGKY equation we obtain a generalized (non Markoffian) kinetic equation for the singlet distribution.

Attempts to solve the doublet cumulant equation, including both the direct binary collision and medium terms, have produced an extensive literature [Gould and DeWitt, 1967; Guernsey, 1964; Kihara and Aono, 1963; Aona, 1968; Honda, Aono and Kihara, 1963]. For the weak coupling limit, $\gamma < 1$, the Landau length, mean particle separation, and Debye lengths are well separated. The bare Coulomb interaction leads to the Landau kinetic equation with a logarithmically divergent upper length. With a Debye screened binary interaction this large distance divergence is avoided. The retention of solely the medium terms leads to the kinetic equation of Balescu, Lenard and Guernsey, with a logarithmic divergence at small distances. We will not go into this problem and will be content to establish equations at the double cumulant level. As was the case for the equilibrium cumulant theory, the standard cumulant truncation cannot be expected to yield a good kinetic theory for strongly coupled plasmas.

How can we estimate the accuracy of the usual cumulant truncations for the short time description of time correlation functions? Let us take the situation where the initial distribution is a general one body additive function N(px), and examine the t^2 response of the singlet distribution function. We need

$$\left.\frac{\partial f_1(1)}{\partial t}\right|_{t=0} \quad \text{and} \quad \left.\frac{\partial^2 f_1(1)}{\partial t^2}\right|_{t=0} .$$

The initial conditions are given as

$$f_1(1;\ 0) = \langle N(1)\ N(\vec{p}\vec{x})\rangle$$

$$f_2(12;\ 0) = \langle N(12)\ N(\vec{p}\vec{x})\rangle$$

$$f_3(123;\ 0) = \langle N(123)\ N(\vec{p}\vec{x})\rangle \tag{3.15}$$

The second time derivative of the singlet response involves the first time derivative of $f_2(12)$ which is computed from the second hierarchy equation. With the correct initial condition for $f_2(12)$, the streaming term is accurate. So the inaccuracy of

$$\left.\frac{\partial^2 f(1)}{\partial t^2}\right|_{t=0}$$

is given by

$$\iint dx_2\, dp_2\, dx_3\, dp_3 \frac{\partial V(\vec{x}_1 - \vec{x}_2)}{\partial \vec{x}_1} \frac{\partial}{\partial \vec{p}_1} \frac{\partial V(\vec{x}_1 - \vec{x}_3)}{\partial \vec{x}_1} \frac{\partial}{\partial \vec{p}_1} \tilde{c}(123;\ t=0) \tag{3.15'}$$

In the usual (small γ) theory $\tilde{c}(123;\ t = 0)$ is set equal to zero, but with the exact microscopic initial condition it is not zero.

The standard cumulant approach gives theories of both the static and time dependent correlation functions. The simple truncations yield a theory for small γ, but the path to a theory valid for $\gamma > 1$ is obscure. One may try a 'memory function' approach, in which the chain of equations starting with

$$\frac{\partial \tilde{c}_2(12;\ t)}{\partial t}$$

is solved formally with the singlet distribution treated as an inhomogeneous term. $\tilde{c}_2(12;\ t)$ is to be inserted in the singlet equation to obtain a non-Markoffian kinetic equation for $f_1(1;\ t)$. This leads to an ugly theory for two reasons. First, the initial values of all of the higher order cumulants enter explicitly. Second, the 'vertex' structure is complicated, since every equation of the cumulant hierarchy involves many lower order cumulants, including the singlet distribution. We will see that both of these difficulties are avoided in the systematic microscopic theory.

We examine what happens when we carry out the standard procedure but use different cumulants. The time dependent Kirkwood superposition approximation [Rice and Gray, 1964; Stillinger and Suplinskas, 1966; Mortimer, 1968; Jordan, 1974] can be used to define a triplet cumulant by

$$D_3 = F_3 - F_2(12)\, F_2(23)\, F_2(31)/F_1(1)\, F_1(2)\, F_1(3) \tag{3.16}$$

There are a number of ways of extending this to define higher order 'Kirkwood' cumulants. The KSA is not nonsensical for both small and large interparticle separations, even for systems with strong short range forces. It is therefore a possibility for a preliminary unified microscopic theory of fluids and plasmas.

When one writes the linearized doublet equation in KSA ($D = 0$), one finds that $S_0(12)$ is replaced by $S(12)$ where $S(12)$ involves $-\ln n_2(\vec{x}_1\vec{x}_2)$ in place of the bare interaction $\psi(\vec{x}_1\vec{x}_2)$. Here $n_2(\vec{x}_1\vec{x}_2)$ is the equilibrium pair distribution in the KSA. The new feature is that the effective interaction is medium dependent, and for a plasma is screened at the Debye length. A second feature is that the medium terms that replace the Vlasov operators are more complicated and are no longer one body additive. When a member of the pair (1,2) interacts with a medium particle the interaction depends on the spatial position of the other member. This is clearly a geometric effect which is entirely reasonable. The new terms do go over to a sum of Vlasov operators when $\gamma \to 0$. We expect that these qualitative features of the KSA will be retained in a correct microscopic theory.

B. Hard Sphere Hierarchy

We now examine the kinetic theory of hard spheres.

The key point is that in the hard sphere limit the duration of a collision vanishes. This is true in the many dimensional phase space description as well as in ordinary space. Of course this simplification only holds in classical physics. In quantum scattering the boundary condition implies nonlocal effects and the distinction between smooth and hard core potentials is not so sharp. We encounter the standard Boltzmann operator for a pair of particles of mass m = 1.

$$\overline{T}(ij) = \delta(|\vec{q}_{ij}| - a)\left| \vec{p}_{ij} \cdot \hat{q}_{ij}\right| [\theta(\vec{p}_{ij}\cdot\hat{q}_{ij})\, b_{ij} - \theta(-\vec{p}_{ij}\cdot\hat{q}_{ij})] \tag{3.17}$$

Here $\theta(x)$ is the step function. $\theta(x) = 0$ for $x < 0$ and $\theta(x) = 1$ for $x > a$. a is the hard sphere diameter.

$$\vec{p}_{ij} \equiv \vec{p}_i - \vec{p}_j \quad , \quad \hat{q}_{ij} = \vec{q}_{ij}/|\vec{q}_{ij}| \tag{3.18}$$

The exchange operator b_{ij} operates according to the rule

$$b_{ij}\, f(\vec{p}_1 \,..\, \vec{p}_i, \vec{q}_i, \,..\, \vec{p}_j, \vec{q}_j, \,..\, \vec{q}_N)$$

$$= f(\vec{p}_1 \,..\, \vec{p}_i^{\,F}\, \vec{q}_i, \vec{p}_j^{\,F}\, \vec{q}_j, \,..\, \vec{q}_N) \tag{3.19}$$

where the final momenta $\vec{p}_i^{\,F}$, $\vec{p}_j^{\,F}$ obey

$$\vec{p}_i^{\,F} = \vec{p}_i - (\vec{p}_{ij} \cdot \hat{q}_{ij})\, \hat{q}_{ij}$$

$$\vec{p}_j^{\,F} = \vec{p}_j + (\vec{p}_{ij} \cdot \hat{q}_{ij})\, \hat{q}_{ij} \tag{3.20}$$

The operator $\overline{T}(ij)$ has a simple intuitive content [Ernst, Dorfman, Hoegy and van Leeuwen, 1969]. The directions of colliding particles are arranged for forward time propagation; there is the standard 'collision cylinder' factor and a factor indicating action only when the spheres are in contact. We then have a pseudo-Liouville equation

$$\left\{\frac{\partial}{\partial t} + \sum_{i=1}^{N} \vec{p}_i \frac{\partial}{\partial \vec{q}_i} - \frac{1}{2} \sum_{i \neq j} \overline{T}(ij)\right\} F_N = 0 \tag{3.21}$$

This is a description that goes back at least to Kirkwood [1946] and Grad [1949]. Its practical importance has emerged forcefully in the papers of Lebowitz, Percus, and Sykes [1969] and of Ernst and Dorfman [1972]. There is an analytically distinct collision operator that propagates events backward in time. This maintains the overall reversibility, but is not important for our considerations, since we are interested in forward propagation.

The pseudo-Liouville hierarchy is very similar to the smooth potential hierarchy. It is

$$\left(\frac{\partial}{\partial t} + \vec{p}_1 \frac{\partial}{\partial \vec{x}_1}\right) f_1(1;\, t) = \overline{T}(1\overline{2})\; f_2(1\overline{2};\, t)$$

$$\left(\frac{\partial}{\partial t} + \vec{p}_1 \frac{\partial}{\partial \vec{x}_1} + \vec{p}_2 \frac{\partial}{\partial \vec{x}_2} - \overline{T}(12)\right) f_2(12;\, t) = [\overline{T}(1\overline{3})$$

$$+ \overline{T}(2\overline{3})]\; f_3(12\overline{3};\, t) \;\ldots \tag{3.22}$$

Introducing standard linearized cumulants, the first non-equilibrium equation becomes

$$\left(\frac{\partial}{\partial t} + \vec{p}_1 \frac{\partial}{\partial \vec{x}_1}\right) \tilde{f}(1;\, t) - \rho_0\, J_B(1)\, \tilde{f}(1;\, t) = \overline{T}(1\overline{2})\; \tilde{c}_2(1\overline{2};\, t)$$

$$J_B(1)\; \tilde{f}(1) \equiv -\overline{T}(1\overline{2})\; \{\phi_1\; \tilde{f}(\overline{2}) + \phi_2\; \tilde{f}(1)\} \tag{3.23}$$

The term on the left hand side, a medium term, is proportional to the density, and replaces the Vlasov term. It is in fact the linearized Boltzmann-Enskog operator. It rearranges velocities and acts even on spatially homogeneous singlet distributions. This is noteworthy, since the most rudimentary approximation, (neglect of the second cumulant), leads to the Boltzmann equation with its long time irreversible behavior, velocity relaxation, hydrodynamical limit, etc. These are things that emerge at the level of the doublet equation, (with neglect of $\tilde{c}_3$ and of medium terms), for the smooth hierarchy. The way in which the Boltzmann-Enskog equation for hard spheres emerges as a limit of the doublet equation for smooth potentials has been studied extensively. The use of the pseudo-Liouville hierarchy yields immediate results and enables us to penetrate more deeply into the structure of the theory.

Remarkable results were obtained by Ernst and Dorfman by proceeding to the next step. The second hierarchy equation, rewritten in terms of cumulants, is

$$\left(\frac{\partial}{\partial t} + \vec{p}_1 \frac{\partial}{\partial \vec{x}_1} + \vec{p}_2 \frac{\partial}{\partial \vec{x}_2} - \overline{T}(12)\right) \tilde{c}_2(12;\, t)$$

$$- \rho_0\, \overline{T}(1\overline{3})\, [\tilde{c}_2(21)\, \overline{\phi}_3 + \tilde{c}_2(2\overline{3})\, \phi_1]$$

$$- \rho_0\, \overline{T}(2\overline{3})[\tilde{c}_2(12)\, \overline{\phi}_3 + \tilde{c}_2(1\overline{3}\, \phi_2] = H_2(12) \qquad (3.24)$$

Note that the medium operators for the doublet cumulant are Boltzmann operators acting between one of the doublet particles and a particle of the medium. In this scheme the position of the other member of the doublet doesn't enter, i.e. we have one body additive operators. In addition, there is a direct interaction $\overline{T}(12)$. So the formal structure is identical to that for smooth potentials. We employ a solution procedure similar to the one that leads to the Balescu equation. But the 'dressing' of each particle interacting with the medium now has irreversible hydrodynamic behavior at long wavelengths. We have

$$H_2(12) = \rho_0 \overline{T}(12)[\phi_1\, \tilde{f}(2) + \phi_2\, \tilde{f}(1)]$$

$$+ \overline{T}(1\overline{3})\, \{c_{eq}(12)\, \tilde{f}_1(\overline{3}) + c_{eq}(\overline{3}1)\, \tilde{f}_1(2)\} + 1 \rightleftarrows 2$$

$$+ [\overline{T}(1\overline{3}) + \overline{T}(2\overline{3})]\, \tilde{c}_3(12\overline{3}) \qquad (3.25)$$

In the lowest order in the density only the first term enters. Ernst and Dorfman use a more rudimentary set of cumulants, and their approximation neglects the terms in c_{eq} and $\tilde{c}_3$. We then write the formal solution as

$$\tilde{c}(12,\, t) = \left[\frac{\partial}{\partial t} + \vec{p}_1 \frac{\partial}{\partial \vec{x}_1} - \rho_0\, J_B(1) + \vec{p}_2 \frac{\partial}{\partial \vec{x}_2} - \rho_0\, J_B(2) - \overline{T}(12)\right]^{-1} * \rho_0\, \overline{T}(12)\, [\phi_1\, \tilde{f}(2) + \phi_2\, \tilde{f}(1)] \qquad (3.26)$$

and insert it into the singlet equation. (The initial condition is ignored.) The density independent $\overline{T}(12)$ in the propagator is treated by iteration. In the lowest approximation where $\overline{T}(12)$ is neglected one has an exact solution in terms of the eigenfunctions of the linearized hard sphere Boltzmann-Enskog equation. The higher order terms may be treated in the same way. Ernst and Dorfman kept one additional term in the series and made contact with earlier ring diagram theories of the long time tails for correlation functions (to the lowest order in density). In addition, by analyzing the implications of the ring terms for hydrodynamic behavior, they uncovered nonanalytic behavior in the dispersion relations for long wavelength collective modes.

The structure of the corrections to the Boltzmann operator is that of second order perturbation theory, just as for the Balescu theory. The situation is reminiscent of the theory of the chemical bond. The weak but inverse power law van der Waals forces arise in second order perturbation theory. The stronger, exponential short range forces arise in first order. The long time effects have little to do with the microscopic preparation question. The important point is that the energy denominators for the hard sphere case exhibit hydrodynamic behavior, including damping. This is not the case for the Balescu theory which uses Vlasov propagators to represent medium interactions.

IV. FUNCTION SPACE AND ALGEBRAIC RELATIONS

A. Function Space and Cumulants

We return to the initial preparation and short time considerations of Chapter II and develop a systematic way of dealing with the problem. We do this in a way that has a clear relation to the conventional schemes of Chapter III.

Start with one body additive functions N(1). Construct a linear function space with unity as the first element and define an inner product; using the Gibbs weight factor

$$\langle A|B\rangle = \int \Phi \, d\Gamma \, A^*B \tag{4.1}$$

The second set of functions is

$$T(1) \equiv \delta N(1) = N(1) - \langle N(1)\rangle \quad , \tag{4.2}$$

and each function is orthogonal to unity. T(1) is parametrized by the variables $\vec{p}_1$ and $\vec{x}_1$. The functions T(1) are not mutually orthogonal. In fact we have

$$\begin{aligned}\langle T(1)\,T(2)\rangle &= \langle N(12)\rangle - \langle N(1)\rangle\langle N(2)\rangle + \delta(1-2)\langle N(1)\rangle \\ &= \left\{\rho_2(\vec{x}_1\vec{x}_2) - \rho_0^2\right\}\phi_1\phi_2 + \delta(1-2)\,\rho_0 \quad .\end{aligned} \tag{4.3}$$

It is of course possible to find linear combinations that are mutually orthogonal, in an infinite number of ways. We would then avoid the use of projection operators needed to maintain consistency.

One simple choice is the one body additive set

$$E_n(\vec{k}) = \sum_{\alpha=1}^{N} h_n(\vec{p}_\alpha)\, e^{i\vec{k}_i \cdot \vec{p}_\alpha}$$

parametrized by the indices n and $\vec{k}$. Here $h_n(\vec{p}_\alpha)$ are normalized Hermite polynomials. (We use a one dimensional notation for simplicity.) We find

$$\langle E_m(\vec{k}')|E_n(\vec{k})\rangle = [\delta_{m,n}\, \rho_0 + \delta_{m,o}\, \delta_{n,o}\, \tilde{\rho}_2(k)]\, \delta(\vec{k},\vec{k}') \, . \tag{4.5}$$

where $\tilde{\rho}_2(k)$ is the Fourier transform of the pair distribution. Thus

$$\frac{E_0(\vec{k})}{\sqrt{\rho_0 + \tilde{\rho}(k)}} \, , \quad \frac{E_1(\vec{k})}{\sqrt{\rho_0}} \, , \quad \frac{E_2(\vec{k})}{\sqrt{\rho_0}} \, , \ \ldots \tag{4.6}$$

is an orthonormal set in the one body additive space. We have only to subtract equilibrium averages for $\vec{k} = o$ to make them orthogonal to unity.

However the T(1) are more convenient for general arguments. Define a projection operator for the functions T(1) by

$$P_1 = |T(\bar{1})\rangle\langle\bar{1}|Z_1|\bar{2}\rangle\langle T(\bar{2})| \tag{4.7}$$

The one body inverse Z_1 is defined by

$$\langle T(1)|T(\bar{2})\rangle\langle\bar{2}|Z_1|3\rangle = \delta(1 - 3) \tag{4.8}$$

This integral equation is readily solved to yield

$$\langle 1|Z_1|2\rangle = [\delta(1 - 2)/\rho_0\, \phi_1] - B(\vec{x}_1 - \vec{x}_2) \tag{4.9}$$

Here $B(\vec{x})$ is the Ornstein-Zernike direct correlation function, defined by the integral equation

$$h(\vec{x}) = B(\vec{x}) + \rho_0 \int B(|\vec{x} - \vec{x}'|)\, h(\vec{x}')\, dx'$$

$$h(\vec{x}) = [\rho_2(\vec{x})/\rho_0^2] - 1 \tag{4.10}$$

We have

$$P_1\, F_N(\vec{p}_1 \ldots \vec{q}_N) = |T(\bar{1})\rangle\langle\bar{1}|Z_1|\bar{2}\rangle\langle T(\bar{2})|F_N\rangle \tag{4.11}$$

Note that the extra term in the inverse lies in the purely spatial sector viz. belongs to the lowest (constant) Hermite polynomial.

The next step is to introduce two body additive functions

$$N(12) = \sum_{\alpha \neq \beta} \delta(\vec{p}_\alpha - \vec{p}_1)\ \delta(\vec{p}_\beta - \vec{p}_2)\ \delta(\vec{q}_\alpha - \vec{x}_1)\ \delta(\vec{q}_\beta - \vec{x}_2) \tag{4.12}$$

which are used to define the usual doublet distribution. We have

$$N(12) = N(1)\ N(2) - \delta(1 - 2)\ N(1)$$

$$\delta N(12) \equiv N(12) - \langle N(12)\rangle \tag{4.13}$$

The two-body additive set contains the one body additive functions, and will be divided into these and an orthogonal part. This very natural mathematical division is the key to finding new cumulants to treat short-time and short-distance behavior. We write

$$T(12) = \delta N(12) - \Lambda(12\overline{3})\ \delta N(\overline{3}) \tag{4.14}$$

and fix $\Lambda(123)$ by requiring that $T(12)$ be orthogonal to $T(3) = \delta N(3)$ for all values of the arguments 1, 2, 3., i.e.

$$\langle T(12)|\delta N(3)\rangle = 0 \tag{4.15}$$

This leads to

$$\Lambda(123) = \langle \delta N(12)|\delta N(\overline{4})\rangle\langle\overline{4}|Z_1|3\rangle \tag{4.16}$$

This has the explicit form

$$\Lambda(123) = K(\vec{x}_1\vec{x}_2\vec{x}_3)\ \phi_1\phi_2 + \left[\frac{\rho_2(\vec{x}_1\vec{x}_2)}{\rho_0}\right] [\phi_2\ \delta(1 - 3) + \phi_1\ \delta(2 - 3)] \tag{4.17}$$

where K is a spatial function involving the triplet static correlation function as well as the direct correlation function. Note again that the main complications occur in the spatial sector.

We need a projection operator P_2 onto the space of the T(12)

$$P_2 = |T(\overline{12})\rangle\langle\overline{12}|Z_2|\overline{34}\rangle\langle T(\overline{34})| \tag{4.18}$$

Here Z_2 is the two body inverse, defined by

$$\langle 12|Z_2|\overline{34}\rangle\langle T(\overline{34})|T(56)\rangle = \frac{1}{2}\ [\delta(5 - 1)\ \delta(6 - 2) + \delta(5 - 2)\ \delta(6 - 1)] \tag{4.19}$$

We will not write down the detailed expressions until we need them.

One can proceed to complete the function space by the Gram-Schmidt procedure. Starting with the three body additive functions N(123) one defines functions T(123) that are orthogonal to unity, T(1) and T(12). There is a three body inverse Z_3 and a projection operator P_3. The detailed formulae are quite complicated and are given elsewhere.

Before proceeding to an examination of the algebraic structure of the theory, we note some properties of the Liouville operator. We reexpress the matrix element

$$\begin{aligned} \langle A|L|B\rangle &\equiv \int \Phi \, d\Gamma \, A^*(\vec{q}_1 \ldots \vec{p}_N)\left(\sum_\alpha \vec{P}_\alpha \frac{\partial}{\partial \vec{q}_\alpha} - \frac{\partial V}{\partial \vec{q}_\alpha} \frac{\partial}{\vec{P}_\alpha}\right) B \\ &= -\frac{1}{\theta} \int d\Gamma \, A^* \sum_\alpha \left(\frac{\partial \Phi}{\partial \vec{p}_\alpha} \frac{\partial}{\partial \vec{q}_\alpha} - \frac{\partial \Phi}{\partial \vec{q}_\alpha} \frac{\partial}{\partial \vec{p}_\alpha}\right) B \\ &= +\frac{1}{\theta} \int d\Gamma \, \Phi \sum_\alpha \left(\frac{\partial A^*}{\partial \vec{p}_\alpha} \frac{\partial B}{\partial \vec{q}_\alpha} - \frac{\partial A^*}{\partial \vec{q}_\alpha} \frac{\partial B}{\partial \vec{p}_\alpha}\right) \\ &= -\frac{1}{\theta} \langle \{A^*, B\}_{P,B} \rangle \end{aligned} \tag{4.20}$$

in terms of the Poisson-Bracket. This form is responsible for the unexplicitly simple form that some of the matrix elements take. Thus operating to the right we have for our singular basis functions

$$\langle T(1)|L|T(2)\langle = \rho_0 \, \delta(1 - 2) \, \phi_2 \, \vec{p}_2 \frac{\partial}{\partial \vec{x}_2} \tag{4.21}$$

$$\langle N(1)|L|N(23)\rangle = -2 \, \rho_2(\vec{x}_2 \vec{x}_3) \, \phi_2 \phi_3 \cdot [\delta(1 - 2) \, L(2|3) + \delta(1 - 3) \, L(3|2)] \tag{4.22}$$

where

$$L(1|2) = \vec{p}_1 \frac{\partial}{\partial \vec{x}_1} + \frac{1}{\theta} \frac{\partial \ln \rho_2(\vec{x}_1 \vec{x}_2)}{\partial \vec{x}_1} \frac{\partial}{\partial \vec{p}_1} \tag{4.23}$$

$$\langle N(12)|L|N(3)\rangle = -2\{\delta(1 - 3) \, \rho_2(\vec{x}_3 \vec{x}_2) \, \phi_3 \phi_2 \, L(3|2) + 1 \rightleftarrows 2\} \tag{4.24}$$

The formulae for <T(1)|L|T(23)>, <T(12)|L|T(3)> are more complicated, since one must add the one body part. The binary collision matrix element is surprisingly simple. We find

$$\langle T(12)|L|T(34)\rangle = [\delta(3-1)\,\delta(4-2)$$

$$+ \delta(3-2)\,\delta(4-1)]\langle N(34)\rangle L(3|4)$$

$$+ \phi_1\phi_2[1+P(12)][1+P(34)]\;\phi_4\;\delta(3-1)$$

$$\cdot \left[R_3(\vec{x}_3\vec{x}_2\vec{x}_4)\;\vec{p}_3\,\frac{\partial}{\partial\vec{x}_3} + \frac{1}{\theta}\,\frac{\partial R_3}{\partial\vec{x}_3}\,\frac{\partial}{\partial\vec{p}_3}\right] \qquad (4.25)$$

$$R_3(\vec{x}_1\vec{x}_2\vec{x}_4) = \rho_3(\vec{x}_1\vec{x}_2\vec{x}_4) - \rho_2(\vec{x}_1\vec{x}_2)\;\rho_2(\vec{x}_1\vec{x}_2)/\rho_0 \quad . \qquad (4.26)$$

A phase space function is expanded as

$$F_N(t) = |T(\overline{1})\rangle\;A_1(\overline{1};\;t) + |T(\overline{12})\rangle\;A_2(\overline{12};\;t)$$

$$+ |T(123)\rangle\;A_3(\overline{123};\;t) + \ldots$$

$$A_n(t=o) \equiv A_n^0 \qquad (4.27)$$

where the $A_n(t)$ are amplitudes. The most common microscopic preparations have only the lowest A_n different from zero initially. We will be interested in cumulants, defined with the T functions

$$\Delta_1(1;\;t) = \langle T(1)|F_N(t)\rangle$$

$$\Delta_2(12;\;t) = \langle T(12)|F_N(t)\rangle$$

$$\Delta_2(123;\;t) = \langle T(123)|F_N(t)\rangle \quad ,\text{etc.} \qquad (4.28)$$

The connection between the amplitudes and cumulants is formally simple in virtue of the orthogonality of the spaces

$$\Delta_1(1;\;t) = \langle T(1)|T(\overline{2})\rangle\;A_1(\overline{2};\;t)$$

$$\Delta_2(12;\;t) = \langle T(12)|T(\overline{34})\rangle\;A_2(\overline{34};\;t) \qquad (4.29)$$

Thus if we want to work entirely with cumulants, the expansion of $F_N(t)$ is

$$F_N(t) = |T(\overline{1})\rangle\langle\overline{1}|Z_1|\overline{2}\rangle\;\Delta_1(2;\;t)$$

$$+ |T(\overline{12})\rangle\langle\overline{12}|Z_2|\overline{34}\rangle\;\Delta_2(\overline{34};\;t) + \ldots \qquad (4.30)$$

Now $\Delta_1(1;\;t)$ may be written as

$$\Delta_1(1;\, t) = \langle T(1)|e^{-Lt}\; F_N^0\rangle$$

$$= \langle T(1)|e^{-Lt}|T(\bar{1}')\rangle\; A_1^0(\bar{1}') + \langle T(1)|e^{-Lt}|T(\bar{1}'\bar{2}')\rangle$$

$$\cdot\; A_2^0(\bar{1}'\bar{2}';\, t) + \ldots \tag{4.31}$$

This introduces time dependent correlation between the T functions. We also have

$$\Delta_2(12;\, t) = \langle T(12)|e^{-Lt}|T(\bar{1}')\rangle\; A_1^0(T')$$

$$+ \langle T(12)|e^{-Lt}|T(\bar{1}'\bar{2}')\rangle\; A_2^0(\bar{1}'\bar{2}';\, t) \tag{4.32}$$

These correlation functions have been introduced by Mazenko [1973; 1974]. The function space point of view, as pointed out by Boley, is a simple way of generating them to any order.

$\Delta_1(t)$ is just the usual singlet distribution and in general depends on all the initial amplitudes. But the simplest microscopic preparations pick out only a few of the correlation functions.

We write the evolution operator for the Liouville equation as $G(t) = e^{-Lt}$. Then the projections $P_n\; G(t)\; F_N^0$ give the cumulants Δ_n and the projections $P_n\; G(t)\; P_m \equiv G_{nm}(t)$ are the correlation functions. We now proceed to use this algebraic notation. It should be noted that the present point of view allows us to see explicitly what the form of $F_N(t)$ is for any approximation. It is thus suitable for use in conjunction with the variational principles discussed in Section A.

B. Algebraic Aspects of the General Theory

Consider the operator equation for $G(t) = e^{-Lt}$

$$\left(\frac{\partial}{\partial t} + L\right) G(t) = 0 \quad , \qquad G(o) = 1 \tag{4.33}$$

and the Laplace transform or resolvent operator $\tilde{G}(S) = (S + L)^{-1}$

$$(S + L)\; \tilde{G}(S) = 1 \tag{4.34}$$

where 1 is the symmetrized identity operator in N body space. Using a projection operator P and its complement Q, P + Q = 1, we operate first with P, then with Q, from the left

$$(S + PLP)\; P\tilde{G} + PLQ \cdot Q\tilde{G} = P$$

$$(S + QLQ)\; Q\tilde{G} + QLP \cdot P\tilde{G} = Q \tag{4.35}$$

Formal elimination of the second equation yields

$$\left\{(S + PLP) - PLQ \frac{1}{S + QLQ} QLP\right\} P\tilde{G} = P - PLQ \frac{1}{S + QLQ} Q \quad (4.36)$$

In the time domain this is the formal Zwanzig-Mori solution as applied for example to the cumulant $\Delta_1(1; t)$.

From these formulae, taking right hand projections we find $(S + PLP + \tilde{M})\ P\tilde{G}P = P$

$$\tilde{M} = -PLQ(S + QLQ)^{-1}\ QLP \quad (4.37)$$

Here $\tilde{M}$ is the memory operator associated with P.

Clearly the utility of the formalism depends on the choice of the projection operators. If P is to be a very small part of the function space, too much is thrown into the determination of the memory operator. The choice of P_1 as the one body additive space is a very natural one in kinetic theory since we then have a self contained singlet equation for free particles. This comes from the 'static' term P_1LP_1 and the memory function is zero. If P_1 only includes hydrodynamic microscopic quantities we have to work with a memory function even in the free particle limit.

Rewrite the preceding formulae as

$$(S + L_{11} + M_{11})\ \tilde{G}_{11} = P_1 \quad ,$$

$$\tilde{M}_{11} = -P_1\, L\, Q_1 (S + Q_1\, L\, Q_1)^{-1}\ Q_1\, L\, P_1 \quad (4.38)$$

The next step is to introduce two body additive functions. These include the space P_1 and a space P_2 orthogonal to P_1. The space Q_2 is orthogonal to both P_1 and P_2. The equation

$$(S + Q_1\, L\, Q_1)\ Q_1\tilde{G} + Q_1\, L\, P_1 \cdot P_1\tilde{G} = Q_1 \quad (4.39)$$

is now broken up into ($Q_2 = Q_1 - P_2$, $P_2\ Q_1 = P_2$, $Q_2\ Q_1 = Q_2$)

$$(S + P_2\ L\ P_2)\ P_2\tilde{G} + P_2\ L\ Q_2\ Q_2\tilde{G} + P_2\ L\ P_1\ P_1\tilde{G} = P_2$$

$$(S + Q_2\ L\ Q_2)\ Q_2\tilde{G} + Q_2\ L\ P_2 \cdot P_2\tilde{G} = Q_2 \quad (4.40)$$

When the Liouville operator is two-body additive we only connect adjacent spaces. So $Q_2\ L\ P_1 = 0$

$$\{S + P_2\ L\ P_2 + \tilde{M}_{22}\}\ P_2\tilde{G} = P_2 - P_2\ L\ P_1 \cdot P_1\tilde{G} - P_2\ L\ Q_2 \frac{1}{S + Q_2\ L\ Q_2}\ Q_2 \quad (4.41)$$

This is the algebraic counterpart of the doublet cumulant equation with the information on higher cumulants hidden in the two body memory function

$$\tilde{M}_{22} = -P_2 \; L \; Q_2 \frac{1}{S + Q_2 \; L \; Q_2} Q_2 \; L \; P_2 \tag{4.42}$$

and in the inhomogeneous term involving Q_2. The singlet cumulant equation is

$$(S + P_1 \; L \; P_1) \; P_1\tilde{G} + P_1 \; L \; P_2 \cdot P_2\tilde{G} = P_1 \quad . \tag{4.43}$$

Let us consider the sequence generated by $P_1\tilde{G}P_1$, multiplying by P_1 from the right. This removes many inhomogeneous terms. We see that the one and two body memory functions are connected by

$$\tilde{M}_{11} = -P_1 \; L \; P_2\{S + P_2 \; L \; P_2 + \tilde{M}_{22}\}^{-1} \; P_2 \; L \; P_1 \tag{4.44}$$

To evaluate the inverse operator is equivalent to solving the doublet cumulant equation. There is a 'static' part, P_2LP_2, not involving the Laplace transform variables S. It contains medium modifications of the direct two body interactions. So this in itself is a difficult problem, i.e. it involves solution of a complicated doublet equation. The memory function $\tilde{M}_{22}$ contains the information relating to the elimination of the higher order cumulants. If one neglects it entirely one has the two body additive approximation which is the same as truncating the BBGKY hierarchy by setting $\Delta_3(t)$ equal to zero at all times.

The preceding argument is easily generalized. One finds a memory operator for the Δ_n cumulant which is connected to the one for Δ_{n+1} by

$$\tilde{M}_{nn} \equiv P_n \; \tilde{M} \; P_n = -P_n \; L \; P_{n+1}\{S + P_{n+1} \; L \; P_{n+1} + \tilde{M}_{n+1,n+1}\}^{-1} \cdot P_{n+1} \; L \; P_n \tag{4.45}$$

This is the algebraic version of Boley's continued fraction representation.

We next turn to the notion of a one body additive operator. Examples are the free streaming operator and an operator that is the sum of Vlasov operators acting separately on the arguments of a function like $\Delta_n(1 \; ... \; n; \; t)$. We define the operator L_0 by first noting its action on a one body additive function

$$L_0|N(1)\rangle = \langle 1|R|\overline{3}\rangle|N(\overline{3})\rangle \tag{4.46}$$

The matrix $\langle 1|R|3\rangle$ completely characterizes the operator L_0. The definition is completed by giving its matrix elements in the entire function space, e.g.

$$L_0|N(1)\ N(2)\rangle = \langle 1|R|\bar{3}\rangle|N(\bar{3})\ N(2)\rangle + \langle 2|R|\bar{3}\rangle|N(1)\ N(\bar{3})\rangle, \text{ etc.} \tag{4.47}$$

A one body additive operator has the property

$$P_m\ L_0\ P_1 = 0 \qquad m \geq 2 \qquad \text{i.e. } Q_1\ L_0\ P_1 = 0 \tag{4.48}$$

$L_0\ P_2$ has components in both P_1 and P_2 but we still have

$$P_m\ L_0\ P_2 = 0 \qquad m \geq 3 \qquad \text{i.e. } Q_2\ L_0\ P_2 = 0 \tag{4.49}$$

We now use some algebraic identities. Starting from the well known

$$\begin{aligned}(S + Q_1\ L_0)^{-1} &= (S + L_0)^{-1}\{1 + P_1\ L_0(S + Q_1\ L_0)^{-1}\} \\ &= \{1 + (S + Q_1\ L_0)^{-1}\ P_1\ L_0\}(S + L_0)^{-1}\end{aligned} \tag{4.50}$$

we find

$$Q_1(S + Q_1\ L_0)^{-1} = Q_1(S + L_0)^{-1} \tag{4.51}$$

With the definitions

$$\tilde{G}_0 = (S + L_0)^{-1} \quad , \qquad \tilde{I}_1 = \{S + Q_1(L_0 + L_1)\}^{-1} \tag{4.52}$$

the same identity yields

$$\tilde{I}_1 = (S + Q_1\ L_0)^{-1}\{1 - Q_1\ L_1\ \tilde{I}_1\} = \{1 - \tilde{I}_1\ Q_1\ L_1\}\{S + Q_1\ L_0\}^{-1} \tag{4.53}$$

This has the formal solution

$$\begin{aligned}Q_1\ \tilde{I}\ Q_1 &= Q_1\{1 + \tilde{G}_0\ Q_1\ L_1\ Q_1\}^{-1}\ \tilde{G}_0\ Q_1 \\ &= Q_1\ \tilde{G}_0\ Q_1[1 + Q_1\ L_1\ Q_1\ \tilde{G}_0]^{-1}\ Q_1\end{aligned} \tag{4.54}$$

So

$$\begin{aligned}\tilde{M}_{11} &= -P_1\ L\ Q_1[1 + \tilde{G}_0\ Q_1\ L_1\ Q_1]^{-1}\ Q_1\ \tilde{G}_0\ Q_1\ L\ P_1 \\ &= -P_1\ L\ Q_1\ \tilde{G}_0\ Q_1[1 + \tilde{G}_0\ Q_1\ L_1\ Q_1]^{-1}\ Q_1\ L\ P_1\end{aligned} \tag{4.55}$$

The same type of expression holds for any n, viz

$$\tilde{M}_{nn} = -P_n\, L\, Q_n [1 + \tilde{G}_0\, Q_n\, L_1\, Q_n]^{-1}\, \tilde{G}_0\, Q_n\, L\, P_n$$

$$= -P_n\, L\, Q_n\, \tilde{G}_0 [1 + Q_n\, L_1\, Q_n\, \tilde{G}_0]^{-1}\, Q_n\, L\, P_n \tag{4.56}$$

Let us now discuss how existing theories fit into this framework. The first class of theories works at the $\tilde{M}_{11}$ level. In the cumulant language this means at the level where $\tilde{\Delta}_2$ is approximated as a linear functional of $\tilde{\Delta}_1$. The kinetic equation is

$$(S + P_1\, L\, P_1 + \tilde{M}_{11})\, P_1 \tilde{G}\, P_1 = P_1 \tag{4.57}$$

where $P_1 L P_1$ is easily computed without any split of L and involves the direct correlation function. Thus neglecting $\tilde{M}_{11}$ entirely yields the one body additive theory i.e. the modified Vlasov equation. The Forster-Martin weak coupling theory [Forster and Martin, 1970; Forster, 1974] takes L_0 equal to the free particle streaming one body additive operator, and retains only the first term of a geometric series, viz

$$\tilde{M}_{11} \sim -P_1\, L\, P_1\, \tilde{G}_0\, P_2\, L\, P_1 \tag{4.58}$$

A primitive dressed particle approximation also uses the first term for $\tilde{M}_{11}$. But L_0 is a modified Vlasov operator. It is chosen as the one body additive extension of the relation

$$P_1\, L\, P_1 = P_1\, L_0\, P_1 \tag{4.59}$$

viz

$$L_0 |N(1)\rangle = \vec{p}_1 \cdot \frac{\partial}{\partial \vec{x}_1} |N(1)\rangle - \frac{\partial B(\vec{x}_1 - \overline{\vec{x}_3})}{\partial \vec{x}_1} \frac{\partial}{\partial \vec{p}_1} |N(\overline{3})\rangle \tag{4.60}$$

This yields a modified Balescu equation. For this type of approximation one can proceed to analyze the geometric series for $\tilde{M}_{11}$ (as in the early days of the Brussels school) [Prigogine, 1962; Balescu, 1963; Resibois, 1966]. Thus the next term has the form

$$+ P_1\, L\, P_2\, \tilde{G}_0\, P_2\, L_1\, P_2\, \tilde{G}_0 \cdot Q_1\, L\, P_1 \tag{4.61}$$

and describes scattering of excitations by the residual L_1.

It is tempting to choose L_0 so that the resolvant $(S + L_0)^{-1}$ has a hydrodynamic pole structure. This leads to theories similar to those first explored by Nelkin [Kim and Nelkin, 1971; Ortoleva and Nelkin, 1969; 1970]. They are semi-phenomenological and one does not know how important the corrections are (although they can be estimated by studying the geometric series). The parameters in such a model can be fixed using the variational approach. This type of approach overemphasizes the coupling to collective excitations. The propagator $(S + L_0)^{-1}$ acts in the two particle space and involves a product of two one body propagators at each time. One can set up a nonlinear integral equation for the time dependent propagator if it is assumed to be the exact S dependent one body streaming plus memory kernel. But it is not clear that the results would be satisfactory even if the equation could be solved.

Another class of theory works at the $\tilde{M}_{22}$ level. The simplest one is the two body additive theory where $\tilde{M}_{22} \approx 0$. One still has to analyse the doublet cumulant kernel given by $(S + P_2 L P_2)^{-1}$ to obtain $\tilde{M}_{11}$. This is itself quite difficult but physically has the virtue that the direct screened binary interaction appears explicitly. Some of the medium terms can be simplified in a manner appropriate to particular systems. We will examine the doublet kernel in more detail in the next section.

The only feasible estimates of $\tilde{M}_{22}$ to date are based on a dressed particle approximation of a type

$$\tilde{M}_{22} \approx -P_2 L P_3 (S + L_0)^{-1} P_3 L P_2 \tag{4.62}$$

which involves the product of three one body propagators at the same time. Again we have to choose L_0 and the same type of considerations already discussed are involved. The most tempting choice is the one body additive extension of

$$P_1 L_0 P_1 = P_1 L P_1 + \tilde{M}_{11} \tag{4.63}$$

This amounts to adding and subtracting L_0 in the equations governing $\tilde{M}_{22}$ (or $\tilde{\Delta}_3$ in cumulant language), treating $L - L_0$ as a perturbation again one obtains very complicated self consistent nonlinear integral equations for L_0. Thus use of exact one body propagators to estimate $\tilde{M}_{22}$ does take into account the fact that when the three particles involved are well separated each one does move in accordance with the fully damped exact one body propagator.

One virtue of the projection formalism (or of the Δ_n cumulant approach) is that the vertices of type $P_2 L P_3$ are exactly known. Other cumulants contain all the lower cumulants on the right hand side of the governing equations.

C. Short Time Behavior

If we use the usual sequence 1BA, 2BA, etc., or $P_1, P_2, P_3, \ldots$ we have only a few general properties. We use the notation $L_{m,n} = P_m L P_n$. Then only $L_{m,m\pm 1}$, $L_{m,m}$ are different from zero. In addition L changes sign under momentum reversal and only contains first derivatives with respect to coordinates and momenta, so that the chain rule of differentiation applies. However we will not make maximal use of these facts.

Let us ask what we need to do in an approximation scheme to ensure that P_1 G P_1 is accurate at short times to a given power t^n. We assume that the initial condition is such that only $P_1 G P_1 \neq 0$ i.e. $F_N(t = o)$ belongs to P_1. Then one computes, using the basic chain

$$\left.\frac{\partial G_{11}}{\partial t}\right|_{t=o} = -L_{11}$$

$$\left.\frac{\partial G_{11}}{\partial t^2}\right|_{t=o} = L_{11}^2 + L_{12} L_{21} \tag{4.64}$$

Now consider the chain of equations for cumulants (in the present projection language).

$$\frac{\partial}{\partial t} P_1 G + L_{11} P_1 G = -L_{12} P_2 G$$

$$\frac{\partial}{\partial t} P_2 G + L_{22} P_2 G = -L_{21} P_1 G - L_{23} P_3 G \tag{4.65}$$

Consider a truncation that neglects P_3G. We see that L_{22} plays no role to order t^2. Any additional replacement of L_{22}, say by a free particle or dressed particle model Liouville operator will do for the term L_{22} P_2G. We do have to be exact for the terms on the right hand side viz L_{12}, L_{21}.

In the next step we have

$$\left.\frac{\partial^3 G_{11}}{\partial t^3}\right|_{t=o} = -L_{11}^3 - L_{12} L_{21} L_{22} - 2L_{12} L_{21} L_{11} \tag{4.66}$$

Thus we have to use a model operator that gives L_{22} exactly to get P_1 G P_1 to order t^3.

We also have

$$\left.\frac{\partial^4 G_{11}}{\partial t^2}\right|_{t=o} = +L_{12} L_{11}^3 + L_{11}^2 L_{21} L_{12} + L_{11}\{2L_{12}^2 L_{21} + L_{12} L_{21} L_{22} + L_{21} L_{22}^2\} + L_{12}^2 L_{21} L_{22} + L_{12}^2 L_{21}^2 + L_{21} L_{12} L_{23} L_{32} \tag{4.67}$$

To have t^4 accuracy for $P_1 \, G \, P_1$ we need to retain the doublet equation in toto. In the triplet equation

$$\left(\frac{\partial}{\partial t} + L_{33} \right) P_3 G = -L_{32} \, P_2 G - L_{34} \, P_4 G \tag{4.68}$$

we can neglect P_4G and do anything we like to L_{33}, provided L_{32} is accurate and is retained in the truncated hierarchy. Of course, the exact L_{22}, L_{21}, L_{11} must be used in the lower order equations.

From the continuity equation, the time derivative of the density is proportional to the longitudinal current density which is one body additive. Thus the t^2 requirements on the phase space correlation function cover the t^4 requirements for the density autocorrelation function.

V. KINETIC EQUATIONS

A. Cumulants

We now study some of the detailed equations implied by the preceding algebraic considerations. The approach is to define modified cumulants as linear combinations of the usual reduced distribution functions. $\Delta_2(12)$ is given by

$$\Delta_2(12) = \langle \delta N(12) \, F_N \rangle \, \frac{\rho_2(\vec{x}_1 \vec{x}_2)}{\rho_0} \, \{\phi_2 \, \Delta_1(1) + \phi_2 \, \Delta_1(2)\}$$

$$- K(\vec{x}_1 \vec{x}_2 \vec{x}_3) \, \phi_1 \, \phi_2 \, \Delta_1(3) \tag{5.1}$$

$$K(\vec{x}_1 \vec{x}_2 \vec{x}_3) = \left[\frac{\rho_3(\vec{x}_1 \vec{x}_2 \bar{\vec{x}}_4)}{\rho_0} - \rho_2(\vec{x}_1 \vec{x}_2) \right] [\delta(\bar{\vec{x}}_4 - \vec{x}_3) - B(\bar{\vec{x}}_4 - \vec{x}_3)]$$

$$- \rho_2(\vec{x}_1 \vec{x}_2) \, [B(\vec{x}_3 - \vec{x}_1) + B(\vec{x}_3 - \vec{x}_2)] \tag{5.2}$$

The third cumulant is quite complicated. It may be written as

$$\Delta_3(123) = \langle \delta N(123) \, F_N \rangle - \Lambda(123|\bar{4}) \, \Delta_1(\bar{4}) - \Lambda(123|\overline{45}) \, \Delta_2(\overline{45}) \tag{5.3}$$

where

$$\Lambda(123|\bar{4}) = \langle \delta N(123) | \delta N(\bar{5}) \rangle \langle \bar{5} | Z_1 | 4 \rangle$$

$$\Lambda(123|\overline{45}) = \langle \delta N(123) | T(\overline{67}) \rangle \langle \overline{67} | Z_2 | 45 \rangle \tag{5.4}$$

It involves static correlation functions as high as ρ_5. The only term which requires discussion is the two body inverse. It is defined by Eq. (4.19). It has three parts A, B, C. The first part is

$$<12|Z_2|34>_A = \frac{\delta(3-1)\,\delta(4-2) + \delta(3-2)\,\delta(4-1)}{4\rho_2(\vec{x}_1\vec{x}_2)\;\phi_1\phi_2} \tag{5.5}$$

The second part is determined by an integral equation driven by the quantity

$$F(\vec{x}_1 - \vec{x}_2||\vec{x}_1 - \vec{x}_3) \equiv \frac{\rho_3(\vec{x}_1\vec{x}_2\vec{x}_3)}{\rho_2(\vec{x}_1\vec{x}_2)\;\rho_2(\vec{x}_1\vec{x}_3)} - \frac{1}{\rho_0} \tag{5.6}$$

The integral equation determines a function D according to

$$D(\vec{y}_2||\vec{y}_5) = -\frac{1}{4}F(\vec{y}_2||\vec{y}_5) - \int F(\vec{y}_2||\vec{y}_3)\;\rho_2(\vec{y}_3)\;D(\vec{y}_3||\vec{y}_5)\;dy_3 \tag{5.7}$$

Then

$$<12|Z_2|34>_B = \{1 + P(12)\}\{1 + P(34)\}\;D(\vec{x}_1 - \vec{x}_2||\vec{x}_1 - \vec{x}_3) \cdot \frac{\delta(1-4)}{\phi_1} \tag{5.8}$$

Note that the A part is of order ρ_0^{-2}, while the B part is of order ρ_0^{-1}. However for weak coupling with

$$h(r) = \frac{\rho_2(r)}{\rho_0^2} - 1$$

we have an additional spatial factor viz.

$$D(\vec{x}_1 - \vec{x}_2||\vec{x}_1 - \vec{x}_3) \rightarrow -\frac{1}{4\rho_0}\,h(\vec{x}_2\vec{x}_3) \tag{5.9}$$

Finally, the C part is purely spatial. It is

$$<12|Z_2|34>_C \equiv <\vec{x}_1\vec{x}_2|U|\vec{x}_3\vec{x}_4> \tag{5.10}$$

where U may be found from the defining relation (or by consulting Gross [1976]). It is a density dependence $\rho_0^0 \sim 1$ and in weak coupling has two spatial factors involving h(r).

B. Singlet Equation

Once one has decided to use the Δ_n cumulants one can adopt the pedestrian approach of reexpressing the hierarchy as cumulant equations. This holds for both smooth potentials and for the pseudo-Liouville hierarchy. Clearly, no approximation is made -- the theory is merely set up so that traditional ideas can be taken over. It is not advisable to use simpler cumulants at the first stage since one would lose the formal structure that makes the algebraic arguments of the preceding section applicable.

A disadvantage of the straightforward method is that a number of simplifications are possible if one invokes the equilibrium hierarchy, and these have to be put in by hand. Use of the orthogonal function machinery for smooth potentials gives a singlet equation

$$\frac{\partial}{\partial t}\Delta(1) + \langle 1|L|\overline{2}\rangle\langle\overline{2}|Z_1|\overline{3}\rangle\,\Delta_1(\overline{3}) = -\langle 1|L|\overline{23}\rangle\langle\overline{23}|Z_2|\overline{45}\rangle\,\Delta_2(\overline{45}) \tag{4.11}$$

The left hand side is simply the modified Vlasov structure. The right hand side seems quite complicated. But we know what it must be from the usual hierarchy, because $F_2(12)$ on the right side gives directly the term in Δ_2 with the straightforward approach. In fact note that

$$\begin{aligned}\langle 1|L|2\rangle &= \langle T(1)|\sum_\alpha \vec{p}_\alpha \frac{\partial}{\partial\vec{q}_\alpha} - \frac{\partial V}{\partial\vec{q}_\alpha}\frac{\partial}{\partial\vec{p}_\alpha}|T(23)\rangle \\ &= -\frac{\partial V(\vec{x}_1 - \vec{x}_4)}{\partial\vec{x}_1}\frac{\partial}{\partial\vec{p}_1}\langle 1\overline{4}|Z_2|23\rangle\end{aligned} \tag{5.12}$$

So we have the identity

$$\begin{aligned}\langle 1|L|\overline{23}\rangle\langle\overline{23}|Z_2|45\rangle = -\frac{1}{2}\Bigg[&\delta(1-5)\frac{\partial V(\vec{x}_1 - \vec{x}_4)}{\partial\vec{x}_1}\frac{\partial}{\partial\vec{p}_1} \\ &+ \delta(1-4)\frac{\partial V(\vec{x}_1 - \vec{x}_5)}{\partial\vec{x}_1}\frac{\partial}{\partial\vec{p}_1}\Bigg]\end{aligned} \tag{5.13}$$

The singlet equation is therefore

$$\begin{aligned}\frac{\partial}{\partial t}\Delta_1(1) + \vec{p}_1\frac{\partial}{\partial\vec{x}_1}\Delta_1(1) - \frac{\partial B(\vec{x}_1 - \vec{x}_{\overline{2}})}{\partial\vec{x}_1}\frac{\partial\phi_1}{\partial\vec{p}_1}\Delta_1(\overline{2}) \\ = \frac{\partial V(\vec{x}_1 - \vec{x}_{\overline{3}})}{\partial\vec{x}_1}\frac{\partial}{\partial\vec{p}_1}\Delta_2(1\overline{3})\end{aligned} \tag{5.14}$$

The orthogonal function approach shows that everything can in principle be expressed in terms of static correlation functions without the bare potentials appearing. But in contrast to our initial point of view [Gross, 1972; Bergeron, Gross and Varley, 1974], it is now clear that the expressions are much simpler with a mixed description.

The neglect of Δ_2 yields the modified Vlasov equation with its superior short time and short distance properties. If we are interested in computing phase space correlation functions where the microscopic initial condition is $F_N = T(3)$, the initial condition is

$$\Delta_1(1, t = 0) = \langle T(1)|T(3)\rangle = \rho_0 \, \phi_1 \, \delta(1 - 3) + \\ + [\rho_2(\vec{x}_1\vec{x}_3) - \rho_0{}^2] \, \phi_1 \, \phi_3$$

$$\Delta_2(12; t = 0) = 0 \qquad (5.15)$$

The singlet equation for hard spheres of diameter a takes the form

$$\left(\frac{\partial}{\partial t} + \vec{p}_1 \frac{\partial}{\partial \vec{x}_1}\right) \Delta_1(1; t) - \rho_0 \frac{\partial \phi_1}{\partial \vec{p}_1} \frac{\partial}{\partial \vec{x}_1} V_{eff}(\vec{x}_1 - \bar{\vec{x}}_2) \, \Delta_1(\bar{2}) \\ - g(a) \, \rho_0 \, J_B(1) \, \Delta(1) \\ = \bar{T}(1\bar{2}) \, \Delta_2(1\bar{2}) \qquad (5.16)$$

Here $J_B(1)$ is the linearized Boltzmann operator and V_{eff} is

$$V_{eff} = - \frac{1}{\theta} \{B(r) + g(a) \, \Theta(a-|r|)\} \qquad (5.17)$$

When $\Delta_2 = 0$ we have the kinetic equation derived by Lebowitz, Percus and Sykes [1969] from considerations of short time behavior based on microscopic initial conditions. The effective potential is continuous with a vanishing linear term in the density. Since we have a Vlasov-like medium term, the solution can be given in terms of the Green's function for the Boltzmann-Enskog equation. The medium term does not disturb the satisfactory hydrodynamic behavior, but corrects the short time, small distance behavior of time correlation functions.

The relation of this equation to other theories has been analyzed by Sykes [1973]. The standard Enskog equation, (different from what we have called the Boltzmann-Enskog equation), can be analyzed in terms of its time reversible and irreversible parts [Gross and Wisnivesky, 1968]. The main defect is in the short distance behavior of the reversible part. Sykes shows however that the LPS equation agrees with the Enskog equation in the limit of low densities. We have the remarkable result that the original Enskog equation in the low density limit gives the correct short time behavior. Of course, one must use the correct initial condition in computing time correlation functions. Sykes also shows that an equation derived by Mazenko is also the Enskog equation at low densities. None of these equations is reliable at higher densities. Detailed analysis of the hydrodynamic limit may be found in the cited papers [This volume; Konijnendjck and van Leeuwen, 1973].

More detailed kinetic modelling of the Enskog equation has been carried out by Mazenko and collaborators. The passage from continuous short range potentials to hard spheres has been studied by Blum and Lebowitz [1969] from the point of view of a binary collision expansion. Mazenko obtained the equation by neglecting medium terms and collision duration effects in the doublet equation and evaluating the Green's function for the doublet cumulant. This is the same procedure that was used in older derivations of the Boltzmann equation.

The short time and short distance behaviors are intimately connected. Theories that are framed to account for short time behavior, (in a non ad-hoc manner), like the LPS theory, give a good account of the short distance behavior.

C. Doublet Equation

The new doublet equation is

$$\frac{\partial}{\partial t}\Delta_2(12) + \langle 12|L|\overline{34}\rangle\langle\overline{34}|Z_2|\overline{56}\rangle\,\Delta_2(\overline{56}) = \langle 2|L|\overline{3}\rangle\langle\overline{3}|Z_1|\overline{4}\rangle\,\Delta_1(\overline{4}) + \left(\frac{\partial V(\vec{x}_1 - \overline{\vec{x}}_3)}{\partial\vec{x}_1}\frac{\partial}{\partial\vec{p}_1} + \frac{\partial V(\vec{x}_2 - \overline{\vec{x}}_3)}{\partial\vec{p}_2}\frac{\partial}{\partial\vec{p}_2}\right)\Delta_3(12\overline{3}) \tag{5.18}$$

Again we have taken advantage of the direct approach to write the Δ_3 contribution in terms of the bare potentials. The singlet contribution to the right hand side simplifies because the direct correlation function part of the one body inverse doesn't contribute. Note that

$$\langle T(12)|L|\sum_\alpha\delta(\vec{q}_\alpha - \vec{x}_3)\rangle = 0$$

$$\langle 12|L|\overline{3}\rangle\langle\overline{3}|Z_1|\overline{4}\rangle\,\Delta_1(\overline{4}) = \langle 12|L|\overline{3}\rangle\,\frac{\Delta_1(\overline{3})}{\rho_0\,\phi_3} \tag{5.19}$$

Using the formula for the matrix element of the Liouville operator, we have

$$\{ \frac{\partial}{\partial t} + L(1|2) + L(2|1)\} \Delta_2(12) + <12|K_2|\overline{34}> \Delta_2(\overline{34})$$

$$= \frac{1}{\theta\rho_0} \frac{\partial\rho_2(\vec{x}_1\vec{x}_2)}{\partial\vec{x}_1} \left(\phi_2 \frac{\partial}{\partial\vec{p}_1} - \phi_1 \frac{\partial}{\partial\vec{p}_2}\right) [\Delta_1(1) + \Delta_1(2)]$$

$$+ \{1 + P(12)\} \frac{\partial V(\vec{x}_1 - \vec{\overline{x}}_3)}{\partial\vec{x}_1} \frac{\partial}{\partial\vec{p}_1} \Delta_3(12\overline{3})$$

$$- \phi_1 \phi_2 S(\vec{x}_1\vec{x}_2\vec{x}_3) \vec{\overline{p}}_3 \frac{\partial}{\partial\vec{x}_3} \Delta_1(\overline{3}) \tag{5.20}$$

$$S(x_1x_2x_3) \equiv \{\rho_3(\vec{x}_1\vec{x}_2\vec{\overline{x}}_4) - \rho_0\ \rho_2(\vec{x}_1\vec{x}_2)\}\{\delta(\vec{\overline{x}}_4 - \vec{x}_3) - \rho_0\ B(\vec{\overline{x}}_4 - \vec{x}_3)\} \tag{5.21}$$

Here we have anticipated the fact that the doublet interaction kernel contains a direct part with the medium dependent potential

$$- \frac{1}{\theta} \ln \rho_2(\vec{x}_1\vec{x}_2) \quad .$$

The term involving $S(\vec{x}_1\vec{x}_2\vec{x}_3)$ drops out for spatially homogeneous problems.

Our main task is now the analysis of the kernel K_2. The Liouville matrix element has two parts, a direct and a medium part

$$<12|L|34>_D = [\delta(3 - 1)\ \delta(4 - 2) + \delta(3 - 2)\ \delta(4 - 1)]$$
$$\cdot <N(34)>\ L(3|4)$$

$$<12|L|34>_M = \phi_1\ \phi_2[1 + P(12)][1 + P(34)]\ \phi_4\ \delta(3 - 1)$$
$$\cdot \left[R_3(x_3x_2x_4)\ \vec{p}_3 \frac{\partial}{\partial\vec{x}_3} + \frac{1}{\theta}\frac{\partial R_3}{\partial\vec{x}_3}\frac{\partial}{\partial\vec{p}_3}\right] \tag{5.22}$$

There are three parts to the two body inverse. The term

$$<12|L|\overline{34}>_D <\overline{34}|Z_2|\overline{56}>_A \Delta_2(\overline{56})$$

yields the direct binary interaction that we have already isolated.

There are five remaining terms that contribute to the kernel K_2. These terms can be analyzed in the limit of weak coupling and yield

$$- \{1 + P(12)\} \rho_0 \vec{p}_1 \phi_1 \frac{\partial}{\partial \vec{x}_1} B(\vec{x}_1 \vec{x}_3) \Delta_2(2\bar{3}) \tag{5.23}$$

which is the sum of one body additive operators.

One can thus show explicitly that in the weak coupling limit, the theory has two of the desirable features that were present in the older theories of the BBGKY hierarchy. First the direct interaction in the doublet equation is given by the effective potential. This was a feature that did not appear in the usual cumulant approximation, but did appear with Kirkwood cumulants. Second the introduction of the Δ cumulants leads to medium terms which reduce in weak coupling to dressed particle (one body) operators. However, just as in the conventional theories, when one is away from the weak coupling limit the medium terms depend on the spatial relations of both members of the pair as well as the relation of each to a medium particle. These spatial relations are even more pronounced for the $\langle 12|Z|34\rangle_C$ contributions, which we have not discussed.

We now introduce the one body additive operators referred to in Chapter IV. A one body operator L_0 is defined by Eq. (4.47). We have

$$\langle 12|L_0|\overline{34}\rangle\langle\overline{34}|Z_2|\overline{56}\rangle \Delta_2(\overline{56}) = -\{1 + P(12)\} \langle 1|R|\bar{3}2\rangle \Delta_2(\bar{3}2) \tag{5.24}$$

If one writes $L = L_0 + (L - L_0)$ on the left hand side, the resulting doublet L_0 equation can be solved exactly. One obvious choice for L_0 is to take a modified Vlasov operator such as the one in the singlet equation. Using the matrix elements of the exact L

$$\langle 1|R|3\rangle = \langle 1|L|\bar{2}\rangle\langle\bar{2}|Z_1|3\rangle \tag{5.25}$$

and

$$\langle 1|R|\bar{3}\rangle \Delta(\bar{3}) = \left\{\delta(\bar{3} - 1) \vec{p}_3 \frac{\partial}{\partial \vec{x}_3} - \rho_0 \vec{p}_1 \phi_1 \frac{\partial B(\vec{x}_1 - \bar{\vec{x}}_3)}{\partial \vec{x}_1}\right\} \Delta(\bar{3})$$
$$\equiv \left\{\vec{p}_1 \frac{\partial}{\partial \vec{x}_1} - V^*(1)\right\} \Delta(1) \tag{5.26}$$

This 'dressed particle' approximation has the same analytic structure as the Balescu-Lenard equation.

In the dressed particle approximation the doublet equation is

$$\left\{\frac{\partial}{\partial t} + \vec{p}_1 \frac{\partial}{\partial \vec{x}_1} - V^*(1) + \vec{p}_2 \frac{\partial}{\partial \vec{x}_2} - V^*(2)\right\} \Delta_2(12) = H_2(12,t) \quad . \tag{5.27}$$

The right hand side contains the singlet and triplet terms as well as the rest of the doublet kernel $L - L_0$ (which might be treated by iteration). In the dressed particle approximation only the singlet is retained.

To solve the equation we need the time dependent propagator for the singlet equation viz.

$$\left\{\frac{\partial}{\partial t} + \vec{p}_1 \frac{\partial}{\partial \vec{x}_1} - V^*(1)\right\} \langle 1|\Gamma(t)|3\rangle = \delta(1 - 3)\ \delta(t)$$

$$\Gamma(t) = 0 \quad \text{for} \quad t < 0 \quad . \tag{5.28}$$

The solution for the doublet is then (with $\Delta_2(t = 0) = 0$)

$$\Delta_2(12;\ t) = \int_0^t dt'\ \langle 1|\Gamma(t - t')|\overline{3}\rangle\langle 2|\Gamma(t - t')|\overline{4}\rangle H_2(\overline{34},t') \tag{5.29}$$

In contrast to the theory of Chapter III there is now no term from the initial condition because of the definition of Δ_2 (with the usual microscopic preparation). It involves the product of two propagators at the same time and is to be inserted into the singlet Eq. (5.14).

As emphasized in the algebraic considerations, the same approximation could be made in the triplet equation. Then Δ_3 involves the product of three Γ's. However careful analysis is needed to decide which of the many medium terms can be neglected to find a tractable approximation for Δ_2. This can be done for low densities and for weak coupling but it is much harder to justify approximations for denser systems.

As noted in Chapter IV, other one body propagators, which themselves contain hydrodynamic damping, are in principle more satisfactory. One sees concretely in the case of the pseudo-Liouville hierarchy what is involved. For smooth potentials one may imagine that the Δ_2 equation is solved explicitly keeping only the direct screened binary interaction. This part of the true Δ_2 (which is a linear functional of Δ_1) is then taken from the right hand side of the singlet and moved to the left hand side to define a new one body operator $\langle 1|R|3\rangle$. The action in the higher parts of the phase space is uniquely defined by the single quantity $\langle 1|R|3\rangle$, taken from the singlet equation.

One can envision a scheme which starts at the Δ_2 level by defining $<1|R|3>$ to be the exact singlet memory operator. Δ_2 is then analyzed in terms of scattering of fully dressed excitations. This is the standard procedure in quantum field theory and many body theory. Mazenko's theory is set up to do this in a natural way. It leads to nonlinear integral equations for the self consistent determination of $<1|R|3>$ or $<1|\Gamma|3>$. However, since close collision effects have been treated crudely, there is no reason to believe that the solution of the nonlinear equations, even if they could be found, would be satisfactory.

To simplify the nonlinear equations for Γ that result when a 'dressed' particle description is used, one can keep for example only the hydrodynamic parts. This intuitive procedure has been used in discussing long time effects (cf. H. Gould and G. F. Mazenko [1975; 1977], and M. Baus [Baus and Wallenborn, 1977; 1975; Baus, 1975; 1977]) in the one component plasma. It is outside the scope of the present work to try to assess the range of validity of such procedures. They take their cues from the analysis of properties of the exact memory function [Forster, 1974; Pomeau and Resibois, 1975; Resibois, 1972].

D. Summary

Let us now summarize:

In order to maintain historical continuity we have emphasized the cumulant approach. The Δ cumulants reduce to ordinary cumulants at large separations but are more satisfactory for the study of problems with microscopic initial conditions and for the short time evolution of the system. There is a surprising bonus in that short distance difficulties in the conventional cumulant approach are overcome and that binary interactions are screened. There is a basic reason that the present theory has these features. It is that when a preparation is made microscopically by adding a term to a Hamiltonian, we are certain that the system is 'smart enough' to arrange itself so that both the short and long range characteristics of distributions are correctly described. The price that must be paid is the presence of very high order static correlation functions.

We stress that it is already at the singlet level that the superiority to the conventional form is seen. The modified Vlasov medium term remains meaningful at small distances even for strong short range forces since the direct correlation function enters. The t^2 behavior of the density autocorrelation is now correct. The defects of the conventional theory are revealed by the fact that the Balescu equation doesn't correct the short time, short distance inadequacies of the Vlasov equation.

Once one has decided on what to choose for the Δ_n, one could use the usual hierarchy. The Δ_n are just linear combinations of the usual distribution functions or cumulants at each time t. The connection involves space dependent kernels that depend on the static correlation functions. It is the underlying function space description that tells us uniquely what combinations to form. These combinations are far from obvious, particularly in the C sector. There are two further advantages to the function space theory approach. First, a systematic continued fraction structure emerges, i.e. there is some clarity as to what has been neglected, and there is a definite structure to the medium corrections retained at a given level of approximation. Second, the straightforward approach of rearranging the BBGKY in terms of the Δ_n leads to many terms involving the bare potentials. One can use the equilibrium hierarchy to simplify the time dependent hierarchy, and in particular to bring out the effective potential in the two body direct interaction. This process occurs automatically with the function space approach for smooth potentials in view of the Poisson Bracket form for the matrix elements of the Liouville operator.

VI. IMPURITY PROBLEM

A. Initial Conditions

We use the general formalism to study the interaction of an impurity of mass M with particles of mass unity that constitute a medium. We treat the case of smooth potentials and cite results for the hard sphere case. To ensure generality the Hamiltonian is written as

$$H = \frac{\hat{\vec{p}}^2}{2M} + \sum_{i=1}^{N} U(\hat{\vec{q}} - \hat{\vec{q}}_i) + \sum_{i=1} \frac{\hat{\vec{p}}_i^2}{2} + \frac{1}{2} \sum_{i=j}^{N} V(\hat{\vec{q}}_i - \hat{\vec{q}}_j) \tag{6.1}$$

with $\hat{\vec{p}}(t)$, $\hat{\vec{q}}$, $\hat{\vec{p}}_1$, ..., $\hat{\vec{q}}_N$ as dynamical variables.

The type of microscopic initial condition considered is [Lebowitz, Percus and Sykes, 1969]

$$F_{N+1}(t = o) = \Phi \frac{N}{\rho_0} W(\vec{\hat{p}},\vec{\hat{q}})/\phi(\vec{\hat{p}}), \iint W(\vec{\hat{p}},\vec{\hat{q}})\, d\hat{p}\, d\hat{q} = 1 \tag{6.2}$$

Here the medium particles are thermalized relative to the impurity. The Gibbs Φ contains a factor $\phi(p)$ which cancels the ϕ in the denominator. A particular case is

$$W(\vec{\hat{p}},\vec{\hat{q}}) = \delta(\vec{\hat{p}} - \vec{p}^{\,*}_0)\, \delta(\vec{\hat{q}} - \vec{x}^{\,*}_0) \equiv N(0*) \tag{6.3}$$

We use stars on the test particle parameters to indicate that N(0*) differs from

$$N(0) = \sum_{i=1}^{N} \delta(\vec{\hat{p}}_i - \vec{p}_0)\,\delta(\vec{\hat{q}}_i - \vec{x}_0) \quad .$$

The initial singlet distribution is

$$f(1^*;\ t = o) \equiv \int N(1^*)\, F_{N+1}(t = o)\, d\Gamma = W(\vec{p}_1^{\,*}, \vec{x}_1^{\,*})$$

$$f(\bar{1}^*) = 1 \quad , \quad f_{eq}(1^*) = \phi(p_1^*)/\Omega \tag{6.4}$$

The equilibrium singlet distribution is spread out over all space with a strength $1/\Omega$. With the sharp initial condition, one is considering strong deviations from equilibrium, which is, strictly speaking, never reached.

The reduced distributions implied by the microscopic initial conditions are

$$f_{n+1}(1^*, 1, \ldots n;\ t = o) \equiv \int N(1^*)\, N(1, \ldots, n)\, F_{N+1}(t = o)\, d\Gamma$$

$$= W(\vec{p}_1^{\,*}\vec{x}_1^{\,*})\ \frac{\rho_{n+1}(\vec{x}_1^{\,*}\vec{x}_1 \ldots \vec{x}_N)\ \phi_1 \ldots \phi_n}{\rho_0}$$

$$f_{n+1,eq}(1^*, 1, \ldots n) = \frac{\rho_{n+1}(\vec{x}_1^{\,*}\vec{x}_1 \ldots \vec{x}_n)\ \phi_1^*\ \phi_1 \ldots \phi_n}{N} \tag{6.5}$$

We divide by N, since there is only one test particle and $\rho_2(\vec{x}_1^{\,*}\vec{x}_1)$ refers to a pair distribution that reduces to the usual one when the impurity is identical to a medium particle.

The first two equations of the BBGKY hierarchy are

$$\left(\frac{\partial}{\partial t} + \frac{\vec{p}_1^{\,*}}{M}\frac{\partial}{\partial \vec{x}^*}\right) f(1^*) = \frac{\partial U(\vec{x}_1^{\,*} - \vec{\bar{x}}_1)}{\partial \vec{x}_1^{\,*}}\ \frac{\partial}{\partial \vec{p}_1^{\,*}}\ f_2(1^*\bar{1})$$

$$\left\{\frac{\partial}{\partial t} + \frac{\vec{p}_1^{\,*}}{M}\frac{\partial}{\partial \vec{x}_1^{\,*}} + \vec{p}_1\frac{\partial}{\partial \vec{x}_1} - \frac{\partial U(\vec{x}^* - \vec{x}_1)}{\partial \vec{x}_1^{\,*}}\left(\frac{\partial}{\partial \vec{p}_1^{\,*}} - \frac{\partial}{\partial \vec{p}_1}\right)\right\} f_2(1^*1)$$

$$= \left(\frac{\partial U(\vec{x}_1^{\,*} - \vec{\bar{x}}_2)}{\partial \vec{x}_1^{\,*}}\ \frac{\partial}{\partial \vec{p}_1^{\,*}} + \frac{\partial V(\vec{x}_1 - \vec{\bar{x}}_2)}{\partial \vec{x}_1}\ \frac{\partial}{\partial \vec{p}_1}\right) f_3(1^*1\bar{2}) \tag{6.6}$$

with the associated equilibrium hierarchy.

B. Construction of the Function Space

We start with

$$T(1^*) \equiv \sqrt{N}\ N(1^*) \quad . \tag{6.7}$$

Strictly speaking, we would have $T(1^*) = \sqrt{N}\ \{N(1^*) - \langle N(1^*)\rangle\}$ if $T(1^*)$ is to be orthogonal to 1 with weight function Φ. But as remarked earlier

$$\langle N(1^*)\rangle = \frac{\phi(p_1^*)}{\Omega} \quad ,$$

and can be neglected. We also note that $\langle N(2)\rangle = \rho_0\ \phi_2$, $\delta N(1) \equiv N(1) - \langle N(1)\rangle$.

The next step is to introduce

$$T(1^*1) = \sqrt{N}\ N(1^*)\ \delta N(1) - \Lambda(1^*1\bar{2}^*)\ T(\bar{2}^*) \tag{6.8}$$

and to choose $T(1^*1)$ to be orthogonal to $T(3^*)$. One finds

$$\Lambda(1^*12^*) = \frac{\phi_1}{\rho_0}\ \delta(2^* - 1^*)\ \{\rho_2(\vec{x}_1^*\vec{x}_1) - \rho_0^2\} \tag{6.9}$$

This generates the first two cumulants

$$\Lambda_1(1^*) \equiv \langle T(1^*)\ F_{N+1}\rangle = \sqrt{N}\ f(1^*)$$

$$\Lambda_2(1^*1) \equiv \langle T(1^*1)\ F_{N+1}\rangle = \sqrt{N}\left\{f_2(1^*1) - f(1^*)\ \phi_1\ \frac{\rho_2(\vec{x}_1^*\vec{x}_1)}{\rho_o}\right\} \tag{6.10}$$

Δ_2 measures the deviation of f_2 from instantaneous medium equilibrium. The first hierarchy equation is

$$\left(\frac{\partial}{\partial t} + \frac{\vec{p}_1^*}{M}\frac{\partial}{\partial \vec{x}_1^*}\right)\ \Delta_1(1^*) = \frac{\partial U(\vec{x}_1^* - \vec{\bar{x}}_1)}{\partial \vec{x}_1^*}\ \frac{\partial}{\partial \vec{p}_1^*}\ \Delta_2(1^*\bar{1}) \tag{6.11}$$

The simplest truncation $\Delta_2(1^*1;\ t) = 0$ just gives free streaming of the impurity, since the assumption implies local equilibrium of the medium.

For the case where the impurity and medium particles are hard spheres of diameter a., one has

$$\left\{\frac{\partial}{\partial t} + \frac{\vec{p}_1^*}{M}\frac{\partial}{\partial \vec{x}_1^*} - \rho_0\ g(a)\ J_B(1^*)\right\}\ \Delta_1(1^*) = \bar{T}(1^*\bar{1})\ \Delta_2(1^*\bar{1}) \tag{6.12}$$

The $\Delta_2 = 0$ approximation now corresponds to a standard Boltzmann theory of the one body phase space correlation function and thus of the velocity autocorrelation function.

To go to the next level we need T(1*12). We will not construct a complete space with T(1), T(12) involving medium particles alone, since these cumulants are not connected directly to $\Delta(1^*)$. Write

$$T(1^*12) = \sqrt{N}\; N(1^*)\;\delta N(12) - \Lambda(1^*12|\overline{2}^*\overline{3})\; T(\overline{2}^*\overline{3}) - \Lambda(1^*12|\overline{2}^*)\; T(\overline{2}^*) \tag{6.13}$$

and chose the Λ functions to insure orthogonality. We have

$$\Lambda\langle 1^*12|\overline{2}^*\rangle\langle T(\overline{2}^*)\; T(3^*)\rangle = \sqrt{N}\;\langle N(1^*)\;\delta N(12)\; T(3^*)\rangle \tag{6.14}$$

The static correlation function

$$\langle T(2^*)\; T(3^*)\rangle = \rho_0\;\phi_2^*\;\delta(3^* - 2^*) \tag{6.15}$$

has the inverse

$$\langle 2^*|Z_1|3^*\rangle = \frac{1}{\rho_0\;\phi_2^*}\;\delta(3^* - 2^*) \quad . \tag{6.16}$$

Hence

$$\Lambda(1^*12|2^*) = \delta(2^* - 1^*)\left(\frac{\rho_3(\vec{x}_1^*\vec{x}_1\vec{x}_2) - \rho_2(\vec{x}_1\vec{x}_2)}{\rho_0}\right)\phi_1\;\phi_2 \tag{6.17}$$

The construction of $\Lambda(1^*12|2^*3)$ is more complicated. We need the static correlation function

$$\langle T(1^*2)\; T(3^*4)\rangle = \delta(3^* - 1^*)\{\delta(4 - 2)\;\phi_1^*\;\phi_2\;\rho_2(\vec{x}_1^*\vec{x}_2) + \phi_1^*\;\phi_2\;\phi_4\; R_3(\vec{x}_1^*\vec{x}_2\vec{x}_4)\} \tag{6.18}$$

with

$$R_3(\vec{x}_1^*\vec{x}_2\vec{x}_4) = \rho_3(\vec{x}_1^*\vec{x}_2\vec{x}_4) - \frac{\rho_2(\vec{x}_1^*\vec{x}_2)\;\rho_2(\vec{x}_1^*\vec{x}_4)}{\rho_0} \tag{6.19}$$

Then the static inverse is defined by

$$\langle T(1^*1)|T(\overline{2}^*\overline{2})\rangle\langle\overline{2}^*\overline{2}|Z_2|3^*3\rangle = \delta(3^* - 1^*)\;\delta(3 - 1) \tag{6.20}$$

Writing this as an integral equation one finds that the solution can be expressed as

$$\langle 1{*}1|Z_2|2{*}2\rangle = \frac{\delta(2^* - 1^*)}{\phi_1^*} \frac{\delta(2-1)}{\phi_2\ \rho_2(x_2^* x_2)} + \langle \vec{x}_1|Z_2|\vec{x}_2\rangle_{\vec{x}_1^*} \tag{6.21}$$

Here the purely spatial function Z_2 obeys the integral equation

$$\langle\vec{x}_1|Z_2|\vec{x}_2\rangle_{\vec{x}_1^*} + \frac{R_3(\vec{x}_1^*\vec{x}_1\vec{\bar{x}}_3)}{\rho_2(\vec{x}_1^*\vec{x}_1)} \langle\vec{\bar{x}}_3|Z_2|\vec{x}_2\rangle_{\vec{x}_1^*}$$

$$= - \frac{R_3(\vec{x}_1^*\vec{x}_1\vec{x}_2)}{\rho_2(\vec{x}_1^*\vec{x}_1)\ \rho_2(\vec{x}_1^*\vec{x}_2)} \tag{6.22}$$

The iteration solution shows that Z_2 starts as $1/\rho_0$ and contains terms ρ_0^n, $n \geq 0$. The first term in the static inverse starts as $1/\rho_0^2$.

After computing $\Lambda(1{*}12|2{*}3)$, we have

$$T(1{*}12) = \sqrt{N}\ N(1^*)\ N(12) - \frac{\rho_3(\vec{x}_1^*\vec{x}_1\vec{x}_2)}{\rho_0}\ \phi_1\ \phi_2\ T(1^*)$$

$$- \rho_3(\vec{x}_1^*\vec{x}_1\vec{x}_2)\left[\phi_2 \frac{T(1{*}1)}{\rho_2(\vec{x}_1^*\vec{x}_1)} + \phi_1 \frac{T(1{*}2)}{\rho_2(\vec{x}_1^*\vec{x}_2)}\right]$$

$$- T(1{*}\bar{5})\ \phi_1\ \phi_2\ X(\vec{x}_1^*\vec{x}_1\vec{x}_2\vec{\bar{x}}_5) \tag{6.23}$$

where X is a complicated function

$$X(\vec{x}_1^*\vec{x}_1\vec{x}_2\vec{x}_5) = \frac{Q_4(\vec{x}_1^*\vec{x}_1\vec{x}_2\vec{x}_5)}{\rho_2(\vec{x}_1^*\vec{x}_5)} + Q_4(\vec{x}_1^*\vec{x}_1\vec{x}_2\vec{\bar{x}}_7)\ \langle\vec{\bar{x}}_7|Z_2|\vec{x}_5\rangle_{\vec{x}_1^*}$$

$$+ \rho_3(\vec{x}_1^*\vec{x}_1\vec{x}_2)\left\{\vec{x}_1|Z_2|\vec{x}_5\rangle_{x_1^*} + \langle\vec{x}_2|Z_2|\vec{x}_5\rangle_{\vec{x}_1^*}\right\} \tag{6.24}$$

$$Q_4(\vec{x}_1^*\vec{x}_1\vec{x}_2\vec{x}_5) = \rho_4(\vec{x}_1^*\vec{x}_1\vec{x}_2\vec{x}_5) - \rho_2(\vec{x}_1^*\vec{x}_5)\ \rho_2(\vec{x}_1\vec{x}_2) + \rho_0^2(\vec{x}_1\vec{x}_2)$$

$$- \rho_3(\vec{x}_1^*\vec{x}_1\vec{x}_2)\ \frac{\rho_2(\vec{x}_1^*\vec{x}_5)}{\rho_0} \tag{6.25}$$

We examine the Z_2 inverse and the cumulants Δ_2, Δ_3 in some limiting cases.

(a) No interaction between medium particles

$$R_3(\vec{x}_1^*\vec{x}_1\vec{x}_2) = 0 \quad , \quad \rho_3(\vec{x}_1^*\vec{x}_1\vec{x}_2) = \frac{\rho_2(\vec{x}_1^*\vec{x}_1)\ \rho_2(\vec{x}_1^*\vec{x}_2)}{\rho_0}$$

$$\langle\vec{x}_1|Z_2|\vec{x}_2\rangle_{\vec{x}_1^*} = 0 \quad , \quad Q_4(\vec{x}_1^*\vec{x}_1\vec{x}_2\vec{x}_5) = -\rho_0^4\ h(\vec{x}_1^*\vec{x}_5)$$

$$X(\vec{x}_1^*\vec{x}_1\vec{x}_2\vec{x}_5) = -\rho_0^4\ h(\vec{x}_1^*\vec{x}_5)/\rho_2(\vec{x}_1^*\vec{x}_5)$$

$$\begin{aligned}\Delta_3(1{*}12) \to \sqrt{N}\left\{f_3(1{*}12) - \frac{\rho_2(\vec{x}_1^*\vec{x}_1)\ \rho_2(\vec{x}_1^*\vec{x}_2)}{\rho_0^2}\right\}\\ - \rho_2(\vec{x}_1^*\vec{x}_2)\ \phi_2\ \Delta_2(1{*}1)\\ - \rho_2(\vec{x}_1^*\vec{x}_1)\ \phi_1\ \Delta_2(1{*}2)\\ + \rho_0^4\ \phi_1\ \phi_2\ \Delta_2(1{*}\bar{5})\ h(\vec{x}_1^*\bar{\vec{x}}_5)/\rho_2(\vec{x}_1^*\bar{\vec{x}}_5)\end{aligned} \tag{6.26}$$

(b) Vanishingly weak interaction between impurity and medium particles

$$\rho_2(\vec{x}_1^*\vec{x}_2) \to \rho_0^2 \quad , \quad \rho_3(\vec{x}_1^*\vec{x}_1\vec{x}_2) \to \rho_0\ \rho_2(\vec{x}_1\vec{x}_2) \quad ,$$

$$R_3(\vec{x}_1^*\vec{x}_1\vec{x}_2) \to \rho_0^3\ h(\vec{x}_1\vec{x}_2)$$

The integral equation for Z_2 is

$$\langle\vec{x}_1|Z_2|\vec{x}_2\rangle_{\vec{x}_1^*} + \rho_0\ h(\vec{x}_1\bar{\vec{x}}_3)\langle\bar{\vec{x}}_3|Z_2|\vec{x}_2\rangle_{\vec{x}_1^*} = -\frac{h(\vec{x}_1\vec{x}_2)}{\rho_0}$$

thus

$$\langle\vec{x}_1|Z_2|\vec{x}_2\rangle_{\vec{x}_1^*} \to -\frac{B(\vec{x}_1\vec{x}_2)}{\rho_0} \tag{6.27}$$

$$Q_4(\vec{x}_1^*\vec{x}_1\vec{x}_2\vec{x}_5) \to \rho_0\{\rho_3(\vec{x}_1\vec{x}_2\vec{x}_5) - \rho_0\ \rho_2(\vec{x}_1\vec{x}_2)\}$$

(c) All interactions weak (first order in correlation functions)

$$X \to -\rho_0^2\ h(\vec{x}_1^*\vec{x}_5)$$

$$\Delta_3(1{*}12) \to \sqrt{N}\ f_3(1{*}12) - \rho_2(\vec{x}_1\vec{x}_2)\ \phi_1\ \phi_2\ \Delta_1(1{*})$$

$$- \rho_0\ \phi_2\ \Delta_2(1{*}1) - \rho_0\ \phi_1\ \Delta_2(1{*}2) \qquad (6.28)$$

This reduces to the conventional cumulants adapted for the impurity problem.

(d) Superposition approximation

The integral equation for Z_2 may be written as

$$\langle\vec{x}_1|Z_2|\vec{x}_2\rangle_{\vec{x}_1^*} + \rho_0\ h(\vec{x}_1\overline{\vec{x}_3})\ \langle\overline{\vec{x}_3}|Z_2|\vec{x}_2\rangle_{\vec{x}_1^*}$$

$$+ \rho_0\ h(\vec{x}_1\overline{\vec{x}_3})\ h(\vec{x}_1^*\overline{\vec{x}_3})\ \langle\overline{\vec{x}_3}|Z_2|\vec{x}_2\rangle_{\vec{x}_1^*}$$

$$= - \frac{h(\vec{x}_1\vec{x}_2)}{\rho_0} \qquad (6.29)$$

If we neglect the third term on the left hand side and use the definition of the O-Z direct correlation function, we find

$$\langle\vec{x}_1|Z_2|\vec{x}_2\rangle_{\vec{x}_1^*} \to - \frac{1}{\rho_0}\ B(\vec{x}_1 - \vec{x}_2)\ , \qquad (6.30)$$

which is the first term of a series expansion.

C. Kinetic Equations

Let us now take up the problem of constructing the kinetic equation for Δ_2. One could use the direct approach of expressing f_3 in terms of Δ_1, Δ_2, Δ_3. The truncation $\Delta_3(1{*}12); t) = 0$ yields

$$f_3(1{*}12;\ t) \approx \Lambda(1{*}12|\overline{2} * \overline{3})\ \Delta_2(\overline{2}{*}\overline{3};\ t)$$

$$+ \frac{\rho_3(\vec{x}_1^*\vec{x}_1\vec{x}_2)}{\rho_0}\ \phi_1\ \phi_2\ f_1(1{*};\ t) \qquad (6.31)$$

We insert this in the second hierarchy equation and obtain an integro-differential equation for $(\partial/\partial t)\ \Delta_2(1^*1;\ t)$ with $f_1(1^*;\ t)$ as an inhomogeneous term. This is what is done in the hard sphere case. However the bare potential and the static correlation functions enter in a complicated way and one has to use the equilibrium hierarchy to effect simplifications. For smooth potentials one can use the orthogonal function machinery instead.

We need matrix elements of the Liouville operator taken between our basis functions. One finds

$$\langle 1^*1|L|\overline{2}^*\rangle\langle\overline{2}^*|Z_1|\overline{3}^*\rangle\ \Delta_1(\overline{3}^*) = \phi_1\ \phi_1^*\ \frac{1}{\theta}\ \frac{\partial\rho_2(\vec{x}_1^*\vec{x}_1)}{\partial\vec{x}_1^*}\ \frac{\partial}{\partial\vec{p}_1^*} \cdot \left(\frac{\Delta_1(1^*)}{\rho_0\ \phi_1^*}\right) \tag{6.32}$$

The exact doublet equation is

$$\frac{\partial}{\partial t}\ \Delta_2(1^*1) + \langle 1^*1|K|\overline{3}^*\overline{3}\rangle\ \Delta_2(\overline{3}^*\overline{3})$$

$$= -\ \frac{\phi_1}{\rho_0}\ \frac{\partial\rho_2(\vec{x}_1^*\vec{x}_1)}{\partial\vec{x}_1^*}\left(\frac{1}{\theta}\ \frac{\partial}{\partial\vec{p}_1^*} + \frac{\vec{p}_1^*}{M}\right)\Delta_1(1^*)$$

$$+\left(\frac{\partial U(\vec{x}_1^* - \vec{\overline{x}}_2)}{\partial\vec{x}_1^*}\ \frac{\partial}{\partial\vec{p}_1^*} + \frac{\partial V(\vec{x}_1 - \vec{\overline{x}}_2)}{\partial\vec{x}_1^*}\ \frac{\partial}{\partial\vec{p}_1}\right)\Delta_3(1^*1\overline{2}) \tag{6.33}$$

The doublet kernel is

$$\langle 1^*1|K|3^*3\rangle = \langle 1^*1|L|\overline{2}^*\overline{2}\rangle\langle\overline{2}^*\overline{2}|Z_2|3^*3\rangle \tag{6.34}$$

Direct calculation shows that the matrix element of L is the sum of two parts. The direct part is

$$\langle 1^*1|L|2^*2\rangle_D = \delta(2 - 1)\ \delta(2^* - 1^*)\{L(2^*|2) + L(2|2^*)\}\langle N(2^*2)\rangle \tag{6.35}$$

The medium part is

$$\langle 1^*1|L|2^*2\rangle_M = \delta(2^* - 1^*)\left\{\frac{\vec{p}_2^*}{M}\ \frac{\partial}{\partial\vec{x}_2^*} + \frac{1}{\theta}\ \frac{\partial\ \ell n\ R_3(\vec{x}_1^*\vec{x}_1\vec{x}_2)}{\partial\vec{x}_2^*}\ \frac{\partial}{\partial\vec{p}_2^*}\right\} \cdot R_3(\vec{x}_1^*\vec{x}_1\vec{x}_2)\ \phi_1^*\ \phi_1\ \phi_2 \tag{6.36}$$

The exact doublet kernel is

$$\langle 1^*1|K|\overline{2}^*\overline{2}\rangle\, \Delta_2(\overline{2}^*\overline{2}) = \{L(1^*|1) + L(1|1^*)\}\, \Delta_2(1^*1)$$
$$+ \vec{p}_1\, \phi_1\, \rho_2(x_1^* x_1)\, \frac{\partial}{\partial \vec{x}_1} \langle \vec{x}_1 | Z_2 | \vec{\overline{x}}_2 \rangle_{\vec{x}_1^*}\, \Delta_2(1^*\overline{2})$$
$$- \phi_1 \frac{\rho_2(\vec{x}_1^* \vec{x}_1)}{\theta} \frac{\partial}{\partial \vec{x}_1^*} \ell n[R_3(\vec{x}_1^* \vec{x}_1 \vec{x}_2)/\rho_2(\vec{x}_1^* \vec{x}_1)]$$
$$\cdot \langle \vec{x}_1 | Z_2 | \vec{\overline{x}}_2 \rangle_{\vec{x}_1^*} \frac{\partial \Delta_2(1^*\overline{2})}{\partial \vec{p}_1^*} \qquad (6.37)$$

This kernel has a direct pair interaction where the effect of the medium is present in the effective potential

$$-\frac{1}{\theta}\, \ell n\, \rho_2(\vec{x}_1^* \vec{x}_1) \quad .$$

For small impurity medium-impurity separations it becomes the bare potential $U(\vec{x}_1^* - \vec{x}_1)$, while at larger separations we encounter the screening effects of the medium.

The behavior of the other terms in the kernel depends on the approximate form of

$$\langle \vec{x}_1 | Z_2 | \vec{\overline{x}}_2 \rangle_{\vec{x}_1^*} \quad .$$

From our earlier discussion a reasonable first approximation is

$$\langle \vec{x}_1 | Z_2 | \vec{x}_2 \rangle_{\vec{x}_1^*} \to -B(\vec{x}_1 - \vec{x}_2)/\rho_0$$

Then the second term in K is

$$-\rho_1\, \phi_1\, \rho_0[1 + h(\vec{x}_1^* \vec{x}_1)]\, \frac{\partial B(\vec{x}_1 - \vec{\overline{x}}_2)}{\partial \vec{x}_1}\, \Delta_2(1^*\overline{2}) \qquad (6.38)$$

When $\vec{x}_1^*$ and $\vec{x}_1$ are well separated this is a modified Vlasov term, and is a one body additive operator on the medium particles. At small separations the one body character is destroyed because of the spatial correlations introduced by $h(x_1^* x_1)$. This is a characteristic result. Recall that in plasma units both B(r) and h(r) are of order γ. So the correction is of higher order in γ. Since $\gamma n_0 = 1/4\pi$ the leading term is independent of γ. (One must of course modify the hierarchy to include the compensating background as in Chapter III.) The collision kernel for the singlet equation is of order γ and has the modified Balescu form. This supplies an explicit justification for the dressed particle approximation.

We now set down the doublet equation for the hard sphere case. It is

$$\left[\frac{\partial}{\partial t} + \frac{\vec{p}_1^*}{M}\frac{\partial}{\partial \vec{x}_1^*} + \vec{p}_1 \frac{\partial}{\partial \vec{x}_1} - \overline{T}(1^*1) - \frac{\rho_3(\vec{x}_1^*\vec{x}_1\vec{\overline{x}}_2)}{\rho_2(\vec{x}_1^*\vec{x}_1)}\right.$$

$$\left.\cdot\left\{\overline{T}(1^*\overline{2}) + \overline{T}(1\overline{2})\right\}\overline{\phi}_2\right]\Delta_2(1^*1)$$

$$- \overline{T}(1\overline{2})\frac{\rho_3(\vec{x}_1^*\vec{x}_1\vec{\overline{x}}_2)}{\rho_2(\vec{x}_1^*\vec{x}_1)}\phi_1\,\Delta_2(1^*\overline{2}) + A - B = H_2(1^*1)$$

(6.39)

where

$$A = -\left\{\frac{\rho_3(\overline{x}_1^*\overline{x}_1\vec{\overline{x}}_2)}{\rho_2(\vec{x}_1^*\vec{\overline{x}}_2)} - \frac{\rho_2(\vec{x}_1^*\vec{\overline{x}}_2)}{\rho_0}\right\}\phi_1\,\overline{T}(1^*\overline{2})\Delta_2(1^*\overline{2})$$

$$B = -X(\vec{x}_1^*\vec{x}_1\vec{x}_2\vec{\overline{x}}_3)\,\overline{T}(1^*\overline{2})\,\overline{\phi}_2\,\phi_1\,\Delta_2(1^*\overline{3}) \qquad (6.40)$$

All of the terms except A and B occur when we set up the kinetic equation for the standard doublet cumulant $\tilde{c}_2(1^*1)$. They reduce to the standard terms at low density, except for geometrical factors involving spatial correlation functions that insure that the collisions are physically reasonable. The theory thus explicitly generates independent Boltzmann operators in the doublet cumulant equation.

The term A is a new medium term. It is proportional to the density, but at low densities a cumulant expansion of the static correlation shows that it also has a factor $h(\vec{x}_1^*\vec{x}_1)$ which is small over most of space. The final term B is proportional to ρ_0^2 and also has a factor $h(\vec{x}_1^*\vec{x}_3)$.

The right hand side is found by straightforward calculation

$$H_2(1^*1) = \frac{\rho_2(\vec{x}_1^*\vec{x}_1)}{\rho_0}\,\overline{T}(1^*1)\,\phi_1\,\Delta_1(1^*)$$

$$+ \frac{R_3(\vec{x}_1^*\vec{x}_1\vec{\overline{x}}_2)}{\rho_0}\,\phi_1\,\overline{T}(1^*\overline{2})\,\overline{\phi}_2\,\Delta_1(1^*)$$

$$+ \left(\frac{\vec{p}_1^*}{M} - \vec{p}_1 \right) \rho_0 \frac{\partial h(\vec{x}_1^* \vec{x}_1)}{\partial \vec{x}_1^*} \phi_1 \Delta_1(1^*)$$

$$+ [\bar{T}(1^*\bar{2}) + \bar{T}(1\bar{2})] \Delta_3(1^*1\bar{2}) \qquad (6.41)$$

Thus far we have only rearranged the first two equations of the BBGKY hierarchy. For the hard sphere case, one may not be too far from a truly microscopic theory. The $\Delta_2 \approx 0$ approximation already yields a Boltzmann equation. At low densities and for large separations of 1* and 1 the doublet kernel reduces to the sum of $-\rho_0 J_B(1^*) \Delta_2(1^*1)$ for the impurity and the L.P.S. operator for the medium particle. This sum of one body additive operators yields a natural dressed particle approximation to $\Delta_2(1^*1)$ which is a short time modification of the Ernst-Dorfman theory. Of course, this is not the best choice of the one body additive operators, since the medium is only described in the L.P.S. approximation. Improvements have been discussed by Resibois and Lebowitz.

It is tempting to use the exact one body propagator for the medium particles in absence of the impurity. An interesting theory of this type has been developed for smooth potentials by Sjölander and Sjögren. Of course, something microscopic or phenomenological has to be said about the medium propagator. Even then it is not clear how much effective mass and close collision effects contribute. The conservative point of view is that one should make the dressed particle approximation at the level of $\Delta_3(1^*12)$. One imagines solving the equations in the 2BA approximation ($\Delta_3 = 0$) to give a better account of close collisions. One then uses the corrected Boltzmann propagator to make a dressed particle approximation to $\Delta_3(1^*12)$. This is hopeless in practice, unless kinetic models or variational methods can be used to analyze the 2BA. In any event, we seem here to be at the limit of controlled approximations for the high density case, and more intuitive considerations take over.

In the smooth potential case and in particular for the strongly coupled plasma, we are far from a convincing microscopic theory. Consider theories that work at the level of the first memory function. From the continued fraction structure we know that the Laplace transform of the doublet cumulant can be written as

$$\tilde{\Delta}_2(1^*1) = \langle 1^*1 | \tilde{W}(S) | (\bar{2}^*2) \rangle \bar{\phi}_2 \frac{\partial \rho_2(\vec{x}_2^* \vec{x}_2)}{\partial \vec{x}_2^*} \frac{1}{\rho_0}$$

$$\cdot \left(\frac{1}{\theta} \frac{\partial}{\partial \vec{p}_2^*} + \frac{\vec{p}_2^*}{M} \right) \tilde{\Delta}_1(\bar{2}^*)$$

to be inserted into

$$\left(S + \frac{\vec{p}_1^*}{M}\frac{\partial}{\partial \vec{x}_1^*}\right)\tilde{\Delta}_1(1^*) = \frac{\partial U(\vec{x}_1^* - \vec{\bar{x}}_1)}{\partial \vec{x}_1^*}\frac{\partial}{\partial \vec{p}_1^*}\tilde{\Delta}_2(1^*\bar{1}) + \Delta_1(1^*, t = 0)$$

The only rigorous result is the 'vertex' function on which $\tilde{W}(S)$ operates. $\tilde{W}(S)$ must incorporate the close collision effect which are already partly taken into account in the singlet equation for the low density hard sphere case. In addition it should describe the coupling to the damped collective modes. As noted many times the dressed particle 1BA for hard spheres is on the same level as the dressed particle 2BA for smooth potentials. It is possible to make intelligent guesses as to the form of $\tilde{W}(S)$, cf. Gould and Mazenko and Baus. It is also possible to use sum rules and short time behavior to control parameters in an assumed form. (cf. the contributions of Singwi, Ichimaru, Totsuji, Golden, Kalman in this volume). It is, however, our feeling that we are still far from a microscopic theory of time dependent correlation functions in strongly coupled plasmas.

One suggestion is that the variational approach (cf. Chapter II) may help to obtain tractable and accurate theories. Recall that the microscopic theory is equivalent to the expansion

$$F_{N+1}(t) = |T(\bar{1}^*)\rangle\langle\bar{1}^*|Z_1|\bar{2}^*)\ \Delta_1(\bar{2}^*;\ t)$$

$$+\ |T(\bar{1}^*\bar{1}\rangle\langle\bar{1}^*\bar{1}|Z_1|\bar{2}^*\bar{2}\rangle\ \Delta_2(\bar{2}^*\bar{2};\ t) + \ldots$$

The form of the exact equations satisfied by Δ_2 can be used to suggest trial forms. This amounts to choosing trial forms for the 1BA memory operator. The variational principle can be used to determine free parameters present in the trial functions.

REFERENCES

Akcasu, A. Z. and J. J. Duderstadt, 1969, Phys. Rev. 188, 479; 1970, Phys. Rev. A1, 905.

Aona, A., 1968, Phys. of Fluids 11, 341.

Balescu, R., 1963, Statistical Mechanics of Charged Particles, John Wiley & Sons, New York.

Baus, 1975, Physica 79A, 377.

Baus, 1977, Phys. Rev. A15, 790.

Baus, M. and J. Wallenborn, 1977, J. of Statistical Physics, 16, 91.

Baus, M. and J. Wallenborn, 1975, Phys. Lett. 55A, 90.

Bergeron, K. D. and E. P. Gross, 1975, J. of Statistical Physics 13, 85.

Bergeron, K. D., E. P. Gross, and R. D. Varley, 1974, J. of Statistical Physics 10, 111.

Berne, B. J., 1971, Chap. 9 in Physical Chemistry, edited by H. Eyring, D. Henderson, W. Jost, Vol. VIII B, Academic Press, New York.

Blum, L. and J. L. Lebowitz, 1969, Phys. Rev. 185, 273.

Boley, C. D., 1975, Phys. Rev. A11, 328.

Boley, C. D., 1974, Annals of Physics 86, 91.

DeWitt, H. E., this Volume.

Dupree, T., 1961, Phys. of Fluids 4, 696.

Ernst, M. H., J. R. Dorfman, W. R. Hoegy and J. M. M. van Leeuwen, 1969, Physica 45, 127.

Ernst, M. H. and J. R. Dorfman, 1972, Physica 61, 157.

Fisher, I. Z., 1964, Statistical Theory of Liquids, U. of Chicago Press.

Forster, D. and P. C. Martin, 1970, Phys. Rev. A2, 1565.

Forster, D., 1974, Phys. Rev. A9, 943.

Frieman, E. A. and D. L. Book, 1963, Phys. of Fluids 6, 1700.

Gotze, W. and M. Lucke, 1975, Phys. Rev. A11, 2173.

Gould, H. and H. E. DeWitt, 1967, Phys. Rev. 155, 68.

Gould, H. and G. F. Mazenko, 1975, Phys. Rev. Lett. 35, 1455.

Gould, H. and G. F. Mazenko, 1977, Phys. Rev. A15, 1274.

Grad, H., 1949, Handbuch der Physik.

Gross, E. P., 1972, Annals of Physics 69, 42.

Gross, E. P., 1973, J. of Statistical Physics 9, 275, 297; 1974, J. of Statistical Physics 11, 503; 1976, J. of Statistical Physics 15, 181.

Gross, E. P. and M. Lindenfeld, to be published.

Gross, E. P. and D. Wisnivesky, 1968, Phys. of Fluids 11, 1387.

Guernsey, R. L., 1964, Phys. of Fluids 7, 1600.

Hansen, J. P., this Volume.

Honda, N., O. Aono and T. Kihara, 1963, J. Phys. Soc. Japan 18, 856.

Hopps, J., 1971, Ph.D. Thesis, Brandeis University.

Hopps, J., 1976, Phys. Rev. A13, 1226.

Jhon, M. S. and D. Forster, 1975, Phys. Rev. A12, 254.

Jordan, P. C., 1974, Phys. Rev. A10, 319.

Keyes, T. and I. Oppenheim, 1973, Phys. Rev. A7, 1384; 1973, Phys. Rev. A8, 937.

Kihara, T. and O. Aono, 1963, J. Phys. Soc. Japan 18, 837.

Kim, K. and M. Nelkin, 1971, Phys. Rev. A4, 2065.

Kirkwood, J. G., 1946, J. Chem. Phys. 14, 180.

Klimontovich, Y. L., 1967, Statistical Theory of Non-Equilibrium Processes in Plasmas, Pergamon Press.

Konijnendyck, H. H. U. and M. J. van Leeuwen, 1973, Physica 64, 342.

Lebowitz, J. L., J. K. Percus and J. Sykes, 1969, Phys. Rev. 188, 487.

Lindenfeld, M., 1975, Thesis, Brandeis University.

Lindenfeld, M., 1977, Phys. Rev. A15, 1801.

Martin, P. C., 1968, in Many Body Physics, edited by C. DeWitt and R. Balian, Gordon and Breach, New York.

Mazenko, G., 1973, Phys. Rev. 7, 209, 222; 1974, Phys. Rev. 9, 360.

Mazenko, G. F. and S. Yip, to be published, Renormalized Kinetic Theory of Dense Fluids in Statistical Mechanics, in Modern Theoretical Chemistry, Part B., edited by B. J. Berne, Plenum Press, New York.

Mortimer, R. G., 1968, J. Chem. Phys. 48, 1023.

Mostellor, R. D. and J. J. Duderstadt, 1974, J. of Statistical Physics 11, 409.

Nossal, R., 1968, Phys. Rev. 166, 81.

Nossal, R. and R. Zwanzig, 1967, Phys. Rev. 157, 120.

O'Neil, T. and N. Rostoker, 1965, Phys. of Fluids 8, 1109.

Ortoleva, P. J. and M. Nelkin, 1969, Phys. Rev. 181, 429; 1970, Phys. Rev. A2, 187.

Phythian, R., 1972, Phys. A-5, 1566.

Pomeau, Y. and P. Resibois, June 1975, Physics Reports 19C, No. 2.

Prigogine, L., 1962, Non-Equilibrium Statistical Mechanics, Interscience.

Resibois, P., 1966, in Physics of Many Particle Systems, edited by E. Meeron, Gordon and Breach.

Resibois, P., 1972, in Irreversibility in the Many Body Problem, edited by J. Biel and J. Rae, Plenum Press, New York.

Resibois, P., 1975, J. of Statistical Physics 13, 393.

Resibois, P., 1976, in Statistical Physics, edited by L. Pal and P. Szepfalusy, North Holland.

Resibois, P. and J. L. Lebowitz, 1975, J. of Statistical Physics 12, 483.

Rice, S. A. and P. Gray, 1964, Stat. Mech. of Simple Liquids, John Wiley & Sons, New York.

Rosenbluth, M. and N. Rostoker, 1960, Phys. of Fluids 3, 1.

Stillinger, F. H. and R. J. Suplinskas, 1966, J. Chem. Phys. 44, 2432.

Sykes, J., 1973, J. of Statistical Physics 8, 279.

Wu, T. I., 1966, Kinetic Equations of Gases and Plasmas, Addison Wesley, Reading, Massachusetts.

Zwanzig, R., 1968, Phys. Rev. 144, 170; 1967, Phys. Rev. 156, 190.

PERTURBATIVE AND NON-PERTURBATIVE METHODS FOR QUANTUM AND CLASSICAL PLASMAS

A. Sjölander

Institute of Theoretical Physics
Chalmers University of Technology
Fack, S-402 20 Göteborg, Sweden

TABLE OF CONTENTS

PERTURBATIVE AND NON-PERTURBATIVE METHODS FOR QUANTUM AND CLASSICAL PLASMAS

A. Sjölander

Institute of Theoretical Physics, Chalmers University of Technology
Fack, S-402 20 Göteborg, Sweden

I. INTRODUCTION

The present lectures are divided into two different parts. In the first one we shall briefly outline the conventional perturbation treatment for the dynamics of a quantum mechanical system of interacting Fermi particles, with particular reference to electrons in metals. Due to the long range of the Coulomb interaction, straightforward expansion in powers of the potential has to be handled with some care. Partial summation of the series becomes necessary and this introduces a screened Coulomb interaction between the electrons. At the same time new collective modes appear, the so-called plasma oscillations. In all these respects everything is very analogous to what occurs in a classical one-component plasma. However, due to the existence of a Fermi surface and other quantum mechanical effects the theoretical treatment is conventionally developed very differently for the quantum plasma than for the corresponding classical system. The basic procedure was developed during the later part of 1950's and the first half of 1960's, and it goes under the name of Green-function technique. For a more comprehensive treatment references are given to some standard textbooks [Schultz, 1964; Raimes, 1972; Fetter and Walecka, 1971; Doniach and Sondheimer, 1974]. Here, only the basic ideas will be presented and a few specific situations considered.

In the second part of the lectures some new results for dense classical plasmas are presented. The calculations are based on a theory [Sjögren and Sjölander, in press; Sjogren, preprint], which was developed for ordinary classical liquids with strong short range interactions, and it has been modified to apply to a one-component plasma by Sjödin and Mitra [Sjödin and Mitra, 1977]. The original theory was used successfully for studying Argon [Sjögren, in press] and Rubidium [Sjögren and Sjölander, in press], and with appropriate modifications it seems to work well also for plasmas.

II. QUANTUM PLASMAS

A. Free Fermions

The characteristic feature of a system of non-interacting fermions is that at most one particle can occupy a single quantum state, and at zero temperature (note that the Fermi temperature in simple metals is 1-10x10^4K) all the states inside the Fermi sphere are occupied and the others are empty. At finite temperatures a smearing of the occupation number occurs near the Fermi surface. A particle with the momentum $\vec{p}$ has the energy $\varepsilon_p = p^2/2m$, where m is the particle mass. The amplitude of a particle wave evolves in time as

$$a_{\vec{p}}(t) = a_{\vec{p}} \exp(-i\varepsilon_p t/\hbar) \exp(-\delta t/\hbar) \quad , \quad t>0 \ (\delta = 0^+) \quad . \quad (2.1)$$

The correlation function

$$G^o_{part}(\vec{p},t) = -i <0|a_{\vec{p}}(t)\ a^+_{\vec{p}}(0)|0>$$

$$= -i <0|a_{\vec{p}}\ a^+_{\vec{p}}|0> \exp(-i\varepsilon_p t/\hbar) \exp(-\delta t/\hbar) \quad , \quad t>0 \quad (2.2)$$

represents the situation where one particle of momentum $\vec{p}$ ($p>p_F, p_F$ = Fermi momentum) is added above the filled Fermi surface and it shows how the amplitude of the particle wave evolves in time. $|0>$ denotes the ground state and an averaging is made over this. Second quantization is used and $a^+_{\vec{p}}$ and $a_{\vec{p}}$ are then creation and annihilation operators, which add and subtract, respectively, one particle. We have

$$<0|a_{\vec{p}}\ a^+_{\vec{p}}|0> = 0 \quad , \quad p < p_F \quad ,$$

$$= 1 \quad , \quad p > p_F \quad . \quad (2.3)$$

For $p<p_F$ we may take one particle out of the system, creating a hole inside the Fermi sphere, and the evolution of this hole is described through

$$G^o_{hole}(\vec{p},t) = i <0|a^+_{\vec{p}}(0)\ a_{\vec{p}}(t)|0>$$

$$= i <0|a^+_{\vec{p}}\ a_{\vec{p}}|0> \exp(-i\varepsilon_p t/\hbar) \exp(\delta t/\hbar) \quad , \ t<0 \ , \quad (2.4)$$

where we here consider negative times. Defined in this way, we can combine the two correlation functions to one quantity, the so-called time-ordered Green-function,

$$G_o(\vec{p},t) = -i \langle 0| a_{\vec{p}}(t)\ a^{+}_{\vec{p}}(0)|0\rangle \quad , \quad t>0 \quad ,$$

$$= i \langle 0| a^{+}_{\vec{p}}(0)\ a_{\vec{p}}(t)|0\rangle \quad , \quad t<0 \quad , \tag{2.5}$$

where

$$\langle 0| a^{+}_{\vec{p}}\ a_{\vec{p}}|0\rangle = n^{o}_{\vec{p}} \quad ,$$

$$\langle 0| a_{\vec{p}}\ a^{+}_{\vec{p}}|0\rangle = 1 - n^{o}_{\vec{p}} \quad , \tag{2.6}$$

$n^{o}_{\vec{p}}$ being the average number of particles in the quantum state $\vec{p}$. One has actually to introduce separate notations for particles with spin up and down, i.e., $\vec{p}\rightarrow(\vec{p},\sigma=\pm 1)$. The Fourier transform of the Green function takes the following simple form:

$$G_o(\vec{p},\omega) = \int_{-\infty}^{\infty} dt\ e^{i\omega t}\ G_o(\vec{p},t) = \frac{i}{\omega - \varepsilon_p + i\ \delta_p} \quad , \tag{2.7}$$

with $\delta_p = 0^+$ or 0^-, depending on whether $p>p_F$ or $p<p_F$.

B. Interacting Fermions

We may now ask what happens, when the interaction between the fermions is turned on. Some of the main changes are summarized below.

(i) The ground state $|0\rangle$ is now a collection of N interacting particles, which are correlated and have a pair correlation function differing from that of a non-interacting system (see Figure 1).

(ii) The occupation number $n_{\vec{p}}$ is modified and even at zero temperature we can find particles with momenta above the Fermi surface (see Figure 2). This is directly reflected in Compton scattering against conduction electrons in metals. We have still a discontinuity at p_F and this has far reaching consequences for the system. So for instance, the specific heat becomes linear in temperature as for free particles and only the proportionality factor is affected by the interaction.

(iii) The wave amplitudes of the particles and the holes are modified and an essential part of it has the form

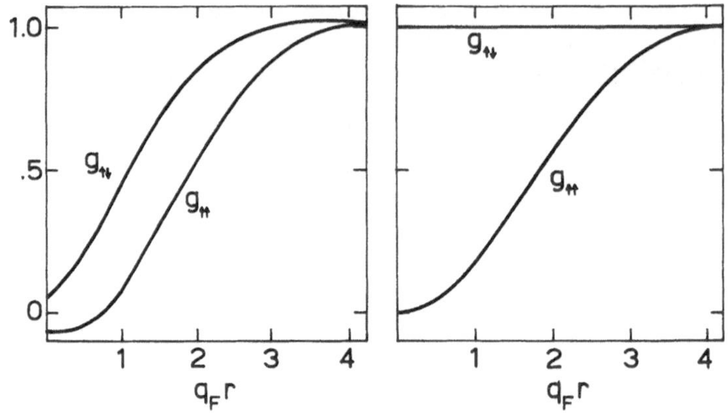

Figure 1. Static pair correlation functions for parallel and anti-parallel spins and for the interacting (left figure) and the non-interacting (right figure) system, respectively. (From Lobo, Singwi and Tosi [1969]). The curves are representative for electrons in Sodium ($r_s = 4$). The negative values of $g_{\uparrow\uparrow}$ for small r is a defect of the theory.

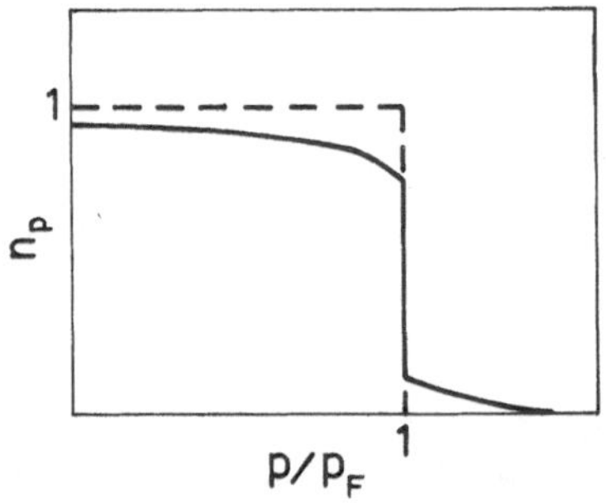

Figure 2. Occupation number n_p, representative for electrons in Sodium. The dotted line shows the corresponding curve for the non-interacting system. (From Lindqvist [1969]).

$$a_{\vec{p}}(t) \sim a_{\vec{p}} \exp(-iE_p t/\hbar) \exp(-\Gamma_p t/\hbar) \quad , \quad t>0 \quad , \tag{2.8}$$

where $E_p \neq \varepsilon_p$ represents the energy of the particle wave and Γ_p is the damping. Due to the Pauli principle we have

$$\Gamma_p \sim (p - p_F)^2 \quad , \quad T=0 \quad , \tag{2.9}$$

for $p \simeq p_F$ and it implies that a particle state and a hole state near the Fermi surface has a very long life time. This has important consequences for the transport properties of metals, for instance. For long wavelength disturbances the system can be considered as built up of particles with energies E_p and with a reduced strength of the interaction. This is precisely the picture, from which Landau developed his famous Fermi liquid theory, and it was done before the modern Green function technique was introduced in solid state physics.

As for free particles, we may through Eq. (2.5) introduce a single particle propagator $G(\vec{p},t)$ and $n_{\vec{p}}$ means now the occupation number for the interacting system. $G(\vec{p},t)$ still represents for $t>0$ the situation where an extra particle of momentum $\vec{p}$ is inserted into the system and it shows how the wave amplitude evolves in time. For negative time it describes the situation where a particle is removed from the system, i.e., a hole is introduced. To begin with, the bare particle surrounds itself with a cloud of other correlated particles and this local configuration of particles behaves to some extent as a unit, a so-called quasi-particle, and it moves with the energy E_p. The quasi-particle has, however, a certain finite life time and this enters through Γ_p in Eq. (2.8). All this can be described quantitatively, if we are able to calculate $G(\vec{p}\omega)$.

C. Perturbation Expansion

Starting from the Schrödinger equation, one can write down an equation for $G(\vec{p},t)$, which contains the interparticle potential $v(r)$ explicitly. We may then expand the solution in powers of $v(r)$ and each term will contain the free particle and hole propagator $G_0(\vec{p},t)$ besides the potential. To make the writing more economic one introduces a diagrammatic language, where a full line with an arrow to the right means a free particle propagator $G^0_{part}(\vec{p},\omega)$, a full line with an arrow to the left means a free hole propagator $G^0_{hole}(\vec{p},\omega)$, and a dashed line represents the interaction potential $v(r)$. Precise rules have been worked out for how to interpret the diagrams in terms of explicit analytic expressions.

Let us now consider the full propagator $G(\vec{p},t)$, for which the first few terms in the diagram expansion is shown in Figure 3. As an illustration the diagrams (a) and (b) are also given analytically;

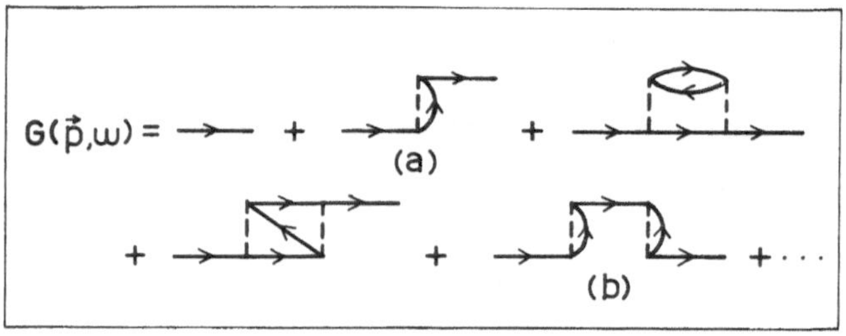

Figure 3. Diagram expansion of $G(\vec{p},\omega)$ to second order in v(q).

$$(a) = [-(2\pi)^{-3}\int v(\vec{p}-\vec{p}')\, n^{o}_{\vec{p}'}\, d\vec{p}'][G_o(\vec{p},\omega)]^2 \quad ,$$

$$(b) = [-(2\pi)^{-3}\int v(\vec{p}-\vec{p}')\, n^{o}_{\vec{p}'}\, d\vec{p}']^2[G_o(\vec{p},\omega)]^3 \quad , \tag{2.10}$$

where $v(\vec{p})$ is the Fourier transform of $v(\vec{r})$. These are the first two terms in a geometric series and we can sum up the whole subseries without any extra effort. Actually, by rearranging the terms we can make a partial summation of the full $G(\vec{p},\omega)$, yielding

$$G(\vec{p}\omega) = [\omega - \varepsilon_p - \Sigma(\vec{p}\omega)]^{-1} \quad . \tag{2.11}$$

The so-called self energy $\Sigma(\vec{p}\omega)$ contains only those diagrams which cannot be split in two disconnected pieces by cutting one particle or one hole line. So for instance, diagram (b) in Figure 3 can be cut in this way and is, therefore, not included in $\Sigma(\vec{p}\omega)$. It is recovered when expanding Eq. (2.11) in powers of $\Sigma(\vec{p}\omega)$. The first few terms in the self energy is shown in Figure 4. The diagrams, denoted here by (a), are of purely quantum mechanical origin and contain exchange effects. Let us consider diagrams (b) in the figure. They represent a situation where a particle of momentum $\vec{p}$ enters from the left and disturbs the surrounding by lifting one particle from a state underneath the Fermi surface to a state above, creating a particle-hole pair. This pair can recombine and create another pair and so on. The original particle continues to move as a free particle with changed momentum and it feels its own disturbance at another time. $\Sigma(\vec{p}\omega)$ can, therefore, be interpreted as an energy and momentum dependent potential, in which the primary particle moves. It can also be considered as a refractive index for the particle wave. Adding diagrams (c), we begin to include self energy corrections to the free particle line. When this is fully done, G_0 in (b) should be replaced by the full G and the corresponding self energy corrections should consequently be dropped. Again, a partial resummation of our original series has been achieved without any real effort.

If, in diagrams (b) of Figure 3, we sum up all terms containing repeated particle-hole pairs (see Figure 5) and also include all kinds of higher order lines inside each bubble, we obtain the full density-density correlation function for the interacting system. We denote this in Figure 6a by a wiggly line. There, we present what we have obtained so far by summing a large number of terms in our original series. The diagram represents the situation where an incoming particle with momentum $\vec{p}$ disturbs its surrounding, continues to move as a real particle and feels its own disturbance at another time. The disturbance is described through the density correlation of the medium. This is actually what we would obtain when calculating the disturbance from the incoming particle to linear response. There are still terms in our expansion which have not been taken care of. These are conventionally included in an effective potential, called vertex corrections by people in the field (see Figure 6b). The effective potential contains the response from the primary particle also to nonlinear order. Due to the momentum dependence of the effective potential the response of the surrounding medium is no longer expressible only through the density correlation. Current correlations and other correlation functions enter as well. We have now sorted out various effects in the self energy arising from the interaction. It is then simply a matter of carrying out the explicit calculations to desired accuracy. Unfortunately, higher order diagrams become very difficult to evaluate numerically and one cannot go very far in this way.

As an illustration, we consider the diagrams in Figure 5, which give an approximate expression for the density correlation. Summing up this series in Fourier space we get

$$\langle n(\vec{r},t)\; n(0,0)\rangle_{q,\omega} = \alpha(q,\omega)/[1 - v(q)\;\alpha(q,\omega)] \quad , \qquad (2.12)$$

where $\alpha(q\omega)$ stands for one particle-hole bubble. The latter represents the density correlation for the non-interacting system, for it contains no interaction lines. It can easily be evaluated and Eq. (2.12) yields the Random phase approximation, which is the quantum mechanical analogue of the Vlasov approximation. The zeros of the denominator give for Coulomb interaction an approximate value for the plasma dispersion curve, and give for short range interaction the dispersion for zero sound.

Another illustration is given in Figure 7. The first term represents two particles interacting through the potential $v(r)$. The higher order terms can be summed up and they lead to a screening of the potential,

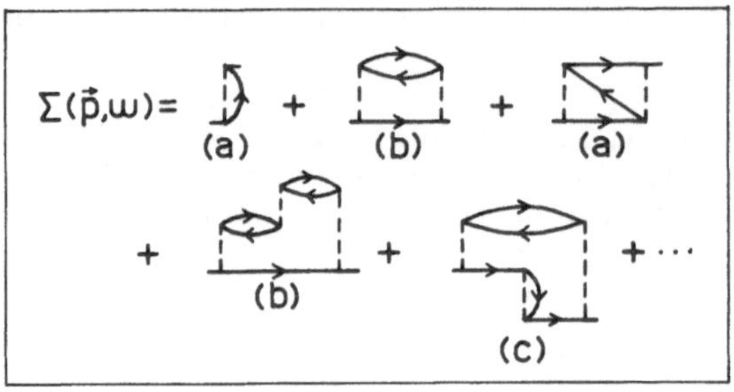

Figure 4. Diagram expansion of the self energy $\Sigma(\vec{p},\omega)$.

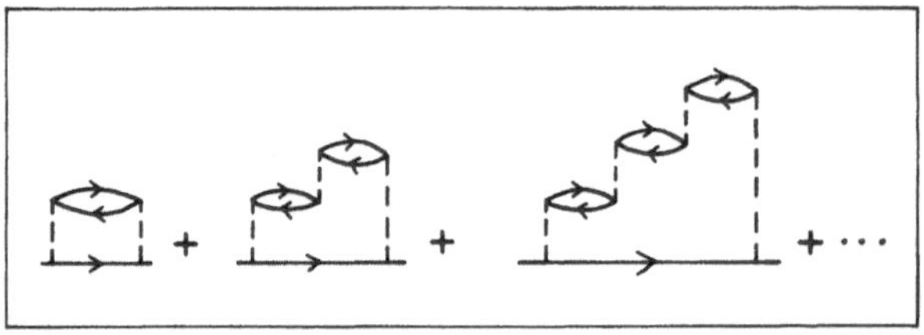

Figure 5. Diagrams contributing to the density-density correlation functions.

$$v_{sc}(q) = v(q)/[1 - v(q)\ \alpha(q,\omega)] \quad . \tag{2.13}$$

The same denominator enters here as in Eq. (2.12) and is the frequency and wavevector dependent dielectric function of the medium. What we learned from this example is that the bare potential is always screened and it leads in many cases to a considerable reduction in the strength of the interaction between the particles.

The above perturbation procedure has been used extensively for explicit calculations and these have confirmed what was said in subsection B. For the details we refer to a recent review article by Hedin and Lundqvist [Hedin and Lundqvist, 1969].

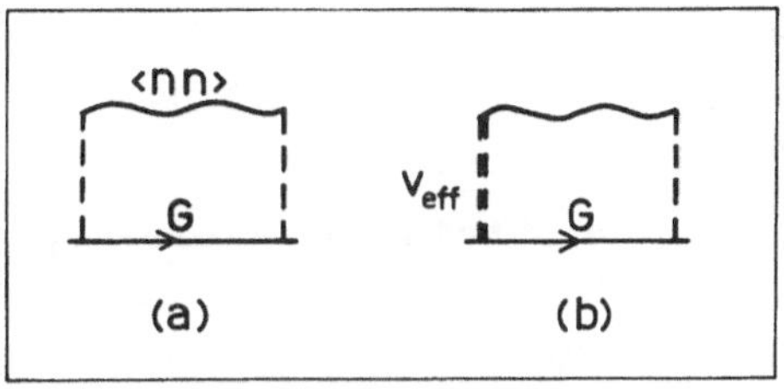

Figure 6. (a) Diagram for $\Sigma(\vec{p},\omega)$ after partial summation to include all self energy terms in G and all terms in the density-density correlation <n n>.
(b) Diagram after including all remaining terms in the vertex correlation v_{eff}. The wiggly line contains now more than the density-density correlation function.

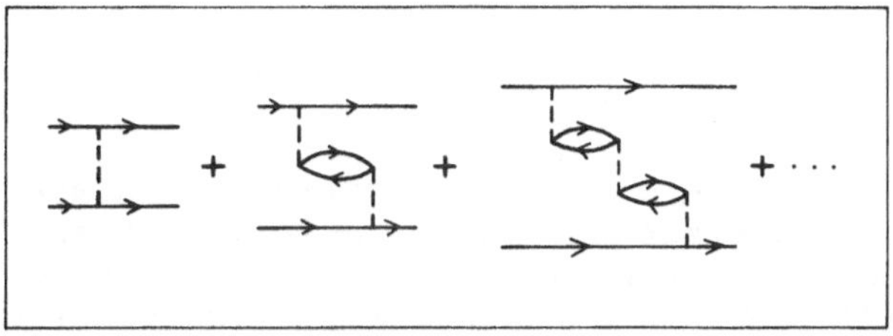

Figure 7. Diagrams yielding a screened interaction between two particles.

D. The Wigner Distribution Function

The quantum mechanical analogue to the classical Klimontovich function

$$f_{cl}(\vec{r}\vec{p}t) = \sum_{\ell=1}^{N} \delta\{\vec{r} - \vec{r}_\ell(t)\} \, \delta\{\vec{p} - \vec{p}_\ell(t)\} \quad , \qquad (2.14)$$

or rather

$$f_{cl}(\vec{q}\vec{p}t) = \sum_{\ell=1}^{N} \exp\{i\vec{q}\cdot\vec{r}_\ell(t)\} \, \delta\{\vec{p} - \vec{p}_\ell(t)\} \quad , \qquad (2.15$$

is

$$f(\vec{q}\vec{p}t) = a^{+}_{\vec{p}+\hbar\vec{q}/2}\, a_{\vec{p}-\hbar\vec{q}/2} \quad . \tag{2.16}$$

Summing over all $\vec{p}$-values, we obtain the microscopic density operator

$$n(\vec{q},t) = \sum_{p} a^{+}_{\vec{p}+\hbar\vec{q}/2}\, a_{\vec{p}-\hbar\vec{q}/2} \quad . \tag{2.17}$$

If we consider a system of non-interacting particles, the Schrödinger equation yields

$$\{-i\hbar\frac{\partial}{\partial t} + [\varepsilon_{\vec{p}+\hbar\vec{q}/2} - \varepsilon_{\vec{p}-\hbar\vec{q}/2}]\}\, f(\vec{q}\vec{p}t) = 0 \tag{2.18}$$

with

$$\varepsilon_{\vec{p}+\hbar\vec{q}/2} - \varepsilon_{\vec{p}-\hbar\vec{q}/2} = \frac{1}{m}\vec{p}\cdot\vec{q} \quad . \tag{2.19}$$

Written in space and time variables, it leads to

$$(\frac{\partial}{\partial t} + \frac{1}{m}\vec{p}\cdot\vec{\nabla})\, f(\vec{r}\vec{p}t) = 0 \quad , \tag{2.20}$$

which is identical to the classical equation for a system of free particles. For an interacting system additional terms enter and in order to apply the conventional Green function technique one considers the equilibrium phase-space correlation function

$$F(\vec{q}t|\vec{p}\vec{p}') = \langle 0|a^{+}_{\vec{p}+\hbar\vec{q}/2}(t)\, a_{\vec{p}-\hbar\vec{q}/2}(t)\, a^{+}_{\vec{p}'-\hbar\vec{q}/2}(0) \cdot a_{\vec{p}'+\hbar\vec{q}/2}(0)|0\rangle \quad . \tag{2.21}$$

Here, we may expand in powers of the interaction potential. The first few terms in the diagrammatic expansion are shown in Figure 8. By rearranging the terms one can derive an equation for $F(\vec{q}t|\vec{p}\vec{p}')$ of the following form:

$$\{\frac{\partial}{\partial t} + \frac{i}{m}(\vec{p}\cdot\vec{q})\}\, F(\vec{q}t|\vec{p}\vec{p}') + \int dt_1 \int d\vec{p}_1\, L(\vec{q},t-t_1|\vec{p}\vec{p}_1)\, F(\vec{q}t_1|\vec{p}_1\vec{p}') = 0 \quad , \tag{2.22}$$

where the quantity $L(\vec{q}t|\vec{p}\vec{p}')$ is given as an infinite series. Then one can derive the Landau transport equation, for instance.

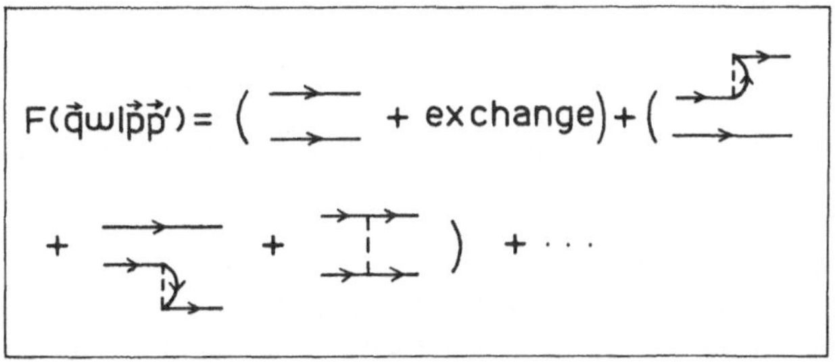

Figure 8. Diagram expansion of the phase-space correlation function $F(\vec{q}\omega|\vec{p}\vec{p})$.

This brief review has hopefully given the flavour of the perturbation treatment and indicated its possibilities and also its limitations. Whenever more complicated diagrams need be evaluated, one is normally in trouble and should try some other way of proceeding. One such possibility is described in Professor Singwi's lectures.

III. CLASSICAL PLASMAS

A new approach for treating dynamics of dense classical plasmas is outlined in this last chapter. The Vlasov equation is known to be the correct kinetic equation to lowest order in the interaction potential and in next order we have the Lenard-Balescu-Guernsey equation [Balescu, 1963; Montgomerey and Tidman, 1964; Ichimaru, 1973] which contains a collision term with a screened Coulomb interaction besides the Vlasov mean field term. Here the expansion parameter is $\Gamma = (Ze)^2/k_BTa$, where Ze is the ionic charge, a is the average interparticle distance, and T is the temperature. This transport equation is valid for $\Gamma<1$. However, for many important applications in astrophysics and also in some fusion problems Γ is considerably larger than unity and the ions become strongly correlated, invalidating the above theories. From computer simulations it is found that this system undergoes a transition to the crystalline phase for $\Gamma \simeq 150$ (more about this is presented by Professor Hansen in his lectures.

Let me first outline the basic idea of the new approach. We focus the attention on one particular ion, let us call it the <u>zeroth particle</u>, and we describe its self motion through a distribution function $f_s(\vec{r}\vec{p}t)$, which should satisfy a generalized Focker-Planck equation. The surrounding particles are disturbed by the zeroth particle and we describe their motions through another distribution $f_d(\vec{r}\vec{p}t)$, which contains information on how the density is modified around the zeroth particle and how a backflow current is built up. We may then calculate various equilibrium correlation functions, such as $\langle f_s(\vec{r}\vec{p}t)\ f_s(0\vec{p}'0)\rangle$, $\langle f_d(\vec{r}\vec{p}t)\ f_s(0\vec{p}'0)\rangle$ etc. Adding f_s and f_d, calculating the corresponding full phase-space correlation function, and then integrating over the momenta and going over to the Fourier space we obtain the dynamical structure factor $S(q\omega)$, which is of particular interest for us.

Referring to the initial values of $\langle f_s\ f_s\rangle$ and $\langle f_d\ f_s\rangle$, we are for t=0 faced with the situation of having the zeroth particle located at the origin with the momentum $\vec{p}'$ and the surrounding particles being in equilibrium around this particle; i.e.,

$$f_d(\vec{r}\vec{p},t=0) = \Phi_M(p)\ ng(r) \quad , \tag{3.1}$$

where $\Phi_M(p)$ is the Maxwellian distribution and $g(r)$ is the static pair correlation function. For $t>0$, the zeroth particle has moved to a different position with a different momentum -- say $\vec{r}_o$ and $\vec{p}_o$ -- and $f_s(\vec{r}_o,\vec{p}_o,t)$ gives the probability for this to occur. The distribution of the surrounding is written as

$$f_d(\vec{r}\vec{p}t) = \Phi_M(p)\ ng(\vec{r}-\vec{r}_o) + \overline{f}_d(\vec{r}\vec{p}t) \quad , \tag{3.2}$$

where the first term represents the situation where the surrounding is in equilibrium around the zeroth particle and $\overline{f}_d(\vec{r}\vec{p}t)$ describes the deviation from this. We can easily understand that $\overline{f}_d$ should be small when the zeroth particle is moving slowly and thus gives time for the other particles to relax essentially to the equilibrium situation. If we simply ignore the effect of $\overline{f}_d$, we can show that

$$S(q\omega) = S(q)\ S^s(q\omega) \quad , \tag{3.3}$$

where $S(q)$ is the static structure factor and $S^s(q\omega)$ is the self part of the density correlation and describes the motion of one single particle. The above approximation was introduced a long time ago by Vineyard [Vineyard, 1958], but it is found to be a rather poor approximation for wavevectors of main interest for us here.

Starting from the BBGKY-hierarchy, one easily finds an equation for $\overline{f}_d$ of the following form:

$$\left\{\frac{\partial}{\partial t} + \frac{1}{m}\vec{p}\cdot\vec{\nabla}_r\right\}\overline{f}_d(\vec{r}\vec{p}t) + \{\text{terms containing higher order correlations, including zeroth particle}\}$$

$$= -n\ \Phi_M(p)\int d\vec{r}_1\ d\vec{p}_1\ g(\vec{r}-\vec{r}_1)\frac{\partial}{\partial t} f_s(\vec{r}_1\vec{p}_1t) \quad . \tag{3.4}$$

The right hand side is the primary source of disturbance from the zeroth particle and it becomes small if this particle is moving slowly. Using the Zwanzig-Mori projection operator technique [Zwanzig, 1961; Mori, 1965], for instance, one can rewrite the equation as

$$f_d(\vec{r}\vec{p}t) = -n \int_0^{\tau} dt' \int d\vec{r}'\, d\vec{p}'\, H(\vec{r}\vec{p}t|\vec{r}'\vec{p}'t')\, \Phi_M(\vec{p}')\, g(\vec{r}'-\vec{r}_1)$$

$$\cdot \frac{\partial}{\partial t'} f_s(\vec{r}_1\vec{p}_1t')\, d\vec{r}_1\, d\vec{p}_1 \quad . \tag{3.5}$$

$H(\vec{r}\vec{p}t|\vec{r}'\vec{p}'t')$ contains information on how the disturbance at $(\vec{r}'\vec{p}')$ at time t' propagates to $(\vec{r}\vec{p})$ at time t, and it is affected by the presence of the zeroth particle. It is evident that this can be ignored when we consider positions far away from the zeroth particle. H(...) is then a characteristic propagator for the undisturbed fluid and it can be expressed in terms of the equilibrium correlation function <f f>. Assuming this for all positions and times, we obtain the following simple result for the density response function:

$$\chi(q\omega) = \chi_s(q\omega)/[1 - v_{eff}(q)\, \chi_s(q\omega)] \quad , \tag{3.6}$$

where $v_{eff}(q) = -k_BT\, c(q)$, c(q) being the direct correlation function and $\chi_s(q\omega)$ is the corresponding response function for a single ion. From here we can easily go over to S(qω). The Vlasov equation yields the same result, but with

$$\chi_s(q\omega) = \chi_{free}(q\omega) \quad , \quad v_{eff}(q) = v(q) \quad , \tag{3.7}$$

v(q) being the bare Coulomb potential. Modifications, as obtained in Eq. (3.6), have been suggested before [Kerr, 1968; Singwi, Sköld and Tosi, 1970] and they gave a considerable improvement for ordinary liquids. One unsatisfactory feature of Eq. (3.6) is that an unknown quantity $\chi_s(q\omega)$ enters. Like any other mean field theory, it lacks the effect of collisions and is, therefore, missing an essential part of the dynamics.

In the following we make improvements on two essential points: (i) we develop a procedure for calculating $\chi_s(q\omega)$ as well, and (ii) we correct for the fact that H(...) is affected by the zeroth particle. This gives rise to a collision term in the corresponding kinetic equation, analogous to that in the Lenard-Balescu-Guernsey equation.

There are no possibilities of going into any derivation in this single lecture and here I will only discuss the results, re-referring to Sjögren and Sjölander [in press] and Sjödin and Mitra [1977] for details. The full phase-space correlation function

$F(\vec{r}\vec{t};\vec{p}\vec{p}') = \langle f(\vec{r}\vec{p}t)\ f(0\vec{p}'0)\rangle$ is found to satisfy the equation

$$\left\{\frac{\partial}{\partial t} + \frac{1}{m}\,\vec{p}\cdot\vec{\nabla}_r\right\} F(\vec{r}t;\vec{p}\vec{p}') - n\,\vec{\nabla}_p\Phi_M(p)$$

$$\cdot \int d\vec{r}_1\, d\vec{p}_1\, \vec{\nabla}_r\{-k_BT\ c(\vec{r}-\vec{r}_1)\}\ F(\vec{r}_1t;\vec{p}_1\vec{p}')$$

$$- \int_0^\tau dt_1 \int d\vec{r}_1\, d\vec{p}_1\, \vec{\nabla}_p\cdot L(\vec{r}\vec{p}t\,|\,\vec{r}_1\vec{p}_1t_1)$$

$$\cdot \left\{\frac{1}{m}\,\vec{p}_1 + k_BT\,\vec{\nabla}_{p_1}\right\} F(\vec{r}_1t_1;\vec{p}_1\vec{p}') = 0 \quad , \tag{3.8}$$

where $c(r)$ is the direct correlation function and the tensor $L(\ldots)$ is a certain memory function, which arises from $\vec{F}_d$ in Eq. (3.2) and describes how a backflow is built up around the zeroth particle. The term containing $c(q)$ is the ordinary mean field term with $(-k_BT)c(q)$ replacing the bare potential. An equation similar to Eq. (3.8) has been obtained before, starting from a different point of view [Mazenko, 1974; Gross, 1976; Boley, 1974a, 1974b]. A corresponding equation is obtained for $F_s(\vec{r}\vec{p}t;0\vec{p}'0)=\langle f_s(\vec{r}\vec{p}t)f_s(0\vec{p}'0)\rangle$; namely

$$\left\{\frac{\partial}{\partial t} + \frac{1}{m}\,\vec{p}\cdot\vec{\nabla}_r\right\} F_s(\vec{r}t;\vec{p}\vec{p}') - \int_0^\tau dt_1 \int d\vec{r}_1\, d\vec{p}_1\, \vec{\nabla}_p\cdot L_s(\vec{r}\vec{p}t\ \vec{r}_1\vec{p}_1t_1)$$

$$\cdot \left\{\frac{1}{m}\,\vec{p}_1 + k_BT\,\vec{\nabla}_{p_1}\right\} F_s(\vec{r}_1t_1;\vec{p}_1\vec{p}') = 0 \tag{3.9}$$

with another memory function $L_s(\ldots)$ entering. This is a generalization of the ordinary Focker-Planck equation for a Brownian particle. The latter emerges, if we assume

$$L_s(\vec{r}\vec{p}t\,|\,\vec{r}_1\vec{p}_1t_1) = \zeta\delta(\vec{r}-\vec{r}_1)\ \delta(\vec{p}-\vec{p}_1)\ \delta(t-t_1) \quad . \tag{3.10}$$

Approximate, but explicit, expressions were given for L and L_s in Sjögren and Sjölander [in press] and Sjödin and Mitra [1977]. There, it was assumed that the disturbance of the fluid around the zeroth particle can be described through the collective variables particle number density and longitudinal current density. This implies that we ignore the possibility of having a transverse component in the blackflow current. With these assumptions, the dynamical structure factor takes the form

$$S(q,\omega) = \frac{1}{\pi}\,\mathrm{Re}\ F(q,z=i\omega) \tag{3.11a}$$

with

$$F(q,z) = S(q) \frac{F_s(q,z) - (z/k_BTq^2) L(q,z) [z F_s(q,z) - 1]}{1 + [n c(q) - (z/k_BTq^2) L(q,z)][z F_s(q,z) - 1]} \tag{3.11b}$$

We recover Eq. (3.6) by assuming L(qz)=0. The explicit expression for L(qt), used by Sjödin and Mitra [Sjödin and Mitra, 1977] in their numerical calculations, is given below. For a detailed discussion of the approximation I refer to Sjögren and Sjölander [in press] and to a forthcoming paper of Sjödin and Mitra.

$$L = L_1 + L_2 - L_s \quad , \tag{3.12a}$$

where

$$L_1(q,t) = -n \int \frac{d\vec{q}\,'}{(2\pi)^3} \{(\hat{q}\cdot\vec{q}\,') v(\vec{q}\,') + [\hat{q}\cdot(\vec{q}-\vec{q}\,')] v(\vec{q}-\vec{q}\,')\} F(\vec{q}\,',t) F(\vec{q}-\vec{q}\,',t) \times [1 - n c(\vec{q}\,')] c(\vec{q}-\vec{q}\,') [(\vec{q}-\vec{q}\,')\cdot\hat{q}] \quad , \tag{3.12b}$$

$$L_2(q,t) = n^2 k_BT q \int \frac{d\vec{q}\,'}{(2\pi)^3} (\hat{q}\cdot\vec{q}\,') c(\vec{q}\,') F(\vec{q}\,',t) F(\vec{q}-\vec{q}\,',t) \times c(\vec{q}\,') c(\vec{q}-\vec{q}\,') \tag{3.12c}$$

and

$$L_s(q,t) = -n \int \frac{d\vec{q}\,'}{(2\pi)^3} (\hat{q}\cdot\vec{q}\,') v(\vec{q}\,') F(\vec{q}\,',t) F_s(\vec{q}-\vec{q}\,',t) \times c(\vec{q}\,') (\vec{q}\,'\cdot\vec{q}) \quad . \tag{3.12d}$$

For the self correlation function $F_s(qz)$, being the Laplace transform of $F_s(qt)$, the so-called Gaussian approximation was employed, i.e.

$$F_s(q,t) = \exp\left[- \frac{q^2}{3} \int_0^t dt' \, (t-t') \, \Phi(t')\right] \quad , \tag{3.13}$$

where $\Phi(t) = \langle\vec{v}(t)\cdot\vec{v}(0)\rangle$ is the velocity auto-correlation function. The latter satisfies the equation

$$\frac{d}{dt} \Phi(t) + \int_0^t dt' \, M(t-t') \, \Phi(t') = 0 \tag{3.14}$$

with

$$M(t) = L_s(q=0,t) = -L(q=0,t) \quad . \tag{3.15}$$

Eqs. (3.11)-(3.15) form a set of nonlinear equations for $F(qz)$ and $F_s(qz)$, and these were solved through iterations for the plasma parameters $\Gamma=1$ and $\Gamma=10$. Some comparisons with available computer results are shown in Figures 9-10. The present theory can be shown to go over to the Lenard-Balescu-Guernsey equation for $\Gamma<1$. It seems to reproduce quite well the main features in $S(q\omega)$ also for larger Γ-values. There are still some interesting discrepancies between the theory and the molecular dynamics results and an important point is now to find the physical reason for this. Calculations for much larger values of Γ would be desirable, but then a straightforward iterative procedure of solving the equations is not practical.

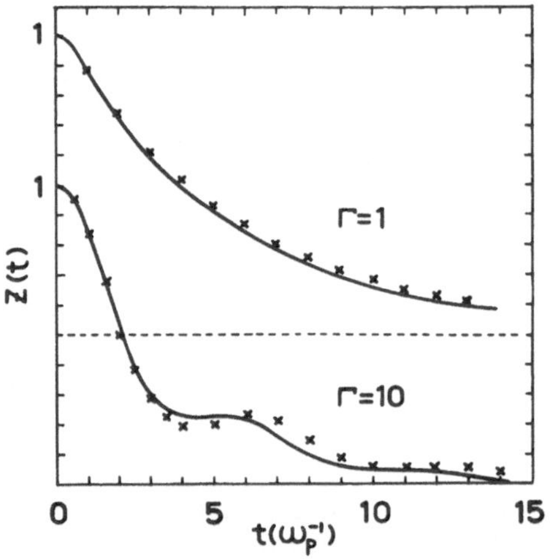

Figure 9. The normalized velocity auto-correlation function $Z(t) = \langle\vec{v}(t)\cdot\vec{v}(0)\rangle/\langle v^2\rangle$ for $\Gamma=1$ and $\Gamma=10$. Present theory (full curve) and molecular dynamics results (crosses) of Hansen, McDonald and Pollock [1975]. Unit of time is the inverse plasma frequency ω_p^{-1}.

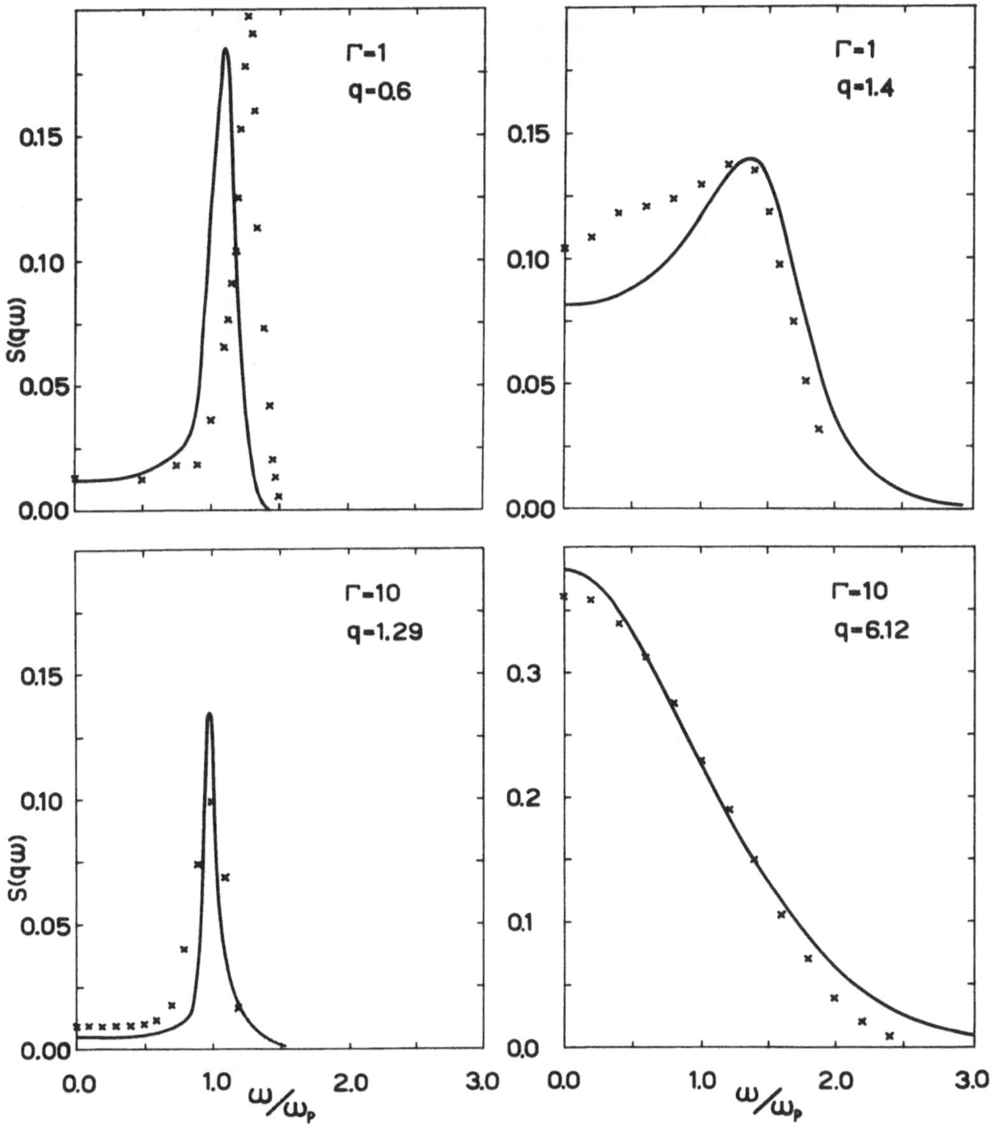

Figure 10. The dynamical structure factor S(q,ω) in units of ω_p^{-1}. The wavenumber is in units of a^{-1}, a being the mean interparticle distance. Present theory (full curve) and molecular dynamics results of Hansen et al (crosses).

ACKNOWLEDGEMENTS

It is my pleasure to acknowledge close collaboration with Drs. Lennart Sjögren and Swapan Mitra and with Mr. Staffan Sjödin on the problems covered in the second part of my lectures. I am grateful for getting permission to present results on plasmas before publication.

REFERENCES

Balescu, R., 1963, Statistical Mechanics of Charged Particles, Interscience, John Wiley & Sons, Ltd., London.

Boley, C. D., 1974, Ann. Phys. (New York), 86, 91.

Boley, C. D., 1974, Phys. Rev. A11, 328.

Doniach, S. and E. H. Sondheimer, 1974, Green's Functions for Solid State Physicists, W. A. Benjamin, Inc., London.

Fetter, A. L. and J. D. Walecka, 1971, Quantum Theory of Many-Particle Systems, McGraw-Hill, New York.

Gross, E. P., 1974, J. Stat. Phys. 11, 503; 1976, J. Stat. Phys. 15, 181.

Hansen, J. P., I. R. McDonald and E. L. Pollock, 1975, Phys. Rev. A11, 1025.

Hedin, L. and S. Lindqvist, 1969, Solid State Physics, Vol. 23, 1-181, Academic Press, New York.

Ichimaru, S., 1973, Basic Principles of Plasma Physics, A Statistical Approach, W. A. Benjamin, Inc., London.

Kerr, W. C., 1968, Phys. Rev. 174, 316.

Lindqvist, B. I., 1969, Physik Kondensierten Materie 7, 117.

Lobo, R., K. S. Singwi and M. P. Tosi, 1969, Phys. Rev. 186, 470.

Mazenko, G. F., 1974, Phys. Rev. A9, 360.

Montgomerey, D. C. and D. C. Tidman, 1964, Plasma Kinetic Theory, McGraw-Hill, New York.

Mori, H., 1965, Progr. Theor. Phys. (Japan), 33, 423; 34, 399.

Raimes, S., 1972, Many-Electron Theory, North Holland, Amsterdam.

Schultz, T. D., 1964, Quantum Field Theory and the Many-Body Problem, Gordon and Breach, New York.

Singwi, K. S., K. Sköld and M. P. Tosi, 1970, Phys. Rev. A1, 454.

Sjödin, S. and S. Mitra, 1977, J. Phys. A.

Sjögren, L., in press, Ann. Phys. (New York).

Sjögren, L. and A. Sjölander, in press, Ann. Phys. (New York).

Vineyard, G. H., 1958, Phys. Rev. 110, 999.

Zwanzig, R., 1961, Lectures in Theoretical Physics, Boulder, edited by W. F. Britton, W. B. Downs and J. Downs, Interscience, New York.

EQUILIBRIUM STATISTICAL MECHANICS OF STRONGLY COUPLED PLASMAS BY NUMERICAL SIMULATION*

Hugh E. DeWitt

Lawrence Livermore Laboratory

Livermore, California 94550

*Work performed under the auspices of the Department of Energy under contract W-7405-Eng-48.

TABLE OF CONTENTS

EQUILIBRIUM STATISTICAL MECHANICS OF STRONGLY COUPLED PLASMAS BY NUMERICAL SIMULATION

Hugh E. DeWitt

Lawrence Livermore Laboratory

Livermore, CA 94550

I. INTRODUCTION

The intent of these lectures is to give a summary of theoretical understanding and computer simulation results for the equilibrium thermodynamic properties of fully ionized light elements at high density. Matter in this state is a "dense" plasma or "strongly coupled" plasma because of the strength of the Coulomb interaction among the point nuclei. At the extreme densities found in white dwarf stars the Coulomb interactions can be so strong that the nuclei are localized into lattice sites, i.e. a Coulomb solid. Most of these lectures, however, will deal with lower densities in which the nuclei form a strongly-correlated fluid. Real matter in this state may be considered to be a mixture of two fluids: the fluid of point nuclei governed by pairwise Coulomb interactions with classical mechanics, and a neutralizing fluid of electrons which is degenerate due to Fermi statistics. In real physical systems such as stellar interiors, interiors of large planets (Jupiter), and laser fusion compression experiments, the two fluids interact with each other chiefly by means of the electron screening effect due to some increase of electron density in the neighborhood of each nucleus. A large fraction of the matter in the universe is in the strongly coupled plasma state, and consequently the physics of this state of matter is of great importance.

It will be useful at the beginning of these talks to specify the appropriate form of the Coulomb coupling parameters. In usual laboratory plasmas at low density or weak coupling the appropriate parameter is normally taken as the inverse of the number of particles in the Debye sphere; for ions this is:

$$\varepsilon = \frac{1}{4\pi n\lambda_D^3} = \frac{Z^2\beta e^2}{\lambda_D} = 2\pi^{1/2} Z^3 e^3 n^{1/2}/(kT)^{3/2} \qquad (1.1)$$

where Z is the nuclear charge, $\beta = 1/kT$, $n = N/V$, the ion number density, and λ_D is the Debye screening length. Weak coupling means $\varepsilon << 1$. This usual plasma parameter can be used to characterize dense plasmas for which $4\pi n\lambda_D^3 < 1$, however the concept of Debye length is no longer useful or appropriate when $\varepsilon >> 1$. It is more convenient to use the form:

$$\Gamma = Z^2\beta e^2 \left(\frac{4\pi}{3} n\right)^{1/3} = Z^2\beta e^2/\bar{r} \qquad (1.2)$$

where

$$\bar{r} = \left(\frac{4\pi}{3} \right)^{-1/3}$$

is the ion sphere radius. Note the relation:

$$\varepsilon = \sqrt{3}\,\Gamma^{3/2}$$

$$\Gamma = \frac{\varepsilon^{2/3}}{3^{1/3}}$$

An ionized gas is strongly coupled when $\Gamma \gtrsim 1$. At temperature and densities for strong coupling the plasma is a fluid with properties similar to liquids, and consequently the description used here will use liquid state physics terminology.

The main advances in understanding of the properties of the strongly coupled Coulomb fluid in the last few years have come from Monte Carlo simulations, or numerical "experiments" from Professor Hansen and his co-workers in Paris and from the group at Livermore. Both groups were inspired by the 1966 pioneering Monte Carlo study of the Coulomb fluid by Brush, Sahlin, and Teller [1966] at Livermore. Typically these Monte Carlo simulations compute the classical Coulomb interactions between nuclei, and treat the electrons as a neutralizing background. This procedure is possible because at high density the small electron mass results in degeneracy since the Fermi energy is large compared to the temperature, $\varepsilon_F >> kT$. The electron pressure is much greater than the ion pressure, but to a large extent the two fluids are decoupled so that the separate contributions are additive. To make this point clearer, one should first consider the real two component system: electrons and point nuclei. The Hamiltonian for this system is:

$$H = K_e + K_i + U_{ee} + U_{ei} + U_{ii}$$

$$= \sum_{\alpha=1}^{N_e} \frac{p_\alpha^2}{2m_e} + \sum_{i=1}^{N} \frac{p_i^2}{2M}$$

$$= \sum_{\alpha>\beta}^{N_e} \frac{e^2}{r_{\alpha\beta}} - \sum_{\substack{\alpha=1 \\ i=1}}^{N_e,N} \frac{ze^2}{r_{\alpha i}} + \sum_{i<j}^{N} \frac{(ze)^2}{r_{ij}} \qquad (1.3)$$

for nuclei of mass M and charge ze. Neutrality for a single element requires

$$N_e = zN$$

and for a fully ionized mixture of elements with z_1 and z_2 the neutrality condition is:

$$N_e = z_1 N_1 + z_2 N_2 = \bar{z} N$$

with $N = N_1 + N_2$. Obviously this system is a quantum mechanical many-body problem which cannot yet be handled by present day computational methods. However, if the electrons are treated as a fluid rather than point particles, the energy of the system can be written as:

$$E = \langle H \rangle = \langle K_e \rangle + \langle K_i \rangle + \langle U_{ee} \rangle + \langle U_{ii} + U_{ei} \rangle$$

$$= \frac{3}{2} NkT + N_e \left(\frac{3}{5} \varepsilon_F + \varepsilon_{ex} + \varepsilon_{corr} \right) + U \qquad (1.4)$$

where the brackets indicate an ensemble average, and U is the internal energy of the nuclear Coulomb interactions ensemble averaged in the presence of the electron fluid. The term (3/2)NkT is obviously the ideal gas kinetic energy of the nuclei. The electron term includes the average kinetic energy, (3/5) ε_F, the exchange energy, and correlation energy, assuming complete electron degeneracy. A second parameter is needed to describe the electron terms and the electron screening effect, namely:

$$r_s = \bar{r}_e / a_0 \qquad (1.5)$$

where $a_0 = h^2/m_e c^2$, and

$$\bar{r}_e = \left(\frac{4\pi}{3} n_e \right)^{-1/3} = \bar{r}/\bar{z}^{1/3}$$

is the electron ion sphere radius, and $n_e = \overline{z}_n$. The Coulomb energy of ions in Eq. (1.4) may be written as a function of the two parameters, and r_s:

$$U = \left\langle \sum_{i<j}^{N} \frac{(ze)^2}{r_{ij}} + \text{background} \right\rangle$$

$$= NkT\, f\, (\Gamma,\, r_s) \tag{1.6}$$

Numerical simulation methods, Monte Carlo and molecular dynamics give U/NkT directly.

Strong coupling usually means densities and temperatures such that $\Gamma > 1$. Debye shielding of the ions by electrons does not apply; to the extent that exponential screening of the nuclei does occur the Thomas-Fermi screening length applies. A few astronomical facts are relevant here. Approximately 80% of the hydrogen in Jupiter is pressure ionized. The Coulomb coupling parameter in the Jovian interior is of the order of 20, and the electron screening parameter is $r_s \sim 1$. For the enormously greater densities in white dwarf stars the two parameters are roughly $\Gamma \sim 100$ to 200 and $r_s \sim 0.01$. Also is should be noted that hydrogenic plasmas in laser fusion pellet compression experiments can include the ranges $\Gamma \sim 0.1$ to 10 and $r_s \sim 0.1$ to 1.

In the limit of $r_s = 0$ the electron density is constant, and electron screening effects disappear. This important limiting situation, classical point charges in a uniform background, is called the one component classical plasma (OCP). Evidently the OCP does not exist in nature, though white dwarf star interiors come close to it. As a mathematical model the OCP has the same importance for real strongly coupled plasmas as the hard sphere fluid has for the understanding of properties of real liquids [Alder and Wainwright, 1957; Rushbrooke, 1968]. The OCP has been the subject of intense research, both theoretical and computational, in recent years. It should also be noted that the OCP can be thought of as the extreme limit of the soft sphere fluid in which the particle interactions are described by inverse power potentials:

$$U(r) = \varepsilon \left(\frac{\sigma}{r} \right)^m \tag{1.7}$$

where $m = \infty$ is the hard sphere system and $m = 1$ is the Coulomb system. Since inverse power fluids have similar behavior [Hoover, Gray and Johnson, 1971] in strong coupling and go into a lattice at high enough density, it seems useful to include a discussion of the inverse power fluids in these lectures.

The modern understanding of the equilibrium properties of

strongly correlated plasmas, at least the OCP, began with the Monte Carlo work of Brush, Sahlin and Teller (BST) [1966]. In their simulation of the canonical ensemble average of the OCP internal energy they adopted the Metropolis procedure [Metropolis, Rosenbluth, A., Rosenbluth, M., Teller, A. and Teller, E., 1953] and they handled the long range of the Coulomb interaction by replicating the basic three-dimensional cell containing N point charges in all directions to infinity. The interactions among charges and with image charges were summed with the Ewald method [Nijboer and DeWette, 1957]. Numerical results were obtained for U/NkT, C_V, and g(r) by averaging over about 10^5 configurations of N = 108 charges for values of Γ from 0.1 (weak coupling) up to 125. Because of numerical inaccuracies for large Γ the BST results for the thermal energy, heat capacity, and location of the bluid-lattice transition are not too reliable.

In 1971 Hubbard and Slattery applied the BST methods to the calculation of the properties of the dense ionized hydrogen in Jupiter [Hubbard and Slattery, 1971]. The BST program was extended to include electron screening effects for non-zero values of r_s by means of linear response theory and the Lindhard dielectric function for completely degenerate electrons. Their code can also handle arbitrary mixture of two nuclear components so that they could compute the thermodynamic properties of mixtures of light elements, assumed to be fully ionized [Hubbard, 1972]. Their first published results were of limited accuracy because computer time limitation required that they use a small number of particles ($N \sim 40$) and average over few configurations (10^4). Later the Hubbard Monte Carlo code was brought to the Livermore Laboratory for use on a larger computer (CDC 7600). At Livermore a large number of runs were made to map out the thermodynamics of one and two nuclear component plasmas including electron screening as functions of Γ and r_s [DeWitt and Hubbard, 1976].

In France Professor J. P. Hansen and his collaborators developed a new and very accurate OCP Monte Carlo code. In 1973 they published results for OCP thermodynamic functions obtained by averaging N = 128 particles over 10^6 configurations for values of Γ from 1 to 160 for the fluid state [Hansen, 1973] and 150 to 300 for the solid state, the Coulomb lattice [Pollock and Hansen, 1973]. Their U/NkT results for the OCP are apparently very accurate and reliable and at the present time they are the standard results available for testing the validity of analytical theories and integral equation results. The Hansen OCP fluid and solid OCP equation of state data can be viewed in the same way as the Alder-Wainwright molecular dynamics data for the fluid and solid hard sphere equation of state in the fluid and solid phases. The numerical experimental data give an accurate mapping of the thermodynamic properties of systems of particles governed by inverse power fluids including the limiting cases, m = 1 for the OCP, and $m = \infty$ for the hard sphere system. Empirical relations have been found from the analysis of this data which hopefully will suggest theoretical models for strongly coupled systems.

From the analysis of the data obtained from the Livermore and Paris Monte Carlo strongly coupled plasma equilibrium results a number of general results have emerged:

(1) The fluid internal energy is a sum of a static energy resembling the energy of particles in a lattice plus a well-defined thermal energy:

$$U = U_0(n) + U_{th}(n,kT)$$

$$C_v = dU_{th}/dT \quad .$$

(2) Certain integral equations give this qualitative feature, namely the hypernetted chain (HNC) equation and the mean spherical approximation.

(3) For arbitrary mixtures of different nuclear charges the ion-sphere charge averaging largely determines the thermodynamic properties so that a one fluid model can be used with

$$\Gamma = z^{5/3}\, \overline{z}^{1/3}\, \Gamma_0$$

$$\Gamma_0 = \beta e^2/\overline{r} \quad .$$

(4) The direct correlation function for distances less than the nearest neighbor distance ($r < 1.7\overline{r}$) has a simple algebraic form dominated by a linear term:

$$c(r) = -\Gamma(a_0 - a_1 x) \quad , \qquad x = r/\overline{r}$$

and this form determines the ion fluid structure factor, S(k).

These four general conclusions are of course all related and are mainly a consequence of the apparent fact that a strongly coupled plasma may be described as a disorder lattice. The static energy, $U_0(n)$, represents the average energy of the system with the strong interparticle correlations keeping the ions in positions that resemble a lattice structure. Since the system is a fluid, the average position of the particles changes slowly with correlation. The magnitude of $U_0(n)$ may be expected to be comparable to the energies of simple cubic, face centered cubic, or body centered cubic lattices, but since the strongly coupled fluid is a disordered array, this static energy is expected to be slightly larger than the Madelung energy of the lattice that gives the minimum Helmholtz free energy for the solid, i.e. the bcc lattice for the Coulomb potential [Pollock and Hansen, 1973]. In addition to the slow shifts of the average particle positions there are also much more rapid short distance movements of the particles around their average positions. These rapid movements are analogous to the harmonic vibrations that give the thermal energy of a real lattice, and for the fluid these movements give rise to the fluid thermal energy, $U_{th}(n,kT)$. The essential difference of the fluid state as compared with the solid state shows up in the form of U_{th}. The Monte Carlo simulations of the strongly coupled plasma fluid state show that the temperature dependence of the fluid U_{th} per particle is approximately $(kT)^{3/4}$ instead of $(3/2)kT$ for a particle vibrating harmonically around a lattice site [DeWitt, 1976].

II. THE OCP IN WEAK AND STRONG COUPLING

The equilibrium properties of the OCP may be rigorously obtained from the canonical partition function for N point charges in a volume V at temperature $\beta = 1/kT$. The thermodynamic limit is assumed:

$$\lim_{\substack{N\to\infty \\ V\to\infty}} \frac{N}{V} = n \quad . \tag{2.1}$$

With classical mechanics the kinetic energy portion of the OCP Hamiltonian is easy to separate from the interaction portion. Including the neutralizing uniform background the interaction Helmholtz free energy of the OCP is obtained from:

$$\beta F_I = \ln \int \cdots \int \frac{\prod_{i=1}^{N} d^3 r_i}{V^N} \exp(-\beta) \left\{ \sum_{i<j}^{N} \frac{(ze)^2}{|\underline{r}_i - \underline{r}_j|} - \frac{N}{V} \sum_{i=1}^{N} \int d^3 r \frac{ze^2}{|\underline{r}_i - \underline{r}|} + \frac{N^2}{2V^2} \iint d^3 r \, d^3 r^1 \frac{e^2}{|\underline{r} - \underline{r}^1|} \right\} \tag{2.2}$$

In principle we would like to evaluate this partition function analytically for all values of Γ. In practice so far the analytical evaluation has been limited to weak coupling ($\Gamma \ll 1$) and intermediate coupling ($0.3 \lesssim \Gamma \lesssim 1$). For potentials that fall to zero faster than $1/r^3$ at large distance a virial expansion in powers of density is possible. In the case of the Coulomb potential the virial coefficients as usually defined from the Mayer cluster expansion all diverge, and a rearrangement of the expansion is necessary to obtain convergent results. This procedure was worked out in 1959 by Abe [Abe, 1959] with the result:

$$\beta \bar{F}_I = N \{- \frac{1}{3} (\sqrt{3}\, \Gamma^{3/2}) + S_2(\Gamma) + S_3(\Gamma) + \ldots\} \qquad (2.3)$$

The first terms in Eq. (2.3) is the sum of the ring diagram, and is equivalent to the Debye-Hückel theory. The higher order terms, S_2, S_3, etc. are functions of Γ, and not coefficients of powers of n. These integrals involve interactions among clusters of particles as in the Mayer expansion, but the Debye screened potential appears:

$$\beta u_s(r) = \frac{\beta e^2}{r} e^{-4/\lambda_D} \qquad (2.4)$$

instead of the bare Coulomb potential, u(r) and e^2/r. The result for S_2 is:

$$S_2 = \frac{n}{2} \int d^3r \left\{ e^{-\beta u_s} - 1 + \beta u_s - \frac{1}{2} (\beta u_s)^2 \right\}$$

$$= \frac{1}{2\varepsilon} \int_0^\infty y^2\, dy \left\{ e^{-\frac{\varepsilon}{y} e^{-y}} - 1 + \frac{\varepsilon}{y} e^{-y} - \frac{1}{2} \left(\frac{\varepsilon}{y} e^{-y}\right)^2 \right\} \qquad (2.5)$$

and has the form $O(\varepsilon^2 \ln \varepsilon)$ or $O(\Gamma^3 \ln \Gamma)$ for $\Gamma \ll 1$. Complicated analytic expansion exist for S_2 and S_3 but they are generally not very useful. A better use of the Abe expansion is simply to numerically evaluate the integrals to sufficient accuracy that numerical derivatives can be taken. The Abe cluster integrals are all monotonic functions of Γ and easy to evaluate for the lower orders [Rogers and DeWitt, 1973]. The internal energy and the pressure are found by temperature and volume derivatives respectively of the free energy expression, Eq. (2.3). One finds:

$$\beta U/N = - \frac{1}{2} (\sqrt{3}\, \Gamma^{3/2}) + \frac{3}{2} \Gamma \frac{d}{d\Gamma} [S_2(\Gamma) + S_3(\Gamma) + \ldots] \qquad (2.6)$$

$$C_v/Nk = \frac{1}{4} (\sqrt{3}\, \Gamma^{3/2}) - \frac{3}{2} \Gamma^2 \frac{d^2}{d\Gamma^2} [S_2(\Gamma) + S_3(\Gamma) + \ldots] \qquad (2.7)$$

The interaction contribution to the pressure, $P_I = P - P_0$ with $P_0 = NkT$, is by the virial theorem:

$$(P - P_0)/n = \frac{1}{3} \beta U/N \qquad (2.8)$$

when there is no electron screening ($r_s = 0$). One notes that in weak coupling, $\Gamma \ll 1$, that the leading contribution is $O(\Gamma^{3/2})$ or

$O(\Gamma)$, thus from Eqs. (2.3), (2.6) and (2.7) we have

$$\beta F_I/N = -\frac{1}{3}\varepsilon \quad ,$$

$$\beta U/N = -\frac{1}{2}\varepsilon \quad ,$$

and

$$C_V/Nk = \frac{1}{4}\varepsilon$$

which are the well-known Debye-Hückel results. The addition of the Abe cluster term in Eqs. (2.3), (2.6) and (2.7) simples reduces the $O(\Gamma^{3/2})$ contribution as Γ increases. In fact it has been shown [Brush, DeWitt and Trulio, 1963] that as $\Gamma \to \infty$ the sum of the Abe integrals identically cancels the $O(\Gamma^{3/2})$ term leaving $O(\Gamma)$ as the leading term. Lower bounds on the interval energy are useful and important. The known lower bounds are:

$$\beta U/N = -\frac{1}{2}(\sqrt{3}\,\Gamma^{3/2}) \quad , \qquad \Gamma << 1 \tag{2.9}$$

$$= -\frac{9}{10}\Gamma \quad , \qquad \Gamma >> 1 \tag{2.10}$$

as will be discussed by Professor Choquard. The change in character of the functional dependence on Γ, i.e. $O(\Gamma^{3/2})$ changing to $O(\Gamma)$, occurs in the region $0.3 \lesssim \Gamma \lesssim 0.75$, and in this intermediate coupling region the Abe expansion is numerically useful for computing values of the thermodynamic functions. The transition value, $\Gamma_t = 0.75$, is an approximate lower limit of the strong coupling regime; this is simply value of Γ where the static energy of the OCP fluid, the $O(\Gamma)$ term becomes apparent in the Monte Carlo data.

For the present discussion a strongly coupled plasma will be taken to mean a coupling parameter range, $\Gamma_t < \Gamma < \Gamma_f$, for which there seems to be a well-defined separation of the potential energy into two parts: $U = U_0 + U_{th}$. This region might also be called the asymptotic Coulomb fluid region since a very simple analytic expression gives the equation of state and leads to all other thermodynamic functions. The upper limit is clearly the fluid-lattice transition which is indicated to be from the Pollock and Hansen Monte Carlo work as $\Gamma_f = 155$. For $\Gamma_t = 0.75$ the potential energy may be obtained accurately with a few terms of the Abe cluster expansion [Rogers and DeWitt, 1973].

The most accurate Monte Carlo data on the potential energy for the strongly coupled Coulomb fluid that is available at the present time is that of Hansen [1973], and some of this data is shown in Table I. In a separate column the values of the thermal energy are given for each Γ value. A striking feature of the Coulomb fluid data is that the thermal energy is only a small

TABLE I

THERMAL EQUATION OF STATE FOR OCP

	Monte Carlo		Hyper-netted Chain		Mean Spherical Model	
Γ	U/NkT	U_{th}/NkT	U/NkT	U_{th}/NkT	U/NkT	U_{th}/NkT
1	-.580	.315	-.570	.331	-.607	.294
2	-1.318	.471			-1.377	.424
3	-2.111	.572	-.2103	.598	-2.180	.522
4	-2.926	.652			-2.999	.579
6	-4.590	.778				
10	-7.996	.950	-7.9355	1.070	-8.053	.952
15	-12.313	1.106			-12.343	1.165
20	-16.667	1.225	-16.538	1.472	-16.667	1.343
30	-25.429	1.409			-25.373	1.642
40	-34.232	1.552	-33.999	2.020	-34.125	1.895
60	-51.936	1.741	-51.597	2.431	-51.710	2.320
80	-69.690	1.879	-69.264	2.774	-69.360	2.680
100	-87.480	1.981	-86.973	3.074	-87.053	2.997
120	-105.284	2.069	-104.713	3.343	-104.775	3.285

fraction of the total potential energy for large ; U_{th}/NkT at $\Gamma = 155$, the fluid-lattice transition, is only 2% of the total energy. The Monte Carlo process necessarily gives U/NkT and not U_{th}/NkT so that it is necessary to obtain U/NkT with great accuracy in order to get dependable results for the thermal energy. Hansen used N = 128 particles and averaged over 10^6 configurations to obtain his results. A statistical analysis of this data indicated that up to Γ of 40 the results are consistent. For $\Gamma > 40$ the data indicate a possible small systematic error. Also it was found that the second moment of g(r) did not satisfy the Stillinger-Lovett condition [Stillinger and Lovett, 1968]. Consequently, the data for Γ in the range from 1 to 40 was used to deduce the quantitative results for the thermal energy. The total energy is written as:

$$U/NkT = (U_0 + U_{th})/NkT$$
$$= -a\Gamma + g(\Gamma) \quad . \tag{2.11}$$

For the Coulomb solid the Monte Carlo data indicated the expected harmonic vibrations with a small anharmonic correction, namely $g(\Gamma) = 3/2 + 3500/\Gamma^2$. The fluid data, however, indicated a very different form, namely $g(\Gamma) = b\Gamma^s - c$, where s is a small power between 0.2 and 0.3. A nonlinear least squares fitting procedure for both the energy and heat capacity data established the best value to be s = 0.25. The total energy was found to be [DeWitt, 1976]:

$$U/NkT = -0.89461\Gamma + 0.8165\Gamma^{1/4} - 0.5012 \quad , \quad 0.75 < \Gamma < 40 \quad . \tag{2.12}$$

The static energy constant obtained by fitting the fluid data is about 0.15% higher than the Madelung constant for the bcc lattice, $a_{bcc} = 0.895929$. This formula must obviously fail at small Γ, and presumably there must be corrections to this asymptotic formula that are inverse powers of Γ. It has not been possible with the present data to find a believable correction to the above result. For the data with $\Gamma > 40$ in the fluid phase the same kind of fitting formula applies. However, the values of the three coefficients are slightly different. Also it was not possible to establish the exponent in the thermal energy to be precisely s = 1/4 as was the case for the data in the region $1 < \Gamma < 40$. The nature of the data suggested a slight systematic error, although this may be an artifact due to the finite number of particles used in Hansen's simulations, N = 128. Hansen's three values obtained with N = 250 particles for Γ = 70, 100 and 140 were all closer to Eq. (2.12) than his N = 128 data. Also it should be mentioned that Hansen's data for a very small number of particles, N = 16, showed a van der Waals loop in the region, $40 < \Gamma < 140$. Some evidence of this apparent van der Waals loop is also evident in his data for N = 54. It is possible that the accuracy of U/NkT for large Γ will be increased by going to much larger systems.

The $\Gamma^{1/4}$ behavior of the thermal energy appears to be quite definite. It is possible, of course that Eq. (2.12) is only a fortunate fitting function. However, the clear appearance of the exponent s = .25 is suggestive of some underlying mechanism governing the thermal energy of the Coulomb fluid. It is a very interesting theoretical challenge to find a theoretical model that can explain the $\Gamma^{1/4}$. It is my opinion that this result for U_{th}/NkT is a fundamental result deduced from valid numerical "experimental" data.

An independent Monte Carlo calculation of the heat capacity should in principle give additional information about the functional form of U_{th}. Both BST and Hansen obtained results for C_V using:

$$C_V/Nk = \beta^2(\langle U^2\rangle - \langle U\rangle^2)/N \tag{2.13}$$

Unfortunately the numerical data for C_V so obtained are less accurate than the data for U/NkT because the computer must work with the difference of very large numbers, $\langle U^2\rangle$ and $\langle U\rangle^2$. This calculation of C_V is further complicated by the fact that the OCP fluid static energy $U_0/NkT = -a\Gamma$ is so much larger than U_{th}/NkT. From Eqs. (2.11) and (2.12) a simple expression for the heat capacity is readily found:

$$\begin{aligned} C_V/Nk &= -\Gamma^2 \frac{d}{d\Gamma}[g(\Gamma)/\Gamma] \\ &= \frac{3}{4} b\Gamma^{1/4} - c \end{aligned} \tag{2.14}$$

This result agrees well with Hansen's numerical data for C_V/Nk.

An asymptotic form of the interaction Helmholtz free energy is readily found by integration:

$$\begin{aligned} F_I/NkT &= \int_{\Gamma_1}^{\Gamma} d\Gamma^1 \frac{(U/NkT)}{\Gamma^1} + \frac{\overline{F}_I(\Gamma_1)}{NkT_1} \\ &= -a\Gamma + 4b\Gamma^{1/4} - c\,\ell n\,\Gamma + d \end{aligned} \tag{2.15}$$

where the entropy constant is found to be d = -2.809 by comparison with the Abe cluster expansion evaluated at $\Gamma = 1$. Since the potential energy is negative (meaning the Coulomb interaction energy) the pressure due to the Coulomb interactions is also negative. For the OCP the interaction pressure is given exactly by the virial theorem, PV/NkT = (1/3)U/NkT. The negative pressure of the Coulomb interactions among the ions is, of course, more than balanced by the large positive pressure of Fermi degenerate electrons for real systems. Similarly the compressibility from the Coulomb interactions alone is negative for the OCP. Working from the Coulombic pressure the compressibility is found to be:

$$\left(\frac{\partial \beta P}{\partial n}\right)_\beta = -\frac{4}{9} a\Gamma + \frac{13}{36} b\Gamma^{1/4} - \frac{c}{3} \quad . \tag{2.16}$$

This result for the compressibility is of considerable importance since it appears in the OCP structure factor [Vieillefosse and Hansen, 1976] as:

$$S(k) = \frac{1}{1 + \frac{3\Gamma}{K^2} + \frac{\partial\beta P}{\partial n}} \quad , \qquad K = k\bar{r} \tag{2.17}$$

for the small k limit of S(k). Compressibility results have been obtained from Hansen's values for S(k), and at least for the $\Gamma < 40$, the results are in good agreement with the above result for $(\partial\beta P/\partial n)$.

The various results quoted in this section completely specify the equilibrium thermodynamic functions for the OCP. The $\Gamma^{1/4}$ form appearing in U_{th}/NkT indicates that the thermal energy itself varies as $T^{3/4}$ in contrast to (3/2)NkT for the thermal energy observed in the solid phase due to harmonic vibrations around lattice sites. It is not surprising that the thermal energy of the OCP fluid differs from the solid thermal energy. An understanding of the underlying mechanism to account for the $T^{3/4}$ behavior in the fluid state would be most helpful for dense plasma theory.

III. STRONG COUPLING RESULTS FOR INVERSE POWER FLUIDS

Fluids governed by repulsive $1/r^m$ potentials (Eq. 1.7) may at first glance seem remote from plasma physics. This may seem particularly so for the m = 12 case which describes the repulsive part of the Lennard Jones 6-12 potential used for describing interaction among molecules of ordinary gases and liquids. However, the Monte Carlo calculation of U/NkT for the cases of m = 4, 6, 9 and 12 indicate a basic similarity to the OCP Monte Carlo results [Hoover, Gray and Johnson, 1971]. One sees the same kind of change of functional form in the internal energy from weak to strong coupling as is seen in the plasma case. In weak coupling, however, there is a virial expansion in powers of density unlike the weak coupling plasma case. The strong coupling parameter for the inverse power potentials can be defined in a manner analogous to Eq. (1.2) for the 1/r potential. The general inverse power potential can be written as:

$$\beta u(r) = \beta\varepsilon(\sigma/r)^m = \beta\varepsilon(\sigma/\bar{r})^m\,(\bar{r}/r)^m$$
$$= \Gamma/x^m \quad , \qquad x = r/\bar{r} \tag{3.1}$$

where

$$\Gamma = \beta\varepsilon(\sigma/\bar{r})^m \propto \frac{n^{m/3}}{kT} \tag{3.2}$$

is the generalization of the OCP strong coupling parameter. The inverse power potential has a particularly useful scaling property which is apparent from the virial theorem:

$$E = E_0 + U$$

$$PV = P_0V + \frac{m}{3} U \tag{3.3}$$

where

$$U = \left\langle \sum_{i<j} U(r_{ij}) \right\rangle = \frac{Nn}{2} \int d^3r \; u(r) \; g(r) \tag{3.4}$$

and $E_0 = (3/2)NkT$ for the ideal gas term. The m/3 factor follows from the virial expression for the pressure:

$$(P - P_0)V = \frac{Nn}{6} \int d^3r \left[-r \frac{du(r)}{dr}\right] g(r) \tag{3.5}$$

For low densities the pressure and energy have an expansion in powers of density, the cluster expansion, and the strong coupling parameter Γ is inappropriate. (Recall in the OCP case that weak coupling expressions behaved linearly with $\varepsilon \sim \Gamma^{3/2}$). It is convenient to express the density coupled with the appropriate power of temperature as:

$$\tilde{n} = \frac{n\sigma^3}{\sqrt{2}}\left(\frac{\varepsilon}{kT}\right)^{3/m} = \left(\frac{3}{4\pi\sqrt{2}}\right) \Gamma^{3/m} \tag{3.6}$$

As with the OCP all thermodynamic quantities are functions of Γ (or equivalent of $\tilde{n}$). The virial expansion in powers of $\tilde{n}$ gives exact results for weak coupling (low density or high temperature) and strong coupling results are available from the Monte Carlo studies of Hoover, et al [1971]. The interaction internal energy has the form:

$$\begin{aligned} U/NkT &= \frac{3}{m} (P - P_0)/nkT \\ &= \frac{3}{m}\left\{\tilde{B}_2\tilde{n} + \tilde{B}_3\tilde{n}^2 + \tilde{B}_4\tilde{n}^3 + \ldots\right\} \\ &= a\Gamma + (b\Gamma^s - c) \quad , \quad \Gamma_t < \Gamma < \Gamma_f \end{aligned} \tag{3.7}$$

The virial expansion is convergent for the entire fluid phase, but at present time only a few of the dimensionless virial coefficients ($\tilde{B}_2$, $\tilde{B}_3$, etc.) have been evaluated; seven are known for the hard sphere fluid, $m = \infty$ [Ree and Hoover, 1967]. The Monte Carlo strong coupling data for m = 4, 6, 9 and 12 can all be fitted with the second form of Eq. (3.7). As described in OCP case the $a\Gamma$ term is the static energy of disordered system, U_0/NkT, and the constant a is expected to be close to or perhaps identical to the Madelung constant for the lattice phase. The f_{cc} lattice has the lowest energy for $m > 3$; the bcc lattice is lowest for the OCP. The remaining piece, $b\Gamma^s - c$, is the fluid thermal energy, U_{th}/NkT.

The actual data for all five available cases, m = 1, 4, 6, 9, 12, is shown in Table II. The number quoted in Table II are for $(U_{MC} - U_0)/NkT$ with U_0/NkT taken to be the appropriate lattice value. Thus in this tabulation the constant, a, in Eq. (3.7), is taken to be exactly the appropriate Madelung constant for the solid phase, rather than being treated as a free parameter. This is an unproved assumption, and one that I regard as questionable. Because of this assumption, the U_{th}/NkT data for the OCP as tken from Hansen's paper differ slightly from the U_{th}/NkT data for the OCP in Table I. Recall that Eq. (2.12) for the fit to Hansen's OCP data for $1 < \Gamma < 40$ used a as free parameter and it came out slightly higher than the lattice value. The interesting aspect of this thermal energy data for the strongly coupled fluids is the uniformity in behavior and magnitude. Recently Hoover, et al [1975] developed a generalized van der Waals equation of state for liquids, and made a simple fit to the thermal energy data with the assumption in Eq. (4.8) that c = 0 and s = 1/3. A more careful analysis with c as a free parameter indicates that the exponent is lower than 1/3, and is probably close to 1/4 as in the OCP case.

For the m = 12 case the virial expansion with four exactly known coefficients is:

$$U/NkT = \frac{3}{12}(P - P_0)/NkT$$

$$= \frac{1}{4}\left\{3.63\tilde{n} + 7.58\tilde{n}^2 + 9.94\tilde{n}^3 + 8.45\tilde{n}^4 + \ldots\right\} \qquad (3.8)$$

and for large Γ the result is:

$$U/NkT = .00493\Gamma + (.516\Gamma^{1/4} - .49) \quad , \quad 200 \leq \Gamma < 538 \qquad (3.9)$$

where a = .00493 is the fcc Madelung constant for the $1/r^{12}$ potential, $\Gamma_f = 538$ is the value of the coupling parameter when the fluid freezes. $\Gamma_t \sim 200$ is an estimate of the point at which the asymptotic strong coupling form appears. Γ_t for m = 12 is far larger than in the OCP case ($\Gamma_t \sim .75$ for m = 1) because of the far shorter range of the $1/r^{12}$ potential. In general it will be true that

$$\lim_{m\to\infty} \frac{\Gamma_f - \Gamma_t}{\Gamma_f}$$

and for the hard sphere gas there is no static lattice energy and no

TABLE II

INVERSE POWER FLUID MONTE CARLO DATA

Hansen, 1973		Hoover, et al, 1971			
m = 1		m = 4		m = 12	
Γ	U_{th}/NkT	Γ	U_{th}/NkT	Γ	U_{th}/NkT
1	.316	.498	.331	.123	.111
2	.473	1.688	.605	1.97	.270
3	.577	4.25	.907	9.98	.476
4	.658	10.72	1.347	31.5	.734
6	.786	18.40	1.644	77.0	1.034
10	.971	27.01	1.887	159.6	1.329
15	1.126	36.37	2.055	219.8	1.506
20	1.252	46.38	2.262	295.7	1.661
30	1.450	56.96	2.412	369.3	1.780
40	1.606	68.06	2.583	432.9	1.870
50	1.703	σ = .013		504.4	1.956
60	1.821	m = 6		σ = .007	
70	1.909	3.51	.247		
80	1.986	2.19	.577		
90	2.066	8.77	1.036		
100	2.116	35.09	1.739		
110	2.195	78.96	2.195		
120	2.230				
125	2.262	m = 9			
130	2.292	.208	.159		
140	2.34	3.25	.469		
155	2.43	25.98	1.086		
160	2.46	207.9	2.199		
σ = .007					

$$U_{th}/NkT = \Delta a\Gamma + b\Gamma^s - c$$

1) Nonlinear l. sq. (s, b, c)
2) Linear l. sq. s = .20, .25, .3, .35

(Δa, b, c)

thermal energy. Table III gives the strong coupling fits to the data for the five known cases assuming that a = Madelung constant, and s = 1/4. For this investigation a nonlinear least squares

TABLE III

FLUID THERMAL ENERGY FOR INVERSE POWER POTENTIALS

$$u(r) = \varepsilon(\sigma/r)^m$$

Data for m = 4, 6, 9, 12 (Hoover, et al) and m = 1 (Hansen) all fit the form:

$$U_{th}/NkT = b\Gamma^{1/4} - c \quad , \quad \text{where } \Gamma = (\beta\varepsilon)(\sigma/\bar{r})^m$$

$$\bar{r} = \left(\frac{1}{\frac{4\pi}{3} N/V}\right)^{1/3}$$

m	$U/NkT = (U_o + U_{th})/NkT$
1	$-.895929\Gamma + .8739\Gamma^{1/4} - .5777$
4	$1.181978\Gamma + 1.150\Gamma^{1/4} - .7413$
6	$.205945\Gamma + .9267\Gamma^{1/4} - .5484$
9	$.0300478\Gamma + .7067\Gamma^{1/4} - .4913$
12	$.0049258\Gamma + .5156\Gamma^{1/4} - .4870$

The $\Gamma^{1/4}$ power law appears best for the m = 1 and m = 12 cases for which there are the most points.

$C_v/Nk = \frac{3}{4} b\Gamma^{1/4} - c$; Hansen's data checks this result very well.

---

fitting routine was used with a, b, c and s as free parameters. The value of the exponent as $s \approx 1/4$ appeared best for m = 1 and m = 12. There are too few data points for m = 4, 6 and 9 to assert confidently that in all cases s = 1/4. However, the results do strongly suggest a universal form of the thermal energy, namely $\Gamma^{1/4}$, for all the inverse power fluids.

IV. INTEGRAL EQUATION RESULTS

Until about 20 years ago most calculations of the equation of state of dense gases and liquids had to rely on the virial expansion with the cluster integrals evaluated for assumed intermolecular potentials. In the 1960's the density expansions were powerfully supplemented by integral equations for g(r) which in effect summed to infinity certain kinds of the integrals in the Mayer irreducible clusters. The Percus-Yevick equation was found to give an excellent description of the hard sphere fluid and the PY equation of state was found to be in near quantitative agreement with the molecular dynamics hard sphere data [Rushbrooke, 1968]. The hypernetted chain (HNC) equation did not work so well for the hard sphere system, but it did appear to be useful for longer range potentials including the inverse power potentials [Hutchinson and Conkie, 1972].

Monte Carlo simulations give presumably nearly exact "experimental" results for the strongly coupled OCP. It is of considerable interest to see what extent integral equations of present day liquid state theory can reproduce the Monte Carlo results. The Percus-Yevick equation has been investigated for the Coulomb potential [Springer, Pokrant and Stevens, 1973] and found to be completely inaccurate for large Γ. However, the hypernetted chain equation (HNC) when solved numerically was found to give results for the total potential energy, U/NkT, for the OCP that are remarkably close to the Monte Carlo results [Springer, Pokrant and Stevens, 1973]. Recently Ng has made an extremely accurate numerical solution of the HNC equation [Ng, 1974], and obtained results for U/NkT for the OCP to seven and eight figure accuracy for values of Γ from 20 to 7000. Although $\Gamma = 7000$ is far beyond any conceivable physical situation, these exact numerical results for the HNC equation allow one to find the functional form of U/NkT with respect to Γ without the difficulty presented by the inevitable noise in the Monte Carlo data. To explain simply the HNC equation one notes that the pair distribution function for the OCP may be written generally as:

$$g(r) = h(r) = 1 = \exp\left\{-\frac{\Gamma}{x} + S(x) + B(x)\right\} \tag{4.1}$$

where g(r) is the pair distribution function, h(r) = g(r) - 1 is the total correlation function, S(x) indicates the sum of all convolution or series of graphs in the cluster expansion of g(r), and B(x) is the sum of all bridge graphs. h(r) is related to the direct correlation function, c(r), by the Ornstein-Zernike equation:

$$h(r) = c(r) + n\int d^3r'\, c(r')h\,(|\underset{\sim}{r} - \underset{\sim}{r}'|) \quad . \tag{4.2}$$

The HNC integral equation is obtained from the above relations by the approximation of neglecting the bridge graph contributions, i.e. assuming B(x) = 0. The resulting equation is nonlinear for h(r) but may be solved with computers to any desired accuracy, and the OCP internal energy is obtained from the integral:

$$U/NkT = \frac{n}{2}\int d^3r \, \frac{\beta e^2}{r} \, [g(r) - 1]$$

$$= \frac{3\Gamma}{2}\int_0^\infty x^2 \, dx \, \frac{1}{x} \, h(x) \tag{4.3}$$

Some of Ng's HNC results for U/NkT are given in Table I to show a comparison with the Monte Carlo results of Hansen for U/NkT. Since Ng's results are of high accuracy and span an extremely wide range in Γ it was possible to find the functional dependence on Γ; the result is [DeWitt, 1976]:

$$(U/NkT)_{HNC} = -0.90047\Gamma + 0.26883\Gamma^{1/2} + 0.0720 \ \ln \Gamma + 0.0533 \tag{4.4}$$

The HNC solution is quite continuous for large Γ even up to Γ = 7000, and consequently it should be be interpreted as an approximation for the fluid branch of the OCP equation of state. There is no indication that any known integral equation can give a second solution that would correspond to the equation of state for the solid phase of the OCP. What is remarkable about the above nearly exact analytic result for the fluid phase potential energy is that it clearly shows a separation of the internal energy into a fluid static energy and a thermal energy portion which is dominated by a $\Gamma^{1/2}$ dependence. The HNC fluid static energy comes out remarkably close to the prediction of the ion sphere model which for the OCP would be:

$$(U_0/NkT)_{\text{ion-sphere}} = \left(-\frac{3}{2} + \frac{3}{5}\right)\frac{\beta(ze)^2}{\bar{r}} = -0.9\Gamma \quad . \tag{4.5}$$

The HNC equation evidently goes to the ion-sphere result in the limit. The Monte Carlo result for the static energy, namely $U_0/NkT = -0.89461\Gamma$, is very close to the bcc lattice value, which differs from the ion-sphere model by only 0.45%. Since the thermal energy portion is only a small fraction of the total potential energy for large , the close agreement of U_0 for HNC and Monte Carlo insures that the HNC results for U/NkT seem to agree well with the Monte Carlo results. This agreement is deceptive since the HNC thermal energy ($\sim\Gamma^{1/2}$) is very different from the presumed exact thermal energy ($\sim\Gamma^{1/4}$) obtained from the Monte Carlo simulations. Evidently this difference is entirely due to the basic HNC approximation of neglecting the bridge graphs. It is an open question as to whether a more exact integral equation that includes one or more of the lower order bridge graph terms could account for the difference in the HNC and the exact thermal energies. It is, in any case, significant that the simple approximation of the HNC equation is sufficient to give the basic qualitative feature of the fluid phase OCP internal energy, namely a division into a static portion and a thermal portion.

Another widely used approximation in liquid state theory is the mean spherical approximation which states that:

$$c(r) = -\beta u(r) \quad , \quad r < \sigma$$

$$h(r) = -1 \quad , \quad r < \sigma \tag{4.6}$$

where σ is an equivalent hard sphere radius, and c(r) and h(r) are connected by the usual Ornstein-Zernike equation. Gillan [1974] has solved the mean spherical model for the OCP by using a judicious choice of σ so that g(r) does not go negative. He obtains g(r) and computes values of U/NkT some of which are shown in Table I. As is the case with the HNC numerical results the absolute values of U/NkT obtained from the mean spherical model are in moderately good agreement with the Monte Carlo results. The functional form obtained from Gillan's numbers is:

$$(U/NkT)_{MS} = -0.9005\Gamma + 0.2997\Gamma^{1/2} + 0.0007 \quad . \tag{4.7}$$

As with the HNC results there is a clear separation of the potential energy into a static portion, U_0, that is very close to the ion-sphere result, and a thermal energy that is dominated by $\Gamma^{1/2}$. In a numerical sence there is little to distinguish the HNC results from the mean spherical model. It is not even clear whether one approximation is better than the other. Both give a thermal energy in clear disagreement with the thermal energy from the Monte Carlo data. However, the qualitative agreement of the mean spherical model with the "exact" OCP results does indicate that the mean spherical model approximation may well be improved by a better guess for the form of the direct correlation function, c(r). The actual form of c(r) is obtainable from the Monte Carlo simulations and will be discussed later.

V. DENSE PLASMA MIXTURES IN STRONG COUPLING

We consider now arbitrary mixtures of two nuclear species, and for the moment $r_s = 0$. This is simply the extension of the OCP to two components [Salpeter, 1954]. The ion sphere model result, Eq. (4.5), is easily extended to two components. With the total ion number density as $n = n_1 + n_2$ for charges z_1 and z_2, the background density of the entralizing electron is:

$$n_e = z_1 n_1 + z_2 n_2 = \bar{z} n$$

and the radius of a sphere around a charge z_1 of sufficient size to neutralize z_1 is:

$$\bar{r}_1 = \left(\frac{z_1}{\bar{z}}\right)^{1/3} \bar{r} = z_1^{1/3}\, \bar{r}_e \qquad (5.1)$$

From this result the static energy for a two-component mixture according to the ion-sphere prescription is:

$$(U/NkT)_{\text{ion-sphere}} = -\frac{9}{10}\left(x_1 z_1^{5/3} + x_2 z_2^{5/3}\right) \bar{z}^{-1/3}\, \beta e^2/\bar{r}$$

$$= -\frac{9}{10}\, \overline{z^{5/3}}\, \bar{z}^{-1/3}\, \Gamma_0 \qquad (5.2)$$

where

$$x_1 = \frac{n_1}{n_1 + n_2} \quad , \qquad x_2 = \frac{n_2}{n_1 + n_2}$$

and

$$\Gamma_0 = \beta e^2/\bar{r} \qquad .$$

Thus the ion-sphere model gives a characteristic simple charge averaging prescription for mixtures of ions, namely

$$\overline{z^{5/3}}\, \bar{z}^{-1/3} \qquad ,$$

that is very different from the $\overline{z^2}$ charge averaging that appears in the Debye result for weak coupling. In view of the remarkable agreement of the OCP ion-sphere result with both the Monte Carlo and the HNC results for the static energy, it is reasonable to expect that the ion-sphere result will be equally good for the static energy for the ion mixtures. Indeed this is the case. The two nuclear component Monte Carlo data from Livermore [DeWitt and Hubbard, 1976] agree perfectly with the ion-sphere charge averaging prescription. More recent (unpublished) and much more extensive Monte Carlo for a variety of mixtures for $z_1 = 1$, $z_2 = 2$ and $z_1 = 1$, $z_2 = 3$ also completely agree with the ion-sphere charge average prescription. Also Hansen and Vieillefosse have solved the coupled hypernetted chain equations for two nuclear components [Hansen and Vieillefosse, 1976], and have found that for the HNC approximation again the ion-sphere charge averaging is satisfied. (Hansen and Vieillefosse use the term "Two-Component Plasma" or TCP in their paper, and of course, mean two components of the same sign; this should not be confused with a two-component plasma in the sense of fully ionized hydrogen, i.e., charges of opposite sign). The recent Monte Carlo data on mixtures from Livermore also suggests that the thermal energy has the same density and temperature dependence as was found for the OCP, namely $\Gamma_0^{1/4}$. Consequently the internal energy for a two-component mixture can be written as:

$$U/NkT = -a\overline{z^{5/3}\,\bar{z}^{1/3}}\,\Gamma_0 + bg(z_1x_1,\ z_2x_2)\,\Gamma_0^{1/4} - c \qquad (5.3)$$

where the function $g(z_1x_1;\ z_2x_2)$ is a charge average for the thermal energy. Clearly for $x_1 = 1$ the value of g is $z_1^{1/2}$ and for $x_1 = 0$ it would be $z_2^{1/2}$. It is tempting to assume a one fluid model for the two-component system and to use the ion-sphere charge average for g, namely:

$$g = (\overline{z^{5/3}}\,\bar{z}^{1/3})^{1/4}$$

While this assumption does not give mixture energies that are badly in error, this one fluid model is definitely not correct. Recent Monte Carlo results from both Livermore and Paris for two components indicate that the above assumption for the thermal energy charge average is definitely outside the noise of the Monte Carlo simulations. The Livermore data shown in Table IV can be fitted reasonably well with the form:

$$g = \bar{z}^{1/2} \qquad (5.4)$$

although it should be noted that this form has no theoretical justification at the present time. Hansen and Vieillefosse note that the ion-sphere charge average is strictly additive when the electron density remains constant. Their solution of the HNC equations for two components suggests that this additive property also holds true for the thermal energy. Thus they suggest that the two-component energy is:

$$U/NkT = x_1\left\{U(z_1^{5/3}\,\Gamma_0^1) + x_2\ U(z_2^{5/3}\,\Gamma_0^1)\right\}/NkT \qquad (5.5)$$

where $\Gamma_0^1 = \bar{z}^{1/3}\,\Gamma_0$. Very recently they have two-component Monte Carlo results which indicate the same additivity [Hansen, Torrie and Vieillefosse, 1976]. The Livermore and Paris Monte Carlo results for mixtures are so close in numerical agreement and also close to the noise level, that at the moment it is difficult to give a final answer for the thermal energy for mixtures. The precise form will probably not be clear until there is a good theoretical model for the $\Gamma^{1/4}$ thermal term.

VI. SCREENING CORRECTIONS

So far in these lectures the electron background has been treated as a rigid neutralizing background with no properties other than the constant densities. This is obviously unrealistic for applications to the interior of Jupiter and stellar interiors. In order to treat the electrons as a responding background the

TABLE IV

TWO NUCLEAR COMPONENT PLASMA MIXTURE FLUID DATA FROM LLL

		Hansen	LLL					
Γ_0	$\tilde{\Gamma}$	(U/NkT)	$(U/NkT)_{\tilde{z}\tilde{z}}$	$(U/NkT)_{1,2}$	$\frac{U_{12} - U_{\tilde{z}\tilde{z}}}{NkT}$	$C_{v,\tilde{z}\tilde{z}}$	$C_{v,1,2}$	$C_{v1,2} - C_{v\tilde{z}\tilde{z}}$
.1875	.3124	-.170	-.135	-.141	-.006	.035	.033	-.002
.375	.6236	-.334	-.323	-.332	-.009	.081	.078	-.003
.6	.9977	-.578	-.579	-.588	-.009	.134	.129	-.05
.75	1.2472	-.754	-.754	-.772	-.018	.161	.147	-.013
1.5	2.2472	-1.707	-1.716	-1.731	-.015	.257	.258	+.001
3.0	4.4944	-3.744	-3.750	-3.779	-.029	.386	.382	-.004
6.0	9.9774	-7.976	-7.980	-8.011	-.031	.601	.586	-.015
12.0	19.9548	-16.627	-16.611	-16.661	-.051	.772	.753	-.019
		128 particles	48 particles		36 with $z_1 = 1$ 12 with $z_2 = 2$			

$$\tilde{\Gamma} = \overline{z^{5/3}}\,\bar{z}^{-1/3}\,\Gamma_0 = \tilde{z}^2\,\Gamma_0$$

$$(U/NkT)_{\tilde{z}\tilde{z}} = \text{OCP with } \tilde{\Gamma}$$

linear response theory has been used to describe the increase in electron density near each ion. With the ion charge density given by:

$$\rho_i(\underline{r}) = ze \sum_{i=1}^{N} \delta(\underline{r} - \underline{r}_i) \tag{6.1}$$

the induced electron density, i.e. the deviation from uniform density, can be written as:

$$\rho_{induced}(k) = - \left[1 - \frac{1}{\varepsilon(k)}\right] \rho_i(k) \tag{6.2}$$

where $\varepsilon(k)$ is the dielectric function of the electrons. In general $\varepsilon(k)$ is a function of the electron temperature, and indeed finite temperature effects may be considerable in the interiors of laser fusion pellets. However, at sufficiently high densities the electron Fermi energy dominates the electron temperature, $kT/\varepsilon_F \ll 1$, and the electrons can be treated as a completely degenerate T=0 fluid. For $r_s \lesssim 1$ the random phase approximation (RPA) is adequate for the electron dielectric function; at T=0 this is given by the Lindhard expression:

$$\varepsilon(k) = 1 + \frac{1}{(k\,\lambda_{TF})^2} f(y) \tag{6.3}$$

where $y = k/k_F$ and k_F is the Fermi momentum wave number, and

$$f(y) = \frac{1}{2} + \frac{1 - y^2}{4y} \ln \left| \frac{1 + y}{1 - y} \right| \tag{6.4}$$

and

$$\bar{r}/\lambda_{TF} = \left(\frac{12z}{\pi}\right)^{1/3} r_s^{1/2} = q_{TF} \tag{6.5}$$

defines the Thomas-Fermi screening length. Hubbard and Slattery [1971] incorporated the electron screening effect into their Monte Carlo code, and with some minor modifications this code has been operated extensively at Livermore. The effect of the electron screening also changes the form of the ion-ion potential. In k space this is:

$$u(k) = \frac{4\pi(ze)^2}{k^2\varepsilon(k)} \tag{6.6}$$

and in r space:

$$u(r) = \frac{(ze)^2}{r} g(r/\lambda_{TF}) \sim \frac{(ze)^2}{r} e^{-r/\lambda_{TF}} \quad , \quad r < \lambda_{TF} \qquad (6.7)$$

The r space representation is not readily useful since the screening function, $g(r/\lambda_{TF})$, exhibits Friedel oscillations for large r. The simple exponential screening is apparent only at short distances. The k space form, Eq. (6.6), however, can be adapted into the calculation of the Ewald potential needed to describe the image charges for the plasma Monte Carlo calculation.

The Monte Carlo code gives results for U/NkT and P/nkT as functions of Γ and r_s. Note that the simple virial theorem factor of 1/3 relating the energy and pressure no longer applies since the ion-ion potential, Eq. (6.7), is no longer pure Coulomb. It is more general and useful to discuss the results by starting with the Helmholtz free energy which can be written as:

$$\beta F/N = \beta[F^{(0)}(\Gamma) + F^{(1)} (\Gamma, r_s)]/N \qquad (6.8)$$

with $F^{(0)}(\Gamma)$ given by Eq. (2.15) for the OCP. A two-component version of $F^{(0)}$ can also be easily worked out using the charge averaging results given in the previous section. The entropy constant for the two-component mixture is given in DeWitt and Hubbard [1976]. Our interest here is in the screening correction, $F^{(1)}$. The Monte Carlo results show that for

$$0 < r_s \lesssim 0.5$$

$F^{(1)}$ is proportional to r_s times a function of Γ [Hubbard and Slattery, 1971]. This numerical result has been well-confirmed by the theoretical analysis of Hansen and Golam [1976] who show that:

$$F^{(1)}/NkT = \frac{1}{3\pi} \int_0^\infty q^2 \, dq \; S^{(0)}(q) \; w(q) \qquad (6.9)$$

with

$$w(q) = \frac{3\Gamma}{q^2} \left[\frac{1}{\varepsilon(q)} - 1\right] \quad , \qquad q = k\bar{r}$$

and $S^{(0)}(q)$ is the OCP structure factor. Evaluation of the integral numerically and fitting to the same form for U/NkT gives:

$$F^{(1)}/NkT = r_s \; (.0579\Gamma + .971\Gamma^{1/4} - .343) \qquad (6.10)$$

The energy and pressure are obtained from:

$$U/NkT = \beta \frac{\partial}{\partial\beta} (\beta F/N) = \Gamma \frac{d}{d\Gamma} (\beta F/N)$$

$$\beta P/n = -V \frac{\partial}{\partial V} (\beta F/N) = \frac{1}{3} \beta U/N - \frac{r_s}{3} \frac{\partial}{\partial r_s} (\beta F^{(1)}/N) \qquad (6.11)$$

The explicit results from the Livermore Monte Carlo data are:

$$\begin{aligned} U/NkT = &-(.8946 + .0543\ r_s)\ \Gamma \\ &+ (.8165 - .1853\ r_s)\ \Gamma^{1/4} \\ &- (.5012 - .0659\ r_s) \end{aligned} \qquad (6.12)$$

and for the pressure:

$$(P - P_0)/nkT = \frac{1}{3} (U^{(0)}/NkT) + .185\ r_s\ \Gamma^{1/4} \qquad (6.13)$$

Similar numerical results are available for mixtures of two elements in Hubbard and Slattery [1971], but will not be quoted here. The two ionic component results are being done now more accurately, so that the free energy of a mixture (e.g. fully ionized hydrogen and helium) can be used for locating the condition for phase separation [Stevenson, 1975].

The largest numerical effect of electron screening is on the static portion of the energy, but even this is only a few percent at $r_s \sim .5$. The pressure of the plasma is affected rather little by electron screening. This fact is apparent from the form of $F^{(1)}$ which is proportional to $r_s\Gamma = \beta e^2/a_0$, and hence independent of volume. Consequently the correction due to screening in Eq. (6.11) comes only from the thermal portion of the free energy.

Where detailed comparisons have been made there is excellent agreement between the Livermore Monte Carlo screening results and the calculation based on Eq. (6.9) reported by Galam and Hansen. For $r_s \geq .5$ there is an additional $O(r_s^{3/2})$ contribution which also shows up in the Monte Carlo data.

VII. PAIR CORRELATION AND DIRECT CORRELATION FUNCTIONS

The pair correlation function g(r) is obtainable from the Monte Carlo calculations but only to a distance of approximately half of the size of all containing the N charges. This distance, $x \sim N^{1/3}/2$, is sufficient to show the major structural features of g(r). An example is shown in Figure 1 for $\Gamma = 40$. The g(r)'s for the OCP resemble the g(r) for the inverse power fluids and even the

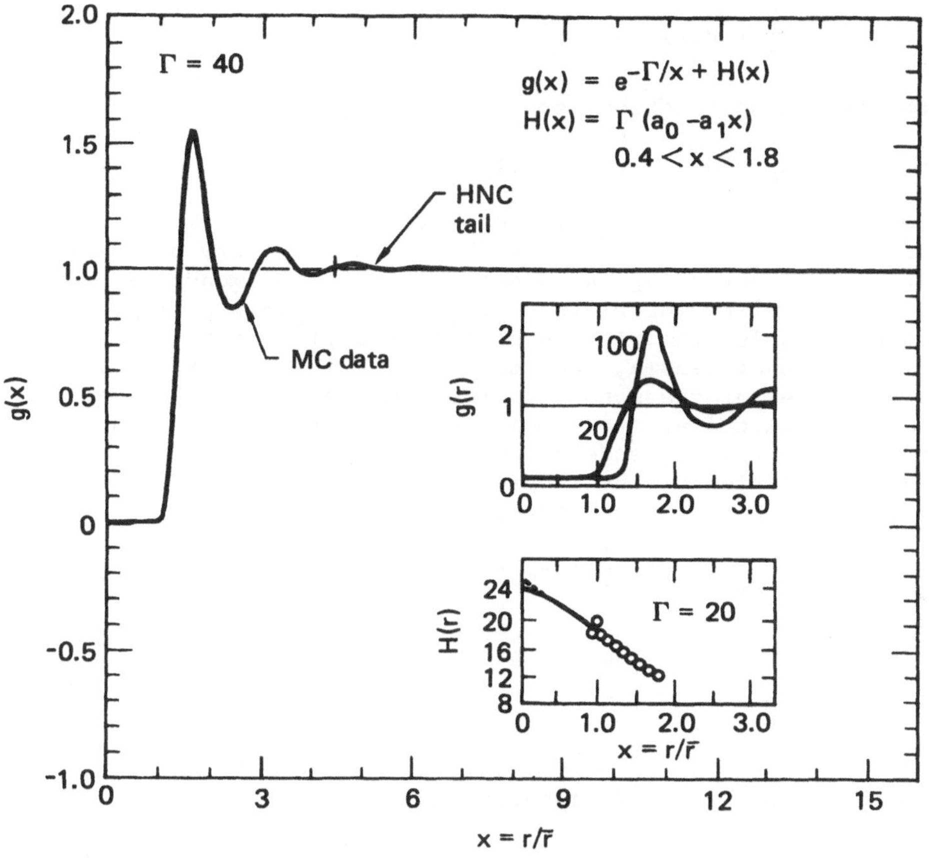

Figure 1. Monte Carlo g(r) for Γ = 40 with HNC extension. The insert shows the linear behavior of the screening function.

hard sphere g(r). This is another indication that the strongly coupled plasma has properties closely related to more ordinary liquids. The typical oscillations of g(r) about 1 appear when $\Gamma \sim 2.5$. The exact transition value is of some interest, and will be discussed in more detail by Dr. Deutsch. Here I will just point out that the transition value from monotonic to oscillatory behavior according to the Monte Carlo calculations is definitely less than $\Gamma = 3.08$ at which the OCP compressibility becomes negative (where Eq. (2.16) gives -1). Although short range order is certainly indicated by the oscillations in g(r), there is no indication of any kind of phase change at $\Gamma \sim 2.5$ or 3.08. The thermodynamic functions are all quite continuous.

A striking feature of the g(r) data is the linear behavior in the screening function in the region below the first peak:

$$g(x) = e^{-\frac{\Gamma}{x} + H(x)} \tag{7.1}$$

with the screening function as:

$$H(x) = \Gamma(a_0 - a_1 x) \quad , \quad .4 < x < 1.8 \tag{7.2}$$

This empirical feature was first observed in 1973 from the BST data [DeWitt, Graboske and Cooper, 1973] and was very useful in calculating the screening enhancement rate of thermonuclear reactions in stellar interiors. Professor Ichimaru will present a discussion of this linearity in H(x) in relation to the lattice model of the dense plasma, and the use of this relation in more recent estimates of thermonuclear reaction rates. The values of the constants a_0 and a_1 are almost independent of Γ, though a slight but important dependence has been found. a_0 and a_1 have a geometrical relation to the location and height of the first peak in g(r). With the empirical form, Eq. (7.2), one has:

$$g(x) = e^{\Gamma f(x)} = e^{\Gamma(-\frac{1}{x} + a_0 - a_1 x)} \tag{7.3}$$

The first peak occurs at $x_m \approx 1.68$, and the height is:

$$g(x_m) = e^{\Gamma\Delta}$$

where Δ is a small number found empirically from the MC data. A little algebra with Eq. (7.3) shows that:

$$a_1 = \frac{1}{x_m^2} = .39 \tag{7.4}$$

and

$$a_0 = 2a_1^{1/2} + \Delta = 1.27 + \Delta \tag{7.5}$$

For large Γ the linear behavior in x of H(x) cannot be seen very far because g(x) is almost zero until $x \sim .8$. For small x a different functional form appears:

$$H(x) = \Gamma(\alpha_0 - \alpha_1 x^2) \quad , \quad x \lesssim .4 \tag{7.6}$$

Recently Professor Jancovici has shown that α_1 should be 1/4 rather than 1/2 as reported by DeWitt, Graboske and Cooper [1973]. The constant α_0 (which is crucial for the reaction rate calculation) is related to the difference of free energies of two charges before and after reaction (DeWitt, Graboske and Cooper [1973] and also Jancovici's work reported at this meeting). Using Eq. (2.15) for the OCP free energy one finds that the theoretical value of α_0 is:

$$\Gamma\alpha_0 = [2F_I(\Gamma) - F_I(2^{5/3}\,\Gamma)]/NkT$$

$$= 1.051\Gamma + 2.172\Gamma^{1/4} - .501\,\ell n\,\Gamma = 2.24 \qquad (7.7)$$

and this seems to agree with Monte Carlo results as far as it can be checked.

In order to find the OCP structure factor S(k) by Fourier transforming g(r) it is necessary to have an accurate "tail" to extend the MC numerical data for g(r) from $x \sim 2.5$ on to ∞. This can be done by fitting the solution of the HNC equation on to the MC data with the help of the Ornstein-Zernicke equation. The procedure is described in some detail by Galam and Hansen [1976]. The assumption of importance here is that the bridge graphs are negligible for $x \gtrsim 2.5$ thus ensuring the validity of the HNC equation. The structure factor is obtained then from:

$$S(k) = 1 + c \int_0^\infty x^2 dx \,\frac{\sin kx}{kx}\, h(x) \qquad (7.8)$$

where $h(x) = g(x) - 1$. A typical result is shown in Figure 2. A further check on the accuracy of S(k) so obtained is a direct calculation of the internal energy from

$$U/NkT = \frac{\Gamma}{\pi} \int_0^\infty dk\; S(k) = \frac{3\Gamma}{2} \int_0^\infty x dx\; h(x) \qquad (7.9)$$

For each value of Γ the evaluation of the energy from the k integration and x integration come out slightly different, but they closely bracket the Monte Carlo results for U/NkT. An additional check on the accuracy of the S(k) data is the compressibility relation that appears in S(k), Eq. (2.17). At least up to $\Gamma = 40$ there is good agreement with Eq. (2.16).

With S(k) known numerically one also has $\tilde{c}(k)$ from:

$$S(k) = \frac{1}{1 + \tilde{c}(k)} \qquad (7.10)$$

so that c(r) can be obtained with a Fourier transform. For $x \gtrsim 1.8$ the direct correlation function goes over to the Coulomb potential:

$$c(x) = -\frac{\Gamma}{x} \qquad (7.11)$$

but for $x < 1.8$ c(x) behaves much like H(x), namely

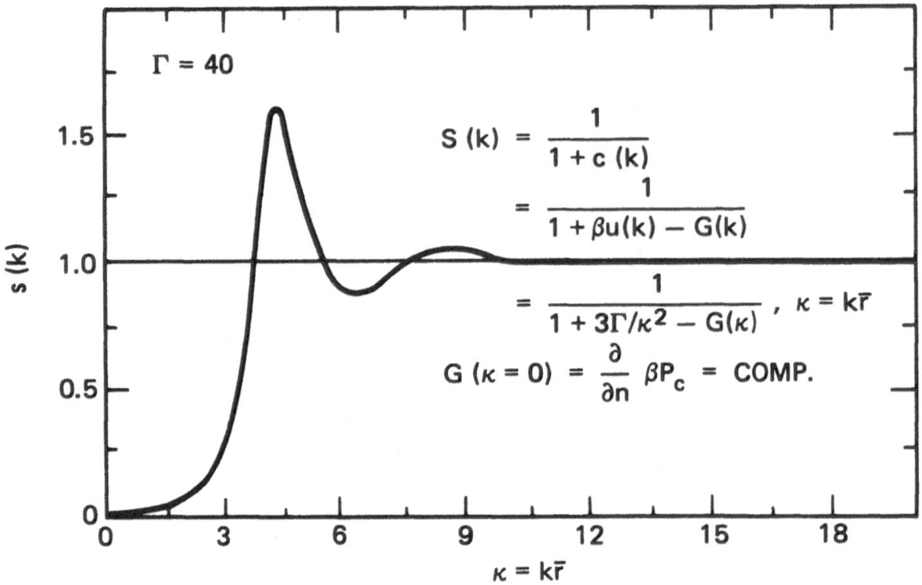

Figure 2. OCP structure factor S(k) for Γ = 40 obtained by Fourier transform of g(x).

$$c(x) = -\Gamma(a_0' - a_1'\, x) \quad , \qquad .4 < x < 1.8$$

$$= -\Gamma(\alpha_0' - \alpha_1'\, x^2) \quad , \qquad x < .4 \qquad (7.12)$$

where the constants a_0', a_1' and α_0', α_1' are slightly different from the values for H(x). This behavior is shown in Figure 3. The calculation of H(x) and c(x) allows one also to find the bridge graph function since:

$$H(x) = T(x) + B(x) \qquad (7.13)$$

where

$$T(x) = \int d^3x' c(x')\, h(|x - x'|) = h - c \qquad (7.14)$$

Since both H(x) and T(x) are now known numerically reasonable results can be obtained for the bridge graph function B(x). This work is in progress.

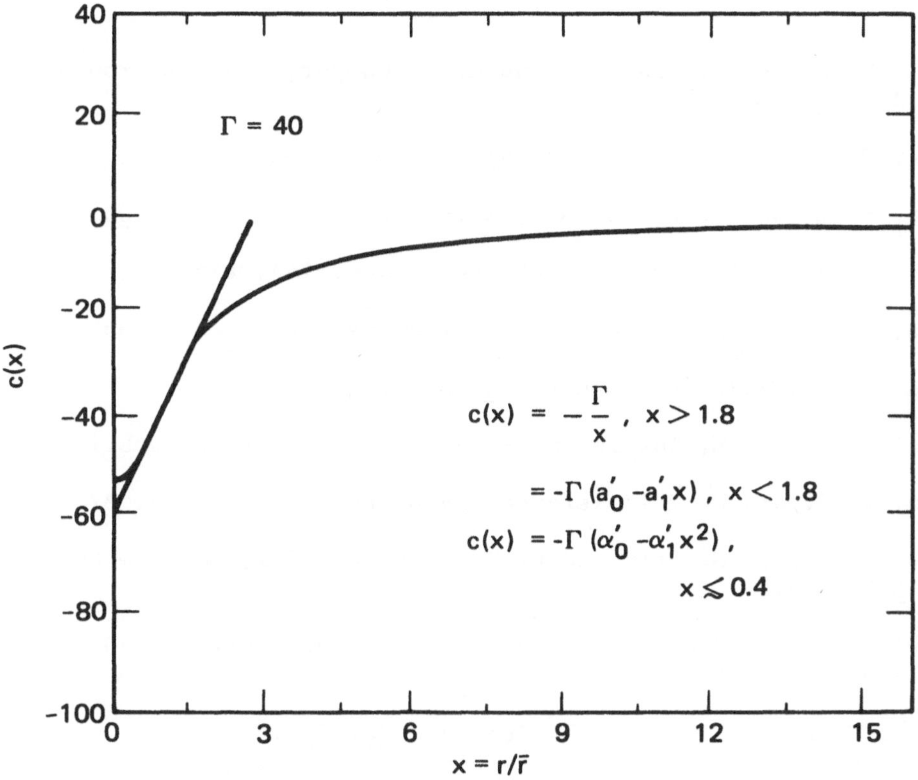

Figure 3. OCP direct correlation function for Γ = 40.

REFERENCES

Abe, R., 1959, Progr. Theor. Phys. 22, 213.

Alder, B. J. and T. E. Wainwright, 1957, J. Chem. Phys. 27, 1208.

Brush, S. G., H. E. DeWitt and J. G. Trulio, 1963, Nuclear Fusion 3, 5.

Brush, S., H. L. Sahlin and E. Teller, 1966, J. Chem. Phys. 45, 2102.

DeWitt, H. E., H. C. Graboske and M. S. Cooper, 1973, Astrophysical J. 181, 439.

DeWitt, H. E., 1976, Phys. Rev. A14, 1290.

DeWitt, H. E. and W. B. Hubbard, 1976, Astrophysical J. 205, 295.

Galam, S. and J. P. Hansen, 1976, Phys. Rev. 14A, 816.

Gillan, M. J., 1965, J. Phys. C: Solid State Physics 7, 353.

Hansen, J. P., 1973, Phys. Rev. A8, 3096.

Hansen, J. P., G. M. Torrie and P. Vieillefosse, to be published.

Hansen, J. P. and P. Vieillefosse, 1976, Phys. Rev. Lett. 37, 391.

Hoover, W. G., S. G. Gray and K. W. Johnson, 1971, J. Chem.Phys. 55, 1128.

Hoover, W. G., G. Stell, E. Goldmark and G. D. Degani, 1975, J. Chem. Phys. 63, 5434.

Hubbard, W. B., 1972, Astrophysical J. 176, 525.

Hubbard, W. B. and W. L. Slattery, 1971, Astrophysical J. 168, 131.

Hutchinson, P. and W. R. Conkie, 1972, Mol. Phys. 24, 567.

Metropolis, N., A. W. Rosenbluth, M. N. Rosenbluth, A. M. Teller and E. Teller, 1953, J. Chem. Phys. 21, 1087.

Nijboer, B. R. A. and F. W. DeWette, 1957, Physica 23, 309.

Ng, K., 1974, J. Chem. Phys. 61, 2680.

Pollock, E. L. and J. P. Hansen, 1973, Phys. Rev. A8, 3110.

Ree, F. H. and W. G. Hoover, 1967, J. Chem. Phys. 46, 4181.

Rogers, F. J., H. E. DeWitt, 1973, Phys. Rev. A8, 1061.

Rushbrooke, G. S., 1968, Equilibrium Theories of the Liquid State in Physics of Simple Liquids, edited by Temperley, Rawlinson and Rushbrooke, North Holland, Amsterdam.

Salpeter, E. E., 1954, Australian J. Phys. 7, 353.

Springer, J. F., M. A. Pokrant and F. A. Stevens, 1973, J. Chem. Phys. 58, 4863.

Stevenson, D. J., 1975, Phys. Rev. B12, 3999.

Stillinger, F. H. and R. Lovett, 1968, J. Chem. Phys. 49, 1991.

Vieillefosse, P. and J. P. Hansen, 1976, Phys. Rev. A12, 1106.

COMPUTER SIMULATION OF COLLECTIVE MODES AND TRANSPORT COEFFICIENTS OF STRONGLY COUPLED PLASMAS

Jean-Pierre Hansen

Laboratoire de Physique Théorique des Liquids*
Université Paris VI
4, Place Jussieu
75230 Paris Cedex 05

*Equipe Associée au CNRS.

TABLE OF CONTENTS

COMPUTER SIMULATION OF COLLECTIVE MODES AND TRANSPORT COEFFICIENTS OF STRONGLY COUPLED PLASMAS

Jean-Pierre Hansen

Laboratoire de Physique Théorique des Liquides
Université Paris VI; 4, Place Jussieu; 75230 Paris
Cedex 05

I. COLLECTIVE MODES AND SINGLE PARTICLE MOTION IN THE ONE COMPONENT PLASMA (OCP)

A. The Model

Consider a periodic system of N point ions of charge Ze and mass M in a rigid, neutralizing uniform background. For a given configuration $\vec{r}^N = (\vec{r}_1, \vec{r}_2, \ldots, \vec{r}_N)$ of the ions, the total potential energy of the system is:

$$V_N = \frac{1}{2V} \sum_{k \neq o} \frac{4\pi (Ze)^2}{k^2} (\rho_{\vec{k}} \rho_{-\vec{k}} - N) \tag{1.1}$$

where:

$$\rho_{\vec{k}} = \sum_{i=1}^{N} e^{i\vec{k}\cdot\vec{r}_i} \tag{1.2}$$

Excess thermodynamic properties, and more generally, all reduced (dimensionless) equilibrium properties depend on the single dimensionless variable:

$$\Gamma = \frac{(Ze)^2}{ak_BT} \tag{1.3}$$

where $a = (3/4\pi\rho)^{1/3}$, $\rho = N/V$. We shall frequently use reduced distances $x = r/a$ and wave numbers $q = k/a$. To describe dynamical (or time-dependent) properties we introduce an additional time variable t which we express in a "natural" unit, equal to the inverse of the plasma frequency:

$$\omega_p = \sqrt{\frac{4\pi\rho(Ze)^2}{M}} \tag{1.4}$$

B. The Physics

The OCP is a model for dense, full ionized matter in which the uniform background is provided by the degenerate electron gas. The model is reasonable if three conditions are fulfilled:

(a) $T << T_F$, i.e. $T << 10^6/r_s^2$

where $r_s = a/a_o$, a_o is the electron Bohr radius and T_F is the electron Fermi temperature; this is the degeneracy condition.

(b) $\lambda_{TF}/a = \left(\frac{\pi}{12Z}\right)^{1/3} r_s^{-1/2} >> 1$, i.e. $r_s << 1$

where λ_{TF} is the Thomas-Fermi screening length; this condition ensures that the electron gas is rigid, i.e. not polarizable by the ionic charge distribution.

(c) $\Lambda/a = \sqrt{h^2/(2\pi M k_B T)}/a = \sqrt{\Gamma\cdot m/(2\pi M r_s)} << 1$

where Λ is the thermal de Broglie wavelength of the ions and m is the electron mass; this condition ensures that the ions can be treated classically.

Applications of the OCP model are essentially astrophysical; Table I lists some orders of magnitude for white dwarf matter and the interior of Jupiter, which show that the three above-mentioned conditions are reasonably well fulfilled in the former case, and are more questionable in the latter. In this latter case, quantum corrections (in powers of Λ^2) [Hansen and Vieillefosse, 1975], and the electron screening corrections (essentially proportional to $1/\lambda_{TF}^2$) [Galam and Hansen, 1976] to the thermodynamic properties of the ions, can be systmatically computed.

The OCP may also be relevant for the determination of the ion equation of state in laser-compressed plasmas (in the imposion zone), where $\Gamma \lesssim 1$.

Finally the OCP is a valuable tool for theorists, since it is the simplest conceivable model incorporatiing Coulomb interactions.

TABLE I

White dwarf	Jupiter (interior)
$T \simeq 10^7 K$	$10^4 K$
$\rho_m \simeq 10^6 - 10^8 gr/cm^3$	$10\ gr/cm^3$
He or C	H (+10% He)
$\Gamma \simeq 6 - 200$	$\Gamma \simeq 50$
$r_s \simeq 10^{-2}$	$r_s \simeq 0.7$
$\Lambda_{He}/a \simeq 0.2$	$\Lambda_H/a \simeq 0.5$
$\lambda_{TF}/a \simeq 4$	$\lambda_{TF}/a \simeq 1$
$T/T_F \simeq 10^{-3}$	$T/T_F \simeq 10^{-2}$
$\omega_p \simeq 5 \times 10^{17}\ sec^{-1}$	$10^{15} sec^{-1}$

C. The Numerical Method

Two computer simulation schemes have been widely used in Statistical Mechanics, in particular in Liquid State Theory, over the last twenty years. The first method, which allows the computation of static ensemble averages, is the Monte Carlo (MC) method which has been reviewed by Professor DeWitt in his lectures. The second method relies on the numerical solution of the coupled equations of motion, for periodic systems of $N \simeq 10^2 - 10^4$ particles, over some finite time interval T. Time averages of various dynamical variables are then taken along the trajectories of the N particles, over the interval T. The system is assumed to be isolated, so that its total energy is fixed; hence the time averages are equivalent to micro-canonical ensemble averages.

The method, called molecular dynamics (MD), was first developed by Alder and Wainwright [Alder and Wainwright, 1959] for hard sphere systems (instantaneous collisions). The MD method was extended to continuous force laws by Rahman [Rahman, 1964] and Verlet [Verlet, 1967]. In this case, the differential equations of motion are replaced by finite difference equations, after the introduction of a finite time increment Δt. The simplest algorithm to solve the equations of motion is that of Verlet [Verlet, 1967]:

$$\vec{r}_i(t + \Delta t) = -\vec{r}_i(t- t) + 2\vec{r}_i(t) + \vec{F}_i(t)(\Delta t)^2 + o(\Delta t^4) \quad (1.5)$$

where $\vec{F}_i(t)$ is the total force acting at time t on particle i from all (N-1) other particles.

The MD estimate of the equilibrium average of any dynamical variable $A(\vec{r}^N, \vec{v}^N)$ is then:

$$<A> = \lim_{T\to\infty} \frac{1}{T} \int_o^T A[r^N(t), v^N(t)]\, dt$$

$$\simeq \frac{1}{\mathcal{N}} \sum_{n=o}^{\mathcal{N}} A[\vec{r}^N(n\cdot\Delta t), \vec{v}^N(n\cdot\Delta t)]$$

where $\mathcal{N} = T/\Delta t$ is the total number of time steps generated in the MD "experiment". Similarly the MD estimate of the time correlation function of two dynamical variables A and B is:

$$C_{AB}(t) = <A(t)B(o)>$$

$$\simeq \frac{1}{(\mathcal{N}-\nu)} \sum_{n=o}^{\mathcal{N}-\nu} A[(n+\nu)\Delta t]\; B(n\cdot\Delta t) \qquad (1.6)$$

In these lectures we review recent applications of the MD method to the study of dynamical properties of the OCP [Hansen, Pollock and McDonald, 1974; Hansen, McDonald and Pollock, 1975]. The systems considered in these computer "experiments" contained generally 250 ions with periodic boundary conditions. The equations of motion were solved over total time intervals $T \simeq 10^3 - 10^4\ \omega_p^{-1}$, for values of the coupling parameter $1 \leq \Gamma \leq 155$. Account was taken of the long range of the Coulomb forces by letting each particle interact not only with the nearest images of the N-1 other ions, but also with the infinite set of periodic images and the background. Ewald techniques were used to calculate the corresponding infinite sums [Brush, Sahlin and Teller, 1966].

D. Some Static Properties of the OCP

The thermodynamic properties and static structure of the OCP, as obtained both from MC [Brush, Sahlin and Teller, 1966; Hansen, 1973] and MD simulations have been extensively reviewed in Professor DeWitt's lectures. Here we only summarize some salient features which will be useful in the following.

The MC excess energy values in the range $\Gamma > 1$ can be very accurately represented by a simple equation of state, proposed by DeWitt [DeWitt, 1976] ($\beta = 1/k_BT$):

$$\frac{\beta U}{N} = a\Gamma + b\Gamma^{1/4} + c \qquad (1.7)$$

with a = -0.895929, b = 0.8739 and c = -0.5777. Note that in the strong coupling limit ($\Gamma >> 1$) the energy (1.7) lies only a few percent above the exact lower bound, -0.9Γ [Lieb and Narnhofer, 1975]. From Eq. (1.7) it can be immediately seen that the isothermal compressibility

$$\chi_T = \frac{\beta}{\rho}\left[\left(\frac{\partial \beta P}{\partial \rho}\right)_T\right]^{-1}$$

becomes <u>negative</u> for $\Gamma > 3$ $\left(\text{note that for the OCP } \frac{\beta P}{\rho} = 1 + \frac{1}{3}\frac{\beta U}{N}\right)$. Nonetheless the OCP has been shown to be still <u>thermodynamically stable</u> against density fluctuations, provided the fluctuations of the local electric field are correctly taken into account [Vieillefosse and Hansen, 1975]. A straightforward fluctuation calculation also yields the $q \to o$ limit of the static structure factor [Vieillefosse and Hensen, 1975]:

$$s(q) = \frac{1}{N} \langle \rho_{\vec{q}}\, \rho_{-\vec{q}} \rangle \underset{q \to o}{\simeq} \frac{1}{\frac{3\Gamma}{q^2} + \frac{\chi_T^o}{\chi_T}} \tag{1.8}$$

where $\chi_T^o = \beta/\rho$ is the ideal gas compressibility.

E. Density Fluctuations, Longitudinal Current Fluctuations and Plasma Oscillations

In addition to the Fourier components (1.2) of the density, we introduce the Fourier components of the particle current:

$$\vec{J}_{\vec{k}}(t) = \sum_{i=1}^{N} e^{i\vec{k}\cdot\vec{r}_i(t)}\, \vec{v}_i(t) \tag{1.9}$$

which satisfy the continuity equation:

$$\frac{\partial}{\partial t} \rho_{\vec{k}}(t) + i\vec{k}\cdot j_{\vec{k}}(t) = o \tag{1.10}$$

We next define the density-density correlation function:

$$F(k,t) = \frac{1}{N} \langle \rho_{\vec{k}}(t)\rho_{-\vec{k}}(o) \rangle \tag{1.11}$$

and its spectrum, the dynamical structure factor:

$$S(k,\omega) = \frac{1}{2\pi}\int_{-\infty}^{+\infty} e^{i\omega t}\, F(k,t)\, dt \tag{1.12}$$

The linear density response function is:

$$\chi(k,\omega) = \frac{1}{N}\int_0^\infty e^{i\omega t} <\{\rho_{\vec{k}}(t),\ \rho_{-\vec{k}}(o)\}> dt \tag{1.13}$$

where { } denote a Poisson bracket; the imaginary part of χ is related to S by the classical limit of the fluctuation-dissipation theorem:

$$S(k,\omega) = -\frac{1}{\pi\beta\rho\omega}\chi''(k,\omega) \tag{1.14}$$

Similarly we introduce the longitudinal current autocorrelation function:

$$C_1(k,t) = \frac{1}{N} <\vec{k}\cdot\vec{j}_{\vec{k}}(t)\ \vec{k}\cdot\vec{j}_{-\vec{k}}(o)> \tag{1.15}$$

whose Fourier transform C_1 is directly related to S via the continuity equation (1.10):

$$\hat{C}_1(k,\omega) = \frac{1}{2\pi}\int_{-\infty}^{+\infty} e^{i\omega t}\ C_1(k,t)\ dt = \omega^2 S(k,\omega) \tag{1.16}$$

The short-time expansion of F(k,t) leads to the frequency moment sum rules of $S(k,\omega)$:

$$<\omega^{2n}> = \int_{-\infty}^{+\infty}\omega^{2n}S(k,\omega)\ d\omega = (-1)^n \left.\frac{d^{2n}F(k,t)}{dt^{2n}}\right|_{t=o} \tag{1.17}$$

Switching to q = ak, we find immediately:

$$<\omega^o> = S(q) \tag{1.18}$$

$$<\omega^2> = \omega_p^2\frac{q^2}{3\Gamma} \equiv \omega_o^2 \tag{1.19}$$

The 4th and 6th moments are given explicitly in Hansen, Pollock and McDonald [1974], Hansen, McDonald and Pollock [1975] and Vieillefosse and Hansen [1975].

In the $q \to o$ limit it can be proved rigorously that [Hansen, Pollock and McDonald, 1974; Hansen, McDonald and Pollock, 1975, Baus, 1975]:

$$\lim_{q\to o}\frac{S(q,\omega)}{S(q)} = \frac{1}{2}\left[\delta(\omega-\omega_p) + \delta(\omega+\omega_p)\right] \tag{1.20}$$

i.e. there is a "plasmon" mode of infinite lifetime (undamped) in the long-wavelength limit. Density fluctuations and plasma oscillations, at finite wavelength have been studied by MD simulations, at $\Gamma \simeq 1$, 10, 110 and 152 [Hansen, Pollock and McDonald, 1974; Hansen, McDonald and Pollock, 1975] and for several wave-numbers ($0.6 < q < 6$); the smallest accessible wave-number is determined by the periodic boundary conditions: $q = (2\pi a)/L = 0.618$ for a cubic box of volume L^3 containing 250 ions. The results for $S(q,\omega)$ are shown in Figures 4, 5, 6 and 7 of Hanzen, Pollock and McDonald [1974] and Hansen, McDonald and Pollock [1975]. The salient features are:

(a) For the smallest q values, $S(q,\omega)$, exhibits sharp peaks around $\omega = \pm\,\omega_p$, showing the existence of a long-lived plasmon mode even at relatively short wavelength.

(b) As Γ increases, the plasmon peak becomes sharper (narrower) for a given q. At $\Gamma = 10$, 110 and 152, $S(q,\omega)$ practically reduces to a pair of δ-functions at $q = 0.618$.

(c) For a given Γ, as q increases, the plasmon peak broadens, as one would intuitively expect; the peak disappears for $q \simeq 3$, which corresponds to a wavelength roughly equal to the mean inter-ionic spacing. For $q > 3$, $S(q,\omega)$ tends rapidly towards its non-interacting gas limit (a gaussian centered on $\omega = o$).

(d) As q increases, the peak position shifts to higher frequencies at $\Gamma = 1$, but to lower frequencies at $\Gamma = 10$, 110 and 152 (negative dispersion).

(e) At the smallest wave numbers, $S(q,\omega)$ shows no trace of a central, diffusive (Rayleigh-type) peak at zero frequency. This finding is in agreement with an exact kinetic theory result by Baus [Baus, 1975] which shows that the strength of the thermal diffusion mode is weaker, by a factor q^2, than the "mechanical" plasmon mode.

In view of the predominance of the plasmon mode, its sharpness at low q, and Eq. (1.20), we can derive a simple phenomenological dispersion relation from the moment relations (1.18) and (1.19) and the small q limit (1.8):

$$\omega^2(q) \simeq \frac{<\omega^2>}{<\omega^o>} = \frac{\omega_p^2 q^2}{3\Gamma S(q)} \simeq \omega_p^2[1 + \delta q^2] + o(q^4) \tag{1.21}$$

where $\delta = \frac{1}{3\Gamma}\, \chi_T^o/\chi_T$. This simple relation then clearly links the observed negative dispersion to the negative compressibility for $\Gamma > 3$.

A standard calculation, based on the linearized Navier-Stokes equations of hydrodynamics, suitably modified to account for the local electric field [Vieillefosse and Hansen, 1975; Balescu, 1975], yields a dynamical structure factor which incorporates all the qualitative features of the MD results at small q, and agrees with the rigorous kinetic theory results of Baus [Baus, 1975; 1977] in the strong coupling limit, where the plasma frequency ω_p becomes small compared to the collision frequency ω_c. The hydrodynamic calculation leads to the following dispersion relation:

$$\omega^2(q) = \omega_p^2 \, [1 + \gamma\delta q^2] + o(q^4) \tag{1.22}$$

where the coefficient of q^2 differs by a factor $\gamma = C_p/C_v$ from the result (1.21). In the strong coupling limit ($\Gamma >> 1$), γ differs very little from 1, so that the hydrodynamic result agrees almost exactly with the phenomenological dispersion relation (1.21).

For shorter wave-lengths, a "generalized hydrodynamics" calculation, which assumes a gaussian form for the positive definite real part of the memory (or "damping") function of $\tilde{\chi}(k,\omega)$ [Kadanoff and Martin, 1963], satisfying a certain number of sum rules, yields excellent agreement with the MD results [Hansen, Pollock and McDonald, 1974; Hansen, McDonald and Pollock, 1975].

F. Transverse Current Fluctuations

In a way similar to Eq. (1.15), we define the transverse current correlation function:

$$C_t(k,t) = \frac{1}{2N} \mathrm{Tr} \, \langle [\vec{k}\Lambda\vec{j}_{\vec{k}}(t)][\vec{k}\Lambda\vec{j}_{-\vec{k}}(o)] \rangle \tag{1.23}$$

and its spectrum:

$$\hat{C}_t(k,\omega) = \frac{1}{2\pi} \int_{-\infty}^{+\infty} C_t(k,t) \, e^{i\omega t} \, dt$$

Defining frequency moments $\langle\omega^{2n}\rangle_l$ and $\langle\omega^{2n}\rangle_t$ for $\hat{C}_t$ and $\hat{C}_l$, as we did for $S(q,\omega)$ (note that, according to Eq. (1.16), $\langle\omega^{2n}\rangle = \langle\omega^{2n-2}\rangle_l$), we immediately obtain the following generalization of the familiar Kohn sum rule for the harmonic Coulomb lattice:

$$\frac{\langle\omega^2\rangle_l + \langle\omega^2\rangle_t}{\omega_o^2} = \omega_p^2 \left[1 + \frac{5q^2}{3\Gamma}\right] \tag{1.24}$$

The MD results for $\hat{C}_t(q,\omega)$ at $\Gamma = 152$ indicate the existence of a well-defined <u>propagating</u> shear mode for $q \gtrsim 0.6$ (cf. Figure 12 of Hansen, Pollock and McDonald [1974] and Hansen, McDonald and Pollock [1975]). For $q \gtrsim 2$, the observed peak splits into two components, the one at the higher frequency being close to ω_p (cf. Figure 13 of Hansen, Pollock and McDonald [1974] and Hansen, McDonald and Pollock [1975]).

G. Single Particle Motion

A more convenient way of studying single-particle (or "self") motion in a fluid is to compute the normalized velocity autocorrelation function:

$$Z(t) = \frac{\langle \vec{v}(t)\cdot\vec{v}(0)\rangle}{\langle v^2\rangle} \tag{1.25}$$

where $\vec{v}(t)$ is the velocity, at time t, of any one of the N particles, and $\vec{v}(o)$ the initial velocity of the same particle. Accurate MD results are available for Γ = 1, 10, 20, 60, 110, 152 (cf. Figure 1 of Hansen, Pollock and McDonald [1974] and Hansen, McDonald and Pollock [1975]). They show that Z(t) exhibits pronounced oscillations, at roughly the plasma frequency, for $\Gamma \gtrsim 10$. In the strong coupling limit (Γ = 110 and 152), these oscillations are very long-lived, extending over many plasma periods. The memory function M(t), defined through the generalized Langevin equation:

$$\dot{Z}(t) = -\int_o^t M(s)\, Z(t-s)\, ds \tag{1.26}$$

exhibits similar, although more damped oscillations. These have been interpreted in terms of a strong coupling of the single particle motion to be collective plasmon mode [Gould and Mazenko, 1977; Gaskell and Chiakvelu, 1977; Varley, 1977].

II. COLLECTIVE MODES IN BINARY IONIC MIXTURES

A. Definitions and General Properties

In this lecture we consider the extension of the results for the OCP, to binary ionic mixtures, (e.g. $H^+ - He^{++}$ or $H^+ - Li^{+++}$) in a rigid, uniform background. Consider a mixture of N_1 ions of charge Z_1 e and mass M_1 and N_2 ions of charge Z_2 e and mass M_2 ($Z_1Z_2 > o$); the concentrations are $x_1 = N_1/N$ and $x_2 = N_2/N$ ($N = N_1 + N_2$); the total number density is $\rho = N/V$ and the charge density is $\rho' = Z_1\rho_1 + Z_2\rho_2 = \overline{Z}\rho$, where $\overline{Z} = x_1Z_1 + x_2Z_2$ is the average charge of the ions. The reduced equilibrium properties depend now on 2 parameters, either $x_1 = 1 - x_2$ and $\Gamma = e^2/(ak_BT)$, or x_1 and $\Gamma' = e^2/(a'k_BT)$, with $a' = (e/4\pi\rho')^{1/3}$ ($\Gamma' = \Gamma\,\overline{Z}^{1/3}$). The interaction hamiltonian reads:

$$V_N = \frac{1}{2V}\sum_{k\neq o}\frac{4\pi e^2}{k^2}\left[\rho'_{\vec{k}}\rho'_{-\vec{k}} - N\overline{Z^2}\right] \tag{2.1}$$

where

$$\rho'_{\vec{k}} = Z_1\rho_{\vec{k}}^{(1)} + Z_2\rho_{\vec{k}}^{(2)} \tag{2.2}$$

$$\rho_{\vec{k}}^{(\nu)} = \sum_{i=1}^{N_\nu} e^{i\vec{k}\cdot\vec{r}_i} \quad , \quad \nu = 1,2 \tag{2.3}$$

Extensive MC computations of the thermodynamic properties and static structure of H^+ - He^{++} mixtures have been carried out both at Livermore [DeWitt and Hubbard, 1976] and in Paris [Hansen, Torrie and Vieillefosse, 1977]. Moreover the coupled HNC integral equations for the three pair distribution functions, g_{11}, g_{12} and g_{22}, have been systematically solved for three concentrations (x_1 = 0.25, 0.5 and 0.75) and for several values of Γ ($0.1 \leq \Gamma \lesssim 10^2$), both for $Z_2/Z_1 = 2$ and $Z_2/Z_1 = 3$ mixtures [Hansen and Vieillefosse, 1976]. All the thermodynamic data can be very accurately represented by a simple linear interpolation at fixed charge density [Hansen and Vieillefosse, 1976]:

$$\frac{\beta U}{N}(\Gamma', x_1) = x_1 \frac{\beta U_{ocp}}{N}(\Gamma_1) + x_2 \frac{\beta U_{ocp}}{N}(\Gamma_2) \tag{2.4}$$

where $\Gamma_1 = \Gamma' Z_1^{5/3}$ and $\Gamma_2 = \Gamma' Z_2^{5/3}$.

This equation of state has been used to establish the phase diagram of H^+ - He^{++} and H^+ - Li^{+++} mixtures at various pressures in the range P > 1Mbar. It is found that at sufficiently high pressures, the mixtures are stable (miscible) at all concentrations [Hansen and Vieillefosse, 1976]. Quantum corrections and electron screening corrections do not modify the calculated phase diagrams very much [Hansen, Torrie and Vieillefosse, 1977].

B. Mass and Charge Density Fluctuation Spectra

We define the Fourier components of the mass (M) and charge (Z) densities:

$$\begin{aligned} \rho_{\vec{k}}^{M} &= M_1 \rho_{\vec{k}}^{(1)} + M_2 \rho_{\vec{k}}^{(2)} \\ \rho_{\vec{k}}^{Z} &= Z_1 \rho_{\vec{k}}^{(1)} + Z_2 \rho_{\vec{k}}^{(2)} \end{aligned} \tag{2.5}$$

Hence we can define three distinct TCF's:

$$F_{XY}(k,t) = \frac{1}{N\overline{XY}} \langle \rho_k^X(t) \rho_{-k}^Y(o) \rangle \tag{2.6}$$

where X, Y = M or Z; their spectra will be represented by $S_{XY}(k,\omega)$. The frequency moments $\langle\omega^{2n}\rangle_{XY}$ of the latter are given by sum rules analogous to Eq. (1.17). In particular for X = Y = Z (charge fluctuations) we find (q = ak):

$$\langle\omega^{o}\rangle_{ZZ} = S_{ZZ}(q) \underset{q\to o}{\simeq} \frac{q^2}{3\Gamma\overline{Z}^2} \tag{2.7}$$

$$\langle\omega^{2}\rangle_{ZZ} = \Omega_p^2 \frac{q^2}{3\Gamma\overline{Z}^2} \tag{2.8}$$

where

$$\Omega_p^2 = 4\pi\rho e^2 \left[x_1 \frac{Z_1^{\;2}}{M_1} + x_2 \frac{Z_2^{\;2}}{M_2}\right] \tag{2.9}$$

is the concentration average of the plasma frequencies of the two components. The $\langle\omega^4\rangle_{ZZ}$ moment [McDonald, Vieillefosse and Hansen, 1977] requires a knowledge of the partial pair distribution functions of the mixture [Hansen, Torrie and Vieillefosse, 1977]. Characteristic frequencies of long-wavelength longitudinal modes can be estimated from ratios of these moments. The ratio

$$\lim_{q\to o} \frac{\langle\omega^{2}\rangle_{ZZ}}{\langle\omega^{o}\rangle_{ZZ}} = \Omega_p^2 \tag{2.10}$$

is independent of the strength of the coupling and hence yields the correct characteristic frequency in the weak coupling limit. This is confirmed by a straightforward solution of the coupled Vlasov (or mean field) equations for the distribution functions of the two components, which leads to the simple dispersion relation:

$$\omega^2(q) = \Omega_p^2 + \delta_V\, q^2 + o(q^4) \tag{2.11}$$

where:

$$\delta_V = \frac{6k_BT}{a^2\Omega_p^2}\left[\frac{\omega_1^2}{M_1} + \frac{\omega_2^2}{M_2}\right]$$

and

$$\omega_\nu^2 = 4\pi\rho e^2\, Z_\nu^2/M_\nu$$

However, contrary to the case of the OCP the ratio $\langle\omega^4\rangle_{ZZ}/\langle\omega^2\rangle_{ZZ}$ yields a different characteristic frequency in the long wavelength limit; we find [McDonald, Vieillefosse and Hansen, 1977]:

$$\lim_{q\to o} \omega^2(q) = \frac{\langle\omega^4\rangle_{ZZ}}{\langle\omega^2\rangle_{ZZ}} = \Omega_p^2 + \omega_p^2 \frac{\dfrac{x_1 x_2 z_1 z_2}{3}\left(\dfrac{z_1}{m_1} - \dfrac{z_2}{m_2}\right)^2}{x_1 \dfrac{z_1^2}{m_1} + x_2 \dfrac{z_2^2}{m_2}} \tag{2.12}$$

where $z_\nu = Z_\nu/\overline{Z}$, $m_\nu = M_\nu/\overline{M}$ and:

$$\omega_p^2 = \frac{4\pi\rho\overline{Z}^2 e^2}{\overline{M}} \leq \Omega_p^2 \tag{2.13}$$

Hence $\omega^2(o) \geq \Omega_p^2 \geq \omega_p^2$; e.g. for a $H^+ - He^{+++}$ mixture with $x_1 = x_2 = 1/2$, we find

$$\omega^2(o)/\omega_p^2 = 1.075 \qquad \text{and} \qquad \Omega_p^2/\omega_p^2 = 1.05 \quad .$$

Note that both ratios $\langle\omega^2\rangle_{ZZ}/\langle\omega^o\rangle_{ZZ}$ and $\langle\omega^4\rangle_{ZZ}/\langle\omega^2\rangle_{ZZ}$ yield, as a function of q, a negative dispersion for $\Gamma \gtrsim 1$, in qualitative agreement with the OCP.

C. The Strong Coupling Limit

A simple hydrodynamic calculation, along the lines sketched in Chapter I, Section C for the case of the OCP, can again be expected to be valid in the limit $\Gamma >> 1$. The results of such a calculation for the long wavelength clarge and mass density fluctuations are given by McDonald, Vieillefosse and Hansen [McDonald, Vieillefosse and Hansen, 1977]. Here we only summarize the salient features of $\sigma_{ZZ}(q,\omega) = S_{ZZ}(q,\omega)/S_{ZZ}(q)$ and $\sigma_{MM}(q,\omega)$:

(a) $\sigma_{ZZ}(q,\omega)$ consists of two conjugate plasmon peaks and a central peak, due to thermal diffusion, and interdiffusion

(b) In the limit $q\to o$, the plasmon peak reduces to

$$\sigma_{ZZ}(o,\omega) = \frac{2\omega^2\omega_d}{\left(\omega^2 \pm \omega_p^2\right)^2 + \omega^2\omega_d^2}$$

with $\omega_d = 4\pi e^2\alpha$, where α is the interdiffusion coefficient. Thus, in contrast to the case of the OCP, the plasmon mode is damped, even at infinite wavelength, by interdiffusion of the two species. Its frequency is ω_p, rather than Ω_p.

(c) The central peak is of intensity q^2, as in the case of the OCP. However, its height is of order q^o, comparable to the finite height of the plasmon peaks; this opens up the possibility of observing the diffusive contribution to the charge fluctuation spectrum in a MD computer "experiment".

(d) $\sigma_{MM}(q,\omega)$ consists of a dominant diffusive central peak, of intensity q^o, which reduces to a δ-function in the limit $q\to o$, and of a pair of plasmon peaks, of intensity q^2, which remain of finite width and are centered on ω_p in the infinite wavelength limit. The integrated intensity of the plasmon peaks is therefore negligible compared to that of the central peak, a behavior which is the exact opposite of that seen in σ_{ZZ}. There is no mode corresponding to a propagating sound wave.

D. MD Results

A MD "experiment" has been performed on a mixture of 125 H^+ and 125 He^{++} ions in a rigid, uniform background, at $\Gamma = 40$ [McDonald, Vieillefosse and Hansen, 1977]. The results for $\sigma_{ZZ}(q,\omega)$ and $\sigma_{MM}(q,\omega)$ are in qualitative agreement with the predictions of the hydrodynamic theory at the smallest wave numbers, except that the extrapolation to $q = o$ of the plasmon peak positions yields an $\omega^2(o)$ in agreement with Eq. (2.12), rather than ω_p^2. The observed dispersion is negative in agreement with the sum rule argument. As q increases, the intensity of the central peak increases rapidly relative to that of the plasmon peaks.

The computed velocity autocorrelation functions $Z_H(t)$ and $Z_{He}(t)$ exhibit marked oscillations, in qualitative agreement with the case of the OCP (cf. Chapter I, Section G). The resulting self-diffusion coefficients are $D^*_{H^+} = D_H/\omega_p a^2 = 0.0086$ and $D_{He^{++}} = 0.0049$ respectively.

A similar MD "experiment" at lower coupling ($\Gamma = 0.4$) is presently under way.

III. LINEAR TRANSPORT COEFFICIENTS OF THE OCP

A. Ionic and Electronic Transport

We consider dense, fully ionized matter in the temperature range:

$$T_d^i << T << T_F^e$$

where T_d^i is the degeneracy temperature of the ions, and T_F^e the Fermi (or degeneracy) temperature of the electrons. In the high density limit the ions and electrons are then essentially decoupled (cf. the discussion in Chapter I, Section B), and it is a reasonable approximation to compute the transport coefficients of both components separately.

Because of the Pauli principle, electron-electron scattering becomes negligible in the high density limit, and each electron is individually scattered by the ions. The electronic transport coefficients can then be calculated in the framework of a Lorentz model, in the formulation due to Ziman [Ziman, 1961] which requires simply a knowledge of the ionic static structure factor. A systematic computation of the electronic transport coefficients using the ionic structure factors of the OCP model has been performed [Minoo, Deutsch and Hansen, 1976]. An advantage of this procedure over previous computations using hard sphere structure factors is the fact that the correct temperature dependence is automatically contained in the OCP structure factors (through Γ) without any adjustable parameter.

The computation of the transport coefficients of the strongly coupled classical ions is more difficult. Recent results are reviewed in this chapter.

B. Ionic Self-Diffusion

For pedagogical reasons we start this chapter with the particularly simple case of the ionic self-diffusion coefficient D. Although, strictly speaking, this is not a genuine transport coefficient associated with dissipation of energy, it is handled in much the same way as true transport coefficients in non-equilibrium Statistical Mechanics.

D is immediately expressible in terms of the velocity autocorrelation function (1.25), through a straightforward reformulation of the Einstein relation [e.g. Hansen and McDonald, 1976]

$$D^* = \frac{D}{\omega_p a^2} = \frac{1}{3\Gamma} \int_0^\infty Z(t)\, dt \tag{3.1}$$

where t is expressed in units of ω_p^{-1}. Z(t) is accurately known from MD computations (cf. Chapter I, Section G) and the integrations lead to D^* values which are fairly well fitted by the simple formula:

$$D^* \simeq 3 \cdot \Gamma^{-4/3} \qquad (\Gamma \simeq 1)$$

We now sketch a simple theoretical calculation [Vieillefosse, 1975] based on the generalized Langevin equation (1.26), the Laplace transform of which reads:

$$[-i\omega + \tilde{M}(\omega)]\ \tilde{Z}(\omega) = 1 \tag{3.2}$$

From Eqs. (3.1) and (3.2) we deduce:

$$D^* = \frac{1}{3\Gamma}\frac{1}{\tilde{M}(o)} \tag{3.3}$$

The short time expansion of M(t) follows from that of Z(t) [Hansen, Pollock and McDonald, 1974; Hansen, McDonald and Pollock, 1975] with the result:

$$M(t) = \omega_{1s}^2\ [1 - \frac{\omega_{2s}^2 - \omega_{1s}^2}{\omega_{1s}^2}\frac{t^2}{2!} + 0(t^4)] \tag{3.4}$$

$$\omega_{1s}^2 = 1/3$$

$$\omega_{2s}^2 = \frac{1}{9}\ [12 I_{-4} + 1 + 9K]$$

$$I_n = \int_o^\infty x^n g(x)\ dx, \quad n < -1$$

$$= \int_o^\infty x^n [g(x) - 1]\ dx, \quad n > -1$$

$$K = \int_o^\infty \frac{dx}{x}\int_o^\infty \frac{dx'}{x'}\int_{-1}^{+1} d(\cos\alpha)\ P_2(\cos\alpha)\ [g_3(x,x',\alpha) - g(x)g(x')]$$

where P_n is the Legendre polynomial of order n, and g_3 is the triplet distribution function. Note that the term 9K in ω_{2s}^2 is missing in papers by Hansen, Pollock and McDonald [Hansen, Pollock and McDonald, 1974; Hansen, McDonald and Pollock, 1975].

We next assume M(t) to be a gaussian satisfying Eq. (3.4):

$$M(t) = \frac{1}{3}\exp\left\{-2(I_{-4} + \frac{3}{4}K)\ t^2\right\} \tag{3.6}$$

This hypothesis is certainly an oversimplification since we pointed out in Chapter I, Section G that M(t) exhibits oscillations in the strong coupling limit. Combining Eqs. (3.3) and (3.6) we now easily obtain:

$$D^* = \frac{1}{\Gamma}\sqrt{\frac{8}{\pi}\; I_{-4} + \frac{3K}{4}} \qquad (3.7)$$

I_{-4} can be computed with the MC or MD pair distribution functions; to compute K one has to make the standard superposition approximation (SA) on g_3; D^* is not too sensitive to the SA, since I_{-4} turns out to be the dominant term. The results are compared to the "exact" MD data in Table II. The agreement is seen to be good only for very large Γ

TABLE II

Γ	D^*_{theory}	D^*_{MD}
0.993	3.5	2.01
9.7	0.078	0.130
19.7	0.031	0.0603
59.1	0.0087	0.0151
110.4	0.0045	0.00511
152.4	0.0033	0.00318

As an illustration, under white dwarf conditions ($\rho_m = 10^6 gr/cm^3$, $T = 10^7 K$, He composition $\rightarrow \Gamma = 5.6$), we find:

$$D \simeq 2 \times 10^{-3}\ cm^2\ sec^{-1}$$

The previous simple calculation has been improved by S. Sjödin and S. K. Mitra [Sjodin and Mitra, 1977] who take into account the coupling of the self motion to the collective plasmon mode.

C. Ionic Shear Viscosity

A simple hydrodynamic calculation, similar to that sketched in Chapter I, Section E, yields the following expression for the Laplace transform of the transverse current correlation function (1.23), valid in the long wavelength, low frequency region:

$$\tilde{C}_t(k,\omega) = \frac{\omega_o^2}{-i\omega + \eta k^2/M\rho} \qquad (3.8)$$

From this we obtain the standard expression for the shear viscosity [Kadanoff and Martin, 1963]:

$$\eta = \lim_{\omega\to o} \lim_{k\to o} \frac{M\rho}{k^2}\left[i\omega + \frac{\omega_o^2}{\tilde{C}_t(k,\omega)}\right] \tag{3.9}$$

which can be recast into a Green-Kubo formula:

$$\eta = \int_o^\infty \eta(t)\, dt$$

where

$$\eta(t) = \frac{M^2}{Vk_BT} \langle A^{\alpha\beta}(t)\, A^{\alpha\beta}(o)\rangle, \quad \alpha \neq \beta = x,y,z \tag{3.10}$$

and $A^{\alpha\beta}$ is an off-diagonal component of the microscopic stress tensor; for the OCP:

$$A^{\alpha\beta} = \sum_{i=1}^{N} v_i^\alpha v_i^\beta + \sum_{i\neq j}\sum \frac{1}{V} \sum_{k\neq o} e^{-i\vec{k}\cdot\vec{r}_{ij}} \frac{4\pi(Ze)^2}{Mk^2}\left[\frac{1}{2}\delta_{\alpha\beta} - \frac{k^\alpha k^\beta}{k^2}\right] \tag{3.11}$$

There exist by now three independent calculations of η for the strongly coupled OCP. The first [Vieillefosse and Hansen, 1975] is based on the known frequency moment of $\tilde{C}_t(k,\omega)$ and is closely related to the calculation of D sketched in Chapter III, Section B. It assumes a gaussian real part of the Laplace transform of the memory function; note that this hypothesis is reasonable, since it is compatible with known exact properties of these functions [Kadanoff and Martin, 1963] which must be positive, even in ω, and decreasing exponentially at high frequency. The final expression for the reduced kinematic shear viscosity $\eta^* = \eta/(M\rho\omega_p a^2)$ is given by Eq. (50) of Vieillefosse and Hansen [Vieillefosse and Hansen, 1975]. It involves integrals over the pair and triplet distribution functions, the latter being approximated by the SA; this introduces an uncertainty in the computed values of η^* (given in Table III) which is largest at high Γ. The results clearly indicate that η^* exhibits a <u>minimum</u> as a function of Γ (i.e. of temperature at constant density) around $\Gamma \simeq 20$. An alternative kinetic calculation by Wallenborn and Baus [Wallenborn and Baus, 1977], which requires only a knowledge of the static structure factor, yields results in semi-quantitative agreement with the previous computation (cf. Table III). It also leads to a minimum in η^*, around $\eta^* = 10$ and it has the advantage of yielding reliable results also in the weak coupling ($\Gamma < 1$) limit.

TABLE III

Γ	η^*_{MD}	η^*_{VH}	η^*_{WB}
1.	1.04	0.35	1.01
10.4	1.085	0.083	0.078
110.4	0.18	0.13 ± 0.05	0.218

Recently the autocorrelation function $\eta(t)$ has been calculated by MD simulations at $\Gamma = 1$, 10 and 100 [Bernu, Vieillefosse and Hansen, 1977]. $\eta(t)$ turns out to be a relatively slowly, monotonically decreasing function of time for $\Gamma = 1$ and 100, while for $\Gamma = 10$, the decay time is much shorter. This behavior is certainly related to the minimum in η^* around that value of Γ.

The results of both theoretical calculations and the MD computations are summarized in Table III.

D. Ionic Bulk Viscosity

The bulk viscosity ζ can also be computed along the lines sketched in the previous section. The longitudinal viscosity $b = \zeta + 4\eta/3$ can be related to the long-wavelength, zero frequency limit of the real part of the Laplace transformed memory function of the longitudinal current correlation function (1.15) [Kadanoff and Martin, 1963]:

$$b^* = \frac{b}{M\rho\omega_p a^2} = \lim_{\omega\to o} \lim_{k\to o} \frac{\tilde{M}'(k,\omega)}{\omega_p a^2 k^2} \tag{3.12}$$

using again a gaussian form for the memory function, b^* can be expressed in terms of integrals over the pair and triplet distribution functions [Vieillefosse and Hansen, 1975]. The resulting b^* turns out to be very close to η^*, so that the bulk viscosity must be smaller than the shear viscosity ($\zeta^* \lesssim 0.02\ \eta^*$).

Equation (3.12) can again be reexpressed in the form of a Green-Kubo relation:

$$\zeta = \int_o^\infty \zeta(t)\ dt \tag{3.13}$$

where $\zeta(t)$ is now equal to the autocorrelation function of the diagonal components of the microscopic stress tensor. In the case of the OCP, $\zeta(t)$ can be further simplified to the expression (valid in the microcanonical ensemble):

$$\zeta(t) = \frac{1}{4Vk_BT} \langle V_N(t)\ V_N(o) - \langle V_N\rangle^2\rangle \tag{3.14}$$

i.e. $\zeta(t)$ is simply the autocorrelation function of the total potential energy of the system.

$\zeta(t)$ has been computed by MD simulations at $\Gamma = 1$, 10 and 100. It is found to be a rapidly oscillating function of time, of frequency $\simeq 2\omega_p$, at all three values of Γ. The values of $\zeta^*(o)$ and of the resulting reduced bulk viscosities ζ^* are summarized in Table IV together with the corresponding data for the shear viscosity and the thermal conductivity. The MD results confirm the theoretical prediction that ζ^* is 2-3 orders of magnitude smaller than η^*, depending on Γ.

TABLE IV

Γ	$\eta^*(o)$	η^*	$\zeta^*(o)$	ζ^*	$K^*(o)$	K^*
1.	0.34	1.04	$4.3\ 10^{-3}$	$2.6\ 10^{-3}$	0.873	2.9
10.4	0.068	0.085	$1.5\ 10^{-3}$	$1.8\ 10^{-3}$	0.354	0.66
100.4	0.041	0.18	$0.27\ 10^{-3}$	$0.21\ 10^{-3}$	0.338	0.88

E. Ionic Thermal Conductivity

The Green-Kubo formula for the thermal conductivity K is [e.g. Hansen and McDonald, 1976]:

$$K = \int_o^\infty K(t)\ dt \tag{3.15}$$

$$K(t) = \frac{\rho^3}{k_B N} \langle j_{\ell s}(k{=}o,t)\ j^*_{\ell s}(k{=}o,t{=}o)\rangle \tag{3.16}$$

where $j_{\ell s}$ is the longitudinal entropy (s) current. In applying Eq. (3.16) to the OCP case, some care must be taken in defining the microscopic entropy current in order to avoid divergences due to the long range of the Coulomb interactions [Bernu, Vieillefosse and Hansen, 1977]. The kth Fourier component of the local internal energy is:

$$\varepsilon(k) = \sum_{i=1}^{N} \frac{1}{2} Mv_i^2\ e^{-i\vec{k}\cdot\vec{r}_i} + \frac{1}{8\pi V} \sum_{k'} \vec{E}_{k'} \cdot \vec{E}_{k-k'} \tag{3.17}$$

where $\vec{E}_k$ is a Fourier component of the local electric field which is related to the charge density by Poisson's equation. From the energy conservation equation we then deduce the k→o limit of the longitudinal energy current:

$$j_{\ell\varepsilon}(k\to o) = \sum_i (\hat{k}\cdot\vec{v}_i)\,\frac{1}{2}\,Mv_i^2$$

$$+ \sum_{i\neq j}\sum \frac{1}{V} \sum_{k'\neq o} e^{-i\vec{k}\cdot\vec{r}_{ij}}\,\vec{v}_i \cdot [\hat{k}-(\hat{k}\cdot\hat{k}')\hat{k}']\,\frac{4\pi(Ze)^2}{k'^2} \tag{3.18}$$

where $\hat{k} = \vec{k}/|\vec{k}|$. A standard thermodynamic relation then links the energy current to the entropy current occuring in Eq. (3.16).

K(t) has also been computed by MD "experiments" at $\Gamma = 1$, 10 and 100; it is an oscillatory function of time, of frequency $\simeq\omega_p$, at the stronger couplings, while it decays essentially monotonically at $\Gamma = 1$ [Bernu, Vieillefosse and Hansen, 1977]. The resulting values for $K^*(o)$ and K^* are summarized in Table IV: K^* is seen to be smallest at $\Gamma = 10$, which indicates that the thermal conductivity exhibits a minimum as a function of Γ just as the shear viscosity.

As an illustration of the orders of magnitude, consider a hydrogen plasma at $T = 10^7$K and $\rho = 5 \times 10^{28}$ ions/cm^3; for this state $\Gamma = 1$, and from $K^* = 2.9$, we find an ionic thermal conductivity $\lambda_i \simeq 2.10^6$ W m^{-1} K^{-1}. The Lorentz-Ziman calculation mentioned in Chapter III, Section A yields for the electronic conductivity $\lambda_e \simeq 2 \times 10^{10}$ W m^{-1} K^{-1} [Minoo, Deutsch and Hansen, 1976], which is, as expected, considerably larger than the ionic conductivity.

ACKNOWLEDGEMENTS

The author wishes to express his gratitude to his collaborators and friends B. Bernu, C. Deutsch, I. R. McDonald, H. Minoo, E. L. Pollock and P. Vieillefosse and acknowledges stimulating discussions with M. Baus and H. E. DeWitt.

REFERENCES

Alder, B. J. and T. E. Wainwright, 1959, J. Chem. Phys. 31, 459.

Balescu, R., 1975, Equilibrium and Non-Equilibrium Statistical Mechanics, Wiley, New York.

Baus, M., 1975, Physica 79A, 377.

Baus, M., 1977, Phys. Rev. A15, 790.

Bernu, B., 1977, P. Vieillefosse and J. P. Hansen, 1977, submitted to Phys. Letters A.

Brush, S. G., H. L. Sahlin and E. Teller, 1966, J. Chem. Phys. 45, 2102.

DeWitt, H. E., 1976, Phys. Rev. A14, 1290.

DeWitt, H. E. and W. B. Hubbard, 1976, Astrophys. J. 205, 295.

Galam, S. and J. P. Hansen, 1976, Phys. Rev. A14, 816.

Gaskell, T. and O. Chiakvelu, 1977, J. Phys. C10, 2021.

Gould, H. and G. F. Mazenko, 1977, Phys. Rev. A15, 1274.

Hansen, J. P., 1973, Phys. Rev. A8, 3096.

Hansen, J. P. and I. R. McDonald, 1976, Theory of Simple Liquids, Academic Press, London.

Hansen, J. P., I. R. McDonald and E. L. Pollock, 1975, Phys. Rev. A11, 1025.

Hansen, J. P., E. L. Pollock and I. R. McDonald, 1974, Phys. Rev. Lett. 32, 277.

Hansen, J. P. G. M. Torrie and P. Vieillefosse, 1977, in press, Phys. Rev. A.

Hansen, J. P. and P. Vieillefosse, 1975, Phys. Lett. 53A, 187.

Hansen, J. P. and P. Vieillefosse, 1976, Phys. Rev. Lett. 37, 391.

Kadanoff, L. P. and P. C. Martin, 1963, Ann. Phys. (New York), 24, 419.

Lieb, E. L. and H. Narnhofer, 1975, J. Stat. Phys. 12, 291.

McDonald, I. R., P. Vieillefosse and J. P. Hansen, 1977, Phys. Rev. Lett. 39, 271.

Minoo, H., C. Deutsch and J. P. Hansen, 1976, Phys. Rev. A14, 840.

Pollock, E. L. and J. P. Hansen, 1973, Phys. Rev. A8, 3110.

Rahman, A., 1964, Phys. Rev. 136, A405.

Sjödin, S. and S. K. Mitra, 1977, preprint.

Varley, R. L., 1977, preprint.

Verlet, L., 1967, Phys. Rev. 159, 98.

Vieillefosse, P. and J. P. Hansen, 1975, Phys. Rev. A12, 1106.

Vieillefosse, P., 1975, thèse de 3ème cycle, Paris.

Wallenborn, J. and M. Baus, 1977, preprint.

Ziman, J. M., 1961, Philos. Mag. 6, 1013.

METHODS AND APPROXIMATIONS FOR STRONGLY COUPLED PLASMAS*

G. Kalman

Department of Physics
Boston College
Chestnut Hill, Massachusetts 02167

*Work partly supported by AFOSR Grant 76-2960, NATO Research Grant No. 1211, and Israel-U.S. BNSF Grant 592.

TABLE OF CONTENTS

METHODS AND APPROXIMATIONS FOR STRONGLY COUPLED PLASMAS

G. Kalman

Department of Physics
Boston College
Chestnut Hill, MA 02167

I. INTRODUCTION

Strongly coupled plasmas are certainly not unique amongst physical systems in the sense of being characterized by a substantial or even overwhelming portion of their energy residing in the form of potential energy. Even if solids are excluded, there are many dense gases and liquids for which such a condition prevails. However, there exist certain features, which do set the strongly coupled plasma problem apart from the rest of the field of dense liquids and other strongly coupled many body systems. First is the long-range character of the coulomb interaction. This feature has been the focal point of attention since the early days that mark the beginning of the investigation of coulomb systems, and plasmas. It is known to be responsible for many unique patterns, both of the physical behavior of such systems, and of the mathematical problems that arise in their treatments (Dr. Baus' lectures [Baus, this Volume] discuss some of these features.) Second, the coulomb potential - in contrast to most of the potentials governing the interaction of neutral fluids - has a simple and well-defined analytic structure. It is probably this, more than any other, feature that motivates the difference in methodology in attacking the problems of plasmas and neutral gases and liquids. For the latter the lack of a simple analytic form or even of precise knowledge of the interaction potential places a premium on approaches that attempt to express relationships between directly measurable quantities and avoid the use of the potential ("fully renormalized kinetic theory", etc.). In the case of plasmas, however, there is an incentive to exploit the simplicity of the coulomb potential and to derive results from first principles. In particular, there is an emphasis on using the powerful formalism of fluctuation-dissipation theorems. Finally, the plasma parameters $\gamma = \kappa^3/(4\pi n)$ ($\kappa^2 = 4\pi e^2 \beta n Z^2$, n = density, β = inverse temperature) or $\Gamma = \beta e^2 Z^2/d$ ($d = [(4\pi/3)n]^{-1/3}$ = ion sphere radius) characterizing the strength of the coupling can vary, for physical systems, over an enormous range of values within which the fundamental physical characteristics of the system remain unchanged. While very little is known, at the present time, about the domain of strong coupling, a great deal of information has accumulated on the properties of weakly coupled plasmas: these results can serve as guidelines, or at least as standards of comparison, for the treatment of strongly coupled plasma systems.

Thus, while many powerful and ingenious methods developed for neutral systems are available and adaptable to plasmas (consult the lectures of Professors Gross, DeWitt and Sjölander [Gross, this Volume; DeWitt, this Volume; Sjölander, this Volume]), some others have been developed with primarily coulomb systems in mind. A review of a group of these which, somewhat arbitrarily, we have selected and grouped as the STLS, TI and GKS methods (see Chapter VII for the references) is the subject of the present lectures. (A quite recent approach, which is not included in the present review, is exposed in Professor Ichimaru's lectures [Ichimaru, this Volume]). Chapter VII deals with the principal ideas and approximations they are based on, with an emphasis of pointing out the common features that unite and the differences that set them apart.

Each approximation relies heavily on the formalism of response functions and fluctuation-dissipation theorems. Some of the features of plasma response functions playing a significant role in this present context, are discussed in Chapter II. Important sum rules are reviewed in Chapter IV. Different kinds of fluctuation-dissipation theorems are derived and exposed in Chapters V and VI.

Although some of the topics on linear response functions and fluctuation-dissipation theorems are fairly standard, what elevates our discussion beyond this level is the equal emphasis we put on linear <u>and</u> nonlinear response functions and fluctuation-dissipation theorems. The reason for this is the central role played by these nonlinear objects in the GKS theory. In particular, the nonlinear fluctuation-dissipation theorem described in Chapter VI is of fairly recent origin.

Most of our discussions and the full discussion of the approximation schemes will be confined to one-component plasmas (ocp) only, primarily for the sake of simplicity and clarity. However, whenever the generalizations of one-component response functions and related objects to multicomponent systems are non-trivial and available, they will be displayed. Chapter II, in particular, is devoted to setting up this generalization. For the ocp, the case of dynamical electrons in an inert background is the more conventional problem, while the opposite case of dynamical ions immersed in a smeared out electron gas, the "inverted plasma" has gained more recent attention and is probably the most significant physical realization of classical strongly coupled plasma systems. Nevertheless, for concreteness, whenever dealing with one-component systems, it will be understood that it is an electron liquid which is under study, with electronic charge $-e = -|e|$.

It will be assumed that the reader is familiar with the elementary properties of response functions. Good reviews on the subject are available [Martin, 1968; Golden and Kalman, 1969; Kalman, 1975].

II. RESPONSE FUNCTIONS

Response functions characterize the behavior of the system under the influence of external perturbations, like electric fields or potentials. (We are concerned with electrostatic coulomb (logitudinal) fields only. Thus the concepts of potential and electric field can be used interchangeably.) However, once an external field, i.e., field generated by external sources, is set up, the response of the system generates additional fields ("polarization" fields), which add to external field. As a result, the particles respond to the total field, and physical response functions relate to the latter. At the same time, it is also extremely useful to retain the somewhat artificial concept of external field and to relate another family of response functions to it.

We will use the notation $\hat{E},\hat{V}$ for the external field and potential (energy), $\check{E},\check{V}$ for the total field, etc., and E,V for the plasma field (polarization field), etc. Evidently

$$E = \check{E} + \hat{E} \tag{2.1}$$

Consider the particle density $\underline{n}$, the (longitudinal) electric current density $\underline{J}$ and the polarization potential itself as responding quantities. Then in a rather symbolic notation

$$n = \chi_1 V + \chi_2 VV + \ldots$$

$$J = \sigma_1 E + \sigma_2 EE + \ldots$$

$$\check{V} = -\underset{1}{\alpha}V - \underset{2}{\alpha}VV - \ldots \tag{2.2}$$

or

$$n = \underset{1}{\hat{\chi}}\hat{V} + \underset{2}{\hat{\chi}}\hat{V}\hat{V} + \ldots \tag{2.3}$$

etc.

The notation can be made more explicit by using Fourier representation. E.g.,

$$n(\vec{k}\omega) = \underset{1}{\chi}(\vec{k}\omega)\ V(\vec{k}\omega) + \sum\!\!\!\!\!\!\diagdown_{\substack{\vec{p}\vec{q}\\ \mu\nu}} \underset{2}{\chi}(\vec{p}\mu,\vec{q}\nu)\ V(\vec{p}\mu)\ V(\vec{q}\nu) + \ldots \tag{2.4}$$

$$\vec{p} + \vec{q} = \vec{k}$$

$$\mu + \nu = \omega$$

$$\sum\!\!\!\!\!\!\diagdown_{\vec{k}\omega} = \frac{1}{2\pi V}\int d\omega \sum_{\vec{k}} \quad (V = \text{volume})$$

$\underset{1}{\chi}$, $\underset{1}{\sigma}$, $\underset{1}{\alpha}$ and their external counterparts $\underset{1}{\hat{\chi}}$, $\underset{1}{\hat{\sigma}}$, $\underset{1}{\hat{\alpha}}$ are linear response functions, while $\underset{2}{\chi}$, etc. are quadratic ones. Obviously, the expansion could be continued to higher orders, but in these lectures we will be mostly content to consider linear and quadratic response functions.

The three response functions (χ, the density response function, σ, the conductivity and α, the polarizability) are obviously not independent of each other. Introducing the Fourier transform of the Coulomb potential

$$\phi_{\vec{k}} = \frac{4\pi\ e^2}{k^2} \tag{2.5}$$

and its quadratic counterpart

$$\phi_{\vec{p}\vec{q}} = \frac{4\pi i e^3}{kpq}$$

the interrelations can be expressed as follows:

$$\alpha_1 = \frac{4\pi i}{\omega} \sigma_1 = -\phi_{\vec{k}} \chi_1$$

$$\alpha_2 = \frac{4\pi i}{\omega} \sigma_2 = -\phi_{\vec{p}\vec{q}} \chi_2 \qquad (2.6)$$

Moreover, the connections between the "external" and total response functions follow in virtue of the relation

$$E(\vec{k}\omega) = \frac{\hat{E}(\vec{k}\omega)}{\varepsilon(\vec{k}\omega)} \qquad (2.7)$$

$$\varepsilon(\vec{k}\omega) = 1 + \alpha(\vec{k}\omega)$$
$$= \{1 - \hat{\alpha}(\vec{k}\omega)\}^{-1}$$

As illustrations, we consider $\hat{\chi}_1$, $\hat{\chi}_2$, $\hat{\chi}_3$

$$\hat{\chi}_1(k) = \frac{\chi_1(k)}{\varepsilon(k)} \qquad (2.8a)$$

$$\hat{\chi}_2(pq) = \frac{\chi_2(pq)}{\varepsilon(p)\ \varepsilon(q)\ \varepsilon(k)} \qquad k = p + q \qquad (2.8b)$$

$$\hat{\chi}_3(pqs) = \frac{1}{\varepsilon(p)\ \varepsilon(q)\ \varepsilon(s)\ \varepsilon(k)} \Bigg\{ \chi_3(pqs)$$
$$+ \frac{2}{3} \Bigg[\phi_{\vec{q}+\vec{s}} \frac{\chi_2(p,\ q+s)\ \chi_2(qs)}{\varepsilon(q+s)} \qquad (2.8c)$$

$$+ \phi_{\vec{s}+\vec{p}} \frac{\chi_2(q, s + p)\ \chi_2(sp)}{\varepsilon(s + p)}$$

$$+ \phi_{\vec{p}+\vec{q}} \frac{\chi_2(s, p + q)\ \chi_2(pq)}{\varepsilon(p + q)} \Bigg]\Bigg\}$$

$$k = p + q + s$$

$$k \equiv \vec{k},\omega \quad , \quad \text{etc.}$$

III. MULTICOMPONENT SYSTEMS

Real plasmas contain, of course, at least two components. Binary ionic mixtures consist of two ion species in a neutralizing background. Electron-hole liquids can be composed of several species. Thus the generalization of the concepts of the foregoing Chapter to multicomponent situations is of great interest. The different species have densities n_A, n_B and each of them gives rise to its own polarization field $\check{E}_A$, $\check{E}_B$, ... Then the corresponding species response functions can be defined by

$$n_A = \chi_{1A}\, V_A + \chi_{2A}\, V_A V_A$$

$$E_A = -\alpha_{1A}\, E - \alpha_{2A}\, EE$$

$$\alpha_{1A} = -\phi_A\, \chi_{1A} \qquad \text{etc.} \tag{3.1}$$

Note that the perturbing V_A depends on the species it is acting on and thus carries the species index, while E is evidently independent of the species. One can also define a full polarizability, relating to the full polarization field $\Sigma\check{E}_A$

$$\alpha_1 = \sum_A \alpha_{1A}$$

$$\alpha_2 = \sum_A \alpha_{2A} \tag{3.2}$$

A similar "full χ" would, however, be void of any physical significance.

The connection between the external species polarizabilities and the full species polarizabilities is now somewhat more involved than before:

$$\underset{1}{\hat{\alpha}}_A(k) = \frac{\underset{1}{\alpha}_A(k)}{\varepsilon(k)}$$

$$\underset{2}{\hat{\alpha}}_A(pq) = \frac{1}{\varepsilon(p)\ \varepsilon(q)\ \varepsilon(k)} \left\{ \left(1 + \sum_{A \neq B} \underset{1}{\alpha}_B(k)\right) \underset{2}{\alpha}_A(pq) - \underset{1}{\alpha}_A(k) \sum_{A \neq B} \underset{2}{\alpha}_B(pq) \right\} \tag{3.3}$$

The relation analogous to Eq. (3.2), however, is still valid:

$$\underset{1}{\hat{\alpha}} = \sum_A \underset{1}{\hat{\alpha}}_A$$

$$\underset{2}{\alpha} = \sum_A \underset{2}{\alpha}_A \tag{3.4}$$

A different extension of the notion of the response function is arrived at by contemplating perturbing fields and potentials which act on one species only. Although such fields are not realizable in normal charged particle systems, they are physically perfectly reasonable. What this kind of perturbation requires is that each species, in addition to its electric charge, be endowed with an extra "species charge" (numerically equal to its electric charge) which can interact with its corresponding field; the latter leaves at the same time, the other species charges unaffected. We will refer to this type of perturbing field by a superscript. Then one has

$$n_A = \sum_B \underset{1}{\hat{\chi}}{}_A^B\, \hat{V}^B + \sum_{BC} \underset{2}{\chi}{}_A^{BC}\, \hat{V}^B\, \hat{V}^C + \dots \tag{3.5}$$

Although it would be possible to fully develop the concept of the fictitious species field by allowing for the generation of corresponding internal polarization fields, such a full generalization doesn't seem to be particularly useful. Instead, we label polarization fields, as previously, by the species they originate from. Thus the relation

$$\check{E}_A = -\hat{\alpha}_A^B \hat{E}^B \tag{3.6}$$

describes the internal polarization field due to species A as a result of the perturbation by the fictitious external field acting on species B.

With the introduction of the interspecies potential ϕ_A^B and

$$\phi_{A\vec{k}}^B = Z_A Z_B \frac{4\pi e^2}{k^2} \tag{3.7}$$

and

$$\phi_{A\vec{p}\vec{q}}^{BC} = Z_A Z_B Z_C \frac{4\pi i e^2}{kpq} \tag{3.8}$$

where $Z_A e$, $Z_B e$, etc. are the ionic charges ($Z_{electron} = -1$),

$$\hat{\alpha}_{1A}^B(k) = -\phi_{A\vec{k}}^B \hat{\chi}_A^B(k)$$

$$\hat{\alpha}_{2A}^{BC}(pq) = -\phi_{A\vec{p}\vec{q}}^{BC} \hat{\chi}_A^{BC}(pq) \tag{3.9}$$

Since the full physical perturbation acts on all species, the following simple relations exist between the full physical species polarizabilities and the partial polarizabilities:

$$\hat{\alpha}_{1^A} = \sum_B \hat{\alpha}^B_{1^A}$$

$$\hat{\alpha}_{2^A} = \sum_{BC} \hat{\alpha}^{BC}_{2^A} \tag{3.10}$$

Finally we note that in the domain of partial response functions (as long as no partial internal fields are introduced), only the external response functions are useful.

Our generalization to multicomponent systems mostly followed the formalisms of Vashishta, Bhattacharyya and Singwi [1974a,b], Tosi, Parinello and March [1974] and Golden and Kalman [1976].

IV. SUM RULES

Response functions obey various constraints which follow from conservation laws and other physical requirements. These constraints are traditionally referred to as "sum rules" although some of them are nothing more but conditions on limiting behaviors with respect to ω or $\vec{k}$. We will discuss some of the sum rules which are of interest here.

A. Compressibility Sum Rules, $\omega = o$

Compressibility sum rules result from the hydrodynamic behavior of the system in the limit $\omega = o$, $k \to o$. Under such conditions the equation of state jointly with the Euler equation is sufficient to determine the limiting behavior of response functions.

Consider now the perturbation of an ocp by a small external electric field under isothermal conditions. Then the equation of state $P = P(n)$ (P = pressure, n = density) can be expanded as [Golden, Kalman and Datta, 1975]

$$P(n) = P\left(n^{(o)}\right) + \frac{\partial P}{\partial n}\bigg|_{n_o} n^{(1)} + \frac{1}{2}\frac{\partial^2 P}{\partial n^2}\bigg|_{n_o} n^{(1)2} + \frac{\partial P}{\partial n} n^{(2)} + \ldots$$

$$= P(n_o) + mc^2\left\{n^{(1)} + n^{(2)}\right\} + \frac{a}{2} n^{(1)2} + \ldots \tag{4.1}$$

Here $n^{(o)} \equiv n_o$ is the unperturbed density, $n^{(1)}$ and $n^{(2)}$ are the first- and second-order density perturbations,

$$c = \left\{ \frac{1}{m} \frac{\partial P}{\partial n}\bigg|_{n_o} \right\}^{1/2} = \frac{1}{(n_o K)^{1/2}} \tag{4.2}$$

is the isothermal sound velocity and K the compressibility; furthermore

$$a = \frac{\partial^2 P}{\partial n^2}\bigg|_{n_o} \tag{4.3}$$

To be sure, the equation of state could depend, in addition to $\underline{n}$ itself, on its derivatives, in the form $P[n, (\nabla n)^2, \nabla^2 n, \ldots]$. It can be shown [Golden, Kalman and Datta, 1975], however, that the inclusion of these terms doesn't affect the leading term in the $k \to o$ limit.

The equilibrium of the system under the effect of a perturbing field is maintained by a pressure gradient, i.e.,

$$-\vec{\nabla} P = -\frac{e}{m} \vec{E} \tag{4.4}$$

where E evidently includes the polarization field. Fourier-analyzing the perturbation, one can identify the first- and second-order terms in E as

$$E^{(1)}_{\vec{k}} = \frac{\hat{E}_{\vec{k}}}{\varepsilon(\vec{k}o)}$$

$$E^{(2)}_{\vec{k}} = -\frac{1}{V} \sum_{\substack{\vec{q} \\ \vec{p}=\vec{k}-\vec{q}}} \frac{\frac{\alpha(\vec{p}o;\, \vec{q}o)}{2}}{\varepsilon(\vec{k}o)\, \varepsilon(\vec{p}o)\, \varepsilon(\vec{q}o)} \hat{E}_{\vec{p}} \hat{E}_{\vec{q}} \tag{4.5}$$

Similarly, the first-order and second-order Fourier-components of $\underline{n}$ can be constructed as

$$-ik\, \phi_{\vec{k}}\, n^{(1)}_{\vec{k}} = -\alpha(\vec{k}o)\, E^{(1)}_{\vec{k}}$$

$$-ik\, \phi_{\vec{k}}\, n^{(2)}_{\vec{k}} = E^{(2)}_{\vec{k}} \tag{4.6}$$

Thus, combining Eqs. (4.1), (4.5) and (4. 6) we obtain the linear relation

$$\alpha(\vec{k}o) = \frac{1}{\beta mc^2}\,\alpha_o(\vec{k}o) = \frac{K}{K_o}\,\alpha_o(\vec{k}o) \tag{4.7}$$

and the additional quadratic relation

$$\underset{2}{\alpha}(\vec{q}o;\ \vec{p}o) = \frac{1}{(\beta mc^2)^2}\left(1 - \frac{n_o a}{mc^2}\right)\underset{2}{\alpha_o}(\vec{q}o;\ \vec{p}o) \tag{4.8}$$

α_o and $\underset{2}{\alpha_o}$ are the Vlasov (or RPA) polarizabilities,

$$\alpha_o(\vec{k}o) = n_o\ \beta\ \phi_{\vec{k}} = \frac{\kappa^2}{k^2}$$

$$\alpha_o(\vec{p}o,\ \vec{q}o) = \frac{1}{2}\,n_o\ \beta\ \phi_{\vec{p}\vec{q}} = \frac{2\pi\ i\ e^3\ n_o\ \beta^2}{kqp} \tag{4.9}$$

In the weak coupling limit the expansion of the equation of state to order $\gamma^2\ \ell n\ \gamma$

$$P = n_o\ \beta^{-1}\left\{1 - \frac{1}{6}\,\gamma - \frac{1}{12}\,\gamma^2\ \ell n\ \gamma\right\} \tag{4.10}$$

$$C = 0.5772\ \ldots\ \text{(Euler constant)}$$

yields the following expressions for α and $\underset{2}{\alpha}$:

$$\alpha(\vec{k}o) = \left\{1 + \frac{1}{4}\,\gamma + \frac{1}{6}\,\gamma^2\ \ell n\ \gamma\right\}\alpha_o(\vec{k}o)$$

$$\underset{2}{\alpha}(\vec{p}o;\ \vec{q}o) = \left\{1 + \frac{5}{8}\,\gamma + \frac{1}{2}\,\gamma^2\ \ell n\ \gamma\right\}\alpha_o(\vec{p}o;\ \vec{q}o) \tag{4.11}$$

The linear compressibility sum rule has been known for a long time. (See, e.g., Pines and Nozieres [1966]; Kalman [1975].) The quadratic one was derived recently [Golden, Kalman and Datta, 1975].

B. High Frequency Sum Rules

The high frequency sum rules provide the coefficients of the inverse powers of ω in the high frequency asymptotic expansion of $\alpha'(\vec{k}\omega)$. They follow from the equations of motion (i.e., conservation laws), via the fluctuation-dissipation theorem and the Kramers-

Kronig relations for the response functions. A quantity of central importance is the dynamical structure factor $S(\vec{k}\omega)$, the Fourier transform of the two-point density correlation function (N is the number of particles in the system):

$$S(\vec{k}\omega) = \frac{1}{2\pi N} \int d\tau \; e^{i\omega\tau} \langle n_{\vec{k}}(t) \; n_{-\vec{k}}(o)\rangle \tag{4.12}$$

Equally important is the static structure factor $S_{\vec{k}}$:

$$\begin{aligned} S_{\vec{k}} &= \frac{1}{N} \langle n_{\vec{k}}(o) \; n_{-\vec{k}}(o)\rangle \\ &= \int d\omega \; S(\vec{k}\omega) \\ &= \frac{1}{N} \left\{ \sum_{ij} \langle e^{-i\vec{k}\cdot(\vec{x}_i - \vec{x}_j)}\rangle - N^2 \, \delta_{\vec{k}} \right\} \\ &= \frac{1}{N} \left\{ N^2 \sum_{i \neq j} \langle e^{-i\vec{k}\cdot(\vec{x}_i - \vec{x}_j)}\rangle + N\langle 1\rangle \right\} - N \, \delta_{\vec{k}} \\ &= N \int d\vec{r} \, \frac{1}{V} \, [1 + g(r)] \, e^{-i\vec{k}\cdot\vec{r}} + 1 - N \, \delta_{\vec{k}} \\ &= 1 + n \, g_{\vec{k}} \end{aligned} \tag{4.13}$$

Here $g(r)$ is the usual pair correlation function and $g_{\vec{k}}$ its Fourier transform, and $n_{\vec{k}}$ is the Fourier component of the microscopic fluctuating density,

$$n_{\vec{k}} = \sum_{i}^{N} e^{-i\vec{k}\cdot\vec{x}_i} - N \, \delta_{\vec{k}} \tag{4.14}$$

The average is taken over the equilibrium ensemble. For the time being we are again restricting ourselves to the ocp situation.

Fluctuation-dissipation theorems (FDT's) will be discussed in greater detail in the next Chapter. The linear FDT relates the linear response functions (say, the polarizability) to the dynamical structure factor:

$$S(\vec{k}\omega) = \frac{1}{\pi \, \phi_{\vec{k}} \, \beta \, n \, \omega} \, \hat{\alpha}''(\vec{k}\omega) \tag{4.15}$$

Now consider the frequency-moments of $\hat{\alpha}''(\vec{k}\omega)$, $\Omega_\ell(\vec{k})$, defined as

$$\Omega_{\ell+1}(\vec{k}) = \frac{1}{\pi} \int d\omega\, \omega^{\ell}\, \hat{\alpha}''(\vec{k}\omega) \tag{4.16}$$

The FDT and the ω-integral allows one to relate the moment to equal-time correlations:

$$\Omega_{\ell+1}(\vec{k}) = \phi_{\vec{k}}\, \beta\, n \int d\omega\, \omega^{\ell+1}\, S(\vec{k}\omega)$$

$$= \frac{\phi_{\vec{k}}\, \beta n}{V} \left(i \frac{d}{d\tau}\right)^{\ell+1} \langle n_{\vec{k}}(\tau)\, n_{-\vec{k}}(o)\rangle \Big|_{\tau=o} \tag{4.17}$$

with the remarkable consequence that these quantities can now be calculated exactly via the equation of motion.

The even moments evidently vanish in virtue of the odd parity of $\hat{\alpha}''(\omega)$. The first moment

$$\Omega_2 = \frac{\phi_{\vec{k}}\, \beta}{V} \left(\frac{d}{d\tau}\right)^2 \langle n_{\vec{k}}(\tau)\, n_{-\vec{k}}(o)\rangle \Big|_{\tau=o}$$

$$= \frac{\phi_{\vec{k}}\, \beta}{V} \langle \dot{n}_{\vec{k}}(o)\, \dot{n}_{-\vec{k}}(o)\rangle \tag{4.18}$$

can be calculated as follows:

$$\dot{n}_{\vec{k}} = -i \sum_i \langle (\vec{k} \cdot \vec{v}_i)\, e^{-i\vec{k}\cdot\vec{x}_i} \rangle \tag{4.19}$$

and

$$\langle \dot{n}_{\vec{k}}(o)\, \dot{n}_{-\vec{k}}(o)\rangle = \sum_{i,j} \langle (\vec{k} \cdot \vec{v}_i)(\vec{k} \cdot \vec{v}_j)\, e^{-i\vec{k}\cdot(\vec{x}_i - \vec{x}_j)} \rangle$$

$$= \sum_i \langle (\vec{k} \cdot \vec{v}_i)^2 \rangle$$

$$= \frac{k^2}{\beta m} N \tag{4.20}$$

The second step follows since different particle velocities are uncorrelated. Thus

$$\Omega_2 = \frac{4\pi e^2 n}{m} = \omega_o^2 \tag{4.21}$$

The third moment can be worked out along the same lines:

$$\Omega_4(\vec{k}) = \frac{\phi_{\vec{k}}\,\beta}{V}\left(i\frac{d}{d\tau}\right)^4 \langle n_{\vec{k}}(\tau)\, n_{-\vec{k}}(o)\rangle\Big|_{\tau=o}$$

$$= \frac{\phi_{\vec{k}}\,\beta}{V} \langle \ddot{n}_{\vec{k}}(o)\, \ddot{n}_{-\vec{k}}(o)\rangle \tag{4.22}$$

The derivatives become

$$\ddot{n}_{\vec{k}} = -\sum_i (\vec{k}\cdot\vec{v}_i)^2 e^{-i\vec{k}\cdot\vec{x}_i} - i\sum_i \vec{k}\cdot\dot{\vec{v}}_i\, e^{-i\vec{k}\cdot\vec{x}_i}$$

$$\dot{v}_i = -\frac{1}{m}\frac{\partial}{\partial\vec{x}_i} H(x,p) \tag{4.23}$$

where H is the Hamiltonian of the system. Equations (4.21) and (4.22) can be combined into

$$\Omega_4(\vec{k}) = \frac{\phi_{\vec{k}}\,\beta}{V}\left\{ \sum_{i,j} \langle \vec{k}\cdot\dot{\vec{v}}_i\, \vec{k}\cdot\dot{\vec{v}}_j\, e^{-i\vec{k}\cdot(\vec{x}_i-\vec{x}_j)}\rangle \right.$$

$$+ \sum_{i,j} \langle (\vec{k}\cdot\vec{v}_i)^2 (\vec{k}\cdot\vec{v}_j)^2\, e^{-i\vec{k}\cdot(\vec{x}_i-\vec{x}_j)}\rangle$$

$$\left. - i\sum_{i,j} \langle (\vec{k}\cdot\vec{v}_i)^2\, \vec{k}\cdot\dot{\vec{v}}_j \left(e^{-i\vec{k}\cdot(\vec{x}_i-\vec{x}_j)} - e^{+i\vec{k}\cdot(\vec{x}_i-\vec{x}_j)}\right)\rangle \right\} \tag{4.24}$$

The first term can be calculated by exploiting the character of the canonical distribution function:

$$\sum_{i,j} \langle \vec{k}\cdot\dot{\vec{v}}_i\, \vec{k}\cdot\dot{\vec{v}}_j\, e^{-i\vec{k}\cdot(\vec{x}_i-\vec{x}_j)}\rangle$$

$$= \frac{1}{m^2} \sum_{ij} \int dx\, dp\, e^{-\beta H(x,p)}\, e^{-i\vec{k}\cdot(\vec{x}_i-\vec{x}_j)}\, \vec{k} \cdot \frac{\partial H}{\partial \vec{x}_i}\, \vec{k} \cdot \frac{\partial H}{\partial \vec{x}_j}$$

$$= \frac{1}{\beta m^2} \sum_{ij} \int dx\, dp\, e^{-\beta H(x,p)} \left\{ \vec{k} \cdot \frac{\partial^2 H}{\partial \vec{x}_i\, \partial \vec{x}_j} \cdot \vec{k} - ik^2\, \vec{k} \cdot \frac{\partial H}{\partial \vec{x}_j} \right\} e^{-i\vec{k}\cdot(\vec{x}_i-\vec{x}_j)}$$

$$= -\frac{1}{\beta m^2} \left\{ \sum_{i \neq j} \langle (\vec{k} \cdot \vec{\nabla})^2\, V(\vec{x}_i - \vec{x}_j)\, e^{-i\vec{k}\cdot(\vec{x}_i-\vec{x}_j)} \rangle - \sum_{i \neq j} \langle (\vec{k} \cdot \vec{\nabla})^2\, V(\vec{x}_i - \vec{x}_j) \rangle \right\}$$

$$+ \frac{k^4}{\beta^2 m^2} \sum_{ij} \langle e^{-i\vec{k}\cdot(\vec{x}_i-\vec{x}_j)} \rangle$$

$$= \frac{n}{\beta m^2} \sum_{\vec{p}}{}' (\vec{k} \cdot \vec{p})^2\, \phi_{\vec{p}} \left\{ (S_{\vec{k}-\vec{p}} + N\delta_{\vec{k}-\vec{p}}) - (S_{\vec{p}} + N\delta_{\vec{p}}) \right\}$$

$$+ \frac{N}{\beta^2 m^2} k^4 S_{\vec{k}} \tag{4.25}$$

The second term yields

$$\sum_{i,j} \langle (\vec{k} \cdot \vec{v}_i)^2 (\vec{k} \cdot \vec{v}_j)^2\, e^{-i\vec{k}\cdot(\vec{x}_i-\vec{x}_j)} \rangle =$$

$$= \sum_{i \neq j} \langle (\vec{k} \cdot \vec{v}_i)^2 \rangle \langle (\vec{k} \cdot \vec{v}_j)^2 \rangle \langle e^{-i\vec{k}\cdot(\vec{x}_i-\vec{x}_j)} \rangle$$

$$+ \sum_i \langle (\vec{k} \cdot \vec{v}_i)^4 \rangle$$

$$= \frac{N}{\beta^2 m^2} k^4 S_{\vec{k}} + 3 \frac{N}{\beta^2 m^2} k^4 \tag{4.26}$$

Finally, the third term becomes

$$i \sum_{i,j}' \langle (\vec{k} \cdot \vec{v}_i)^2 \, \vec{k} \cdot \dot{\vec{v}}_j \, e^{-i\vec{k}\cdot(\vec{x}_i - \vec{x}_j)} \rangle =$$

$$= -\frac{i}{m} \sum_{i,j} \int dx\, dp\, e^{-\beta H(x,p)} (\vec{k} \cdot \vec{v}_i)^2$$

$$\cdot\, e^{-i\vec{k}\cdot(\vec{x}_i - \vec{x}_j)} \, \vec{k} \cdot \frac{\partial H}{\partial \vec{x}_j}$$

$$= \frac{1}{\beta m} \sum_{i,j} \int dx\, dp\, e^{-\beta H(x,p)} (\vec{k} \cdot \vec{v}_i)^2 \, k^2 \, e^{-i\vec{k}\cdot(\vec{x}_i - \vec{x}_j)}$$

$$= \frac{k^2}{\beta m} \sum_{i,j} \langle (\vec{k} \cdot \vec{v}_i)^2 \rangle \langle e^{-i\vec{k}\cdot(\vec{x}_i - \vec{x}_j)} \rangle$$

$$= \frac{N}{\beta^2 m^2} k^4 S_{\vec{k}} \tag{4.27}$$

Combining Eqs. (4.25), (4.26) and (4.27) according to Eq. (4.24), one finds

$$\Omega_4(\vec{k}) = \omega_o^4 \left\{ 1 + 3 \frac{k^2}{\kappa^2} + \frac{1}{N} \sum_{\vec{p}} \frac{(\vec{k} \cdot \vec{p})^2}{k^2 p^2} (S_{\vec{p}-\vec{k}} - S_{\vec{p}}) \right\}$$

$$\kappa^2 = 4\pi e^2 n \beta \tag{4.28}$$

The Ω_6 moment can also be calculated [Forster, Martin and Yip, 1968; Ichimaru and Tange, 1970; Ichimaru, Totsuji, Tange and Pines, 1975] but will not be discussed here in detail.

As a result of their causal behavior, the response functions obey Kramers-Kronig relations (see, e.g., Martin [1968], Kalman [1975]). This fact allows one to convert the result into a high frequency expansion for the <u>real</u> part of $\hat{\alpha}(\vec{k}\omega)$,

$$\hat{\alpha}'(\vec{k}\omega) = -\frac{1}{\pi}\int \frac{d\omega'}{\omega - \omega'}\,\hat{\alpha}''(\vec{k}\omega')$$

$$\simeq -\frac{1}{\pi\omega}\int d\omega' \left(1 + \frac{\omega'}{\omega} + \frac{\omega'^2}{\omega^2} + \ldots\right)\hat{\alpha}(\vec{k}\omega')$$

$$\simeq -\sum_{\ell=1} \frac{\Omega_{\ell+1}(\vec{k})}{\omega^{\ell+1}} \tag{4.29}$$

with only the even $\ell+1$ powers contributing. The expansion of $\hat{\alpha}$ can also be converted into an expansion of α. Since $\alpha = \hat{\alpha}/(1 - \hat{\alpha})$, one has, to lowest order,

$$\hat{\alpha}'(\vec{k}\omega) = -\left\{\frac{\Omega_2}{\omega^2} + \frac{\Omega_4}{\omega^4} + \frac{\Omega_6}{\omega^6}\right\}$$

$$\alpha'(\vec{k}\omega) = -\left\{\frac{\Omega_2}{\omega^2} + \frac{\Omega_4 - \Omega_2^2}{\omega^4} + \frac{\Omega_6 - 2\Omega_2\,\Omega_4 + \Omega_2^3}{\omega^6}\right\} \tag{4.30}$$

Assume that we have a small expansion parameter -- it could be γ, or k^2, or both. Then it is useful to write the moments as

$$\Omega_2 = \omega_o^2$$

$$\Omega_4 = \omega_o^4(1 + \Delta_4)$$

$$\Omega_6 = \omega_o^6(1 + \Delta_6) \tag{4.31}$$

and the substitution of this form into the second line of Eq. (4.29) yields

$$\alpha'(\vec{k}\omega) = -\left\{\frac{\omega_o^2}{\omega^2} + \Delta_4\,\frac{\omega_o^4}{\omega^4} + (\Delta_6 - 2\Delta_4)\,\frac{\omega_o^6}{\omega^6}\right\} \tag{4.31a}$$

Finally, we quote the values of Δ_4 and Δ_6 to order k^2 and γ:

$$\Delta_4 = \left\{3 + \frac{1}{V}\sum_{\vec{p}} \frac{(\vec{k}\cdot\vec{p})^2}{k^2p^2}\,(g_{\vec{k}-\vec{p}} - g_{\vec{p}})\right\}\frac{k^2}{\kappa^2}$$

$$= \left\{ 3 + \frac{1}{V} \sum_{\vec{p}} \left(\frac{(\vec{k} \cdot [\vec{k} - \vec{p}])^2}{k^2 (\vec{k} - \vec{p})^2} - \frac{(\vec{k} \cdot \vec{p})^2}{k^2 p^2} \right) g_{\vec{p}} \right\} \frac{k^2}{\kappa^2}$$

$$= \left\{ 3 + \frac{4}{15} \beta E_c \right\} \frac{k^2}{\kappa^2} + 0(\frac{k^4}{\kappa^4}) \qquad (4.32)$$

where

$$E_c = \frac{1}{2} \int d\vec{r} \, \frac{4\pi e^2}{r} g(r) \qquad (4.33)$$

is the correlation energy per particle. At the same time [Forster, Martin and Yip, 1968; Ichimaru and Tange, 1974]

$$\Delta_6 = 0(k^4, \gamma^2) \qquad (4.34)$$

For small γ one has

$$\Delta_4 = 0(\gamma) + 3\frac{k^2}{\kappa^2} \simeq (3 - \frac{2}{15}\gamma) \frac{k^2}{\kappa^2} \qquad (4.35)$$

and

$$\Delta_6 = 0(\gamma^2) + 15\frac{k^4}{\kappa^4} + 6 \frac{k^2}{\kappa^2} \qquad (4.36)$$

The results for Ω_4 moment for arbitrary interaction were first derived by Placzek [1952] and then rederived by deGennes [1959]; for coulomb interaction the corresponding formula has been given by Pathak and Vashishta [1973]; a full presentation both of the Ω_4 and of the Ω_6 moments for arbitrary interaction is due to Forster, Martin and Yip [1968]; the application to coulomb systems has been given by Ichimaru and Tange [1974] and Ichimaru, Totsuji, Tange and Pines [1975]. (See also Ichimaru [this Volume].)

V. FLUCTUATION-DISSIPATION THEOREMS

Fluctuation-dissipation theorems connect averages of correlations between different space - time points, on the one hand, and response functions, on the other hand. Traditionally, the response functions are linear ones and the averages refer to the equilibrium ensemble. However, a much more general approach is possible and will be followed here. It also turns out to be extremely useful.

The principal idea behind the fluctuation-dissipation theorem is that if the system is subject to a small perturbation, its response to the perturbation can be used to infer correlations that existed in the system before the perturbation was applied. The mathematical formulation of this statement is made possible by connecting the time evolution of the phase space distribution function $\Omega(x,p;\ t)$ with the evolution of dynamical variables through the time evolution operator formalism.

In the following we consider, most of the time, an ocp. The equilibrium system is described by the Hamiltonian

$$H^{(o)} = \sum_i \frac{p_i^2}{2m} + \frac{1}{2} \sum_{i \neq j} V(|\vec{x}_i - \vec{x}_j) \tag{5.1}$$

while perturbation adds to it a term depending on the perturbing potential $\hat{V}_{\vec{k}}(t)$:

$$H^{(1)} = \frac{1}{V} \sum_{\vec{k}} \sum_i \hat{V}_k(t)\ e^{i\vec{k}\cdot\vec{x}_i}$$

$$= \frac{1}{V} \sum_{\vec{k}} \hat{V}_{\vec{k}}(t)\ n_{-\vec{k}} \tag{5.2}$$

$$H = H^{(o)} + H^{(1)}$$

The Hamiltonian generates the Liouville-operators

$$L^{(o)} = -i\ [H^{(o)},\ \ldots]$$

$$L^{(1)} = -i\ [H^{(1)},\ \ldots]$$

$$L = L^{(o)} + L^{(1)} \tag{5.3}$$

and the time evolution operator pertaining to the equilibrium system

$$U(t,\ t') = e^{-iL_o(t-t')} \equiv U(t - t') \tag{5.4}$$

The state of the system is characterized by the phase space distribution function in the 6N-dimensional phase space. In equilibrium

$$\Omega^{(o)} = Z^{-1}\ e^{-\beta H^{(o)}} \tag{5.5}$$

while its evolution under the influence of the perturbation is governed by the Liouville equation,

$$\frac{\partial \Omega}{\partial t} = -i\ L\ \Omega \tag{5.6}$$

Now the formal solution of the Liouville equation can be written down immediately by using the perturbation expansion

$$\Omega(t) = \sum_{n=o} \Omega^{(n)}(t) \tag{5.7}$$

in the perturbing potential $\hat{V}_{\vec{k}}$:

$$\Omega^{(1)}(t) = -i \int_{-\infty}^{t} dt_1\ U(t,\ t_1)\ L^{(1)}(t_1)\ \Omega^{(o)} \tag{5.8}$$

$$\Omega^{(2)}(t) = (-i)^2 \int_{-\infty}^{t} dt_1 \int_{-\infty}^{t_1} dt_2\ U(t,\ t_1)\ L^{(1)}(t_1)\ U(t_1,\ t_2)$$

$$\cdot\ L^{(1)}(t_2)\ \Omega^{(o)} \tag{5.9}$$

First we concentrate on $\Omega^{(1)}$. The closed expression for $\Omega^{(1)}$ can be worked out in a series of simple steps:

$$\Omega^{(1)}(t) = -i \int_{-\infty}^{t} dt'\ U(t,\ t')\ L^{(1)}(t')\ \Omega^{(o)}$$

$$= -i \int_{o}^{\infty} d\tau\ U(t)\ L^{(1)}(t-\tau)\ \Omega^{(o)}$$

$$= -\frac{1}{V} \sum_{\vec{k}} \int_{o}^{\infty} d\tau\ U(t)\ [n_{-\vec{k}},\ \Omega^{(o)}]\ \hat{V}_{\vec{k}}(t-\tau)$$

$$= -\frac{\beta}{V} \sum_{\vec{k}} \int_{o}^{\infty} d\tau\ U(t)\ \frac{dn_{-\vec{k}}}{dt}\ \hat{V}_{k}(t-\tau)$$

$$= -\frac{i\beta}{V}\ \Omega^{(o)} \sum_{\vec{k}} \int_{o}^{\infty} d\tau\ \vec{k}\cdot\vec{j}_{-k}(t-\tau)\ \hat{V}_{\vec{k}}(t-\tau) \tag{5.10}$$

We have used the property of the time evolution operator $\mathcal{U}(t)$ that it shifts the time argument of the dynamical variables by $-\tau$ rather than τ. $\vec{j}_{\vec{k}}$ is the particle current

$$\vec{j}_{\vec{k}} = \sum_i \vec{v}_i \, e^{-i\vec{k}\cdot\vec{x}_i} \tag{5.11}$$

We now can use Eq. (5.10) to evaluate averages of dynamical quantities in the perturbed system. Consider the average of the longitudinal particle current $j_{\vec{k}}$. Because of spatial uniformity, only $\vec{k}$, $-\vec{k}$ combinations contribute.

$$\langle j_{\vec{k}}\rangle^{(1)}(t) = \int dp \, dx \, \Omega^{(1)}(t) \, j_{\vec{k}}$$

$$= i \frac{\beta k}{V} \int_0^{\infty} d\tau \, \langle j_{\vec{k}}(\tau) \, j_{-\vec{k}}(\tau - t)\rangle^{(o)} \, \hat{V}_k(\tau - t) \tag{5.12}$$

We shift to Fourier transform language and trade $j_{\vec{k}}(\omega)$ for $n_{\vec{k}}(\omega)$. Then

$$\langle n_{\vec{k}}(\omega)\rangle^{(1)} = -i \frac{\beta\omega}{V} \int_{-\infty}^{+\infty} d\omega' \int_{-\infty}^{+\infty} d\tau \, e^{i\omega'\tau} \langle n_{\vec{k}}(\tau) \, n_{-\vec{k}}(o)\rangle^{(o)}$$

$$\cdot \, \delta_+(\omega - \omega') \, \hat{V}_{\vec{k}}(\omega) \tag{5.13}$$

Taking now the real part of Eq. (5.13), and using the definitions of Eq. (4.12) and (2.2) we find

$$S(\vec{k}\omega) = - \frac{1}{\pi \, \beta \, n \, \omega} \hat{\chi}''(\vec{k}\omega) \tag{5.14}$$

This is the dynamical linear fluctuation-dissipation theorem. The static FDT is obtained by integrating over ω. Recalling Eq. (4.13) we obtain one of the equivalent forms,

$$\hat{\chi}(\vec{k}o) = -\beta n S_{\vec{k}} = -\beta n \, \{1 + n g_{\vec{k}}\} \tag{5.15}$$

$$S_{\vec{k}} = \frac{\hat{\alpha}(\vec{k}o)}{\alpha_o(\vec{k})} = \frac{\alpha(\vec{k}o)}{\alpha_o(\vec{k})} \frac{1}{\varepsilon(\vec{k}o)} \tag{5.16}$$

$$\alpha(\vec{k}o) = \frac{S_{\vec{k}}}{(k/\kappa)^2 - S_{\vec{k}}} \tag{5.17}$$

The above relations should be understood with the $k \neq o$ qualification;

$$\alpha_o(\vec{k}) = \frac{\kappa^2}{k^2}$$

is the Vlasov (RPA) static polarizability.

Equation (5.17) in conjunction with the sum rule (4.7) can be used to infer the $k \to o$ behavior of the pair correlation function $g_{\vec{k}}$,

$$ng_{\vec{k}} = -1 + (k/\kappa)^2 - \frac{K_o}{K}(k/\kappa)^4 + 0\,(k^6) \tag{5.18}$$

Note that for the above expression to be correct to order k^4, one does not need the (unknown) coefficient of the k^o term in $\alpha(\vec{k}o)$. Equation (5.18) can also be converted into moment conditions in configuration space:

$$n\int d\vec{r}\; g(r) = -1$$

$$n\int d\vec{r}\; r^2\, g(r) = -6/\kappa^2 \tag{5.19}$$

etc.

The second relation is the fairly well known Stillinger-Lovett condition (Stillinger and Lovett [1968]; DeWitt [1978]).

The linear FDT has been derived and rederived innumerable times. The original formulation is due to Kubo [1957]; some more modern discussions in a language more akin to the above presentation are given by Martin [1968], Golden and Kalman [1969] and Kalman [1975].

The generalization of the ocp result to multicomponent systems can proceed in two different ways. First, we can calculate the response of a given species to the physical perturbation. Then Eq. (5.16), for example, becomes

$$\hat{\alpha}_A(\vec{k}o) = \alpha_{oA}(\vec{k})\left\{1 + \sum_B \frac{Z_B}{Z_A} n_B\, g_{AB}\right\}$$

$$\alpha_{oA}(\vec{k}) = \frac{\kappa_A^2}{k^2} \qquad \kappa_A^2 = 4\pi\, e^2\, Z_A^2\, n_A\, \beta \tag{5.20}$$

This is of limited usefulness, since it connects one response function with several correlation functions.

A more useful generalization is arrived at by using the concept of partial response functions, introduced (with this purpose in mind) in Chapter II. We also introduce now the partial dynamical and static structure functions [March and Tosi, 1976] S_{AB} (cf. Eqs. (4.12) and (4.13)):

$$S_{AB}(\vec{k}\omega) = \frac{1}{2\pi \sqrt{N_A N_B}} \int d\tau e^{i\omega\tau} \langle n_{A,\vec{k}}(\tau)\, n_{B,-\vec{k}}(o)\rangle^{(o)} \tag{5.21}$$

$$S_{AB,\vec{k}} = \frac{1}{\sqrt{N_A N_B}} \{\delta_{AB} + \sqrt{n_A n_B}\, g_{AB,\vec{k}}\} \tag{5.22}$$

g_{AB} is the pair correlation function between members of species A and B. Now Eqs. (5.14) and (5.15) are replaced by

$$\hat{\chi}_A^{B\prime\prime}(\vec{k}\omega) = -\pi\,\beta \sqrt{n_A n_B}\ \omega\ S_{AB}(\vec{k}\omega) \tag{5.23}$$

$$\begin{aligned} \hat{\chi}_A^{B}(\vec{k}o) &= -\beta \sqrt{n_A n_B}\ S_{AB,\vec{k}} \\ &= -\beta\sqrt{n_A n_B}\ \{\delta_{AB} + \sqrt{n_A n_B}\ g_{AB,\vec{k}}\} \end{aligned} \tag{5.24}$$

Similarly, Eq. (5.16) becomes [Golden and Kalman, 1976]

$$\begin{aligned} \alpha_A^B(\vec{k}o) &= \beta\ \phi^B_{A,\vec{k}}\ \sqrt{n_A n_B}\ \{\delta_{AB} + \sqrt{n_A n_B}\ g_{AB,\vec{k}}\} \\ &= \alpha_{oA}(\vec{k})\{\delta_{AB} + \frac{Z_B}{Z_A}\, n_B\, g_{AB,\vec{k}}\} \\ \phi^B_{A,\vec{k}} &= Z_A\ Z_B\ \phi_{\vec{k}} \end{aligned} \tag{5.25}$$

Summing over B, and taking into account Eq. (5.20), one verifies Eq. (2.18).

In a two-component plasma (not binary ion mixture) one can exploit the charge neutrality condition $|Z_A| n_A = |Z_B| n_B$. Then Eq.

(5.25) can be rewritten as [Golden and Kalman, 1976]

$$\alpha_A^B(\vec{k}o) = \alpha_{Ao}(\vec{k})\left\{\delta_{AB} + \frac{Z_A Z_B}{|Z_A Z_B|} n_A g_{AB,\vec{k}}\right\} \tag{5.26}$$

and

$$\alpha_A(\vec{k}o) = \alpha_{Ao}(\vec{k})\{1 + n_A(g_{AA,\vec{k}} - g_{AB,\vec{k}})\}$$
$$A \neq B \tag{5.27}$$

VI. NONLINEAR FLUCTUATION-DISSIPATION THEOREM

The generalization of the concepts introduced in the previous Chapter can proceed in two directions. One is to relate nonlinear response functions to higher-order correlations. Only the next step beyond the linear stage is available [Golden, Kalman and Silevitch, 1972; Kalman, 1975]; it consists of relating quadratic response functions to three-point functions. The algebra involved on these manipulations is substantially more complicated than that of the corresponding linear case. In the following the outlines of the derivation will ge given, while the second way of nonlinear generalization will be discussed later.

The formal expression for $\Omega^{(2)}(t)$, the second-order perturbed phase space distribution function, has been given in Eq. (5.9). It evolves around the expression

$$\mathcal{U}(t, t_1)\, \mathcal{L}^{(1)}(t_1)\, \mathcal{U}(t_1, t_2)\, \mathcal{L}^{(1)}(t_2)\, \Omega^{(o)}$$

$$= -\frac{1}{\mathcal{V}^2} \sum_{\vec{k}_1,\vec{k}_2} \mathcal{U}(t - t_1)\left[n_{-\vec{k}_1}, \mathcal{U}(t_1 - t_2)\left[n_{-\vec{k}_2}, \Omega^{(o)}\right]\right]$$

$$\cdot\ \hat{V}_{k_1}(t_1)\, \hat{V}_{k_2}(t_2) \tag{6.1}$$

Using the identity

$$[X,YZ] = Y[X,Z] + Z[X,Y] \tag{6.2}$$

and shifting the time variables of $n_{\vec{k}_1}$, $n_{\vec{k}_2}$ from the arbitrary reference time t to t_1 and t_2, the above expression becomes

$$+ \frac{1}{V^2} \sum_{\vec{k}_1,\vec{k}_2} \Bigg\{ \beta\Omega^{(o)} \left[n_{-\vec{k}_1}(t_1), \left[n_{-\vec{k}_2}(t_2), H^{(o)} \right] \right]$$

$$+ \beta^2 \Omega^{(o)} \left[n_{-\vec{k}_2}(t_2), H^{(o)} \right] \left[n_{-\vec{k}_1}(t_1), H^{(o)} \right] \Bigg\} \hat{V}_{\vec{k}_1}(t_1)\, \hat{V}_{\vec{k}_2}(t_2)$$

$$= \Omega^{(o)} \frac{1}{V^2} \sum_{\vec{k}_1\vec{k}_2} \Bigg\{ i\beta k_2 \left[n_{-\vec{k}_1}(t_1), j_{-\vec{k}_2}(t_2) \right]$$

$$- \beta^2 k_1 k_2\, j_{-\vec{k}_1}(t_1)\, j_{-\vec{k}_2}(t_2) \Bigg\} \hat{V}_{\vec{k}_1}(t_1)\, \hat{V}_{\vec{k}_2}(t_2) \tag{6.3}$$

Now we evaluate $\langle j_{\vec{k}} \rangle^{(2)}(t)$.

$$\langle j_{\vec{k}} \rangle^{(2)}(t) = \int dx\, dp\, \Omega^{(2)}(t)\, j_{\vec{k}}$$

$$= - \frac{1}{V^2} \sum_{\vec{k}_1,\vec{k}_2} \int_{-\infty}^{t} dt_1 \int_{-\infty}^{t_1} dt_2 \Bigg\{ i\beta k_2 \left\langle j_{\vec{k}}(t) \left[n_{-\vec{k}_1}(t_1), j_{-\vec{k}_2}(t_2) \right] \right\rangle$$

$$- \beta^2 k_1 k_2 \langle j_{\vec{k}}(t)\, j_{-\vec{k}_1}(t_1)\, j_{-\vec{k}_2}(t_2) \rangle \Bigg\} \hat{V}_{\vec{k}_1}(t_1)\, \hat{V}_{\vec{k}_2}(t_2) \tag{6.4}$$

We observe that spatial homogeneity imposes the conservation law $\vec{k} = \vec{k}_1 + \vec{k}_2$. This will be understood in the sequel.

Equation (6.4) can be brought into a more appealing form by a number of simple cosmetical operations. First, the asymmetry in the time variables t_1 and t_2 can be eliminated by inverting the order of integration and then symmetrizing the resulting expression. Next, the time variables of the phase-averaged products can be shifted as long as time differences are preserved. Finally, we introduce the abbreviations

$$Q(120) = \langle j_{-\vec{k}_1}(-\tau_1)\, j_{-\vec{k}_2}(-\tau_2)\, j_{\vec{k}}(o) \rangle$$

$$Z(120) = -\frac{i}{k_1} < \left[n_{-\vec{k}_1}(-\tau_1),\ j_{-\vec{k}_2}(-\tau_2) \right],\ j_{\vec{k}}(o) > \tag{6.5}$$

Now Eq. (6.4) can be written as

$$\hat{\sigma}(\vec{k}_1,\tau_1;\ \vec{k}_2,\tau_2) = -\frac{\beta}{2} \ominus (\tau_1) \ominus (\tau_2)$$

$$\cdot \left\{ \beta Q(120) + \ominus (\tau_2 - \tau_1)\ Z(120) + \ominus (\tau_1 - \tau_2)\ Z(210) \right\} \tag{6.6}$$

$\hat{\sigma}$ above is the quadratic, longitudinal, external conductivity in the time representation, i.e.

$$j_{\vec{k}}^{(2)}(t) \equiv <j_{\vec{k}}>^{(2)}(t)$$

$$= \frac{1}{V} \sum_{\vec{k}_1} \int d\tau_1 \int d\tau_2\ \hat{\sigma}(\vec{k}_1,\tau_1;\ \vec{k}_2,\tau_2)$$

$$\cdot\ \hat{E}_{\vec{k}_1}(t - \tau_1)\ \hat{E}_{\vec{k}_2}(t - \tau_2); \tag{6.7}$$

$\ominus(\tau)$ is the step-function.

The problem with Eq. (6.6) lies in the presence of the unwieldy Poisson-bracket terms. Further progress can be made only after their elimination. This can be accomplished through the following steps. First, observe that Z and Q are related to each other by

$$Z(120) + Z(102) = -\beta Q(120)$$

$$Z(210) + Z(201) = -\beta Q(210)$$

$$= -\beta Q(120) \tag{6.8}$$

Substituting Eq. (6.8) into Eq. (6.6), one finds

$$\hat{\sigma}(\vec{k}_1,\tau_1;\ \vec{k}_2,\tau_2) = -\frac{\beta}{2} \ominus(\tau_1) \ominus (\tau_2)$$

$$\cdot \left\{ \ominus(\tau_2 - \tau_1)\ Z(102) + \ominus (\tau_1 - \tau_2)\ Z(201) \right\} \tag{6.9}$$

Re-labelling the argument of $\hat{\sigma}$ in Eq. (6.9) in two different ways and exploiting the time translation invariance property of Z, one can construct

$$\hat{\sigma}(\vec{k}_1, \tau_2 - \tau_1; -\vec{k}, \tau_2) + \hat{\sigma}(-\vec{k}, \tau_1; \vec{k}_2, \tau_1 - \tau_2)$$

$$= -\frac{\beta}{2}\Big\{\Theta(\tau_1)\,\Theta(\tau_2)\;\Theta(\tau_2 - \tau_1)\,Z(120)$$

$$+\,\Theta(\tau_1 - \tau_2)\,Z(210)$$

$$+\,\Theta(-\tau_1)\,\Theta(\tau_2)\,\Theta(\tau_2 - \tau_1)\,Z(021)$$

$$+\,\Theta(\tau_1)\,\Theta(-\tau_2)\,\Theta(\tau_1 - \tau_2)\,Z(012)\Big\} \tag{6.10}$$

which when combined with Eq. (6.6) cancels all the Z-terms except the ones multiplied by $\Theta(-\tau_1)$ and $\Theta(-\tau_2)$, respectively. These latter can, however, be eliminated by projecting out the $\tau_1 > 0$, $\tau_2 > 0$ causal part of the combined response functions. This leads to the desired result, which can conveniently be formulated in Fourier transform language. The central object is the causal symmetrized combination of quadratic response functions,

$$\Xi(\vec{k}_1\omega_1; \vec{k}_2\omega_2) = \omega\hat{\chi}(\vec{k}_1\omega_1; \vec{k}_2\omega_2)$$

$$- \omega_1 \int d\mu\, \delta_+(\omega_1 - \mu)\, \hat{\chi}(-\vec{k}_1 - \mu; \vec{k}\,\omega_2 + \mu)$$

$$- \omega_2 \int d\mu\, \delta_+(\omega_2 - \mu)\, \hat{\chi}(\vec{k}\,\omega_1 + \mu; -\vec{k}_2 - \mu) \tag{6.11}$$

which is related to the quadratic dynamical structure function, defined by

$$S(\vec{k}_1\omega_1; \vec{k}_2\omega_2) = \frac{1}{4\pi^2 N}\int e^{i(\omega_1\tau_1+\omega_2\tau_2)} \langle n_{\vec{k}}(o)\, n_{-\vec{k}_1}(-\tau_1)\, n_{-\vec{k}_2}(-\tau_2)\rangle \cdot d\tau_1\, d\tau_2 \tag{6.12}$$

The result of the above manipulations is

$$\Xi(\vec{k}_1\omega_1;\ \vec{k}_2\omega_2) = -\frac{\beta^2}{2}\int d\mu \int d\omega\ \delta_+\ (\omega_1 - \mu)\ \delta_+\ (\omega_2 - \nu)$$

$$\mu\nu\omega S(\vec{k}_1\mu;\ \vec{k}_2\nu) \tag{6.13}$$

To obtain an explicit relation for S, one can first take the real part of Eq. (6.13) which leads to an expression in terms of repeated Hilbert transforms,

$$H[\omega\mu]\ F(\mu) \equiv \frac{1}{\pi} P \int \frac{d\mu}{\omega - \mu} F(\mu)$$

$$\Xi'(\omega_1,\omega_2) = -\frac{\beta^2}{8}\{1 - H[\omega_1\mu]\ H[\omega_2\nu]\}\ \mu\nu\omega S(\mu\nu) \tag{6.14}$$

In order to invert this relation, we exploit the property of the Hilbert transform operator that its eigenvalue is $-i$ ($+i$) associated with a plus-function (minus-function) eigenfunction:

$$HF^+ = -iF^+$$

$$HF^- = +iF^-$$

$$HHF^{+-} = F^{+-} \tag{6.15}$$

Thus the inversion of Eq. (6.14) can be cast in the form

$$-\frac{\beta^2}{8}\ \omega_1\omega_2\omega\ S(\omega_1,\omega_2) = Re[\Xi(\omega_1,\omega_2) + \Sigma a_i\ X_i(\omega_1,\omega_2)]\} \tag{6.16}$$

where X_i is an arbitrary plus-minus function of its two arguments. However, prescribing that S satisfy the triangle symmetry requirements which obviously follow from Eq. (6.12),

$$S(\vec{k}_1\omega_1;\ \vec{k}_2\omega_2) = S(\vec{k}_2\omega_2;\ -\vec{k}-\omega) = S(-\vec{k}-\omega;\ \vec{k}_1\omega_1)$$

is sufficient to eliminate the ambiguity [Golden, Kalman, Silevitch, 1972] with the final result

$$S(\vec{k}_1\omega_1;\ \vec{k}_2\omega_2) = -\frac{2}{\beta^2}\left\{\frac{\hat{\chi}'(\vec{k}_1\omega_1;\ \vec{k}_2\omega_2)}{\omega_1\omega_2} - \frac{\hat{\chi}'(\vec{k}\omega;\ -\vec{k}_1-\omega_1)}{\omega_2\omega} - \frac{\hat{\chi}'(\vec{k}\omega;\ -\vec{k}_2-\omega_2)}{\omega_1\omega}\right\} \tag{6.17}$$

Equation (6.17) is the quadratic analogue of the linear (5.14).

An alternative way of writing Eq. (6.17) is

$$S(\vec{k}_1\omega_1;\ \vec{k}_2\omega_2) = \frac{1}{\alpha_o(\vec{k}_1\vec{k}_2)}\, Im\left\{\frac{1}{\varepsilon(\vec{k}_1\omega_1)\ \varepsilon(\vec{k}_2\omega_2)\varepsilon^*(\vec{k}\omega)} \cdot \left[\frac{\alpha(\vec{k}_1\omega_1;\ \vec{k}_2\omega_2)}{\omega_1\omega_2} - \frac{\alpha(-\vec{k}-\omega;\ \vec{k}_1\omega_1)}{\omega\omega_1} - \frac{\alpha(\vec{k}_2\omega_2;\ -\vec{k}-\omega)}{\omega_2\omega}\right]\right\} \tag{6.18}$$

$\alpha_o(\vec{p}\vec{q})$ is the absolute value of Vlasov static quadratic polarizability

$$\alpha_o(\vec{p}\vec{q}) = \frac{2\pi\ e^3\ n\ \beta^2}{kqp} \tag{6.19}$$

One can proceed now to evaluate the static limit of Eq. (6.18) by integrating over ω_1 and ω_2. The apparent singularities at $\omega_1 = o$, $\omega_2 = o$ and $\omega_1 = -\omega_2$ are spurious and the integral can be shown to behave regularly. We also recall that $S_{\vec{k}} = 1 + ng_{\vec{k}}$. The analogous quadratic relation is

$$S_{\vec{p}\vec{q}} = 1 + ng_{\vec{p}} + ng_{\vec{q}} + ng_{\vec{k}} + n^2\ h_{\vec{p}\vec{q}} \tag{6.20}$$

We now can assemble the quadratic equivalents of the linear (5.15) and (5.16):

$$\hat{\chi}(\vec{p}o;\ \vec{q}o) = \beta^2 n S_{\vec{p}\vec{q}}$$

$$= \beta^2 n \left(1 + n g_{\vec{p}} + n g_{\vec{q}} + n g_{\vec{k}} + n h_{\vec{p}\vec{q}}\right) \tag{6.21a}$$

$$S_{\vec{p}\vec{q}} = \frac{\hat{\alpha}(\vec{p}o;\ \vec{q}o)}{\alpha_o(\vec{p}\vec{q})} = \frac{\alpha(\vec{p}o;\ \vec{q}o)}{\alpha_o(\vec{p}\vec{q})} \frac{1}{\varepsilon(\vec{p}o)\ \varepsilon(\vec{q}o)\ \varepsilon(\vec{k}o)} \tag{6.21b}$$

Note that $S_{\vec{p}\vec{q}}$ is completely symmetric in its three basic wave vector arguments $\vec{p}$, $\vec{q}$, $-\vec{k}$; so is $h_{\vec{p}\vec{q}}$; therefore, it follows that $\alpha(\vec{p}o;\ \vec{q}o)$ apart from the obvious $\vec{p}\leftrightarrow\vec{q}$ symmetry possesses also the higher triangle symmetry $\vec{p}\leftrightarrow\vec{q}\leftrightarrow-\vec{k}$.

We see that while the static linear FDT allows one to determine the pair correlation function from the knowledge of the response function, the static quadratic FDT provides information on the triplet correlation function (which is not an easily calculable object otherwise). One might easily convince oneself that this escalation of correlations is a general feature of the chain of nonlinear FDT-s. For example, the static cubic FDT should be expressible in terms of the cubic structure function $S_{\vec{p}\vec{q}\vec{t}}$, which is related to the Fourier transforms of the correlation functions g(12), h(123), i(1234) by

$$\begin{aligned} S_{\vec{p}\vec{q}\vec{t}} = 1 &+ n(g_{\vec{p}} + g_{\vec{q}} + g_{\vec{t}} + g_{\vec{p}+\vec{q}} + g_{\vec{q}+\vec{t}} \\ &+ g_{\vec{t}+\vec{p}} + g_{\vec{p}+\vec{q}+\vec{t}}) + n^2(h_{\vec{p}\vec{q}} + h_{\vec{q}\vec{t}} \\ &+ h_{\vec{t}\vec{p}} + h_{\vec{p},\vec{q}+\vec{t}} + h_{\vec{q},\vec{t}+\vec{p}} + h_{\vec{t},\vec{p}+\vec{q}}) \\ &+ n^3\ i_{\vec{p}\vec{q}\vec{t}} \end{aligned} \tag{6.22}$$

A further structural feature of the chain of nonlinear FDT-s seems to be that to <u>lowest order</u> in the coupling they are of the form

$$\begin{aligned} S_{\vec{k}} &= \hat{a}_o(\vec{k}) \\ S_{\vec{p}\vec{q}} &= \hat{a}_o(\vec{p},\vec{q}) \\ S_{\vec{p}\vec{q}t} &= \hat{a}_o(\vec{p},\vec{q},\vec{t}) \end{aligned} \tag{6.23}$$

with

$$\hat{a}_o \equiv \hat{\alpha}_o/\alpha_o \equiv \hat{\chi}_o/\chi_o$$

for a response function of any rank. Equation (6.23) together with Eq. (6.22) allows one to determine the lowest order triplet, quadruplet, etc. correlation functions (which are of order γ^2, γ^3, etc.) virtually by inspection. Invoking Eq. (2.8b) one finds the well-known O'Neil-Rostoker expansion for the triplet correlation function [O'Neil and Rostoker, 1965], while making use of Eq. (2.8c) leads to a cluster expansion for the quadruplet correlation function [Yatom, 1977; Shima, Yatom, Golden and Kalman, to be published].

For multicomponent systems the generalization of the static quadratic FDT is fairly straightforward. First we define the partial static structure function

$$S_{ABC,\vec{p}\vec{q}} = \frac{1}{(N_A\ N_B\ N_C)^{1/3}} \langle n_{A,\vec{p}}\ n_{B,\vec{q}}\ n_{C,-\vec{k}} \rangle \tag{6.24}$$

in terms of which the multispecies equivalent of Eq. (6.21a) reads [Golden and Kalman, 1976],

$$\begin{aligned} \chi_C^{AB}(\vec{p}o;\ \vec{q}o) = \beta^2\ (n_A\ n_B\ n_C)^{1/3} \Bigg\{ & \delta_{AB}\ \delta_{BC} + \left(\frac{n_A^2\ n_B^2}{n_C}\right)^{1/3} g_{AB,\vec{p}}\ \delta_{BC} \\ & + \left(\frac{n_B^2\ n_C^2}{n_A}\right)^{1/3} g_{BC,\vec{q}}\ \delta_{CA} + \left(\frac{n_C^2\ n_A^2}{n_B}\right)^{1/3} g_{CA,\vec{k}}\ \delta_{AB} \\ & + (n_A\ n_B\ n_C)^{2/3}\ h_{CAB,\vec{p}\vec{q}} \Bigg\} \end{aligned} \tag{6.25}$$

The second generalization of the conventional linear FDT concerns itself with averages of two-point functions over the perturbed rather than over the equilibrium ensemble. We can consider for example the equal-time non-equilibrium two-point function $\langle n_{\vec{k}-\vec{q}}\ n_{\vec{q}} \rangle^{(1)}(t)$. In analogy with the relation

$$\langle n_{\vec{k}} \rangle^{(1)}(\omega) = \hat{\chi}(\vec{k}\omega)\ \hat{V}(\vec{k}\omega) \tag{6.26}$$

which defines χ, we may introduce the response function, say, of "second kind" $K(\vec{q}\ ,\vec{k}\omega)$ by

$$\langle n_{\vec{k}-\vec{p}}\ n_{\vec{p}} \rangle^{(1)}(\omega) = \hat{K}(\vec{p},\vec{k}\omega)\ \hat{V}(\vec{k}\omega) \tag{6.27}$$

Using Eq. (5.10), and manipulations similar to what led to Eq. (5.13), to evaluate the left-hand-side, we find (for details see Appendix A of Professor Golden's lectures [Golden, this Volume])

$$\hat{K}(\vec{p};\ \vec{k}\omega) = -\beta n \left\{ 2\pi\ i\ \omega \int d\mu \int d\nu\ \delta_{+}(\omega - \mu - \nu) \cdot S(\vec{p}\mu;\ \vec{k}-\vec{p}\nu) + S_{\vec{p},\vec{k}-\vec{p}} \right\} \tag{6.28}$$

The quadratic FDT (6.17) now allows one to express S in terms of quadratic response functions and thus to relate the non-equilibrium flucuation spectrum to the relatively easily obtainable conventional response functions [Golden and Kalman, to be published]. In particular, in the static situation

$$\hat{K}(\vec{p},\vec{k}) = -\beta n\ S_{\vec{p},\vec{k}-\vec{p}} \tag{6.29}$$

which, via Eqs. (6.20) and (6.27), yields an expression for the perturbed two-point function in terms of equilibrium three-point function,

$$\langle n_{\vec{k}-\vec{p}},\ n_{\vec{p}}\rangle^{(1)} = -\frac{\beta}{V} \langle n_{\vec{k}-\vec{p}}\ n_{\vec{p}}\ n_{-\vec{k}}\rangle^{(o)}\ \hat{V}_{\vec{k}} \tag{6.30}$$

The significance of these relationships (Eqs. (6.27) through (6.30)) in building up a self-consistent approximation scheme will be discussed in the next Chapter and in Professor Golden's lectures [Golden, this Volume].

VII. APPROXIMATION SCHEMES

In this Chapter we compare, from the formal point of view, three leading approximation schemes for strongly coupled plasmas. All of them share the philosophy that they rely on the FDT-s and on the concept of response function to generate self-consistent approximations. Two of the schemes, the one due to Singwi, Tosi, Land and Sjölander [1968] [see also Singwi, Sjölander, Tosi and Land, 1969, 1970] and the one originated by the present author and Golden and Silevitch [Golden, Kalman and Silevitch, 1974; Kalman, Datta and Golden, 1975; Golden and Kalman, 1976] attempt to calculate the dielectric response function in terms of correlation functions and then employ FDT-s to render the relations self-consistent. The way the dielectric response function is expressed depends on the approximation used and this is the point where the STLS and GKS schemes deviate from each other: the former radically truncates the two-particle correlation function while the latter relegates the truncation to higher correlations. The difference manifests itself also in the order of the FDT that one has to evoke.

The STLS scheme relies on the linear FDT solely, while the GKS scheme evokes also the quadratic FDT. The third approximation scheme due to Totsuji and Ichimaru [1973, 1974] (see also Professor Ichimaru's lectures [Ichimaru, this Volume]) although departs from seemingly very different grounds will be shown to amount to an approximation very similar to that of STLS. We will consider only the ocp versions of all the theories, although extension of each of them to two-component systems is possible [Golden and Kalman, 1976].

First we introduce the concept of effective static potential $\psi_{\vec{k}}$ and the static screening function $u_{\vec{k}}$

$$\psi_{\vec{k}} = (1 + u_{\vec{k}})\ \phi_{\vec{k}} \tag{7.1}$$

and assume that the effect of correlations can be accounted for by using this - so far unknown - effective potential to describe the interaction between the particles. Next, using this assumption, we calculate the dielectric function of the system. Since all the correlational effects are assumed to be included in $\psi_{\vec{k}}$, the use of the modified linearized Vlasov equation

$$-i(\omega - \vec{k} \cdot \vec{v})\ F^{(1)}(\vec{k}\omega;\ \vec{v}) - \frac{i}{m} \frac{\partial F^{(o)}(v)}{\partial \vec{v}} \cdot \vec{k}\psi_{\vec{k}}\ n^{(1)}(\vec{k}\omega)$$

$$+ \frac{e}{m}\ \hat{E}(\vec{k}\omega) \cdot \frac{\partial F^{(o)}(v)}{\partial \vec{v}} = 0$$

$$n^{(1)}(\vec{k}\omega) = \int d\vec{v}\ F^{(1)}(\vec{k}\omega;\ \vec{v}) \tag{7.2}$$

is appropriate. From the above expression the polarizability can be calculated in the standard way, with the result

$$\alpha(\vec{k}\omega) = \frac{\alpha_o(\vec{k}\omega)}{1 + u_{\vec{k}}\ \alpha_o(\vec{k}\omega)} \tag{7.3}$$

where $\alpha_o(\vec{k}\omega)$ is the dynamical Vlasov polarizability. It is convenient to rewrite Eq. (7.3) as

$$\alpha(\vec{k}\omega) = \alpha_o(\vec{k}\omega)\ \{1 + v(\vec{k}\omega)\} \tag{7.4}$$

$$v(\vec{k}\omega) = -\frac{u_{\vec{k}}\ \alpha_o(\vec{k}\omega)}{1 + u_{\vec{k}}\ \alpha_o(\vec{k}\omega)} \tag{7.5}$$

or

$$\hat{\alpha}(\vec{k}\omega) = \alpha_o(\vec{k}\omega)\ \{1 + \hat{v}(\vec{k}\omega)\} \tag{7.6}$$

$$\hat{v}(\vec{k}\omega) = \frac{(1 + u_{\vec{k}})\ \alpha_o(\vec{k}\omega)}{1 + (1 + u_{\vec{k}})\ \alpha_o(\vec{k}\omega)} \tag{7.7}$$

which define $v(\vec{k}\omega)$ and $\hat{v}(\vec{k}\omega)$, the coupling factors. In the static limit one can employ the linear FDT to Eq. (7.2) in order to express $u_{\vec{k}}$ or $\psi_{\vec{k}}$ in terms of the pair correlation function. One finds

$$\psi_{\vec{k}} = -\frac{c_{\vec{k}}n}{\beta} \tag{7.8}$$

$$c_{\vec{k}} = \frac{g_{\vec{k}}}{1 + ng_{\vec{k}}} \tag{7.9}$$

Inspecting Eq. (7.9) we recognize that $c_{\vec{k}}$ is the Ornstein-Zernike direct correlation function. The fact that $-n/\beta$ times the direct correlation function can serve as an effective interaction has been recognized, through different arguments, by many people [Nelkin and Ranagathan, 1967; Lebowitz, Percus and Sykes, 1969].

Now we can argue that if we have an independent method for the determination of $\psi_{\vec{k}}$ as a functional of the correlations, Eq. (7.8) will serve as a self-consisting criterion.

A. STLS Scheme

We turn now to the STLS method. In order to assess $\psi_{\vec{k}}$, we write down the full first BBGKY equation in the presence of an external perturbation

$$\left(\frac{\partial}{\partial t} + \vec{v}_1 \cdot \frac{\partial}{\partial \vec{x}_1}\right) F(1) - \frac{e}{m}\,\hat{E}(1) \cdot \frac{\partial}{\partial \vec{v}_1} F(1) = \frac{1}{m}\frac{\partial}{\partial \vec{v}_1} \cdot \int d2\ K(12)\ G(12) \tag{7.10}$$

where

$$\vec{K}(12) = -i\,\frac{1}{V}\sum_{\vec{k}} \vec{k}\phi + \psi_{\vec{k}}\ e^{-i\vec{k}\cdot(\vec{x}_1 - \vec{x}_2)} \tag{7.11}$$

and G(12) is the usual two-particle distribution function. Linearization leads to the following decomposition:

$$F(1) = F^{(o)}(v_1) + F^{(1)}(1), \text{ etc.} \tag{7.12}$$

$$G^{(o)}(12) = F^{(o)}(v_1)\, F^{(o)}(v_2)\, \{1 + g(12)\} \tag{7.13}$$

$$G^{(1)}(12) = \{F^{(o)}(v_1)\, F^{(1)}(2) + F^{(o)}(v_2)\, F^{(1)}(2)\} \cdot \{1 + g(12)\} + G^{(1)}_{irr}(12) \tag{7.14}$$

$G^{(1)}_{irr}$, the irreducible part of the perturbed correlations, is not an easily accessible quantity. The STLS approximation consists of entirely ignoring this term. Thus with

$$G^{(1)}_{irr}(12) \simeq 0 \tag{7.15}$$

one can rewrite Eq. (7.10) as

$$-i(\omega - \vec{k} \cdot \vec{v})\, F^{(1)}(\vec{k}\omega; \vec{v}) - \frac{i}{m} \frac{\partial F^{(o)}(v)}{\partial \vec{v}} \cdot \left\{ \vec{k}\, \phi_{\vec{k}} + \frac{1}{V} \sum_{\vec{q}} \vec{q}\, \psi_{\vec{q}}\, g_{\vec{k}-\vec{q}} \right\} n^{(1)}(\vec{k}\omega) - \frac{e}{m} \hat{E}(\vec{k}\omega) \cdot \frac{\partial F^{(o)}(v)}{\partial \vec{v}} = 0 \tag{7.16}$$

Comparison with Eq. (7.2) yields

$$u_{\vec{k}} = \frac{1}{V} \sum_{\vec{q}} \frac{\vec{k} \cdot \vec{q}}{q^2}\, g_{\vec{k}-\vec{q}} \tag{7.17}$$

which jointly with Eq. (7.9) provide an integral equation for the unknown $g_{\vec{k}}$:

$$n c_{\vec{k}} = - \frac{\kappa^2}{k^2} (1 + u_{\vec{k}}) \tag{7.18}$$

or

$$n g_{\vec{k}} = - \frac{1 + u_{\vec{k}}}{1 + u_{\vec{k}} + k^2/\kappa^2} \tag{7.19}$$

The combined Eqs. (7.17)-(7.19) can, in principle, be solved by numerical iteration. Then $u_{\vec{k}}$ can be determined, and substitution into Eq. (7.3) yields $\alpha(\vec{k}\omega)$.

The STLS method was historically the first to introduce the ingenious idea of generating a self-consistent scheme by linking $\varepsilon(\vec{k}\omega)$ to $g_{\vec{k}}$ both by a kinetic equation and by the FDT. However, it has also been recognized that it has some serious problems. First, it violates both the static compressibility sum rule and the high frequency $1/\omega^4$ sum rule. This is easy to see by expanding Eq. (7.17) for $k \to o$ which yields

$$u_{\vec{k}} = - Z \frac{k^2}{\kappa^2} \tag{7.20}$$

where Z is determined by the integral

$$Z = - \frac{\kappa^2}{6\pi^2} \int_o^\infty dq \; g_{\vec{q}} \tag{7.21}$$

Then one finds from Eq. (7.3) in the $\omega = o$ and $\omega \to \infty$ limits

$$\alpha(\vec{k}o) \simeq \frac{1}{1 - Z} \frac{\kappa^2}{k^2}$$

$$\alpha(\vec{k}\omega\to\infty) \simeq - \left\{ \frac{\omega_o^2}{\omega^2} + \frac{\omega_o^4}{\omega^4} (3 - Z) \, k^2 \right\} \tag{7.22}$$

These two relations are unable to simultaneously satisfy the sum rules (4.7) and (4.32). In particular, for $\gamma \ll 1$, Eq. (4.7) would require $Z = (1/4)\gamma$, while Eq. (4.32) $Z = (2/15)\gamma$; actually it turns out to be $(1/3)\gamma$. The incorrect $\omega = o$ behavior can be remedied by an *ad hoc* modification of the theory, proposed by Vashishta and Singwi [1972]. However, the wrong high-frequency behavior indicates a more serious problem: the theory implies that dynamical properties can be extrapolated from the static behavior; as it is obvious from Eq. (7.3), the only genuine dynamical contribution comes from $\alpha_o(\vec{k}\omega)$ which contains no long time correlational or collisional contribution. The origin of this defect must be sought in the neglect of dynamical correlations contained in G_{irr}.

A further problem arises in connection with the basic integral equation (7.17)-(7.19) and its possible solution. It is found [Bakshi, this Volume; Bakshi, Kalman and Silevitch, to be published] that the solution g(r) for small r has a pathological behavior. This fact has some further consequences from the point of view of of the mathematical structure and mathematical consistency of the equation. These points are further discussed in Professor Bakshi's seminar [Bakshi, this Volume].

B. TI Scheme

Now we turn to the second approach, that of Totsuji and Ichimaru [1973, 1974], [Ichimaru, 1970]. As we have stated, the point of departure is rather different. We consider the second BBGKY equation for the equilibrium system. It can be cast in the form

$$ng_{\vec{k}} = -\frac{\kappa^2}{\kappa^2 + k^2}\left\{1 + \frac{1}{V}\sum_{\vec{q}}\frac{\vec{k}\cdot\vec{q}}{q^2}\, g_{\vec{k}-\vec{q}} + \frac{n}{V}\sum\frac{\vec{k}\cdot\vec{q}}{q^2}\, h_{\vec{q},\vec{k}-\vec{q}}\right\} \tag{7.23}$$

Following the time honored notions of kinetic theory, one adopts a decomposition approximation in terms of pair distribution functions to h. (In a way similar to the old Kirkwood superposition approximation). The paradigm, however, is chosen to conform to the long-range character of the coulomb forces, which would suggest that good agreement for small k (and not for small r) is essential. The chosen structure is the O'Neil-Rostoker solution [O'Neil and Rostoker, 1965; Lie and Ichikawa, 1966] which is exact for small γ and small k:

$$h_{\vec{p}\vec{q}} = g_{\vec{p}}\, g_{\vec{q}} + g_{\vec{q}}\, g_{\vec{k}} + g_{\vec{k}}\, g_{\vec{p}} + n g_{\vec{p}}\, g_{\vec{q}}\, g_{\vec{k}}$$

$$\vec{k} = \vec{p} + \vec{q} \tag{7.24}$$

In configuration space this corresponds to the cluster decomposition

$$h(123) = g(12)\, g(23) + g(23)\, g(31) + g(31)\, g(12) + \int d4\, g(14)\, g(24)\, g(34) \tag{7.25}$$

Substituting Eq. (7.24) into Eq. (7.23), one immediately finds the solution in a form identical to Eq. (7.18)

$$nc_{\vec{k}} = -\frac{\kappa^2}{k^2}(1 + u_{\vec{k}}) \tag{7.26a}$$

but with $u_{\vec{k}}$ being given by

$$u_{\vec{k}}^{TI} = \frac{1}{V}\sum_{\vec{q}} \frac{\vec{k}\cdot\vec{q}}{q^2} g_{\vec{k}-\vec{q}}(1 + ng_{\vec{k}}) \quad , \tag{7.26b}$$

We see that this is quite similar to the STLS model, except for the appearance of the additional screening factor $1 + ng_{\vec{k}}$ ($\sim 1/\varepsilon(\vec{k}o)$). This raises the question whether the STLS result couldn't be derived in a similar fashion, from the equilibrium BBGKY. This is indeed possible, as noted by Yatom and Shima [1978]. If instead of Eq. (7.24) one chooses

$$h_{\vec{p}\vec{q}} = g_{\vec{p}}\, g_{\vec{q}} + g_{\vec{q}}\, g_{\vec{k}} + g_{\vec{k}}\, g_{\vec{p}} \tag{7.27}$$

one recovers $u_{\vec{k}}^{STLS}$.

Evidently, since the TI scheme is built on static, equilibrium concepts, it doesn't lend itself directly to the construction of dynamical response functions. However, one can adopt the philosophy that once $c_{\vec{k}}$ is determined, $-nc_{\vec{k}}/\beta$ can be used as the effective potential, as before, with a result formally identical to Eq. (7.3),

$$\alpha(\vec{k}\omega) = \frac{\alpha_o(\vec{k}\omega)}{1 + u_{\vec{k}}^{TI}(\vec{k}\omega)} \tag{7.28}$$

For static properties, the TI scheme is expected to be superior to the STLS scheme, and indeed it is. It exactly satisfies the compressibility sum rule for $\gamma \ll 1$. For higher γ-s it also seems to give a good agreement between the compressibility calculated from the sum rule and directly from the equation of state. For the description of the dynamical properties of the system the TI model would be plagued with the same problems as the STLS one, since it is void of any genuine dynamical correlations as well. As to the mathematical structure of the integral equation (7.26), the problems found previously probably also prevail here [Bakshi, this Volume; Bakshi, Kalman and Silevitch, to be published].

C. GKS Scheme

The GKS scheme was originated in order to improve the dynamical results of the previous schemes by incorporating the genuine dynamical correlations, while preserving the powerful idea of the FDT generated self-consistency [Kalman, 1975; Golden, Kalman and Silevitch, 1974]. In order to do this the so far ignored G_{irr} has to be salvaged. This can be accomplished by building the theory on the quadratic, rather than the linear FDT. We give here only a brief sketch of the principal structure of the scheme, since its details are given in Professor Golden's lectures [Golden, this Volume].

In order to convert G_{irr} into a tractable object, the velocity average approximation (VAA) is introduced, which transforms the nonequilibrium two-particle distribution function G(12) as follows:

$$G(\vec{x}_1,\vec{v}_1;\ \vec{x}_2,\vec{v}_2) \rightarrow \frac{1}{2}\left\{\frac{F(\vec{x}_1,\vec{v}_1)}{n(\vec{x}_1)}\int d\vec{v}_1'\ G(\vec{x}_1,\vec{v}_1';\ \vec{x}_2,\vec{v}_2) + \frac{F(\vec{x}_2,\vec{v}_2)}{n(\vec{x}_2)}\int d\vec{v}_2'\ G(\vec{x}_1,\vec{v}_1;\ \vec{x}_2,\vec{v}_2')\right\} \tag{7.29}$$

The above expression substituted in the linearized perturbed first BBGKY equation results in

$$-i(\omega - \vec{k}\cdot\vec{v})\ F^{(1)}(\vec{k}\omega;\ \vec{v}) - \frac{i}{m}\frac{\partial F^{(o)}(v)}{\partial\vec{v}} \cdot \frac{1}{N}\sum_{\vec{q}} \vec{q}\ \psi_{\vec{q}}\ \langle n_{\vec{k}-\vec{q}}\ n_{\vec{q}}\rangle^{(1)}(\omega) - \frac{e}{m}\hat{E}(\vec{k}\omega)\cdot\frac{\partial F^{(o)}(v)}{\partial\vec{v}} = 0 \tag{7.30}$$

This equation should be compared with Eq. (7.16) in order to appreciate the difference in the screening structure.

Using Eq. (6.27), the expression for the polarizability can be written down as

$$\hat{\alpha}(\vec{k}\omega) = \alpha_o(\vec{k}\omega)\ \{1 + \hat{v}(\vec{k}\omega)\}$$

$$\hat{v}(\vec{k}\omega) = -\frac{1}{N}\sum_{\vec{p}} \frac{\vec{k}\cdot\vec{p}}{p^2}\,\phi_{\vec{k}}\,\hat{K}(\vec{p};\,\vec{k}\omega) \tag{7.31}$$

where $\hat{K}$, the "response function of the second kind" is defined by Eq. (6.27) and given by Eq. (6.28); Eq. (7.31) should be compared with Eq. (7.7).

Further progress can be made by using the quadratic FDT (6.17) to relate $\hat{K}$ to the quadratic response functions. Rather lengthy algebra leads to the result [Golden, this Volume; Golden and Kalman, to be published]

$$\hat{v}(\vec{k}\omega) = \frac{2}{\beta N}\,\phi_{\vec{k}} \sum \frac{\vec{k}\cdot\vec{p}}{p^2}\int d\mu\ \delta_-(\mu)$$
$$\cdot\ [\hat{\chi}(\vec{p}\mu;\ \vec{k}-\vec{p},\ \omega-\mu) + \hat{\chi}(\vec{p}\omega-\mu;\ \vec{k}-\vec{p}\mu)] \tag{7.32}$$

In the static limit the above expression becomes

$$v_{\vec{k}} = \frac{2}{\beta N}\,\phi_{\vec{k}} \sum_{\vec{p}} \frac{\vec{k}\cdot\vec{p}}{p^2}\,\frac{\chi(\vec{p}o;\ \vec{k}-\vec{p}o)}{\varepsilon(\vec{p}o)\ \varepsilon(\vec{k}-\vec{p}o)} \tag{7.33}$$

Equation (7.33) is equivalent to the exact second BBGKY equation (7.23) [Kalman, Datta and Golden, 1975]. Thus on the static level the VAA is exact.

The scheme as it stands, is not self-consistent; a further approximation is called for in order to express the quadratic response functions in terms of linear response functions. This can be done a number of different ways. For example, the static TI approximation is recovered if

$$\chi(\vec{p}o;\ \vec{q}o) \sim \chi(\vec{p}o)\ \chi(\vec{q}o)\ \chi(\vec{k}o) \tag{7.34}$$

is set. The STLS approximation doesn't have a unique equivalent: if the quadratic response function is calculated, following the philosophy stipulated for the linear one, namely by ignoring the irreducible part of $G^{(2)}$, one obtains a decomposition similar to Eq. (7.34); this, however, is not compatible with Eq. (7.27).

Even a rather crude decomposition of the dynamical expression (7.32) reproduces most of the long-time collisional and correlational dynamical effects in a plasma [Golden and Kalman, to be published]. Thus there is no doubt that as a dynamical theory, the GKS scheme is much superior to its predecessors.

As to testing the model through sum rules, one finds that all the known relevant sum rules, the linear <u>and</u> quadratic compressibility sum rules (4.7) and (4.8), and the $1/\omega^4$ sum rule (4.28) are satisfied; in the small γ limit the correct numerical coefficients, i.e., $(1/4)\gamma$, $(5/8)\gamma$ and $(2/15)\gamma$ emerge, respectively [Golden and Kalman, to be published; Golden, this Volume].

REFERENCES

Bakshi, P., this Volume.

Bakshi, P., G. Kalman and M. B. Silevitch, to be published.

deGennes, P. G., 1959, Physica 25, p. 825.

DeWitt, H. E., this Volume.

Forster, D., P. C. Martin and S. Yip, 1968, Phys. Rev. 170, 155.

Golden, K. I., this Volume.

Golden, K. I. and G. Kalman, 1969, J. Stat. Phys. 1, 415.

Golden, K. I. and G. Kalman, 1975, Phys. Rev. A11, 2147; 1976, Phys. Rev. A14, 1802; 1978, Phys. Rev. A17, 390.

Golden, K. I. and G. Kalman, to be published.

Golden, K. I., G. Kalman and T. Datta, 1975, Phys. Rev. A11, 2147.

Golden, K. I., G. Kalman and M. B. Silevitch, 1972, J. Stat. Phys. 6, 87.

Golden, K. I., G. Kalman and M. B. Silevitch, 1974, Phys. Rev. Lett. 33, 1544.

Gross, E. P., this Volume.

Ichimaru, S., 1970, Phys. Rev. A2, 494.

Ichimaru, S. and T. Tange, 1974, Phys. Rev. Lett. 32, 102.

Ichimaru, S., H. Totsuji, T. Tange and D. Pines, 1975, Progr. Theor. Phys. 54, 1077.

Ichimaru, S., this Volume.

Kalman, G., 1975, in Plasma Physics - Les Houches, 1972, edited by C. DeWitt, Gordon and Breach, New York.

Kalman, G., T. Datta and K. I. Golden, 1975, Phys. Rev. A12, 1125.

Kubo, R., 1957, J. Phys. Soc. Japan 12, 570.

Kubo, R., 1966, Rep. Progr. Phys. 29, 263.

Lebowitz, J. L., J. K. Percus and J. Sykes, 1969, Phys. Rev. 188, 1487.

Lie, T. J. and Y. H. Ichikawa, 1966, Rev. Mod. Phys. 38, 680.

March, N. H. and M. P. Tosi, 1976, Atomic Dynamics in Liquids, John Wiley & Sons, New York.

Martin, P. C., 1968, in Many Body Physics - Les Houches 1967, edited by C. DeWitt, Gordon and Breach, New York.

Nelkin, M. and S. Ranganathan, 1967, Phys. Rev. 164, 222.

O'Neil, T. and N. Rostoker, 1965, Phys. Fluids 8, 1109.

Pathak, K. N. and P. Vashishta, 1973, Phys. Rev. B7, 3649.

Placzek, G., 1952, Phys. Rev. 86, 377.

Pines, D. and Ph. Nozières, 1966, Theory of Quantum Fluids, Benjamin, Inc., New York.

Shima, Y., H. Yatom, K. I. Golden and G. Kalman, to be published.

Singwi, K. S., A. Sjölander, M. P. Tosi and R. H. Land, 1969, Solid State Commun. 7, 1503.

Singwi, K. S., A. Sjölander, M. P. Tosi and R. H. Land, 1970, Phys. Rev. B1, 1044.

Singwi, K. S., M. P. Tosi, R. H. Land and A. Sjölander, 1968, Phys. Rev. 176, 589.

Sjölander, A., this Volume.

Stillinger, F. H. and R. Lovett, 1968, Chem. Phys. 49, 1991.

Tosi, M. P., M. Parinello and N. H. March, 1974, Nuovo Cimento B23, 135.

Totsuji, H. and S. Ichimaru, 1973, Progr. Theor. Phys. 50, 753; 1974, Progr. Theor. Phys. 52, 42.

Vashishta, P., P. Battacharyya and K. S. Singwi, 1974, Phys. Rev. B10, 5102.

Vashishta, P., P. Battacharyya and K. S. Singwi, 1974, Nuovo Cimento B23, 172.

Vashishta, P. and K. S. Singwi, 1972, Phys. Rev. B6, 875.

Yatom, H., Seminar given at this Institute.

Yatom, H. and Y. Shima, 1978, Phys. Rev. A, to appear.

THEORETICAL APPROACHES TO STRONGLY COUPLED PLASMAS*

Setsuo Ichimaru

Department of Physics
University of Tokyo
Bunkyo-ku, Tokyo 113, Japan

*Work supported by the Japan Society for the Promotion of Science and the National Science Foundation under the Japan-United States Cooperative Science Program.

TABLE OF CONTENTS

THEORETICAL APPROACHES TO STRONGLY COUPLED PLASMAS

Setsuo Ichimaru

Department of Physics
University of Tokyo
Bunkyo-ku, Tokyo 113, Japan

I. INTRODUCTION

In this series of lectures, I should like to review some of the theoretical contributions made by our research groups at the University of Tokyo and at the University of Illinois during the past several years, toward understanding of the static and dynamic properties of various versions of the strongly coupled plasmas. The work which I shall describe consists of the results of collaboration with my colleagues: N. Itoh, T. Nakano, D. Pines, T. Tange and H. Totsuji.

Let us begin with a definition of the strongly coupled plasmas. We consider a one-component plasma (OCP) with a uniform neutralizing background, obeying classical statistics; it may be characterized by electric charge e and mass m of a particle, number density n, and temperature T (in energy units). The dimensionless plasma parameter describing discreteness of the system [Rostoker and Rosenbluth, 1960; Ichimaru, 1973] is

$$\varepsilon = (4\pi n\lambda_D{}^3)^{-1} = (4\pi n)^{1/2}\, e^3 T^{-3/2} \quad , \tag{1.1}$$

where $\lambda_D = (T/4\pi ne^2)^{1/2}$ is the Debye length.

The correlation energy density E_c, defined as the statistical average of the interaction Hamiltonian per unit volume, may be calculated in terms of the static form factor $S(\vec{k})$ or the pair correlation function $g(\vec{r})$ as [e.g., Ichimaru, 1973]

$$\frac{E_c}{nT} = \frac{2\pi e^2}{T} \sum_{\vec{k}} \frac{1}{k^2} [S(\vec{k}) - 1] = \frac{ne^2}{2T} \int d\vec{r} \frac{1}{r} g(\vec{r}) \quad . \tag{1.2}$$

The static form factor is related to the spectral function of the density fluctuations in the wave-vector space. For a relatively dilute plasma in thermodynamic equilibrium such that $\varepsilon \lesssim 1$, it is known that Eq. (1.2) takes on a value of the order of the plasma parameter ε, i.e.,

$$\frac{|E_c|}{nT} \simeq \varepsilon \qquad \text{(in thermodynamic equilibrium)} \quad . \tag{1.3}$$

For $n = 10^{13} cm^{-3}$ and $T = 10^6 K$, one finds $\varepsilon \simeq 8 \times 10^{-7}$ so that the ratio (1.3) takes on an extremely small number for an ordinary gaseous plasma.

We define a strongly coupled plasma as that which satisfies

$$\frac{|E_c|}{nT} \simeq 1 \quad . \tag{1.4}$$

For a plasma in thermodynamic equilibrium, the condition (1.4) can be satisfied only when the density is increased to an extent such that $\varepsilon \simeq 1$; such a system may be regarded as a high-density classical plasma or an electron liquid. For a dilute plasma with $\varepsilon << 1$, the condition (1.4) can still be satisfied in a nonequilibrium state when strong plasma turbulence is excited through onset of plasma-wave instabilities; in this case the denominator in Eq. (1.4) is to be replaced by an average density of the kinetic energy.

We thus take the point of view that the problems involved in the theory of plasma turbulence are quite analogous to those in the theory of electron liquids; both systems are characterized by the condition (1.4). In a strongly turbulent plasma the pair correlation function and higher-order correlation functions take on magnitudes of the zeroth order in the discreteness parameters such as ε; those remain finite even in the fluid limit, $\varepsilon \to 0$ [Rostoker and Rosenbluth, 1960], so that strong correlations persist with macroscopic intensities [Ichimaru, 1970a]. In an electron liquid, the pair correlation function, the triple correlation and so on scale as ε, ε^2 and so on at typical interparticle distances of the order of the Debye length; since $\varepsilon \simeq 1$, all of those correlation functions again play important parts in the description of system properties. In either of those cases we are faced with a strong-coupling problem in that we cannot regard fluctuations or correlations as meaningful expansion parameters. Such absence of a systematic expansion scheme may be an essential feature in any theory of strongly coupled plasmas.

II. STATIC PROPERTIES OF ELECTRON LIQUIDS

An important theoretical method for the analysis of the strongly coupled plasmas has been provided by a self-consistent approach. It has been applied to turbulent plasmas [Ichimaru and Nakano, 1967, 1968; Ichimaru, 1969] and to degenerate electron liquids at metallic densities [Hubbard, 1967; Singwi, Tosi, Land and Sjölander, 1968; Singwi, Sjölander, Tosi and Land, 1969; Vashishta and Singwi, 1972]. In this approach, one first calculates a linear response function (e.g., the dielectric response function) as a functional, not only of the single-particle distribution function, but also of the pair correlation function or the static and dynamic form factors; the latter quantities are unknown at this stage and to be determined later in a self-consistent way. One then establishes a relation between the response function and the pair correlation function via, e.g., the fluctuation-dissipation theorem; this equation is to be solved for the pair correlation function. Naturally, various approximations are involved in the calculation of the response function for a strongly coupled plasma.

Ichimaru [1970b] then noted that the second BBGKY equation [e.g., Bogoliubov, 1962; Rostoker and Rosenbluth, 1960; Ichimaru, 1973] should be equivalent in physical content to the self-consistent equation mentioned above, in the sense that both are equations to determine the pair-correlation function. Triple correlation functions are involved in the second BBGKY equation; a proper treatment of the triple correlation functions would enable one to truncate the BBGKY hierarchy at this stage.

An ansatz for the triple correlation function proposed by Ichimaru [1970b] is

$$h(r_{12},r_{23},r_{31}) = g(r_{12})\, g(r_{23}) + g(r_{23})\, g(r_{31}) + g(r_{31})\, g(r_{12}) + n \int d\vec{r}_4\, g(r_{14})\, g(r_{24})\, g(r_{34}) \quad , \qquad (2.1)$$

where

$$r_{ij} \equiv |\vec{r}_i - \vec{r}_j| \quad .$$

O'Neil and Rostoker [1965] have in fact shown that for a dilute plasma ($\varepsilon << 1$), the BBGKY long-range solution of the triple correlation function is expressed in the form (2.1) where the pair correlation function takes on the Debye-Hückel values,

$$g(r) = -\frac{e^2}{Tr} \exp\left(-\frac{r}{\lambda_D}\right) \quad . \qquad (2.2)$$

By adopting the ansatz (2.1) in the second BBGKY solution, one thus automatically guarantees accuracy of the resulting long-range solution of the pair correlation function to the first two terms in the ε expansion.

Substitution of Eq. (2.1) in the second BBGKY equation then yields an integral equation for the static form factor,

$$S(k) = \frac{k^2}{k^2 + k_D^2[1 + w(k)]} \quad , \tag{2.3}$$

$$w(k) = \frac{1}{n} \sum_{\vec{q}} \frac{\vec{k} \cdot \vec{q}}{q^2} S(q)[S(|\vec{k} - \vec{q}|) - 1] \quad , \tag{2.4}$$

where $k_D \equiv 1/\lambda_D$. Connection between the integral equation (2.3) and the foregoing self-consistent approach may be established by setting the (longitudinal) dielectric response function as

$$\varepsilon(k,\omega) = 1 + \frac{\chi_0(k,\omega)}{1 + w(k)\ \chi_0(k,\omega)} \quad , \tag{2.5}$$

where

$$\chi_0(k,\omega) = \frac{4\pi e^2 n}{mk^2} \int d\vec{v} \frac{1}{\omega - \vec{k} \cdot \vec{v}} \vec{k} \cdot \frac{\partial f(\vec{v})}{\partial \vec{v}} \tag{2.6}$$

is the free electron polarizability, and $f(\vec{v})$ is the velocity distribution function normalized to unity.

Totsuji and Ichimaru [1973, 1974] subsequently examined the accuracy and validity of this theory in the light of the following criteria:

(1) The theory should reproduce the correct analytic expressions of the ε expansion for the thermodynamic quantities (e.g., the correlation energy).

(2) It should satisfy the compressibility sum rule [e.g., Pines and Nozières, 1966].

(3) It should describe correct behavior in the short-range parts of the correlation functions, where strong correlations are involved even in a low-density system with small ε.

(4) The theoretical results should agree with the results of Monte Carlo experiments on electron liquids [Brush, Sahlin and Teller, 1966; Hansen, 1973; Pollock and Hansen, 1973].

The correlation energy E_c or the pressure P of a dilute plasma has been calculated by many investigators [e.g., Abe, 1959; Bowers and Salpeter, 1960; O'Neil and Rostoker, 1965]. Its first few terms in the ε expansion have been exactly determined:

$$\frac{E_c}{nT} = 3\left(\frac{P}{nT} - 1\right) = -\frac{\varepsilon}{2} - \varepsilon^2\left[\frac{1}{4}\ \ln 3\varepsilon + \frac{\gamma}{2} - \frac{1}{3}\right] \quad . \tag{2.7}$$

where $\gamma = 0.57721...$ is Euler's constant. It is then possible to show [Totsuji and Ichimaru, 1973] that the pair correlation function calculated from the solution of Eq. (2.3) has the correct long-range and short-range behaviors so that the resulting correlation energy reproduces Eq. (2.7) exactly. When $\varepsilon \simeq 1$, Eq. (2.3) may be solved numerically [Totsuji and Ichimaru, 1974]; the theoretical results exhibit excellent agreement with the Monte Carlo results as Figure 1 illustrates. Consequently, Eq. (2.4) satisfies the criteria (1) and (4) for the correlation energy.

The isothermal sound velocity c of the electron liquid may be calculated in two different ways: one based on the thermodynamic relation,

$$\frac{c^2}{T/m} = \frac{1}{T}\left(\frac{\partial P}{\partial n}\right)_T \quad , \tag{2.8}$$

and the other involving the long-wavelength behavior of the density-density correlation function [e.g., Pines and Nozières, 1966],

$$\frac{c^2}{T/m} = \lim_{k\to 0}\left(\frac{k_D}{k}\right)^4\left\{\left(\frac{k}{k_D}\right)^2 - S(k)\right\} \quad . \tag{2.9}$$

The compressibility sum rule requires that Eqs. (2.8) and (2.9) should agree on the correct value.

Equation (2.3) has been examined in terms of this compressibility sum rule. In the ε expansion, the isothermal sound velocity calculated according to Eq. (2.8) is

$$\frac{c^2}{T/m} = 1 - \frac{1}{4}\,\varepsilon - \varepsilon^2\left(\frac{1}{6}\ \ln 3\varepsilon + \frac{\gamma}{3} - \frac{13}{72}\right) \quad , \tag{2.10}$$

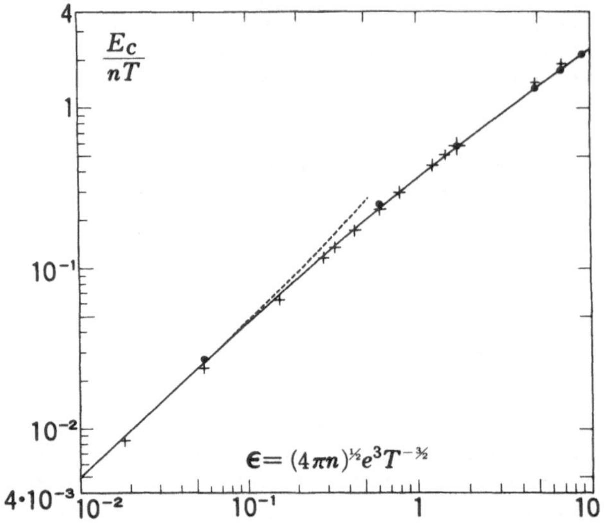

Figure 1. The correlation energy vs. the plasma parameter. The solid line represents the result obtained from the solution of Eq. (2.3); the dotted line, Eq. (2.7). The closed circles denote experimental values due to Brush, Sahlin and Teller [1966]; the crosses, theoretical ones due to Berggren [1970] based on the scheme of Singwi, Tosi, Land and Sjölander [1968].

while Eq. (2.9) gives

$$\frac{c^2}{T/m} = 1 - \frac{1}{4}\,\varepsilon - \varepsilon^2\left(\frac{1}{6}\,\ell n 3\varepsilon + \frac{\gamma}{3} - \frac{11}{54}\right) \quad . \tag{2.11}$$

Since Eq. (2.7) is the correct expression for the pressure, Eq. (2.10) represents the correct evaluation of the sound velocity. Equation (2.11) then shows a slight discrepancy from Eq. (2.10), starting with a term on the order of ε^2. A similar tendency has been noted in the numerical solutions of Eq. (2.3), as Figure 2 illustrates; the numerical solution for the sound velocity obtained from Eq. (2.8) (the solid line in Figure 2) agrees almost completely with that obtained from Eq. (2.9) (the dashed line) for $\varepsilon < 1$, but a notable discrepancy begins to appear for $\varepsilon \geq 1$. Room for improvement on the theory is thus indicated.

The pair correlation function calculated from numerical solution of Eq. (2.3) is shown in Figure 3. In the vicinity of $r = 0$, the

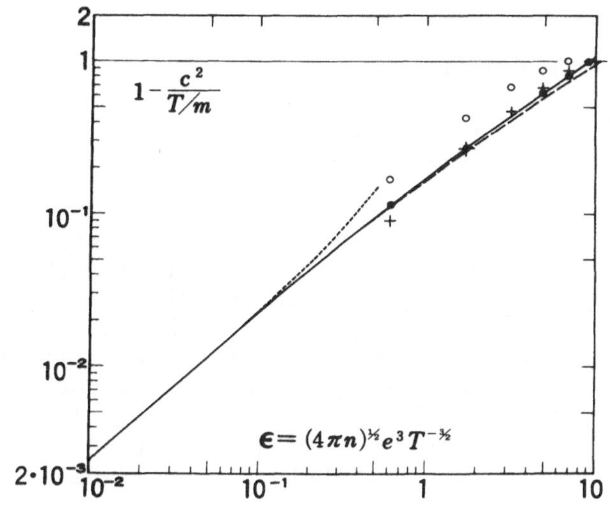

Figure 2. The isothermal sound velocity vs. the plasma parameter. The solid and dashed lines represent the results in the scheme (2.3), evaluated in accord with Eqs. (2.8) and (2.9), respectively. The dotted line describes the result, Eq. (2.10), of the plasma-parameter expansion analysis. The closed circles denote experimental values [Brush, Sahlin and Teller, 1966]. The crosses and open circles denote the theoretical values of Berggren [1970], obtained from Eqs. (2.8) and (2.9), respectively.

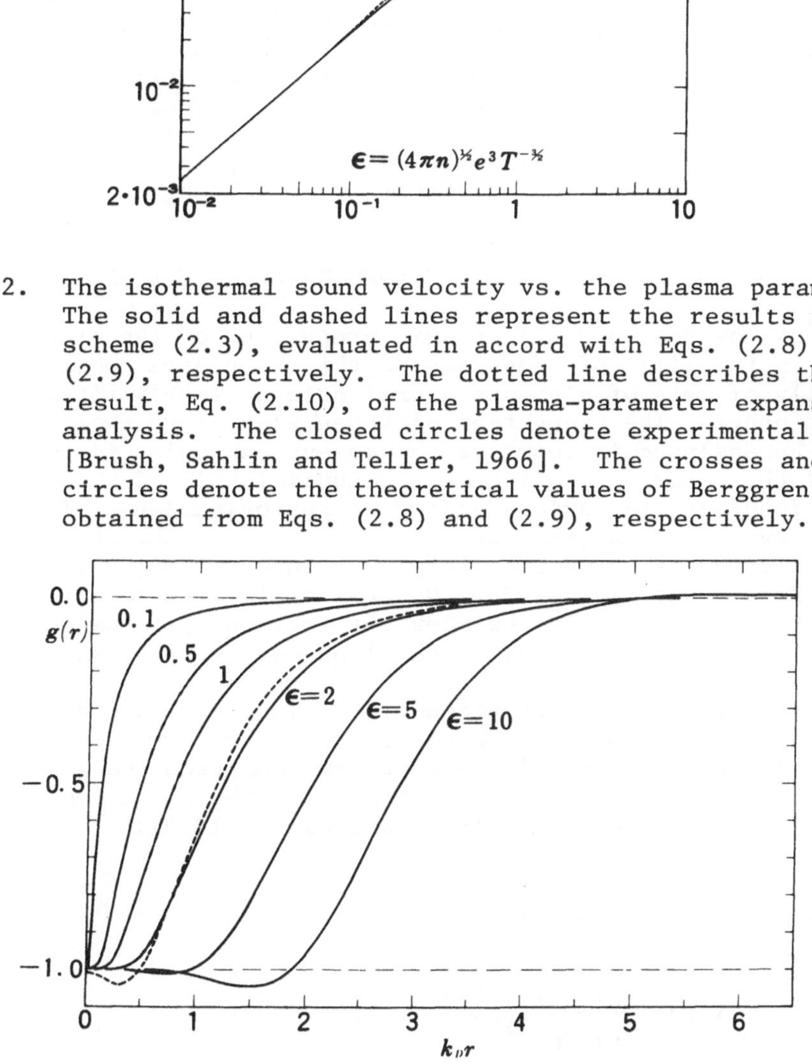

Figure 3. The radial correlation function vs. k_Dr. The dotted line shows a theoretical result of Berggren [1970] for $\varepsilon = \sqrt{3}$.

correlation function takes the form,

$$g(r) = -1 + \exp\,(-e^2/Tr) \quad , \tag{2.12}$$

and generally $g(r) \geq -1$ must be satisfied. For $\varepsilon \geq 5$, the latter condition is slightly violated, another indication of the possibility of theoretical improvement.

As we observe in Figure 2, the isothermal compressibility diverges at $\varepsilon \simeq 10$; a possibility of an onset of a thermodynamic instability is indicated. When the charge density of the neutralizing background is kept uniform, however, the space-charge effect acts to prevent the onset of such an instability. If, on the other hand, the background adjusts itself in such a way as to cancel the space-charge field produced by a fluctuation, the instability may take place [Totsuji and Ichimaru, 1974]. The possibility of this instability thus depends on the nature of the background.

III. SUM-RULE ANALYSIS OF THE LONG-WAVELENGTH EXCITATIONS

Let us now consider the dynamic properties of the electron liquid. In the limit of long wavelengths the plasma consists essentially of sets of elementary excitations, weakly interacting with each other. The properties of those elementary excitations can then be analyzed in terms of the moment sum rules in the frequency domain of the dynamic form factor, or equivalently of the dielectric response function, to various orders of approximations. The sum rules are evaluated exactly with the knowledge of the static correlation functions of the system [e.g., deGennes, 1959; Puff, 1965; Forster, Martin and Yip, 1968]. We thus employ the results of the investigation described in the previous section, for the examination of the elementary excitations in the long-wavelength limit.

To investigate the properties of the long-wavelength excitations in the electron liquids, we express the dynamic form factor in that domain as a superposition of the contributions from the plasma oscillations, the single-particle excitations and the collisional excitations [Ichimaru, Totsuji, Tange and Pines, 1975]:

$$S(k,\omega) = \frac{1}{2} X(k)[\delta(\omega - \omega_k) + \delta(\omega + \omega_k)]$$

$$+ Y(k)\, \frac{y[\omega/k(T/m)^{1/2}]}{k(T/m)^{1/2}} + Z(k)\, \frac{z(\omega/\overline{\omega})}{\overline{\omega}} \quad . \qquad (3.1$$

Here, ω_k, $k(T/m)^{1/2}$ and $\overline{\omega}$ are the characteristic frequencies of the plasma oscillations, the single-particle excitations and the collisional excitations. The collisional excitations are the classical counterpart to the multipair excitations in the degenerate plasma [Pines and Nozières, 1966]; it is important to recognize that $\overline{\omega}$ remains finite in the long-wavelength limit $k \to 0$. Equation (3.1) does not take into account the collisional broadening of the spectral function for the plasma waves; this neglect is justified in the long-wavelength domain with which the present sum-rule analysis is concerned. The spectral functions, $y[\omega/k(T/m)^{1/2}]$ and $z(\omega/\overline{\omega})$, are normalized so that

$$\int_{-\infty}^{\infty} dx\ y(x) = 1 \qquad \text{and} \qquad \int_{\infty}^{\infty} dx\ z(x) = 1 \tag{3.2}$$

The functions, $X(k)$, $Y(k)$ and $Z(k)$, thus represent the strengths of the respective excitations; these are even functions with respect to k.

The frequency moments of the dynamic form factor are defined and calculated according to

$$\langle\omega^{\ell}\rangle = \int_{-\infty}^{\infty} d\omega\ \omega^{\ell}\ S(k,\omega) \quad . \tag{3.3}$$

The moment at $\ell = 0$ is proportional to $S(k)$; the $\ell = 2$ term yields the f-sum rule; the evaluation of the $\ell = 4$ moment involves the pair correlation function; the $\ell = 6$ moment involves the triple correlation function.

Various functions in Eq. (3.1) are expanded in the long-wavelength domain as

$$X(k) = X_0(k/k_D)^2 + X_1(k/k_D)^4 + \ldots$$

$$Y(k) = Y_0(k/k_D)^2 + Y_1(k/k_D)^4 + \ldots$$

$$Z(k) = Z_0(k/k_D)^2 + Z_1(k/k_D)^4 + \ldots$$

$$\omega_k = \omega_p[1 + \delta(k/k_D)^2] + \ldots \tag{3.4}$$

where $\omega_p = (4\pi ne^2/m)^{1/2}$. Comparing the terms proportional to $(k/k_D)^2$ in the sum rules, we find

$$X_0 = 1 \quad , \quad Y_0 = Z_0 = 0 \quad . \tag{3.5}$$

The contributions of the plasma oscillations exhaust the entire strength of the dynamic form factor in the long-wavelength limit.

The dispersion δ in the plasma-wave frequency may be analyzed directly from a sum-rule analysis of the dielectric response function. If the existence of the collisional excitations is totally neglected, one can complete the sum-rule analysis up to the $\ell = 4$ term in Eq. (3.3) [Ichimaru and Tange, 1974]; the result is

$$\delta = \frac{3}{2} + \frac{2}{15}\frac{E_c}{nT} \quad . \tag{3.6}$$

The presence of collisional excitations becomes significant as we proceed to take into account the next $\ell = 6$ term as well in Eq. (3.3); we then obtain

$$\delta = \frac{3}{2} - \frac{1}{2}\tau(\varepsilon) - \frac{2}{15}\frac{E_c}{nT} \quad . \tag{3.7}$$

where

$$\tau(\varepsilon) \equiv -\frac{3ne^2}{10\pi T}\int d\vec{r}\int d\vec{r}\,'h(\vec{r},\vec{r}\,')\left\{\frac{(\vec{r}\cdot\vec{r}\,')^3}{(r\;r')^5} - \frac{\vec{r}\cdot\vec{r}\,'}{(r\;r')^3}\right\} \tag{3.8}$$

and $h(\vec{r},\vec{r}\,')$ is the triple correlation function expressed as a function of relative coordinates, $r = \vec{r}_1 - \vec{r}_3$ and $\vec{r}\,' = \vec{r}_2 - \vec{r}_3$.

Numerical values of δ - (3/2) computed from Eqs. (3.7) and (3.8) are plotted in Figure 4 as functions of ε. A remarkable difference exists between these two sets of results: The values of δ - (3/2) based on Eq. (3.7) always remain negative. According to Eq. (3.8), however, δ - (3/2) starts to take on positive values when ε is small; this quantity changes its sign around $\varepsilon = 1$ and then goes over to negative values with increasing magnitude as ε increases.

In the limit of $\varepsilon \to 0$, the Vlasov description applies [e.g., Ichimaru, 1973], and one finds $\delta = 3/2$. As ε increases, the collisional effects in the plasma act to modify the dielectric response function and thereby produce deviation of δ from the Vlasov value. When $\varepsilon \ll 1$, one can analytically calculate the collisional contribution to the imaginary part of the dielectric function [DuBois, Gilinsky and Kivelson, 1962; Totsuji, unpublished],

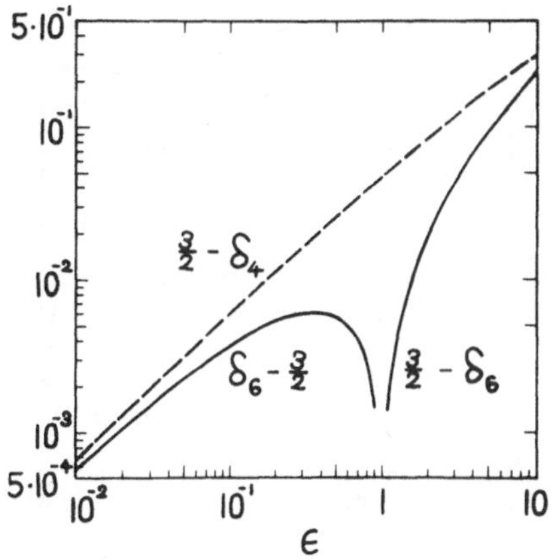

Figure 4. The parameter δ for the plasma-wave dispersion vs. the plasma parameter. The values δ_4 are computed according to Eq. (3.6); δ_6, Eq. (3.7).

$$\text{Im}\ \varepsilon(k,\omega) = -(8/15\pi^{1/2})(\omega_p/\omega)^5(k/k_D)^2\ \varepsilon\ \ell n\ \varepsilon \quad , \tag{3.9}$$

for ω larger than both $\varepsilon\omega_p$ and $k(T/m)^{1/2}$. The real part may then be assessed with the aid of the Kramers-Kronig relations. The result of such an investigation shows that $\delta - (3/2) > 0$ for $\varepsilon \ll 1$, in agreement with the prediction of Eq. (3.7).

In the domain of finite ε, the theoretical results may also be compared with those obtained from the molecular dynamics computations [Hansen, Pollock and McDonald, 1974; Hansen, McDonald and Pollock, 1975]. The molecular dynamics results in fact indicate that $\delta - (3/2) > 0$ for $\varepsilon < \varepsilon_0$ and $\delta - (3/2) < 0$ for $\varepsilon > \varepsilon_0$, where ε_0 falls somewhere between 1 and 10. In addition, it has been pointed out that δ would vanish at $\varepsilon \simeq 52$. These indications are again in agreement with the results of Eq. (3.7). We thus find that inclusion of the collisional excitations in the long-wavelength domain plays a vital part in satisfying the moment sum rules and in securing agreement with those known boundary conditions.

Finally, to determine the parameters X_1, Y_1 and Z_1, we compare the terms proportional to $(k/k_D)^4$ in the moment sum rules (3.3) up to $\ell = 6$; the spectral function $z(x)$ is assumed to be a Gaussian. The numerical results of the solution are shown in Figure 5. With increase of ε, the collisional excitations increase, while the collective and single-particle excitations decrease. The strength of the collisional excitations becomes comparable to that of the single-particle excitations when $\varepsilon \gtrsim 1$.

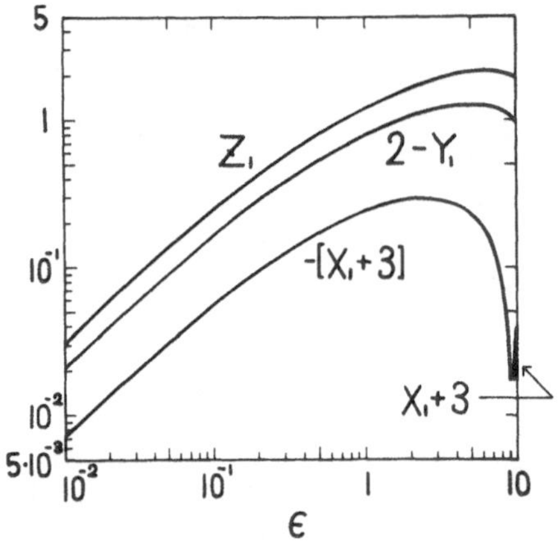

Figure 5. The parameters, X_1, Y_1 and Z_1, describing the strengths of the long-wavelength excitations vs. the plasma parameter.

IV. STRONGLY COUPLED PLASMAS IN A NONEQUILIBRIUM STATIONARY STATE

Thus far we have been concerned with the properties of the electron liquids in thermodynamic equilibrium. Let us now extend the scope and develop a dynamic theory of a strongly coupled plasma in a nonequilibrium state [Ichimaru, 1977]. The theory should thus be relevant directly to description of plasmas in a strongly turbulent state as well. We begin with elucidation of the criteria and requirements for such a theory.

The system under consideration is assumed to be in a nonequilibrium stationary state. An external source of free energy is provided so that it feeds energy through excitation of turbulence in the system; the same amount of energy is removed from the system through dissipative behavior. A stationary flow of energy is thereby established in the system. In such a situation we expect in general that strong correlations or fluctuations co-exist with a single-particle distribution of the particles. A first requirement for the theory is that (a) it provide a formalism in which those quantities may be determined in a nonequilibrium state.

An important physical effect in a turbulent plasma is a statistical modification of single-particle orbits by fluctuating fields. The idea is basically due to Dupree [1966, 1967, 1968], who argued that the dominant nonlinear effect of low-frequency instabilities is an incoherent scattering of particle orbits by waves, which causes particle diffusion and appears in the theory as an enhanced viscosity. We thus require that (b) the theory correctly incorporate such an effect of orbit modifications.

Another aspect of strong-correlation effect in a plasma is modification of effective interactions between particles [Ichimaru, 1970a]. In fact, this has been the major effect taken into account in the theories of electron liquids [Hubbard, 1967; Singwi, Tosi, Land and Sjölander, 1968; Ichimaru, 1970b]. In those theories, however, static approximations are basically involved in describing the modification of effective interactions. Here we adopt a requirement that (c) both static and dynamic aspects in the modification of effective interactions be properly taken into account. The dynamic aspects should physically correspond to creation and destruction of clumps [Dupree, 1970, 1972; Kadomtsev and Pogutse, 1970], bunched particles, and coherent waves.

A fourth requirement is related to the high-frequency response of the system. Generally, for a reflectionally symmetric system, the high-frequency asymptotic expansion of the frequency and wave-vector dependent, longitudinal dielectric response function $\varepsilon(\vec{k},\omega)$ takes the form [Ichimaru and Tange, 1974],

$$\varepsilon(\vec{k},\omega) = 1 - \frac{\omega_p^2}{\omega^2} - A_3 \frac{\omega_p^4}{\omega^4} - \dots \tag{4.1}$$

$$A_3 = \frac{3\langle(\vec{k}\cdot\vec{v})^2\rangle}{\omega_p^2} + \frac{1}{n}\sum_{\vec{q}} \frac{(\vec{k}\cdot\vec{q})^2}{k^2 q^2}\left[S(\vec{k}-\vec{q}) - S(\vec{q})\right] \quad . \tag{4.2}$$

In the long-wavelength limit, Eq. (4.2) reduces to

$$A_3 \rightarrow \frac{k^2}{k_D^2}\left\{3 + \frac{1}{nT}\sum_{\vec{q}} (1 - 5\mu^2 + 4\mu^4)\ \xi_{\vec{q}}\right\} \quad . \tag{4.3}$$

Here, $\xi_{\vec{q}}$ is the energy density contained in the fluctuations with wave vector $\vec{q}$; μ is the direction cosine between $\vec{k}$ and $\vec{q}$. For a turbulent plasma, T is to be replaced by an appropriate average of kinetic energy per particle. For an isotropic system such as an electron liquid in thermodynamic equilibrium, the last term in the curved brackets of Eq. (4.3) reduces to $(4/15)\ E_c/nT$, where

$$E_c = \sum_{\vec{q}} \xi_{\vec{q}} \quad .$$

We thus require that (d) the theory be consistent with Eqs. (4.1) and (4.2).*

Finally we require that (e) when the theory is applied to a particular case of a system in thermodynamic equilibrium, it must reproduce correct calculations of thermodynamic properties of the electron liquid, as outlined in Section II.

Kono and Ichikawa [1973] have shown that the criterion (b) is satisfied if one sums the most secular terms in each order of perturbation-theoretical expansion with respect to fluctuations. Criterion (d) implies that the theory is basically correct in describing the short-time behavior of the system. Criterion (e), on the other hand, is concerned with long-time, hydrodynamic behavior of the system, such as sound propagation, through correct evaluation of compressibilities. By securing these two criteria rigorously applicable in both ends of the frequency domain, we may be reasonably confident about accuracy of the theory in describing the system properties in the intermediate frequency domain as interpolation of those two limiting behaviors.

To obtain a formalism which satisfies these criteria, we must sum not only the most secular terms but also the next most secular terms with inclusion of vertex renormalization [Ichimaru, 1977]. We thus calculate the dielectric response function according to renormalization scheme as described in Figure 6**. Here a thin straight line corresponds to the free-particle propagator,

$$G_0(\vec{k},\omega;\ \vec{v}) = i/(\omega - \vec{k}\cdot\vec{v}) \qquad ; \qquad (4.4)$$

a thick straight line, the renormalized propagator of $G(\vec{k},\omega;\ \vec{v})$; and a shaded double line, the screened propagator $G(\vec{k},\omega;\ \vec{v})/\varepsilon(\vec{k},\omega)$. A wavy line with both ends terminated by vertices represents the contribution of the potential fluctuations $\langle|\phi^2|(\vec{k},\omega)\rangle$; an open wavy line describes the potential field $\phi(\vec{k},\omega)$ stemming from summed contribution of the external and induced potentials. When a wavy line with wave vector $\vec{q}$ is terminated at a filled-circle vertex, the vertex gives rise to a differential operator,

---

*Golden, Kalman and Silevitch [1974] calculated the dielectric response function for an electron liquid in thermodynamic equilibrium. A calculational error was involved in their assessment of Eq. (4.3), so that it appeared that their result did not satisfy the requirement (d). When this error is corrected, however, one finds that their dielectric response function satisfies this requirement for a system in thermodynamic equilibrium.

**Figure 1 in Ichimaru [1977] contained certain overcounting of diagrams; this overcounting has been corrected here.

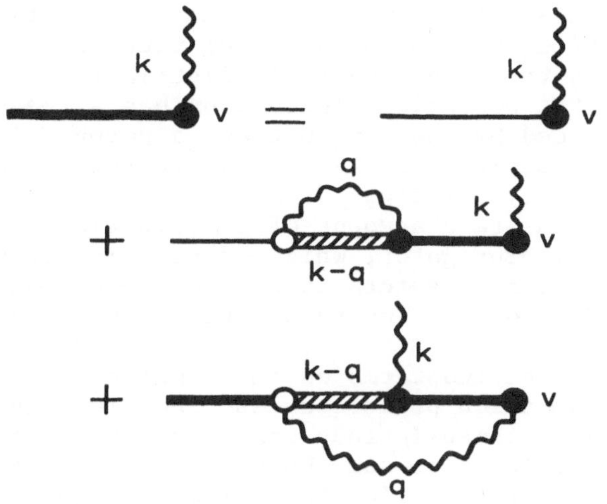

Figure 6. Diagrammatic representation of the renormalized dielectric propagator.

$$\frac{e}{m} i\vec{q} \cdot \frac{\partial}{\partial \vec{v}} \qquad . \tag{4.5}$$

When a wavy line with wave vector $\vec{q}$ and a straight line with wave vector $\vec{k} - \vec{q}$ merge at an open-circle vertex, it produces an operator of (4.5) in which $\vec{q}$ is replaced by

$$\vec{Q} \equiv \vec{q} + \frac{q^2}{|\vec{k} - \vec{q}|^2} (\vec{k} - \vec{q}) \qquad . \tag{4.6}$$

Involvement of the open-circle vertex is related to conservation of the total momentum in a single-component plasma; $\vec{Q}$ defined in Eq. (4.6) vanishes in the limit of $\vec{k} \to 0$.

The necessity of screening only the intermediate propagators with wave vector $\vec{k} - \vec{q}$ in Figure 6 is clear: An outgoing propagator arising from coupling between a potential fluctuation and a particle with velocity $\vec{v}$ in the single-particle distribution $f(\vec{v})$ need not be screened because the potential fluctuation has already been screened. A propagator produced by coupling between a potential fluctuation and a density fluctuation, however, is characterized by a wave vector and a frequency different from those of the parent fluctuations; one must thus take account of polarization processes induced by this new propagator, which results in its screening. Finally the outgoing propagators at the left ends must be left unscreened in order to avoid overcounting of the screening processes.

Clearly the first two terms on the right-hand-side of Figure 6 contain all the summation processes described in Kono and Ichikawa [1973]; in addition, those include screening of the intermediate propagators and enable us to sum the next most secular terms without vertex corrections. The last term of Figure 6 takes account of partial effects of vertex corrections. All of those modifications are necessary to ensure the criteria set forth earlier in this section.

The dielectric response function is calculated from the renormalization scheme of Figure 6 as

$$
\begin{aligned}
\varepsilon(\vec{k},\omega) &= 1 + \chi_2(\vec{k},\omega) \\
&= 1 - \frac{\omega_p^2}{k^2} \int d\vec{v} G(\vec{k},\omega;\ \vec{v})\ i\vec{k} \cdot \frac{\partial f(\vec{v})}{\partial \vec{v}} \\
&= 1 - \frac{\omega_p^2}{k^2} \int d\vec{v} G_0(\vec{k},\omega;\ \vec{v})\ i\vec{k} \cdot \frac{\partial f(\vec{v})}{\partial \vec{v}} \\
&\quad - i \frac{\omega_p^2}{k^2} \frac{e^2}{m^2} \Sigma_{\vec{q}} \int dx \int d\vec{v} \langle |\phi^2| (\vec{q},x) \rangle\ G_0(\vec{k},\omega;\ \vec{v})\ \vec{Q} \cdot \frac{\partial}{\partial \vec{v}} \\
&\quad \cdot \frac{G(\vec{k} - \vec{q},\omega - x;\ \vec{v})}{\varepsilon(\vec{k} - \vec{q},\omega - x)}\ \vec{q} \cdot \frac{\partial}{\partial \vec{v}} G(\vec{k},\omega;\ \vec{v})\ \vec{k} \cdot \frac{\partial f(\vec{v})}{\partial \vec{v}} \\
&\quad - i \frac{\omega_p^2}{k^2} \frac{e^2}{m^2} \Sigma_{\vec{q}} \int dx \int d\vec{v} \langle |\phi^2| (\vec{q},x) \rangle\ G(\vec{k},\omega;\ \vec{v})\ \vec{Q} \cdot \frac{\partial}{\partial \vec{v}} \\
&\quad \cdot \frac{G(\vec{k} - \vec{q},\omega - x;\ \vec{v})}{\varepsilon(\vec{k} - \vec{q},\omega - x)}\ \vec{k} \cdot \frac{\partial}{\partial \vec{v}} G(-\vec{q},-x;\ \vec{v})\ \vec{q} \cdot \frac{\partial f(\vec{v})}{\partial \vec{v}} ,
\end{aligned} \tag{4.7}
$$

where $<|\phi^2|(\vec{k},\omega)>$ represents the spectral function of potential fluctuations [e.g., Ichimaru, 1973].

Let us examine the dielectric function in the light of the criterion (c) concerning modification of the effective interactions. The propagator $G(\vec{k},\omega;\ \vec{v})$ consists of summation of an infinite number of diagrams constructed out of particle lines, field lines, and vertices. It is important to note that each one of such diagrams ends with a line representing $G_0(\vec{k},\omega;\ \vec{v})$. For an arbitrary integrable function $F(\vec{v})$ of velocity vanishing at infinity, we have an identity

$$\int d\vec{v}G_0(\vec{k},\omega;\ \vec{v})\ \vec{p}\cdot\frac{\partial F(\vec{v})}{\partial\vec{v}} = \frac{\vec{P}\cdot\vec{k}}{k^2}\int d\vec{v}G_0(\vec{k},\omega;\ \vec{v})\ \vec{k}\cdot\frac{\partial F(\vec{v})}{\partial\vec{v}} \tag{4.8}$$

We can thus rewrite Eq. (4.7) exactly in the form,

$$\varepsilon(\vec{k},\omega) = 1 - \frac{\omega_p^2}{k^2}\int d\vec{v}G_0(\vec{k},\omega;\ \vec{v})\ i\vec{k}\cdot\frac{\partial\tilde{f}(\vec{k},\omega;\ \vec{v})}{\partial\vec{v}} \quad . \tag{4.9}$$

We now define a function $t(\vec{k},\omega)$ via

$$t(\vec{k},\omega) \equiv \int dv\ \tilde{f}(\vec{k},\omega;\ \vec{v}) - 1 \quad . \tag{4.10}$$

We then introduce a <u>velocity average approximation</u> in the sense that

$$\tilde{f}(\vec{k},\omega;\ \vec{v}) \simeq [1 + t(\vec{k},\omega)]\ f(\vec{v}) \quad . \tag{4.11}$$

The equality of Eq. (4.11) becomes exact by virtue of Eq. (4.10) only after the velocity integration is carried out; as a function of velocity, Eq. (4.11) is an approximation. Adopting this approximation, we may further rewrite Eq. (4.9) as

$$\varepsilon(\vec{k},\omega) = 1 - \frac{\omega_p^2[1 + t(\vec{k},\omega)]}{k^2}\int d\vec{v}G_0(\vec{k},\omega;\ \vec{v})\ i\vec{k}\cdot\frac{\partial f(\vec{v})}{\partial\vec{v}} \quad . \tag{4.12}$$

It is in this form that we identify the effects of modification of particle interactions brought about by the presence of strong correlations in a plasma. When $t(\vec{k},\omega) = 0$, Eq. (4.12) reduces to the linear dielectric function of an uncorrelated charged-particle system, or the Vlasov dielectric function. For the renormalized dielectric function Eq. (4.7) of the strongly correlated plasma,

$t(\vec{k},\omega) \neq 0$; the factor k^{-2} in the Vlasov dielectric function arising from the bare Coulomb potential between two point particles is now effectively modified by a wave-vector and frequency dependent factor and replaced by $[1 + \bar{t}(\vec{k},\omega)]/k^2$. The resulting expression may then be interpreted as representing Coulomb interaction between particles which have both spatial and temporal structures. These aspects may presumably be related to those physical ideas on creation and destruction of clumps, bunching of particles, and coherent waves, set forth as theoretical models of strong plasma turbulence.

Applying the velocity average approximation successively to the terms in Eq. (4.7) and with the aid of Eq. (4.8), we obtain

$$\varepsilon(\vec{k},\omega) = 1 + \frac{\chi_0(\vec{k},\omega)}{1 + \psi(\vec{k},\omega)} \tag{4.13}$$

where

$$\psi(\vec{k},\omega) = \sum_{\vec{q}} \int dx \frac{q^2 \langle |\phi^2| (\vec{q},x) \rangle}{(4\pi ne)^2} \frac{\chi(\vec{k} - \vec{q},\omega - x)}{\varepsilon(\vec{k} - \vec{q},\omega - x)} \vec{k} \cdot \vec{Q}(\vec{k} - \vec{q})$$

$$\cdot \left[\frac{\vec{q}}{q^2} \chi_0(\vec{k},\omega) - \frac{\vec{k}}{k^2} \chi(-\vec{q},-x) \right] \quad . \tag{4.14}$$

Comparison between Eqs. (4.12) and (4.13) yields

$$t(\vec{k},\omega) = - \frac{\psi(\vec{k},\omega)}{1 + \psi(\vec{k},\omega)} \quad . \tag{4.15}$$

In connection with the criterion (d), we examine the high-frequency asymptotic expansion of the response functions for a system with reflectional symmetry. We first note that $\chi_0(\vec{k},\omega)$ for such a system has an expansion,

$$\chi_0(\vec{k},\omega) = - \frac{\omega_p^2}{\omega^2} - 3\left(\frac{k^2}{k_D^2}\right)\frac{\omega_p^4}{\omega^4} - \cdots \quad . \tag{4.16}$$

Let us then write

$$\chi(\vec{k},\omega) = - \frac{\chi_1}{\omega^2} - \frac{\chi_2}{\omega^4} - \cdots \quad . \tag{4.17}$$

Substituting Eqs. (4.16) and (4.17) in (4.13) and (4.14), we find

$$\chi_1 = \omega_p^2 \tag{4.18}$$

$$\chi_2 = \omega_p^4 \frac{k^2}{k_D^2} \left\{ 3 - \sum_{\vec{q}} \int dx \frac{\vec{k}\cdot\vec{Q}(\vec{k}-\vec{q})\cdot\vec{k}}{k^4} \frac{q^2 \langle|\phi^2|(\vec{q},x)\rangle \chi(-\vec{q},-x)}{4\pi nT} \right\} \tag{4.19}$$

Equation (4.18) indicates that the renormalized dielectric function correctly satisfies the f-sum rule.

The last term of Eq. (4.19) stems from the contribution of the last term in Eq. (4.14) or the last diagram in Figure 6; contributions of diagrams like the second one on the right-hand-side of Figure 6 do not influence the high-frequency asymptotic behavior of the response function to the order of ω^{-4}. To establish a connection between Eq. (4.2) and Eq. (4.19), we note that the Fourier components of the induced charge-density fluctuations in the presence of the potential field are given by

$$en(\vec{k},\omega) = -\chi(\vec{k},\omega)(k^2/4\pi)\ \phi(\vec{k},\omega) \quad . \tag{4.20}$$

Then, since

$$S(\vec{k}) - 1 = -\frac{k^4}{(4\pi e)^2 n} \int d\omega \ \langle|\phi^2|(\vec{k},\omega)\rangle\ \chi(-\vec{k},-\omega) \quad , \tag{4.21}$$

we see that Eq. (4.13) exactly satisfies the asymptotic requirement (4.1) to the order of ω^{-4}.

Let us finally examine the thermodynamic properties of the electron liquids predicted in the framework of the present theory. The static form factor, which plays a central part in determining those properties, may be calculated from Eq. (4.13) with the aid of the fluctuation-dissipation theorem and the Kramers-Kronig relations. One thus funds

$$S(k) = -\frac{k^2}{k_D^2}\left[\frac{1}{\varepsilon(k,0)} - 1\right] = \frac{k^2}{k_D^2 + k^2[1 + \psi(k,0)]} \quad . \tag{4.22}$$

With the aid of Eq. (4.21) and the known property $S(k) \to 1$ at large wave numbers, it is possible to prove that

$$\psi(k,0) = (k_D/k)^2\ w(k) \quad , \tag{4.23}$$

to a good degree of accuracy [Ichimaru, 1977]. Hence, the static theory of the electron liquid described in Section II is contained in the dielectric response function (4.13). Incidentally, if the screening of the intermediate propagator is not taken into account, one would obtain

$$\psi(k,0) = \frac{1}{n}\left(\frac{k_D}{k}\right)^2 \sum_{\vec{q}} \frac{\vec{k}\cdot\vec{q}}{q^2} [S(|\vec{k}-\vec{q}|) - 1] \quad , \tag{4.24}$$

which is the earliest version of the theories proposed by Singwi et al.

We have thus shown the dielectric response function (4.13) satisfies all the criteria set forth earlier. A self-consistent equation for determination of the turbulence spectrum may then be established from such a formalism [Ichimaru, 1977].

V. POLARIZATION POTENTIAL MODEL

The polarization potential model of condensed matter, proposed originally by Pines [1966], has recently been applied successfully to the description of elementary excitations in He II and in liquid helium-3 [Aldrich, Pethick and Pines, 1976]. The model is related closely to the phenomenological approach in the Landau Fermi-liquid theory [e.g., Legget, 1965]. In this section we consider the connection between the results, Eqs. (4.13) and (4.14), and the polarization potential model.

In this model, the restoring forces responsible for the collisionless part of the excitation spectrum are described by two kinds of self-consistent fields: a scalar polarization potential,

$$\Phi_{pol}(k,\omega) = f^s(k) \quad \langle n(k,\omega)\rangle \quad , \tag{5.1}$$

which couples directly to the density fluctuations in the system; and a vector polarization potential,

$$\vec{A}(k,\omega) = f^v(k) \quad \langle\vec{J}(k,\omega)\rangle \quad , \tag{5.2}$$

which couples to the particle current density. Here $\langle n(k,\omega)\rangle$ and $\langle\vec{J}(k,\omega)\rangle$ are the particle and current density fluctuations induced by an external scalar probe. The density-density response function takes the form

$$\chi(k,\omega) = \frac{\chi_{sc}(k,\omega)}{1 - [f^2(k) + (\omega^2/k^2)\ f^v(k)](k^2/4\pi e^2)\ \chi_{sc}(k,\omega)} \quad , \tag{5.3}$$

where $\chi_{sc}(k,\omega)$ is the response of the density fluctuations to the external field plus the induced polarization potentials, Eqs. (5.1) and (5.2).

We first note that comparison between Eq. (5.1) and Eq. (4.13) yields

$$\chi_{sc}(k,\omega) = \frac{\chi_0(k,\omega)}{1 + \psi(k,\omega) + [f^s(k) + (\omega^2/k^2)\ f^v(k)](k^2/4\pi e^2)\chi_0(k,\omega)} \quad . \tag{5.4}$$

We then determine $f^s(k)$ and $f^v(k)$ in such a way that they absorb the first two terms of $\psi(k,\omega)$ expanded in power series of ω^2; we thus find with the aid of Eqs. (2.4), (4.14), (4.21) and (4.23)

$$f^s(k) = -\frac{4\pi e^2}{k^2}\frac{\psi(k,0)}{\chi_0(k,0)} = -\frac{4\pi e^2}{nk^2}\sum_{\vec{q}}\frac{\vec{k}\cdot(\vec{k}-\vec{q})}{|\vec{k}-\vec{q}|^2} \cdot S(|\vec{k}-\vec{q}|)[S(q)-1] \quad , \tag{5.5}$$

$$f^v(k) = -4\pi e^2 \lim_{\omega\to 0}\frac{\partial}{\partial\omega^2}\left[\frac{\psi(k,\omega)}{\chi_0(k,\omega)}\right]$$

$$= -\lim_{\omega\to 0}\frac{\partial}{\partial\omega^2}\sum_q \frac{q^2\langle|\phi^2|(q)\rangle}{4\pi n^2}\frac{\chi(\vec{k}-\vec{q},\omega)}{\varepsilon(\vec{k}-\vec{q},\omega)}\ \vec{k}\cdot\vec{Q}(\vec{k}-\vec{q}) \cdot \left[\frac{\vec{q}}{q^2} - \frac{\vec{k}}{k^2}\frac{\chi(-\vec{q},0)}{\chi_0(k,\omega)}\right] \quad . \tag{5.6}$$

The response function obtained from substitution of (5.4) ∿ (5.6) into (5.3) satisfies all the criteria and sum rules discussed earlier.

As we did in Section III, it is meaningful physically to consider the screened density-density response function (5.4) as a summation of contributions from two kinds of particle-like excitations: single-particle excitations and collisional excitations:

$$\chi_{sc}(k,\omega) = \chi^{(s)}(k,\omega) + \chi^{(c)}(k,\omega) \quad . \tag{5.7}$$

The single-particle contribution $\chi^{(s)}(k,\omega)$ may be taken to be the free-particle polarizability (2.6) in which the particle mass is replaced by a wave-number dependent effective mass,

$$m^*(k) = m + nf^v(k) \quad , \tag{5.8}$$

another manifestation of the strong-correlation effects. The explicit expression for the collisional contribution $\chi^{(c)}(k,\omega)$ may then be obtained from the difference between $\chi_{sc}(k,\omega)$ and $\chi^{(s)}(k,\omega)$.

VI. LATTICE MODEL FOR A STRONGLY COUPLED ONE-COMPONENT PLASMA

When $\varepsilon > 1$, the dimensionless parameter,

$$\Gamma \equiv e^2/aT \quad , \tag{6.1}$$

more appropriately describes the system properties than ε itself, where

$$a = [(4/3)\pi n]^{-1/3} \tag{6.2}$$

represents the radius of the spherical volume occupied on average by a particle; this sometimes is referred to as the ion sphere radius. In this domain, the Debye length loses its meaning as a screening radius; the ratio (1.3) now scales as [DeWitt, 1976]

$$|E_c|/nT \simeq \Gamma \quad , \tag{6.3}$$

which basically represents the contribution of the Madelung energy.

Properties of the plasma in such a high-density domain have been investigated by the Monte Carlo method [Brush, Sahlin and Teller, 1966; Hansen, 1973]. One of the most remarkable features that those Monte Carlo investigations have revealed is an accurate linearity of the screening potential in the liquid phase over wide ranges of the distance and the plasma parameter Γ. The screening potential is defined by the relationship,

$$g(r) = \exp\left\{-\frac{1}{T}\left[\frac{e^2}{r} - V_s(r)\right]\right\} - 1 \quad , \tag{6.4}$$

Analyzing the result of Brush, Sahlin and Teller [1966], DeWitt, Graboske and Cooper [1973] found that apart from the vicinity of $r = 0$ the screening potential $V_s(r)$ in the liquid phase has a linear form,

$$V_s(r) = \frac{e^2}{a}\left[c_0 - c_1\left(\frac{r}{a}\right)\right] \quad , \tag{6.5}$$

and that the coefficients c_0 and c_1 satisfy the relationship,

$$c_1 = (c_0/2)^2 \quad . \tag{6.6}$$

In the vicinity of $r = 0$, the screening potential is expanded as $\alpha_0 - \alpha_2 r^2 + \ldots$ [Widom, 1963]. On applying Hansen's result [1973], Itoh, Totsuji and Ichimaru [1977] found the values,

$$c_0 = 1.25 \quad , \qquad c_1 = 0.39 \quad , \tag{6.7}$$

which satisfy the relationship (6.6). The linear screening potential (6.5) with the coefficients (6.7) fits Hansen's result for $4 \lesssim \Gamma \lesssim 160$ with errors less than four percent; the linearity is observed for $0.40 \lesssim r/a \lesssim 2.0$ in the case of $\Gamma = 4$, and for $1.3 \lesssim r/a \lesssim 1.8$ in the case of $\Gamma = 160$. The precise values of the screening potential at short distances are very difficult to deduce from the computed values of the radial correlation function; $g(r) + 1$ takes on extremely small values at short distances because of the Coulomb interaction term e^2/r.

Salient features of the screening potential which the Monte Carlo computations have revealed may thus be summarized in the following three aspects: its apparent linearity (6.5) over wide ranges of parameters, the relationship (6.6) and the absolute magnitude of c_0 or c_1. It has been pointed out by Itoh and Ichimaru [1977] that those empirical features are in fact intimately related to the physical notion that the short-range order in the classical one-component plasma in its liquid phase is already very much like that in its lattice phase. A lattice model for a classical charged liquid is thereby proposed which provides an explicit expression for the short-range correlation function.

To elucidate the content and relevance of the lattice model, let us recall the model calculations of the screening potential in the charged liquid based on an orderly lattice configuration of the bcc type, carried out originally by Salpeter and Van Horn [1969]. They proposed two methods of calculating such a screening potential: a rigid lattice model and a relaxed lattice model. In the rigid lattice model, they consider a situation in which a pair of nearest-neighbor ions approach each other with their center of mass and other surrounding ions fixed rigidly at their equilibrium positions. Along the line passing through the nearest-neighbor lattice points, they carry out the lattice sum numerically, and thereby obtain a result for the screening potential. In the relaxed lattice model, they impose the condition that the screening potential tends to that of the harmonic lattice near the lattice points and to one obtained from the ion-sphere model near the zero separation. The result is

$$V_{rel}(\eta) = \frac{e^2}{b}\,[1.1547 + 1.1602(1-\eta) - 1.0394(1-\eta)^2 + 2.5690(1-\eta)^3 - 1.6971(1-\eta)^4] \quad , \tag{6.8}$$

where b is the lattice constant, and η is the separation between the two ions measured in units of the nearest-neighbor distance d; for the bcc lattice, d = 0.8660b = 1.7589a. In these units the empirical screening potential, (6.5) and (6.7), for the charged liquid reads

$$V_{liquid}(\eta) = \frac{e^2}{b}\,[1.14 + 1.40(1-\eta)] \quad . \tag{6.9}$$

A close examination shows that the relaxed lattice model gives a screening potential which is reasonably close to that for the charged liquid; the discrepancy is less than 10 percent for all η between 0.2 and 1.0. The screening potential for the charged liquid almost coincides with the prediction of the bcc lattice models, $1.1547 = 2/\sqrt{3} = b/d$, at $\eta = 1.0$, the nearest-neighbor distance. Furthermore, we note that the screening potential in the relaxed lattice model is almost linear for $0.1 \leq \eta \leq 1.0$.

We have thus seen that the relaxed lattice model well accounts for linearity and the value at the nearest-neighbor distance of the screening potential in the charged liquid, two of the three empirical features earmarked earlier. To obtain an additional account of the remaining point (6.6), we further investigate the harmonic-oscillator potential model in the vicinity of the nearest-neighbor distance.

In the harmonic-oscillator potential model, the effective potential between two particles, $V_{eff}(r) = (e^2/r) - V_s(r)$, satisfies the conditions,

$$V_{eff}(d) = 0 \quad , \qquad [dV_{eff}(r)/dr]_{r=d} = 0 \quad , \tag{6.10}$$

at the equilibrium position, r = d. The former condition arises from the assumption of the perfect screening at the nearest-neighbor distance; the latter implies that the potential takes on an extremum (minimum) value there. Assuming linearity (6.5) for the screening potential, which has been substantiated in the relaxed lattice model, we obtain from Eq. (6.10)

$$c_1 = (c_0/2)^2 \quad , \qquad c_0 = 2a/d \quad . \tag{6.11}$$

The former is identical to Eq. (6.6); the latter gives $c_0 = 1.137$ for a bcc lattice and $c_0 = 1.241$ for a simple cubic lattice. Those values are again reasonably close to Eq. (6.7).

We have thus seen that the short-range order observed in the classical charged liquid is already very close to that predicted in a harmonic-lattice model for Γ as low as 4, as manifested by the similarity between the screening potential in the charged liquid and that in the lattice model.

VII. ENHANCEMENT OF THERMONUCLEAR REACTION RATE DUE TO STRONG SCREENING

Enhancement of thermonuclear reaction rate arising from Coulomb correlations in strongly coupled plasmas has important consequences in various aspects of stellar evolution such as carbon ignition in degenerate cores. In his pioneering work, Salpeter [1954] presented an analytic treatment of such an effect in a low-density, high-temperature plasma such that $\Gamma < 1$, and introduced the ion-sphere model to describe the effects of interparticle correlations in the strongly coupled regime, $\Gamma > 1$. Later, Salpeter and Van Horn [1969] carried out detailed calculations based on the ion-sphere model.

As we noted in the previous section, the Monte Carlo method has been a powerful tool in the study of Coulomb correlations in strongly coupled plasmas. DeWitt, Graboske and Cooper [1973] developed a generalized statistical-mechanical theory to describe the effects of plasma screening on nuclear reactions; they thereby investigated the effects of strong screening with the aid of the numerical result obtained by Brush, Sahlin and Teller [1966].

Both sets of calculations of the enhancement factor mentioned above are based on evaluation of the screening function at zero separation. For justification of this procedure, it may be argued that the classical turning radii for those particles with relative velocities in the vicinity of the Gamow peak are much smaller than the mean ionic distance; hence, the screening potential may be replaced effectively by its value at zero separation.

Basically, however, the nuclear reaction rate depends on the probability of particles tunneling through the repulsive Coulomb barrier; to evaluate the latter probability one must carry out a relevant WKB integration inside the turning radius. It is therefore expected that the spatial dependence of the screening function will play a crucial part in such an integration.

Itoh, Totsuji and Ichimaru [1977] thus calculated the enhancement factor for the nuclear reaction rate, by taking explicit account of the spatial dependence of the correlation function as obtained by Hansen [1973]. Two alternative methods were proposed: One follows a conventional approach in which only the pair correlations are taken into consideration. The effective interaction between two particles is taken to be

$$V_{eff}(r) = \frac{(Ze)^2}{r} - V_s(r) \quad , \tag{7.1}$$

where $V_s(r)$ is the screening potential given by Eqs. (6.5) and (6.7). [In this section we explicitly write the charge number Z of an ion, so that the dimensionless parameter $\Gamma = (Ze)^2/aT$.]

Substituting Eq. (7.1) in place of the bare Coulomb interaction $(Ze)^2/r$ in a standard calculation of the nuclear reaction rate, we find that the enhancement factor due to strong screening is given by $\exp\,[\tau - Q(\rho_0)]$, where

$$Q(\rho) \simeq \frac{\tau}{3}\left[\frac{1}{\rho} + 2\rho^{1/2} - \frac{3\Gamma}{\tau}c_0 + \left(\frac{3\Gamma}{\tau}\right)^2 c_1\rho - \frac{1}{4}\left(\frac{3\Gamma}{\tau}\right)^2 c_1\rho^{5/2}\right] , \tag{7.2}$$

$$\tau = \left[\left(\frac{27\pi^2}{4}\right)\frac{M(Ze)^4}{T\hbar^2}\right]^{1/3} , \tag{7.3}$$

M is the mass of an ion, and ρ_0 is the value of ρ at which $Q(\rho)$ is minimized; a $(3\Gamma/\tau)\,\rho_0$ represents the classical turning radius at the Gamow peak.

The other method of calculation adopted by Itoh, Totsuji and Ichimaru [1977] takes account of the possibility that the triple correlation function may play a significant part in describing the effects of screening inside the turning radius. Involvement of triple correlation is expected from the consideration that when two reacting particles are separated at a given distance the effective potential between them is determined by the statistical distribution of all the other "field" particles, which may be regarded as "third" particles. It should be remarked that the calculation leading to Eq. (7.2) has implicitly assumed that the statistical distribution of the field particles would adjust "adiabatically" as the reacting particles penetrate through the potential barriers.

The position that we may now adopt is to assume that a wave packet incident onto a potential barrier would tunnel instantantaneously and thus the field particles do not change their relative configuration in the process of tunneling; the field particles are regarded as frozen at their distribution determined when the colliding pair is separated at the distance of the classical turning radius. The expression for the effective potential in the classically forbidden region involves the triple correlation function in this scheme.

At the moment, explicit information on the triple correlation function in the strongly coupled regime is not available from the results of the Monte Carlo investigations. Instead, we note the relationship between the pair and triple correlation functions provided by the Yvon-Born-Green equations [e.g., Rice and Gray, 1965] and certain symmetry properties. We thus seek to find the simplest expression for the triple correlation function that satisfies those requirements and is consistent with the information contained in Eq. (6.5); the effective potential in the classically forbidden region is thereby determined.

The result of such a calculation yields

$$Q(\rho) \simeq \frac{\tau}{3}\left[\frac{1}{\rho} + 2\rho^{1/2} - \frac{3\Gamma}{\tau}c_0 + \left(\frac{3\Gamma}{\tau}\right)^2 c_1\rho - \frac{3}{16}\left(\frac{3\Gamma}{\tau}\right)^2 c_1\rho^{5/2}\right] \tag{7.4}$$

in place of Eq. (7.2). The computed values of $\tau - Q(\rho_0)$ at the Gamow peak ρ_0 from Eq. (7.4) can be reproduced by the following formula within errors less than one percent:

$$\tau - Q(\rho_0) = 1.25\Gamma - 0.10\tau(3\Gamma/\tau)^2 \quad . \tag{7.5}$$

We note that the difference between Eq. (7.2) and Eq. (7.4) is not substantial. The true values of the enhancement factor should lie somewhere between those two estimations. For a carbon plasma ($Z = 12$) at $T = 10^8$K with mass density 10^9g/cm^3, the enhancement factor computed from Eq. (7.2) is $\sim 5 \times 10^{16}$; that computed from Eq. (7.4) is $\sim 4 \times 10^{16}$.

VIII. ELECTRON LIQUID IN A TWO-DIMENSIONAL LAYER

The two-dimensional layer of electrons trapped on the surface of liquid helium [Cole and Cohen, 1969; Cole 1974] offers the cleanest example of strongly coupled classical plasmas hitherto realized in the laboratory. Recently, Zipfel, Brown and Grimes [1976] measured the velocity autocorrelation time τ_c of an electron in such electron liquids with Γ ranging from 9 to 36. In the two-dimensional system, Γ is defined by Eq. (6.1) with

$$a = (\pi n)^{-1/2} \tag{8.1}$$

representing the radius of the characteristic disk area occupied on average by an electron; n is the areal number density of electrons. The measurements have revealed a strikingly close correspondence between τ_c^{-1} and

$$\omega_0 = 2.1(e^2/m)^{1/2}\ n^{3/4} \quad , \tag{8.2}$$

the harmonic-oscillator frequency for the electrons forming a triangular lattice. This observation, providing important clues about the motion of electrons in a liquid state, has given a great impact on the theoretical study of such a two-dimensional one-component plasma.

The static properties of such a system have been investigated by Hockney and Brown [1975] with the aid of molecular dynamics computations, and by Totsuji [1978] through the Monte Carlo method. The detailed features of the radial correlation function clarified in these experiments are found to be closely related again to the physical notion that the short-range order in the two-dimensional classical plasma in its liquid phase is already very much like that in its lattice phase [Itoh, Ichimaru and Nagano, 1978], as we demonstrated in Section VI for a three-dimensional situation.

On analyzing the raw data of g(r) obtained by Hockney and Brown [1975], Itoh, Ichimaru and Nagano [1978] find that except in the vicinity of r = 0, the screening potential $V_s(r)$ defined by Eq. (6.4) is conspicuously expressed in the linear form (6.5) with the coefficients satisfying Eq. (6.6). The values,

$$c_0 = 1.13 \quad , \qquad c_1 = 0.32 \quad , \tag{8.3}$$

fit Hockney and Brown's data for $33.1 \leq \Gamma \leq 1875.8$ within errors less than five percent. Totsuji [1978] finds

$$c_0 = 1.18 \quad , \qquad c_1 = 0.33 \tag{8.4}$$

for $5 \leq \Gamma \leq 50$. These empirical findings can be correlated with a lattice model as we described in Section VI. As we show in the following, the observed features of the velocity autocorrelation time can also be accounted for in terms of such a lattice model.

The velocity autocorrelation time may be calculated from a sum-rule analysis of the dynamic structure factor $S_{inc}(\vec{k},\omega)$

associated with the self-motion of a "tagged" electron,

$$S_{inc}(\vec{k},\omega) = \frac{1}{2\pi}\int_{-\infty}^{\infty} dt \, \langle \exp\{-i\vec{k}\cdot[\vec{r}_1(t) - \vec{r}_1(0)]\}\rangle \exp(i\omega t) \quad , \tag{8.5}$$

where $\vec{r}_1(t)$ is the position of the electrons; < > denotes a statistical average. Defining the frequency moments of $S_{inc}(\vec{k},\omega)$ by

$$\overline{\omega^{\ell}_{inc}} \equiv \int_{-\infty}^{\infty} d\omega \, \omega^{\ell} S_{inc}(\vec{k},\omega) \quad , \tag{8.6}$$

one calculates with the aid of the rigorous equation of motion in the many-particle system [deGennes, 1959; Ichimaru, Totsuji, Tange and Pines, 1975]

$$\Omega^2 \equiv \frac{\langle \dot{v}_{1x}^2\rangle}{\langle v_{1x}^2\rangle} = \frac{\overline{\omega^4_{inc}}}{\overline{\omega^2_{inc}}} - 3\overline{\omega^2_{inc}} = -\frac{ne^2}{m}\int d\vec{r}\left(\frac{\partial}{\partial x}\frac{1}{r}\right)\left[\frac{\partial}{\partial x}g(r)\right] . \tag{8.7}$$

here the x-axis is chosen in the direction of $\vec{k}$; v_{1x} and $\dot{v}_{1x}$ are the velocity and the acceleration of the tagged electron in the x-direction. It is clear from the definition of Eq. (8.7) that Ω^{-1} corresponds to the velocity autocorrelation time of an electron.

To examine the validity of the use of Eq. (8.7) for comparison with the experimental values [Zipfel, Brown and Grimes, 1976], we carry out numerical integration of Eq. (8.7) by substituting the exact molecular-dynamics and Monte-Carlo values of g(r). The result is shown in Table I. The measured values (τ_c) of the velocity autocorrelation time have been interpolated or slightly extrapolated to make a comparison at the same values of Γ where the molecular-dynamics or Monte-Carlo data exist. The agreement is within the experimental error ($\pm$10%). Hence this comparison provides an additional confirmation that the τ_c of Zipfel, Brown and Grimes [1976] corresponds to Ω^{-1} defined by Eq. (8.7).

To obtain an analytical expression of Eq. (8.7) for a two-dimensional electron liquid in the harmonic-lattice model, we use the radial correlation function (6.4) as given by Eqs. (6.5) and (6.11) for the short-range domain, $r \leq d = 1.9046a$. For $r > d$, the radial correlation function generally exhibits a damped-oscillatory behavior around zero. Contributions from the peaks and troughs of

TABLE I

Comparison between the interpolated or extrapolated experimental values (τ_c) of the velocity autocorrelation time and the molecular-dynamics and Monte-Carlo values based on the fundamental relationship (8.7). Temperature is fixed at 1.2K.

Γ	τ_c (10^{-11} s)	Ω^{-1} (10^{-11} s)
7.1	16.0	17.5
15.8	5.1	5.5
22.4	3.2	3.3
33.1	1.9	1.9
46.8	1.00	1.08
50.0	0.95	1.01

g(r) tend to cancel each other in the integration of Eq. (8.7). We may thus take

$$g(r) = \begin{cases} \exp\left[-\frac{e^2}{Td}\left(\frac{d}{r} - 2 + \frac{r}{d}\right)\right] - 1 , & (r \leq d) \\ 0 , & (r > d) \end{cases} \tag{8.8}$$

as an approximate expression, to be substituted in Eq. (8.7); the result for $\Gamma >> 1$ is

$$\Omega^2 = \frac{\pi e^2 n^{3/2}}{\alpha m}\left[1 + \frac{\alpha^{1/2}\pi^{3/4}}{2\Gamma^{1/2}}\right]$$

$$= 0.66\ \omega_0^2(1 + 1.223\Gamma^{-1/2}) , \tag{8.9}$$

where $\alpha = 1.0746$.

The theoretical values Ω^{-1} of the velocity autocorrelation

TABLE II

Comparison between the measured values of (τ_c) of the velocity autocorrelation time and the theoretical values (Ω^{-1}) based on a harmonic-lattice model (8.9).

n ($10^8 cm^{-2}$)	Γ	τ_c (10^{-11} s)	Ω^{-1} (10^{-11} s)
0.15	9	9.9	13.1
0.20	11	7.2	10.6
0.51	17	4.8	5.4
1.4	29	2.1	2.7
2.2	36	1.4	1.9

time are computed from Eq. (8.9) for the values of the electron density studied by Zipfel, Brown and Grimes [1976]; the results are listed in Table II together with the measured values of τ_c. We here observe a good correspondence between the two sets of values. It is also proved that the velocity autocorrelation time in an electron liquid is in fact intimately related to the harmonic-oscillator frequency ω_0.

REFERENCES

Abe, R., 1959, Progr. Theor. Phys. 21, 475.

Aldrich, C. H., C. J. Pethick and D. Pines, 1976, Phys. Rev. Lett. 37, 845.

Gerggren, K. F., 1970, Phys. Rev. A1, 1783.

Bogoliubov, N. N., 1962, in Studies in Statistical Mechanics, translated by E. K. Gora, edited by J. deBoer and G. E. Uhlenbeck, Vol. I, 1, North Holland, Amsterdam.

Bowers, D. L. and E. E. Salpeter, 1960, Phys. Rev. 119, 1180.

Brush, S. G., H. L. Sahlin and E. Teller, 1966, J. Chem. Phys. 45, 2102.

Cole, M. W., 1974, Rev. Mod. Phys. 46, 451.

Cole, M. W. and M. H. Cohen, 1969, Phys. Rev. Lett. 23, 1238.

deGennes, P. G., 1959, Physica 25, 825.

DeWitt, H. E., 1976, Phys. Rev. A14, 1290.

DeWitt, H. E., H. C. Graboske and M. S. Cooper, 1973, Astrophys. J. 181, 439.

DeBois, D. F., V. Gilinsky and M. G. Kivelson, 1962, Phys. Rev. Lett. 8, 419.

Dupree, T. H., 1966, Phys. Fluids 9, 1773; 1967, Phys. Fluids 10, 1049; 1968, Phys. Fluids 11, 2680; 1972, Phys. Fluids 15, 334.

Dupree, T. H., 1972, Phys. Rev. Lett. 25, 789.

Forster, D., P. C. Martin and S. Yip, 1968, Phys. Rev. 170, 155.

Golden, K. I., G. Kalman and M. B. Silevitch, 1974, Phys. Rev. Lett. 33, 1544.

Hansen, J. P., 1973, Phys. Rev. A8, 3096.

Hansen, J. P., I. R. McDonald and E. L. Pollock, 1975, Phys. Rev. A11, 1025.

Hansen, J. P., E. L. Pollock and I. R. McDonald, 1974, Phys. Rev. Lett. 32, 277.

Hockney, R. W. and T. R. Brown, 1975, J. Phys. C8, 1813.

Hubbard, J., 1967, Phys. Lett. 25A, 709.

Ichimaru, S., 1969, in Statistical Physics of Charged Particle Systems, edited by R. Kubo and T. Kihara (Syokabo, Tokyo), 69.

Ichimaru, S., 1970a, Phys. Fluids 13, 1560.

Ichimaru, S., 1970b, Phys. Rev. A2, 494.

Ichimaru, S., 1973, Basic Principles of Plasma Physics, W. A. Benjamin, Reading, Massachusetts.

Ichimaru, S., 1977, Phys. Rev. A15, 744.

Ichimaru, S. and T. Nakano, 1967, Phys. Lett. 25A, 163.

Ichimaru, S. and T. Nakano, 1968, Phys. Rev. 165, 231.

Ichimaru, S. and T. Tange, 1974, Phys. Rev. Lett. 32, 102.

Ichimaru, S., H. Totsuji, T. Tange and D. Pines, 1975, Phys. Rev. Lett. 54, 1077.

Itoh, N. and S. Ichimaru, 1977, in press, Phys. Rev. A16.

Itoh, N., S. Ichimaru and S. Nagano, 1978, to be published, Phys. Rev. B.

Itoh, N., H. Totsuji and S. Ichimaru, 1977, in press, Astrophys. J. 218.

Kadomtsev, B. B. and O. P. Pogutse, 1970, Phys. Rev. Lett. 25, 1155.

Kono, M. and Y. H. Ichikawa, 1973, Progr. Theor. Phys. 49, 754.

Leggett, A. J., 1965, Phys. Rev. 140, A1869.

O'Neil, T. and N. Rostoker, 1965, Phys. Fluids 8, 1109.

Pines, D., 1966, in Quantum Fluids, edited by D. F. Brewer, North Holland, Amsterdam, 257.

Pines, D. and P. Nozières, 1966, The Theory of Quantum Liquids, W. A. Benjamin, New York.

Pollock, E. L. and J. P. Hansen, 1973, Phys. Rev. A8, 3110.

Puff, R. D., 1965, Phys. Rev. 137, A406.

Rise, S. A. and P. Gray, 1965, The Statistical Mechanics of Simple Liquids, Interscience, New York.

Rostoker, N. and M. N. Rosenbluth, 1960, Phys. Fluids 3, 1.

Salpeter, E. E., 1954, Austr. J. Phys. 7, 373.

Salpeter, E. E. and H. M. Van Horn, 1969, Astrophys. J. 155, 183.

Singwi, K. S., A. Sjölander, M. P. Tosi and R. H. Land, 1969, Solid State Commun. 7, 1503.

Singwi, K. S., M. P. Tosi, R. H. Land and A. S. Sjölander, 1968, Phys. Rev. 176, 589.

Totsuji, H., 1978, to be published, Phys. Rev. A.

Totsuji, H. and S. Ichimaru, 1973, Progr. Theor. Phys. 50, 753; 1974, Progr. Theor. Phys. 52, 42.

Vashishta, P. and K. S. Singwi, 1972, Phys. Rev. B6, 875.

Widom, B., 1963, J. Chem. Phys. 39, 2803.

Zipfel, C. L., T. R. Brown and C. C. Grimes, 1976, Phys. Rev. Lett. 37, 1970.

GENERALIZED RESPONSE FUNCTION APPROACH TO STRONGLY COUPLED PLASMAS*

Kenneth I. Golden

Department of Electrical Engineering
Northeastern University
Boston, Massachusetts 02115

*Research partially supported by the United States Air Force Office of Scientific Research Grant No. AFOSR-76-2960.

TABLE OF CONTENTS

GENERALIZED RESPONSE FUNCTION APPROACH TO STRONGLY COUPLED PLASMAS

Kenneth I. Golden

Department of Electrical Engineering, Northeastern University, Boston, Massachusetts 02115, U.S.A.

I. INTRODUCTION

For nearly a decade, theorists in the area of strongly coupled plasmas have been in search of approximation schemes for determining the physical characteristics of the system without recourse to the usual perturbation expansion in the plasma parameter γ ($\gamma = k_D^3/4\pi n$, k_D^{-1} is the Debye length) when γ is not small. The more successful schemes which have emerged from this search can be classified according to whether they are based on the first Bogoliubov-Born-Green-Kirkwood-Yvon (BBGKY) kinetic equation or on the second BBGKY static equation. The former approach, basically a dynamical one, contemplates the introduction of a weak electric field perturbation into the equilibrium system. In this approach, the wave vector- and frequency-dependent dielectric response function can be considered to be the central object and one arrives at an expression for it by combining the first BBGKY kinetic equation with one or more fluctuation-dissipation theorems (FDTs). This method has been pursued in different ways by Singwi et al (STLS) [Singwi Tosi, Land and Sjolander, 1968; Singwi, Sjolander, Tosi and Land, 1969, 1970; Vashishta and Singwi, 1973] and by Golden, Kalman and Silevitch (GKS) [1974]. The latter approach, basically a static one, does not contemplate the introduction of a weak field perturbation into the plasma. Here the system is always in a state of equilibrium and the central object is the equilibrium pair correlation function. In this method, one starts from the second BBGKY static equation relating the pair and triplet correlation functions. Self consistency is then guaranteed by assuming that the triplet correlation can be decomposed into clusters of the pair correlation functions. This is the method pursued by Totsuji and Ichimaru (TI) [1973, 1974].

While the relative merits and defects of all the above listed approaches have already been discussed in Gabor Kalman's lectures, it should nevertheless be stated that these approximation schemes have achieved important results: The equilibrium pair correlation function has been calculated by numerically solving integral equations which result from the theory; equations of state have been computed; conditions for phase transition have been cited; and the theories have been refined to the point that original inconsistencies concerning the satisfaction of sum-rule requirements can be reduced or removed.

In our approach to the problem of the strongly coupled one-component plasma (ocp), we follow a philosophy similar to that of Singwi and his colleagues. The main feature of the STLS method is that the linear dielectric function is calculated by combining the first BBGKY kinetic equation with the linear FDT relating the static polarizability and equilibrium pair correlation function. Self consistency is then guaranteed by supposing that the correlational part of the two-particle distribution function, G(12), involves only the equilibrium pair correlation function. This is the essence of the STLS approximation; their scheme ignores the equally important proper nonequilibrium part of G(12). Unlike Singwi et al, we do take account of the proper nonequilibrium part of G(12) and we do it in a way which avoids the use of the unwieldly second BBGKY kinetic equation. This approach is made possible by introducing into the approximation scheme a relatively new element, the dynamical quadratic FDT derived earlier by us [Golden, Kalman and Silevitch, 1972]. The essence of this nonlinear FDT is that it connects quadratic polarizability response functions to a single equilibrium three-point correlation. By further relating the equilibrium three-point function to G(12), one can formulate the problem in terms of a self-consistent calculation for the combined set of linear and quadratic response functions. We shall see (in Chapter 3) that by invoking the so-called velocity-average-approximation (VAA) [Golden, Kalman and Silevitch, 1974], i.e., replacement of G(12) by its velocity average, it is possible to effect the transition from G(12) language ultimately to equilibrium three-, four-, ... point correlation function language. The VAA is the main assumption of the GKS scheme.

The ingredients of the GKS scheme are then the following:

(i) use of the VAA to convert G(12) on the right-hand-side of the first BBGKY kinetic equation into a nonequilibrium two-point (density-density) correlation function;

(ii) development of first and second order perturbation expansions (in the perturbing external field $\hat{E}$) of the VAA kinetic equation;

(iii) establishment of the relationships between the non-equilibrium two-point functions and equilibrium three- and four-point functions by use of statistical mechanical perturbation theory;

(iv) linking the three- and four-point functions through equilibrium nonlinear FDTs with the quadratic and cubic polarizabilities; and

(v) guaranteeing self-consistency by an appropriate decomposition of the cubic polarizability in terms of the linear and quadratic polarizabilities. The choice of the cubic response function should be made in such a way that the linear and quadratic functions satisfy known long and short wavelength requirements.

In these lectures, both the static ($\omega=0$: Chapters II-VII) and dynamical ($\omega\neq 0$: Chapters IX, X) GKS theories are presented. The plan of these lectures can now be sketched as follows: In Chapter II, the relevant polarizability response functions are defined and relations between the external (responding to an external field) and internal (responding to a total field) polarizabilities are derived. In Chapter III, we introduce the velocity-average-approximation and formulate the VAA kinetic equation. In Chapters IV and V, we calculate the average density response and derive the first two equations in the static GKS hierarchy of coupled polarizability equations. These polarizability equations are then analyzed in the long wavelength limit in Chapter VI. In Chapter VII, we cite short wavelength requirements and in VIII, the GKS pair correlation function is examined in the weak coupling limit. In Chapter IX, we derive an expression for the GKS wave vector- and frequency-dependent dielectric response function and analyze its high frequency behavior in Chapter X. Summary and conclusions are in Chapter XI.

II. ELECTRODYNAMIC RESPONSE FUNCTIONS

External and internal polarizabilities are the principal response functions for the GKS description of the strongly coupled classical ocp. In the sequel, only the longitudinal projections of their tensors will be of interest since we contemplate driving the system (of volume V) only with the weak external longitudinal electric field $\hat{E}(\vec{r},t) = -(i/V)\Sigma_{\vec{k}}\vec{k}\hat{\phi}(\vec{k},t)\exp(i\vec{k}\cdot\vec{r})$; magnetic fields are considered to be entirely absent.

The induced electric field response E of the plasma particles to $\hat{E}$ is described by the so-called external polarizabilities defined through the following relations:

$$E^{(1)}(\vec{k},\omega) = -\underset{1}{\hat{\alpha}}(\vec{k},\omega)\ \hat{E}(\vec{k},\omega) \quad , \tag{2.1}$$

$$E^{(2)}(\mathbf{k},\omega) = -\frac{1}{2\pi V}\Sigma_{\vec{q}}\int_{-\infty}^{\infty} d\mu\ \underset{2}{\hat{\alpha}}(\vec{q},\mu;\ \vec{k}-\vec{q},\omega-\mu)\ \hat{E}(\vec{q},\mu)\ \hat{E}(\vec{k}-\vec{q},\omega-\mu) \quad , \tag{2.2}$$

$$\mathcal{E}^{(3)}(\vec{k},\omega) = -\frac{1}{(2\pi V)^2} \Sigma_{\vec{q}}\Sigma_{\vec{p}} \int_{-\infty}^{\infty} d\mu \int_{-\infty}^{\infty} d\nu \; \hat{\alpha}_3(\vec{q},\mu;\ \vec{p},\nu;\ \vec{k}-\vec{q}-\vec{p},\ \omega-\mu-\nu)$$

$$X\ \hat{E}(\vec{q},\mu)\ \hat{E}(\vec{p},\nu)\ \hat{E}(\vec{k}-\vec{q}-\vec{p},\ \omega-\mu-\nu) \quad , \tag{2.3}$$

where, in an obvious notation, the (n) superscript refers to an nth order response to the perturbation of strength $\hat{E}^n$.

The internal polarizabilities, on the other hand, connect $\mathcal{E}$ to the total field $E = \hat{E} + \mathcal{E}$. They are defined through the following relations:

$$\mathcal{E}^{(1)}(\vec{k},\omega) = -\alpha_1(\vec{k},\omega)\ E^{(1)}(\vec{k},\omega) \quad , \tag{2.4}$$

$$\mathcal{E}^{(2)}(\vec{k},\omega) = -\frac{1}{2\pi V} \Sigma_{\vec{q}} \int_{-\infty}^{\infty} d\mu \; \frac{\alpha_2(\vec{q},\mu;\ \vec{k}-\vec{q},\ \omega-\mu)}{\varepsilon(\vec{k},\omega)}\ E^{(1)}(\vec{q},\mu)\ E^{(1)}(\vec{k}-\vec{q},\omega-\mu) \tag{2.5}$$

$$\mathcal{E}^{(3)}(\vec{k},\omega) = -\frac{1}{(2\pi V)^2} \Sigma_{\vec{q}}\Sigma_{\vec{p}} \int_{-\infty}^{\infty} d\mu \int_{-\infty}^{\infty} d\nu \left[\frac{\alpha_3(\vec{q},\mu;\ \vec{p},\nu;\ \vec{k}-\vec{q}-\vec{p},\ \omega-\mu-\nu)}{\varepsilon(\vec{k},\omega)} - \frac{2\alpha_2(\vec{q},\mu;\ \vec{k}-\vec{q},\ \omega-\mu)}{\varepsilon(\vec{k},\omega)}\ \frac{\alpha_2(\vec{p},\nu;\ \vec{k}-\vec{q}-\vec{p},\ \omega-\mu-\nu)}{\varepsilon(\vec{k}-\vec{q},\ \omega-\mu)} \right]$$

$$X\ E^{(1)}(\vec{q},\mu)\ E^{(1)}(\vec{p},\nu)\ E^{(1)}(\vec{k}-\vec{q}-\vec{p},\ \omega-\mu-\nu) \quad , \tag{2.6}$$

where $\varepsilon(\vec{k},\omega) = 1 + \alpha_1(\vec{k},\omega)$ is the wave vector- and frequency-dependent dielectric response function. Noting from (2.4) that

$$E^{(1)}(\vec{k},\omega) = \hat{E}(\vec{k},\omega)/\varepsilon(\vec{k},\omega) \quad , \tag{2.7}$$

one can readily derive from Eqs. (2.1) to (2.6) the following useful relations between the external and internal polarizabilities:

$$\hat{\alpha}_1(\vec{k},\omega) = \frac{\alpha_1(\vec{k},\omega)}{\varepsilon(\vec{k},\omega)} \quad , \tag{2.8}$$

$$\hat{\alpha}_2(\vec{q},\mu;\ \vec{k}-\vec{q},\ \omega-\mu) = \frac{\alpha_2(\vec{q},\mu;\ \vec{k}-\vec{q},\ \omega-\mu)}{\varepsilon(\vec{q},\mu)\ \varepsilon(\vec{k}-\vec{q},\ \omega-\mu)\ \varepsilon(\vec{k},\omega)} \quad , \tag{2.9}$$

$$\hat{\alpha}_3(\vec{q},\mu;\ \vec{p},\nu;\ \vec{k}-\vec{q}-\vec{p},\ \omega-\mu-\nu)$$

$$= \frac{1}{\varepsilon(\vec{k},\omega)\ \varepsilon(\vec{q},\mu)\ \varepsilon(\vec{p},\nu)\ \varepsilon(\vec{k}-\vec{q}-\vec{p},\ \omega-\mu-\nu)}$$

$$X \left[\alpha_3(\vec{q},\mu;\ \vec{p},\nu;\ \vec{k}-\vec{q}-\vec{p},\ \omega-\mu-\nu) - \frac{2\alpha_2(\vec{q},\mu;\ \vec{k}-\vec{q},\ \omega-\mu)\ \alpha_2(\vec{p},\nu;\ \vec{k}-\vec{q}-\vec{p},\ \omega-\mu-\nu)}{3\varepsilon(\vec{k}-\vec{q},\ \omega-\mu)} - \frac{2\alpha_2(\vec{p},\nu;\ \vec{k}-\vec{p},\ \omega-\nu)\ \alpha_2(\vec{q},\mu;\ \vec{k}-\vec{q}-\vec{p},\ \omega-\mu-\nu)}{3\varepsilon(\vec{k}-\vec{p},\ \omega-\nu)} - \frac{2\alpha_2(\vec{q},\mu;\ \vec{p},\nu)\ \alpha_2(\vec{q}+\vec{p},\ \mu+\nu;\ \vec{k}-\vec{q}-\vec{p},\ \omega-\mu-\nu)}{3\varepsilon(\vec{q}+\vec{p},\ \mu+\nu)} \right] . \tag{2.10}$$

It is clear that $\hat{\alpha}_2$, α_2, $\hat{\alpha}_3$, α_3 must remain invariant under interchange of any two of their wave vector-frequency arguments; this is certainly borne out by Eqs. (2.9) and (2.10).

Finally, from Poisson's equation connecting the average density response n to E,

$$n(\vec{k},\omega) = -\frac{ik}{4\pi e} E(\vec{k},\omega) \tag{2.11}$$

(valid to all orders in $\hat{E}$), one obtains from (2.1) and (2.2) the relations

$$n^{(1)}(\vec{k},\omega) = \frac{ik}{4\pi e}\ \hat{\alpha}_1(\vec{k},\omega)\ \hat{E}(\vec{k},\omega) \quad , \tag{2.12}$$

$$n^{(2)}(\vec{k},\omega) = \frac{ik}{4\pi e}\ \frac{1}{2\pi V}\ \Sigma_{\vec{q}} \int_{-\infty}^{\infty} d\mu\ \hat{\alpha}_2(\vec{q},\mu;\ \vec{k}-\vec{q},\ \omega-\mu)\ \hat{E}(\vec{q},\mu)\ \hat{E}(\vec{k}-\vec{q},\ \omega-\mu), \tag{2.13}$$

which will be useful later on. In writing Eq. (2.11), we have adopted the electron liquid ($n_{ion}\to\infty$, $Z_i=0$) as our ocp model for these lectures. It is, of course, understood that the GKS theory will equally well describe the behavior of the polarizability response functions for the inverted plasma ocp model.

III. VELOCITY AVERAGED KINETIC EQUATION

Following the procedure outlined in the Introduction, one can calculate the linear dielectric response function and higher order polarizabilities from successive perturbation expansions (in $\hat{E}$) of the first BBGKY kinetic equation. This equation must, however, be suitably prepared for such an undertaking first by converting it into its velocity averaged form. This we do in the present Chapter.

We begin by writing the general first BBGKY equation

$$\left[\frac{\partial}{\partial t} + \vec{v} \cdot \frac{\partial}{\partial \vec{x}} - \frac{e}{m} \hat{\vec{E}}(\vec{x},\vec{t}) \cdot \frac{\partial}{\partial \vec{v}}\right] F(\vec{x},\vec{v};\ t)$$

$$= -\frac{1}{m}\frac{\partial}{\partial \vec{v}} \cdot \int d^3\vec{v}' \int d^3\vec{x}' \vec{K}(|\vec{x} - \vec{x}'|)\, G(\vec{x},\vec{v};\ \vec{x}',\vec{v}';\ t) \tag{3.1}$$

for the classical electron liquid in the presence of the weak perturbing field $\hat{E}$ and with magnetic fields assumed to be entirely absent; F and G are one- and two-particle distribution functions (normalized to N and N(N-1), N being the total number of electrons) and $K(|\vec{x}-\vec{x}'|)=(\partial/\partial\vec{x})(e^2/|\vec{x}-\vec{x}'|)$ is the interaction force between the field particle (at x) and a typical source particle (at $\vec{x}'$).

In order to be able to express the right-hand-side of (3.1) in terms of nonequilibrium two-point functions (binary correlations of microscopic charge densities), we assume that G is equal to its velocity average, i.e.,

$$G(\vec{x},\vec{v};\ \vec{x}',\vec{v}';\ t) = (1/2)[f(\vec{x},\vec{v};\ t)\int d^3\vec{v}''G(\vec{x},\vec{v}'';\ \vec{x}',\vec{v}';\ t)$$

$$+ f(\vec{x}',\vec{v}';\ t)\int d^3\vec{v}''G(\vec{x},\vec{v};\ \vec{x}',\vec{v}'';\ t)] \quad , \tag{3.2}$$

where $f(\vec{x},\vec{v};\ t) \equiv F(\vec{x},\vec{v};\ t)/n(\vec{x},t)$, $\int d^3\vec{v}\ f(\vec{x},\vec{v};\ t) = 1$. This is the so-called VAA and is the main assumption of the GKS scheme. The resulting double velocity space integral term $f(\vec{x},\vec{v};\ t)\int d^3\vec{v}'\int d^3\vec{v}'' \times G(\vec{x},\vec{v}';\ \vec{x}',\vec{v}'';\ t)$ which replaces $\int d^3\vec{v}'G(\vec{x},\vec{v};\ \vec{x}',\vec{v}';\ t)$ in (3.1) can then, in turn, be expressed in terms of the nonequilibrium density-density correlation function, $\langle n(\vec{x})n(\vec{x}')\rangle(t)$, in virtue of the relation

$$\int d^3\vec{v}\int d^3\vec{v}'G(\vec{x},\vec{v};\ \vec{x}',\vec{v}';\ t) = \langle n(\vec{x})n(\vec{x}')\rangle(t) - \delta(\vec{x} - \vec{x}')n(\vec{x},t) \ ,$$

$$n(\vec{x}) = \sum_{i=1}^{N} \delta(\vec{x} - \vec{x}_i) \quad . \tag{3.3}$$

This follows from the definitions of G and <nn>, i.e.,

$$G(\vec{x}_1,\vec{v}_1;\ \vec{x}_2,\vec{v}_2;\ t) = N(N-1) \prod_{i=3}^{N} \int d^3\vec{x}_i\ d^3\vec{v}_i \Omega(\Gamma,t) \quad ,$$

$$\langle n(\vec{x})n(\vec{x}')\rangle(t) = \prod_{i=1}^{N} \int d^3\vec{x}_i \int d^3\vec{v}_i \Omega(\Gamma,t)n(\vec{x})n(\vec{x}') \quad ,$$

where $\Omega(\Gamma,t)$ is the nonequilibrium Liouville distribution function (normalized to unity) at time t and at the phase point Γ defined to be the extremity of a position vector in the phase space spanned by the 6N position and velocity coordinates of the electrons. Then upon combining Eqs. (3.1) to (3.3) and taking the Fourier transform of the result, one obtains

$$-i(\omega-\vec{k}\cdot\vec{v})\ F(\vec{k},\vec{v};\ \omega) - \frac{e}{2\pi Vm} \Sigma_{\vec{q}} \int_{-\infty}^{\infty} d\mu\ \hat{\vec{E}}(\vec{q},\mu) \cdot \frac{\partial F(\vec{k}-\vec{q},\vec{v};\ \omega-\mu)}{\partial \vec{v}}$$

$$= \frac{i}{2\pi V^2 m} \Sigma_{\vec{q}} \vec{q}\phi(q) \cdot \Sigma_{\vec{p}} \int_{-\infty}^{\infty} d\nu\ \langle n_{\vec{k}-\vec{p}-\vec{q}} n_{\vec{q}} \rangle(\omega-\nu)\ \frac{\partial f(\vec{p},\vec{v};\ \nu)}{\partial \vec{v}} \quad , \tag{3.4}$$

where $n_{\vec{q}} = \sum_{i=1}^{N} \exp(-i\vec{q}\cdot\vec{x}_i)$, $\phi(q) = 4\pi e^2/q^2$.

The VAA kinetic equation (3.4) is valid to all orders in $\hat{E}$. In the sequel, we shall suppose that the classical electron liquid is initially in a state of equilibrium characterized by the Maxwellian distribution $F°(v) = n(m\beta/2\pi)^{3/2}\exp(-\beta mv^2/2)$, where $n = N/V$ and β^{-1} is the temperature in energy units. The introduction of the external perturbation $\hat{E}$ will then perturb $F°(v)$ by amount $\overline{F}$, i.e.

$$F(\vec{k},\vec{v};\ \omega) = 2\pi V\delta_{\vec{k}}\delta(\omega)\ F°(v) + \overline{F}(\vec{k},\vec{v};\ \omega) \quad , \tag{3.5a}$$

$$f(\vec{k},\vec{v};\ \omega) = 2\pi V\delta_{\vec{k}}\delta(\omega)(1/n)\ F°(v) + \overline{f}(\vec{k},\vec{v};\ \omega) \quad . \tag{3.5b}$$

Finally, if one contemplates only static driving fields, i.e.,

$$\hat{\vec{E}}(\vec{q},\mu) = 2\pi\delta(\mu)\ \hat{\vec{E}}(\vec{q}) \quad , \tag{3.6}$$

where $\hat{\vec{E}}(q) \equiv \hat{\vec{E}}(\vec{q},t=0)$, Eq. (3.4) becomes (in view of (3.5a,b)

$$i\vec{k}\cdot\vec{v}\ \overline{F}(\vec{k},\vec{v};\ t=0) - (e/m)\ \hat{\vec{E}}(\vec{k}) \cdot \frac{\partial F°(v)}{\partial \vec{v}}$$

$$- (e/mV)\ \Sigma_{\vec{q}} \hat{\vec{E}}(\vec{q}) \cdot \frac{\partial \overline{F}(\vec{k}-\vec{q},\vec{v};\ t=0)}{\partial \vec{v}}$$

$$= (i/nmV)\ \Sigma_{\vec{q}}\phi(q)\vec{q} \cdot \frac{\partial F°(v)}{\partial \vec{v}} \langle n_{\vec{k}-\vec{q}} n_{\vec{q}} \rangle(t=0)$$

$$+ (i/mV^2)\ \Sigma_{\vec{q}}\phi(q)\vec{q} \cdot \Sigma_{\vec{p}} \frac{\partial \overline{f}(\vec{p},\vec{v};\ t=0)}{\partial \vec{v}} \langle n_{\vec{k}-\vec{p}-\vec{q}} n_{\vec{q}} \rangle(t=0) \tag{3.7}$$

again valid to all orders in $\hat{E}$.

IV. STATIC DENSITY RESPONSE

We turn now to the calculation of the first and second order average density responses to the static perturbation (3.6). From Eq. (3.7),

$$\overline{F}^{(1)}(\vec{k},\vec{v}) = -(ie/mk)\ \hat{E}(\vec{k})\ \frac{\vec{k}\cdot\partial F^{\circ}(v)/\partial\vec{v}}{\vec{k}\cdot\vec{v}} + (1/Nm)\ \frac{\partial F^{\circ}(v)/\partial\vec{v}}{\vec{k}\cdot\vec{v}}\cdot\Sigma_{\vec{q}}\ \phi(\vec{q})\langle n_{\vec{k}-\vec{q}}n_{\vec{q}}\rangle^{(1)}(t=0) \tag{4.1}$$

$$\begin{aligned}\overline{F}^{(2)}(\vec{k},\vec{v}) = {} & -(ie/mV)\ \Sigma_{\vec{q}}\hat{E}(\vec{k}-\vec{q})\cdot\frac{\partial\overline{F}^{(1)}(\vec{q},\vec{v})/\partial\vec{v}}{\vec{k}\cdot\vec{v}} \\ & + (1/mNV)\ \Sigma_{\vec{q}}\frac{\partial\overline{F}^{(1)}(\vec{q},\vec{v})/\partial\vec{v}}{\vec{k}\cdot\vec{v}} \\ & \cdot\ \Sigma_{\vec{p}}\vec{p}\phi(p)\langle n_{\vec{k}-\vec{q}-\vec{p}}n_{\vec{p}}\rangle^{(1)}(t=0) \\ & - (1/mN^2)\ \Sigma_{\vec{q}}n^{(1)}(\vec{q})\ \frac{\partial F^{\circ}(v)/\partial\vec{v}}{\vec{k}\cdot\vec{v}} \\ & \cdot\ \Sigma_{\vec{p}}\vec{p}\phi(p)\langle n_{\vec{k}-\vec{q}-\vec{p}}n_{\vec{p}}\rangle^{(1)}(t=0) \\ & + (1/mN)\ \frac{\partial F^{\circ}(v)/\partial\vec{v}}{\vec{k}\cdot\vec{v}}\cdot\Sigma_{\vec{q}}\vec{q}\phi(q)\langle n_{\vec{k}-\vec{q}}n_{\vec{q}}\rangle^{(2)}(t=0)\ ,\end{aligned} \tag{4.2}$$

where

$$\overline{F}(\vec{k},\vec{v};\ t=0) = \Sigma_{s\geq 1}\overline{F}^{(s)}(\vec{k},\vec{v})\ ,$$

$$n(\vec{k},t=0) = \Sigma_{s\geq 1}n^{(s)}(\vec{k})\ ,$$

with $O(\hat{E}^s)$ smallness implied by the (s) superscript. Similarly, the $\langle\ldots\rangle^{(s)}$ brackets denote averaging with respect to the sth order perturbed ensemble.

Now, the nonequilibrium two-point functions are linked to equilibrium three- and four-point functions: from statistical mechanical perturbation theoretic calculations (Appendix A), we obtain

$$\langle n_{\vec{k}-\vec{q}}n_{\vec{q}}\rangle^{(1)}(t=0) = \frac{i\beta e}{Vk}\ \hat{E}(k)\langle n_{\vec{k}-\vec{q}}n_{\vec{q}}n_{-\vec{k}}\rangle^{(0)}\ , \tag{4.3}$$

$$\langle n_{\vec{k}-\vec{q}}\, n_{\vec{q}} \rangle^{(2)}(t=0) = -\frac{\beta^2 e^2}{2V^2} \Sigma_{\vec{p}} \frac{1}{p|\vec{k}-\vec{p}|}$$

$$\cdot\ \hat{E}(\vec{p})\ \hat{E}(\vec{k}-\vec{p}) \langle n_{-\vec{p}}\, n_{\vec{p}-\vec{k}}\, n_{\vec{k}-\vec{q}}\, n_{\vec{q}} \rangle^{(0)} , \qquad (4.4)$$

where the $\langle \ldots . \rangle^{(0)}$ brackets denote averaging with respect to the unperturbed ensemble.

Upon combining Eqs. (4.1) and (4.3) on the one hand, and Eqs. (4.1) to (4.4) on the other, and integrating the results over velocity space, one ultimately obtains the density responses

$$n^{(1)}(\vec{k}) = \frac{ik}{4\pi e} \alpha_0(k)\ \hat{E}(\vec{k}) \left[1 - \frac{\alpha_0(k)}{N^2} \Sigma_{\vec{q}} \frac{\vec{k}\cdot\vec{q}}{q^2} \langle n_{\vec{k}-\vec{q}}\, n_{\vec{q}}\, n_{-\vec{k}} \rangle^{(0)} \right] , \qquad (4.5)$$

$$n^{(2)}(\vec{k}) = \frac{1}{2N} \sum_{\vec{q}} n^{(1)}(\vec{q})\ n^{(1)}(\vec{k}-\vec{q})$$

$$+ \frac{ik}{4\pi e} \frac{\alpha_0(k)}{N^2} \frac{1}{V} \sum_{\vec{q}} \alpha_{20}(\vec{q},\vec{k}-\vec{q})\ \hat{E}(\vec{q})\ \hat{E}(\vec{k}-\vec{q})$$

$$X \left[\frac{\hat{\alpha}(q)}{1}{\over \alpha_0(q)} \sum_{\vec{p}} \frac{\vec{k}\cdot\vec{p}}{p^2} \langle n_{\vec{k}-\vec{q}-\vec{p}}\, n_{\vec{p}}\, n_{\vec{q}-\vec{k}} \rangle^{(0)} \right.$$

$$+ \frac{\hat{\alpha}(\vec{k}-\vec{q})}{1}{\over \alpha_0(|\vec{k}-\vec{q}|)} \sum_{\vec{p}} \frac{\vec{k}\cdot\vec{p}}{p^2} \langle n_{\vec{q}-\vec{p}}\, n_{\vec{p}}\, n_{-\vec{q}} \rangle^{(0)}$$

$$\left. - \sum_{\vec{p}} - \frac{\vec{k}\cdot\vec{p}}{p^2} \langle n_{-\vec{q}}\, n_{\vec{q}-\vec{k}}\, n_{\vec{k}-\vec{p}}\, n_{\vec{p}} \rangle^{(0)} \right] , \qquad (4.6)$$

where $\alpha_0(k) = 4\pi\beta ne^2/k^2$ and $\alpha_{20}(\vec{q},\vec{k}-\vec{q}) = 2\pi i\beta^2 ne^3/kq|\vec{k}-\vec{q}|$ are the static linear and quadratic Vlasov polarizabilities.

V. POLARIZABILITY RESPONSE FUNCTIONS

The static versions of Eqs. (2.12) and (2.13) are, in virtue of (3.6),

$$n^{(1)}(\vec{k}) = \frac{ik}{4\pi e} \hat{\alpha}_1(\vec{k})\ \hat{E}(\vec{k}) , \qquad (5.1)$$

$$n^{(2)}(\vec{k}) = \frac{ik}{4\pi e} \frac{1}{V} \sum_{\vec{q}} \hat{\alpha}_2(\vec{q},\vec{k}-\vec{q})\ \hat{E}(\vec{q})\ \hat{E}(\vec{k}-\vec{q}) , \qquad (5.2)$$

where we have adopted the more compact notation

$$\hat{\alpha}_1(\vec{k}) \equiv \hat{\alpha}_1(\vec{k},\omega = 0) \quad , \quad \hat{\alpha}_2(\vec{q},\vec{k}-\vec{q}) \equiv \hat{\alpha}_2(\vec{q},\mu = 0; \vec{k}-\vec{q},\omega-\mu = 0) \quad .$$

Then from Eqs. (4.5), (4.6), (5.1) and (5.2), one readily obtains the polarizability relations

$$\hat{a}_1(\vec{k}) = 1 - \frac{\alpha_0(k)}{N^2} \sum_{\vec{p}} \frac{\vec{k}\cdot\vec{p}}{p^2} \langle n_{\vec{k}-\vec{p}} n_{\vec{p}} n_{-\vec{k}} \rangle^{(0)} \quad , \tag{5.3}$$

$$\begin{aligned} \hat{a}''_2(\vec{q},\vec{k}-\vec{q}) = {} & \hat{a}_1(\vec{q})\, \hat{a}_1(\vec{k}-\vec{q}) \\ & + \frac{\alpha_0(k)}{N^2} \Bigg[\hat{a}_1(\vec{q}) \sum_{\vec{p}} \frac{\vec{k}\cdot\vec{p}}{p^2} \langle n_{\vec{k}-\vec{q}-\vec{p}} n_{\vec{p}} n_{\vec{q}-\vec{k}} \rangle^{(0)} \\ & + \hat{a}_1(\vec{k}-\vec{q}) \sum_{\vec{p}} \frac{\vec{k}\cdot\vec{p}}{p^2} \langle n_{\vec{q}-\vec{p}} n_{\vec{p}} n_{-\vec{q}} \rangle^{(0)} \\ & - \sum_{\vec{p}} \frac{\vec{k}\cdot\vec{p}}{p^2} \langle n_{-\vec{q}} n_{\vec{q}-\vec{k}} n_{\vec{k}-\vec{p}} n_{\vec{p}} \rangle^{(0)} \Bigg] \quad , \end{aligned} \tag{5.4}$$

where $\hat{a}_1(\vec{k}) \equiv \hat{\alpha}_1(k)/\alpha_0(k)$, $\hat{a}_2(\vec{q},\vec{k}-\vec{q}) \equiv i\hat{\alpha}_2(\vec{q},\vec{k}-\vec{q})/\alpha_{2}^{0}(\vec{q},\vec{k}-\vec{q})$, $\hat{a}''_2 = \mathrm{Im}\, \hat{a}_2$. Anticipating the use of fluctuation-dissipation theorems (FDTs) to replace the right-hand-side equilibrium three- and four-point functions in (5.3) and (5.4) by polarizabilities, we next expand these correlation functions as follows:

For Eq. (5.3) with $\vec{k} \neq 0$,

$$\langle n_{\vec{k}-\vec{p}} n_{\vec{p}} n_{-\vec{k}} \rangle^{(0)} = N(\delta_{\vec{p}} + \delta_{\vec{k}-\vec{p}}) \langle n_{\vec{k}} n_{-\vec{k}} \rangle^{(0)} + \langle n_{\vec{k}-\vec{p}} n_{\vec{p}} n_{-\vec{k}} \rangle^{(0)} \Big|_{\vec{p},\vec{k}-\vec{p}\neq 0} \quad ; \tag{5.5}$$

for Eq. (30) with $\vec{q},\vec{k}-\vec{q},\vec{k}\neq 0$,

$$\begin{aligned} \langle n_{\vec{k}-\vec{q}-\vec{p}} n_{\vec{p}} n_{\vec{q}-\vec{k}} \rangle^{(0)} = {} & N(\delta_{\vec{p}} + \delta_{\vec{k}-\vec{q}-\vec{p}}) \langle n_{\vec{k}-\vec{q}} n_{\vec{q}-\vec{k}} \rangle^{(0)} \\ & + \langle n_{\vec{k}-\vec{q}-\vec{p}} n_{\vec{p}} n_{\vec{q}-\vec{k}} \rangle^{(0)} \Big|_{\vec{p},\vec{k}-\vec{q}-\vec{p}\neq 0} \quad , \end{aligned} \tag{5.6}$$

$$\langle n_{\vec{q}-\vec{p}} n_{\vec{p}} n_{-\vec{q}} \rangle^{(0)} = N(\delta_{\vec{p}} + \delta_{\vec{q}-\vec{p}}) \langle n_{\vec{q}} n_{-\vec{q}} \rangle^{(0)} + \langle n_{\vec{q}-\vec{p}} n_{\vec{p}} n_{-\vec{q}} \rangle^{(0)} \Big|_{\vec{p},\vec{q}-\vec{p}\neq 0} \tag{5.7}$$

$$\begin{aligned} \langle n_{-\vec{q}} n_{\vec{q}-\vec{k}} n_{\vec{k}-\vec{p}} n_{\vec{p}} \rangle^{(0)} &= N(\delta_{\vec{p}} + \delta_{\vec{k}-\vec{p}}) \langle n_{-\vec{q}} n_{\vec{q}-\vec{k}} n_{\vec{k}} \rangle^{(0)} \\ &+ (\delta_{\vec{k}-\vec{q}-\vec{p}} + \delta_{\vec{q}-\vec{p}}) \langle n_{-\vec{q}} n_{\vec{q}} \rangle^{(0)} \langle n_{\vec{q}-\vec{k}} n_{\vec{k}-\vec{q}} \rangle^{(0)} \\ &+ \langle n_{-\vec{q}} n_{\vec{q}-\vec{k}} n_{\vec{k}-\vec{p}} n_{\vec{p}} \rangle^{(0)} \Big|_{\vec{p}\neq 0,\vec{k},\vec{q},\vec{k}-\vec{q}} \end{aligned} \tag{5.8}$$

Note that in (5.8), we have assumed the decomposition

$$\langle n_{-\vec{q}} n_{\vec{q}} n_{\vec{q}-\vec{k}} n_{\vec{k}-\vec{q}} \rangle^{(0)} = \langle n_{-\vec{q}} n_{\vec{q}} \rangle^{(0)} \langle n_{\vec{q}-\vec{k}} n_{\vec{k}-\vec{q}} \rangle^{(0)} \quad . \tag{5.9}$$

The replacement of the two-, three-, and four-point equal-time correlations in (5.5) to (5.8) by external polarizability functions is then effected by use of the classical static FDT relations [Golden, Kalman and Silevitch, 1972]

$$\langle n_{\vec{k}} n_{-\vec{k}} \rangle^{(0)} \Big|_{\vec{k}\neq 0} = N\hat{a}_1(\vec{k}) \quad , \tag{5.10}$$

$$\langle n_{\vec{k}-\vec{q}} n_{\vec{q}} n_{-\vec{k}} \rangle^{(0)} \Big|_{\vec{q},\vec{k}-\vec{q},\vec{k}\neq 0} = N\hat{a}''_2(\vec{q},\vec{k}-\vec{q}) \quad , \tag{5.11}$$

$$\langle n_{-\vec{q}} n_{\vec{q}-\vec{k}} n_{\vec{k}-\vec{p}} n_{\vec{p}} \rangle^{(0)} \Big|_{\substack{\vec{q},\vec{k}-\vec{q}\neq 0 \\ \vec{p}\neq 0,\vec{k},\vec{q},\vec{k}-\vec{q}}} = N\hat{a}_3(\vec{p}-\vec{k},\vec{k}-\vec{q},\vec{q}) \quad , \tag{5.12}$$

where

$$\hat{a}_3(\vec{p}-\vec{k},\vec{k}-\vec{q},\vec{q}) \equiv \hat{\alpha}_3(\vec{p}-\vec{k},\vec{k}-\vec{q},\vec{q})/\alpha_{3}^{0}(\vec{p}-\vec{k},\vec{k}-\vec{q},\vec{q})$$

and

$$\alpha_{3}^{0}(\vec{p}-\vec{k},\vec{k}-\vec{q},\vec{q}) = -\frac{2\pi\beta^3 ne^4}{3qp|\vec{p}-\vec{k}|\,|\vec{k}-\vec{q}|} = \frac{2\alpha_{2}^{0}(\vec{k}-\vec{p},\vec{p})\ \alpha_{2}^{0}(\vec{k}-\vec{q},\vec{q})}{3\alpha_{0}(\vec{k})} \tag{5.13}$$

is the Vlasov value of the static cubic polarizability.

Finally, one makes the transition from external to internal

polarizability response functions by means of the static relations (cf. Eqs. (2.8) to (2.10)):

$$\hat{a}_1(k) = \frac{a_1(k)}{\varepsilon(k)} \quad , \tag{5.14}$$

$$\hat{a}_2(\vec{q},\vec{k}-\vec{q}) = \frac{a_2(\vec{q},\vec{k}-\vec{q})}{\varepsilon(q)\ \varepsilon(|\vec{k}-\vec{q}|)\ \varepsilon(k)} \quad , \tag{5.15}$$

$$\hat{a}_3(\vec{p}-\vec{k},\vec{k}-\vec{q},\vec{q}) = \frac{1}{\varepsilon(|\vec{k}-\vec{p}|)\ \varepsilon(|\vec{k}-\vec{q}|)\ \varepsilon(q)\ \varepsilon(p)} \Bigg[a_3(\vec{p}-\vec{k},\vec{k}-\vec{q},\vec{q}) - \frac{\alpha_0(k)}{\varepsilon(k)}\, a_2(\vec{k}-\vec{p},\vec{p})\ a_2(\vec{k}-\vec{q},\vec{q}) - \frac{\alpha_0(|\vec{k}-\vec{q}-\vec{p}|)}{\varepsilon(|\vec{k}-\vec{q}-\vec{p}|)}\, a_2(\vec{k}-\vec{p}-\vec{q},\vec{q})\ a_2(\vec{k}-\vec{q}-\vec{p},\vec{p}) - \frac{\alpha_0(|\vec{q}-\vec{p}|)}{\varepsilon(|\vec{q}-\vec{p}|)}\, a_2(\vec{q}-\vec{p},\vec{k}-\vec{q})\ a_2(\vec{q}-\vec{p},\vec{p}) \Bigg] \quad , \tag{5.16}$$

where $\quad a_1(k) \equiv \alpha_1(\vec{k},\omega=0)/\alpha_0(k) \quad ,$

$$a_2(\vec{q},\vec{k}-\vec{q}) \equiv i\alpha_2(\vec{q},0;\ \vec{k}-\vec{q},0)/\alpha_{20}(\vec{q},\vec{k}-\vec{q}) \quad ,$$

$$a_3(\vec{p}-\vec{k},\vec{k}-\vec{q},\vec{q}) \equiv \alpha_3(\vec{p}-\vec{k},0:\ \vec{k}-\vec{q},0;\ \vec{q},0)/\alpha_{30}(\vec{p}-\vec{k},\vec{k}-\vec{q},\vec{q}) \quad ,$$

and $\quad \varepsilon(k) \equiv \varepsilon(\vec{k},\omega=0) = 1 + \alpha_1(\vec{k},0) \quad .$

From Eqs. (5.3) to (5.16), one then obtains the coupled internal polarizability equations

$$a_1(k) = 1 + v(k) \tag{5.17}$$

$$\hat{a}''_2(\vec{q},\vec{k}-\vec{q}) = \frac{\vec{k}\cdot\vec{q}}{k^2}\, a_1(|\vec{k}-\vec{q}|) + \frac{\vec{k}\cdot(\vec{k}-\vec{q})}{k^2}\, a_1(q)$$

$$- z(\vec{k}-\vec{q},\vec{q}) - z(\vec{q},\vec{k}-\vec{q}) + w(\vec{k}-\vec{q},\vec{q}) \quad , \tag{5.18}$$

where

$$v(k) = -\frac{\alpha_0(k)}{N} \sum_{\vec{p}\neq\vec{0},\vec{k}} \frac{\vec{k}\cdot\vec{p}}{p^2\varepsilon(p)\,\varepsilon(|\vec{k}-\vec{p}|)}\, a''_2(\vec{p},\vec{k}-\vec{p}) \quad , \tag{5.19}$$

$$z(\vec{k}-\vec{q},\vec{q}) = -\frac{\alpha_0(k)}{N} \sum_{\vec{p}\neq\vec{0},\vec{k},\vec{q},\vec{k}-\vec{q}} \frac{\vec{k}\cdot\vec{p}}{p^2\varepsilon(p)\,\varepsilon(|\vec{k}-\vec{p}|)} \frac{\alpha_0(|\vec{k}-\vec{q}-\vec{p}|)}{\varepsilon(|\vec{k}-\vec{q}-\vec{p}|)}$$

$$\times\ a''_2(\vec{k}-\vec{p}-\vec{q},\vec{q})\ a''_2(\vec{k}-\vec{q}-\vec{p},\vec{p}) \quad , \tag{5.20}$$

$$w(\vec{k}-\vec{q},\vec{q}) = -\frac{\alpha_0(k)}{N} \sum_{\vec{p}\neq\vec{0},\vec{k},\vec{q},\vec{k}-\vec{q}} \frac{\vec{k}\cdot\vec{p}}{p^2\varepsilon(p)\,\varepsilon(|\vec{k}-\vec{p}|)}$$

$$\times\ a_3(\vec{p}-\vec{k},\vec{k}-\vec{q},\vec{q}) \quad . \tag{5.21}$$

Eqs. (5.17) and (5.18) are the principal equations in a hierarchy of static internal polarizability equations. We remind the reader that this hierarchy is generated from successive perturbation expansions (in static $\hat{E}$) of the parent VAA <u>first</u> BBGKY kinetic equation coupled with the hierarchy of static FDTs. In particular, Eq. (5.17) is identical to the <u>second</u> BBGKY static equation

$$ng(k) = -\frac{k_D^2}{k^2 + k_D^2} \{1 + \frac{1}{V} \sum_{p} \frac{\vec{k}\cdot\vec{p}}{p^2} [g(|\vec{k}-\vec{p}|) + nh(\vec{p},\vec{k}-\vec{p})]\} \ , \tag{5.22}$$

$$k_D^2 = 4\pi\beta n e^2 \quad ,$$

relating the pair and triplet correlation functions g and h. To see this, one need only apply to (5.17) the linear and quadratic FDTs

$$S(k) = 1 + ng(k) = \hat{a}_1(k) \quad , \tag{5.23}$$

$$S(\vec{p},\vec{k}-\vec{p}) = 1 + ng(k) + ng(p) + ng(|\vec{k}-\vec{p}|) + n^2h(\vec{p},\vec{k}-\vec{p}) = \hat{a}''_2(\vec{p},\vec{k}-\vec{p}), \tag{5.24}$$

(written also in terms of the structure factors $S(k)$ and $S(\vec{p},\vec{k}-\vec{p})$ and) readily derived from (5.10) and (5.11). Evidently, the VAA is exact at the level of the first GKS static polarizability equation (5.17) [Kalman, Datta and Golden, 1975].

While Eq. (5.18) manifestly satisfies the $\vec{q}\leftrightarrow\vec{k}-\vec{q}$ interchange symmetry requirement, it nevertheless fails to satisfy the triangle symmetry requirement (cf. Eq. (5.11))

$$a_2(\vec{q},\vec{k}-\vec{q}) = a_2(-\vec{k},\vec{k}-\vec{q}) = a_2(\vec{q},-\vec{k}) \quad . \tag{5.25}$$

This defect is apparently due to the VAA and is rectified simply by replacing (5.18) by its symmetrized version

$$\begin{aligned} a''_2(\vec{q},\vec{k}-\vec{q}) = (1/3)\Bigg[& \vec{k}\cdot\vec{q}\left(\frac{1}{k^2}+\frac{1}{q^2}\right) a_1(|\vec{k}-\vec{q}|) \\ & + \vec{k}\cdot(\vec{k}-\vec{q})\left(\frac{1}{k^2}+\frac{1}{|\vec{k}-\vec{q}|^2}\right) a_1(q) \\ & - \vec{q}\cdot(\vec{k}-\vec{q})\left(\frac{1}{q^2}+\frac{1}{|\vec{k}-\vec{q}|^2}\right) a_1(k) \\ & - z(\vec{k}-\vec{q},\vec{q}) - z(\vec{q},\vec{k}-\vec{q}) - z(-\vec{k},\vec{q}) - z(\vec{q},-\vec{k}) \\ & - z(\vec{k}-\vec{q},-\vec{k}) - z(-\vec{k},\vec{k}-\vec{q}) + w(\vec{k}-\vec{q},\vec{q}) + w(-\vec{k},\vec{q}) \\ & + w(\vec{k}-\vec{q},-\vec{k})\Bigg] \quad . \end{aligned} \tag{5.26}$$

The question then arises: If upon applying to (5.26) the FDT relations (5.23), (5.24) and (cf. Eq. (5.12)),

$$\begin{aligned} S(\vec{q}-\vec{k},\vec{k}-\vec{p},\vec{p}) = {} & 1 + n[g(|\vec{k}-\vec{q}|) + g(|\vec{k}-\vec{p}|) + g(p) + g(|\vec{p}-\vec{q}|) \\ & + g(|\vec{k}-\vec{q}-\vec{p}|) + g(q) + g(k)] + n^2[h(\vec{q}-\vec{k},\vec{k}-\vec{p}) \\ & + h(\vec{q}-\vec{k},\vec{p}) + h(\vec{k}-\vec{p},\vec{p}) + h(\vec{q}-\vec{k},\vec{k}) \\ & + h(\vec{k}-\vec{p},\vec{p}+\vec{q}-\vec{k}) + h(\vec{p},\vec{q}-\vec{p})] \\ & + n^3 i(\vec{q}-\vec{k},\vec{k}-\vec{p},\vec{p}) = \hat{a}_3(\vec{q}-\vec{k},\vec{k}-\vec{p},\vec{p}) \quad , \end{aligned} \tag{5.27}$$

will the resulting equation be identical to the third BBGKY static equation involving the quartic correlation function i? This question, as yet unanswered, amounts to asking if the VAA is still exact at the level of the second GKS static polarizability equation (5.26).

Eqs. (5.17) and (5.26) are a pair of coupled nonlinear summational equations featuring a_1, a_2, and a_3 as unknowns. Self-consistency is guaranteed by an appropriate decomposition of a_3 entirely in terms of a_1 and a_2. Thus there remains the problem of choosing a_3 before (5.17) and (5.26) can be solved numerically. The choice must certainly be made in such a way that the correct long-wavelength behavior (discussed below) of a_1 and a_2 will be preserved and that a_1 and a_2 will satisfy known short-wavelength asymptotic requirements.

VI. LONG WAVELENGTH LIMIT

At long wavelengths, Eqs. (5.17) and (5.26) simplify to

$$a_1(k\to 0)\Big|_{GKS} = 1 + v \quad , \tag{6.1}$$

$$a_2''(\vec{q}\to 0, \vec{k}-\vec{q}\to 0)\Big|_{GKS} = 1 + v + w - z \quad , \tag{6.2}$$

where v, w and z are wave vector-independent constants. Consequently, the GKS a_1 and a_2 structurally satisfy the compressibility sum rules [Golden, Kalman and Datta, 1975]

$$a_1(k\to 0) = [\beta(\partial p/\partial n)_\beta]^{-1} \quad , \quad \left(mc^2 \equiv (\partial p/\partial n)_\beta\right) \tag{6.3}$$

$$a_2''(\vec{q}\to 0, \vec{k}-\vec{q}\to 0) = [\beta(\partial p/\partial n)_\beta]^{-2}[1 - (n/mc^2)(\partial^2 p/\partial n^2)_\beta] \quad , \tag{6.4}$$

for arbitrary values of the plasma parameter γ. The accuracy of our approximation scheme can be further tested by also going to the weak coupling limit ($\gamma \ll 1$). In this limit, one sets $\varepsilon = 1 + \alpha_0$, $a_2'' = a_3'' = 1$ and (consequently) $w(\vec{k}-\vec{q},\vec{q}) = v(k)$ in (5.19) to (5.21), so that at long wavelengths,

$$v(k\to 0) = \gamma/4$$

$$z(\vec{k}-\vec{q}\to 0, \vec{q}\to 0) = (\gamma/4) - (5\gamma/12) + (5\gamma/24)\,\frac{\vec{k}\cdot\vec{q}}{k^2} \quad .$$

From Eq. (5.17), one then obtains for the linear polarizability

$$a_1(k\to 0)\Big|_{GKS} = 1 + \frac{\gamma}{4} \quad , \qquad (\gamma \ll 1) \tag{6.5}$$

in complete agreement with its compressibility sum rule to $O(\gamma)$. Next, upon combining all terms in (5.26), one obtains for the quadratic polarizability

$$a''_2(\vec{q}\to 0, \vec{k}-\vec{q}\to 0)\Big|_{GKS} = 1 + \frac{5}{8}\gamma \quad , \qquad (\gamma \ll 1) \qquad (6.6)$$

likewise in complete agreement to $O(\gamma)$ with its recently established compressibility sum rule [Golden, Kalman and Datta, 1975]

$$a''_2(\vec{q}\to 0, \vec{k}-\vec{q}\to 0) = 1 + \frac{5}{8}\gamma + \frac{\gamma^2}{2}\left(\ln\gamma + .5772\ldots + \frac{83}{49}\right) \quad . \qquad (6.7)$$
$$(\gamma \ll 1)$$

Recently, we examined the long wavelength limit of the quadratic polarizabilities calculated from the three leading ocp approximation schemes [Golden and Kalman, 1977]. We found that while all three approximation schemes satisfy the linear compressibility sum rule, only the GKS scheme satisfies both the linear and quadratic rules. The significance of the satisfaction of the compressibility sum rules discussed in this section is probably related to the asymptotic $r_{12}\to\infty$, $r_{23}\to\infty$, $r_{31}\to\infty$ behavior of the pair and triplet correlation functions $g(r_{12})$ and $h(r_{12}, r_{23}, r_{31})$. This is implied by the FDT relations (5.23) and (5.24) connecting $g(k\to 0)$ to $a_1(k\to 0)$ and $h(\vec{p}\to 0, \vec{k}-\vec{p}\to 0)$ to $a_2(\vec{p}\to 0, \vec{k}-\vec{p}\to 0)$. The correct small wave number behavior of the GKS correlation functions suggests that their large r behavior is also correct.

VII. SHORT WAVELENGTH LIMIT

The requisite behavior of a_1 at short wavelengths ($k\to\infty$) can be determined by combining the linear FDT relation (5.23) with the fact that

$$g(k\to\infty) = -\frac{4\pi\gamma}{k_D k^2}\,\mathrm{Re}\,K_2\left(\frac{4i\gamma k}{k_D}\right)^{1/2} \simeq -Ae^{-B\sqrt{k}} \quad , \qquad (7.1)$$

$$A \equiv \frac{2\pi^{3/2}\gamma}{k_D k^2}\left(\frac{k_D}{\gamma k}\right)^{1/4}\left[0.9239\cos\sqrt{\frac{2\gamma k}{k_D}} + 0.3827\sin\sqrt{\frac{2\gamma k}{k_D}}\right] \quad ,$$

$$B \equiv (2\gamma/k_D)^{1/2} \quad .$$

One readily obtains

$$a_1(k\to\infty) \simeq \frac{1}{1-\alpha_0(k\to\infty)}\left[1 - \frac{nAe^{-B\sqrt{k}}}{1-\alpha_0(k\to\infty)}\right] \quad . \qquad (7.2)$$

Eqs. (5.17), (5.19) and (7.2), in turn, impose the following large k requirement on $\mathfrak{a}_2$:

$$\lim_{k\to\infty} \frac{1}{N} \sum_{\vec{p}} \frac{\vec{k}\cdot\vec{p}}{p^2\varepsilon(p)\ \varepsilon(|\vec{k}-\vec{p}|)}\ a''_2(\vec{p},\vec{k}-\vec{p})$$

$$= -\frac{1}{1-\alpha_0(k\to\infty)}\left[1-\frac{nAe^{-B\sqrt{k}}}{\alpha_0(k\to\infty)[1-\alpha_0(k\to\infty)]}\right] . \qquad (7.3)$$

Now, the $k\to\infty$ behavior of the GKS approximation scheme has yet to be assessed. In any case, the choice of $\mathfrak{a}_3$ (in terms of $\mathfrak{a}_1$ and $\mathfrak{a}_2$ for self-consistency; see Chapter 5) probably should be made in such a way that the large k requirement (7.3) is met while keeping intact at the same time the long wavelength formulae (6.2) and (6.6).

VIII. PAIR CORRELATION FUNCTION IN THE WEAK COUPLING LIMIT

Eqs. (5.17) and (5.23) combine to give the pair correlation formula

$$g(k) = -\frac{1}{n}\,\frac{\alpha_0(k) + v(k)[\alpha_0(k) - 1]}{1 + \alpha_0(k)[1 + v(k)]} . \qquad (8.1)$$

Setting $\varepsilon = 1 + \alpha_0$ and $\mathfrak{a}_2 = 1$ in (5.19) and expanding Eq. (8.1) in powers of γ, i.e., $g(k) = g_1(k) + g_2(k) + \ldots$, one readily obtains the well-known Debye-Hückel and O'Neil-Rostoker expressions [O'Neil and Rostoker, 1965]

$$g_1(k) = -\frac{1}{n}\,\frac{k_D^2}{k^2 + k_D^2} , \qquad (8.2)$$

$$g_2(k) = -\frac{1}{N}\,\frac{k^2k_D^2}{(k^2 + k_D^2)^2} \sum_{\vec{p}} \frac{\vec{k}\cdot\vec{p}}{p^2\varepsilon(p)}\, g_1(|\vec{k}-\vec{p}|) . \qquad (8.3)$$

These results are, of course, not at all surprising in view of the fact that Eq. (8.1) is the second BBGKY static equation.

IX. DYNAMICAL DIELECTRIC RESPONSE FUNCTION

In the remaining chapters, we develop the GKS expression for the full wave vector- and frequency-dependent dielectric response function and examine its behavior in the high frequency limit.

We begin by introducing into the VAA kinetic equation (3.4) the perturbing field $\hat{E}(\vec{k},\omega\neq 0)$. Then following a procedure similar to that in Chapters 3, 4 and 5, we obtain the GKS dynamical external polarizability

$$\hat{\alpha}_1(\vec{k},\omega) = \hat{\alpha}_0(\vec{k},\omega)[1 + R'(\vec{k},\omega) + iR''(\vec{k},\omega)] \quad , \tag{9.1}$$

where

$$\hat{\alpha}_0(\vec{k},\omega) = \alpha_0(\vec{k},\omega)/\varepsilon_0(\vec{k},\omega) \quad ,$$

$$\alpha_0(\vec{k},\omega) = \varepsilon_0(\vec{k},\omega) - 1 = (4\pi e^2/mk^2)\int d^3\vec{v}\, \frac{\vec{k}\cdot\partial F^\circ(v)/\partial\vec{v}}{\omega-\vec{k}\cdot\vec{v}}$$

$$R'(\vec{k},\omega) = -\frac{\alpha_0(k)}{N}\sum_{\vec{p}}\frac{\vec{k}\cdot\vec{p}}{p^2}\left[S(\vec{p},\vec{k}-\vec{p}) - \frac{\omega}{(2\pi)^2}P\int_{-\infty}^{\infty}d\mu\int_{-\infty}^{\infty}d\nu \right.$$

$$\left. \times\ \frac{S(\vec{p},\mu;\ \vec{k}-\vec{p},\nu)}{\omega-\mu-\nu}\right] \tag{9.2}$$

and

$$R''(\vec{k},\omega) = -\frac{\beta\omega e^2}{k^2V}\sum_{\vec{p}}\frac{\vec{k}\cdot\vec{p}}{p^2}\int_{-\infty}^{\infty}d\mu\ S(\vec{p},\mu;\ \vec{k}-\vec{p},\omega-\mu) \quad ; \tag{9.3}$$

$S(\vec{p},\mu;\ \vec{k}-\vec{p},\nu)$ and $S(\vec{p},\vec{k}-\vec{p})$ are dynamical and static structure factors, i.e.,

$$2\pi\delta(\omega-\mu-\nu)\ S(\vec{p},\mu;\ \vec{k}-\vec{p},\nu) = (1/N)\langle n_{\vec{p}}(\mu)\ n_{\vec{k}-\vec{p}}(\nu)\ n^*_{\vec{k}}(\omega)\rangle^{(0)}\Big|_{\vec{p},\vec{k}-\vec{p},\vec{k}\neq 0} \tag{9.4}$$

$$S(\vec{p},\vec{k}-\vec{p}) = \frac{1}{(2\pi)^2}\int_{-\infty}^{\infty}d\mu\int_{-\infty}^{\infty}d\nu\ S(\vec{p},\mu;\ \vec{k}-\vec{p},\nu) = (1/N)\langle n_{\vec{p}}n_{\vec{k}-\vec{p}}n_{-\vec{k}}\rangle^{(0)}\Big|_{\vec{p},\vec{k}-\vec{p},\vec{k}\neq 0} \tag{9.5}$$

with $n_{\vec{p}}(\mu) = \sum_{i=1}^{N}\int_{-\infty}^{\infty} dt\ \exp\{i[\mu t-\vec{p}\cdot\vec{x}_i(t)]\}$.

(Note that when $\omega=0$, one recovers precisely the result of combining the static Eqs. (5.3), (5.5) and (5.10).)

The replacement of the structure factors by quadratic

polarizabilities is effected by use of (5.11) and our previously established dynamical nonlinear FDT [Golden, Kalman and Silevitch, 1972]

$$-(1/2)\ S(\vec{p},\mu;\ \vec{k}-\vec{p},\nu) = \mathrm{Im}\left[\frac{\underset{2}{\hat{a}}(\vec{p},\mu;\ \vec{k}-\vec{p},\nu)}{\mu\nu} - \frac{\underset{2}{\hat{a}}(\vec{p},\mu;\ -k,-\mu-\nu)}{\mu(\mu+\nu)} - \frac{\underset{2}{\hat{a}}(-\vec{k},-\mu-\nu;\ \vec{k}-\vec{p},\nu)}{\nu(\mu+\nu)}\right] , \tag{9.6}$$

where

$$\underset{2}{\hat{a}}(\vec{p},\mu;\ \vec{k}-\vec{p},\nu) \equiv \frac{i\underset{2}{\hat{\alpha}}(\vec{p},\mu;\ \vec{k}-\vec{p},\nu)}{\underset{2}{\alpha_0}(\vec{p},\vec{k}-\vec{p})} . \tag{9.7}$$

Now, the passage from (9.1) to an expression for $\varepsilon(\vec{k},\omega)$ is somewhat involved due to the appearance of $\varepsilon(\vec{k},\omega)$ in the formula for R" after application of (9.6) to (9.3). We proceed as follows: From Eqs. (9.3) and (9.6) and in virtue of the Kramers-Kronig formula for $\underset{2}{\hat{a}}$,

$$R''(\vec{k},\omega) = \frac{\alpha_0(k)\omega}{2\pi N}\sum_{\vec{p}}\frac{k\cdot p}{p^2}\,\mathrm{Im}\int_{-\infty}^{\infty}\frac{d\mu}{\mu(\omega-\mu)}\,\underset{2}{\hat{a}}(\vec{p},\mu;\ \vec{k}-\vec{p},\omega-\mu)$$

$$-\frac{\alpha_0(k)}{2N}\sum_{\vec{p}}\frac{\vec{k}\cdot\vec{p}}{p^2}\,\mathrm{Re}[\underset{2}{\hat{a}}(\vec{p},0;\ -\vec{k},-\omega) + \underset{2}{\hat{a}}(-\vec{k},-\omega;\ \vec{k}-\vec{p},0)] . \tag{9.8}$$

Letting

$$u(\vec{k},\omega) = \frac{\alpha_0(k)\omega}{2\pi N}\sum_{\vec{p}}\frac{\vec{k}\cdot\vec{p}}{p^2}\int_{-\infty}^{\infty}\frac{d\mu}{\mu(\omega-\mu)}\,\frac{\underset{2}{a}(\vec{p},\mu;\ \vec{k}-\vec{p},\omega-\mu)}{\varepsilon(\vec{p},\mu)\ \varepsilon(\vec{k}-\vec{p},\omega-\mu)} , \tag{9.9}$$

$$v(\vec{k},\omega) = -\frac{\alpha_0(k)}{2N}\sum_{\vec{p}}\frac{\vec{k}\cdot\vec{p}}{p^2}\left[\frac{\underset{2}{a}(\vec{p},0;\ -\vec{k},-\omega)}{\varepsilon(p)\ \varepsilon^*(\vec{k}-\vec{p},\omega)} + \frac{\underset{2}{a}(-\vec{k},-\omega;\ \vec{k}-\vec{p},0)}{\varepsilon(|\vec{k}-\vec{p}|)\ \varepsilon^*(\vec{p},\omega)}\right] , \tag{9.10}$$

where

$$\underset{2}{a}(\vec{p},\mu;\ \vec{k}-\vec{p},\omega-\mu) \equiv \frac{i\underset{2}{\alpha}(\vec{p},\mu;\ \vec{k}-\vec{p},\omega-\mu)}{\underset{2}{\alpha_0}(\vec{p},\vec{k}-\vec{p})} , \tag{9.11}$$

Eq. (9.8) can be written in the more convenient form

$$R''(\vec{k},\omega) = \text{Im}\,\frac{u(\vec{k},\omega)}{\varepsilon(\vec{k},\omega)} + \text{Re}\,\frac{v(\vec{k},\omega)}{\varepsilon^*(\vec{k},\omega)}$$

$$\equiv \frac{1}{2|\varepsilon(\vec{k},\omega)|^2}\{i[u^*(\vec{k},\omega)\,\varepsilon(\vec{k},\omega) - u(\vec{k},\omega)\,\varepsilon^*(\vec{k},\omega)]$$

$$+ [v(\vec{k},\omega)\,\varepsilon(\vec{k},\omega) + v^*(\vec{k},\omega)\,\varepsilon^*(\vec{k},\omega)]\} \quad . \tag{9.12}$$

From Eqs. (9.1) and (9.12), one obtains

$$\varepsilon\varepsilon^* = \frac{(\varepsilon_0 + \alpha_0 u/2 + i\alpha_0 v^*/2)\,\varepsilon^* + (i\alpha_0 v/2 - \alpha_0 u^*/2)\,\varepsilon}{1 - \alpha_0 R'} \quad , \tag{9.13}$$

where it is to be understood that every quantity in (9.13) carries the argument $(\vec{k},\omega)$. Then, in virtue of the reality of the right-hand-side of (9.13),

$$\varepsilon(\vec{k}\omega) = \frac{|\varepsilon_o(\vec{k}\omega)|^2 + \text{Re}\,[\alpha_o(\vec{k}\omega)\,\varepsilon_o^*(\vec{k}\omega)(u + iv^*)]}{A^*(\vec{k}\omega)} \quad , \tag{9.14}$$

where

$$A(\vec{k},\omega) \equiv [\varepsilon_0(\vec{k},\omega) + \alpha_0(\vec{k},\omega)\,u/2 + i\alpha_0(\vec{k},\omega)\,v^*/2]$$

$$\times\,[1 - \alpha_0^*(\vec{k},\omega)\,R'(\vec{k},\omega)]$$

$$+ [i\alpha_0^*(\vec{k},\omega)\,v^*/2 + \alpha_0^*(\vec{k},\omega)\,u/2][1 - \alpha_0(\vec{k},\omega)\,R'(\vec{k},\omega)] \;, \tag{9.15}$$

again with the understanding that u, u^*, v, v^*, A and A^* in (9.14) and (9.15) are functions of $\vec{k}$ and ω. Eq. (9.14) is the desired result.

The collective mode behavior of the strongly coupled electron liquid in the wave number-frequency range $(3k^2/\beta m) \ll \omega_p^2 \equiv (4\pi ne^2)/m \simeq \omega^2$ can be determined by setting $\varepsilon(\vec{k},\omega)$ equal to zero giving $\omega_{\vec{k}} = \omega(\vec{k})$. Such a study is currently in progress and our findings will be reported at a later date.

X. HIGH FREQUENCY BEHAVIOR

Our assessment of the high frequency behavior of $\varepsilon(\vec{k},\omega)$ begins with the evaluation of $R'(\vec{k},\omega\to\infty)$. Upon expanding the principal part denominator of (9.2) and taking account of (9.5), one obtains

$$R'(\vec{k},\omega\to\infty) \simeq \frac{\alpha_0(k)}{N\omega^2} \frac{1}{(2\pi)^2} \sum_{\vec{p}} \frac{\vec{k}\cdot\vec{p}}{p^2} \int_{-\infty}^{\infty} d\mu \int_{-\infty}^{\infty} d\nu (\mu+\nu)^2 \, S(\vec{p},\mu;\ \vec{k}-\vec{p},\nu)$$

$$= -\frac{\alpha_0(k)}{N\omega^2} \sum_{\vec{p}} \frac{\vec{k}\cdot\vec{p}}{p^2} \left[\left(\frac{\partial}{\partial t'} + \frac{\partial}{\partial t''}\right)^2 S(\vec{p},t';\ \vec{k}-\vec{p},t'')\right]_{t'=t''=0}$$

$$= \frac{\alpha_0(k)}{N^2\omega^2} \sum_{\vec{p}} \frac{\vec{k}\cdot\vec{p}}{p^2} [\vec{p} \cdot \langle \vec{J}_{\vec{p}} n_{\vec{k}-\vec{p}} \vec{J}_{-\vec{k}} \rangle^{(0)} \cdot \vec{k}$$

$$+ (\vec{k}-\vec{p}) \cdot \langle \vec{J}_{\vec{k}-\vec{p}} n_{\vec{p}} \vec{J}_{-\vec{k}} \rangle^{(0)} \cdot \vec{k}]_{\vec{p},\vec{k}-\vec{p},\vec{k}\neq 0} \quad , \tag{10.1}$$

where

$$\vec{J}_{-\vec{k}} = \sum_{i=1}^{N} \vec{v}_i e^{i\vec{k}\cdot\vec{x}_i} \quad .$$

From the Appendix B calculation,

$$\frac{1}{N} \sum_{\vec{p}} \frac{\vec{k}\cdot\vec{p}}{p^2} [\vec{p} \cdot \langle \vec{J}_{\vec{p}} n_{\vec{k}-\vec{p}} \vec{J}_{-\vec{k}} \rangle^{(0)} \cdot \vec{k}$$

$$+ \quad (\vec{k}-\vec{p}) \cdot \langle \vec{J}_{\vec{k}-\vec{p}} n_{\vec{p}} \vec{J}_{-\vec{k}} \rangle^{(0)} \cdot \vec{k}]_{\vec{p},\vec{k}-\vec{p},\vec{k}\neq 0}$$

$$= k^2 (n/\beta m) \sum_{\vec{p}} \chi^2 [g(|\vec{k}-\vec{p}|) - g(p)] \quad , \tag{10.2}$$

where $\chi \equiv \vec{k}\cdot\vec{p}/kp$. The resulting expression for R',

$$R'(\vec{k},\omega\to\infty) \simeq \frac{\omega_p^2}{\omega^2} \frac{1}{N} \sum_{\vec{p}} \chi^2 [ng(|\vec{k}-\vec{p}|) - ng(p)] \quad , \tag{10.3}$$

then indicates that the high frequency expansion of the Eq. (9.2) denominator implies $\omega^2 \gg \omega_p^2$. Moreover, at long wavelengths ($k^2 \ll k_D^2$),

$$\frac{1}{N}\sum_{\vec{p}} \chi^2[ng(|\vec{k}-\vec{p}|) - ng(p)] \simeq k^2 \sum_{\vec{p}} \frac{1}{p^2} (1 - 4\chi^2)(1 - \chi^2)\, ng(p)$$

$$= - \frac{4\gamma I}{15\pi} \frac{k^2}{k_D^2} \quad , \tag{10.4}$$

where

$$I \equiv \left|\int_0^\infty dy\; ng(y)\right| \ , \quad y \equiv p/k_D .$$

The combination of Eqs. (10.3) and (10.4) therefore gives

$$R'(\vec{k},\omega) \simeq - \frac{4\gamma I}{15\pi} \frac{k^2}{k_D^2} \frac{\omega_p^2}{\omega^2} \quad , \quad (\omega^2 >> \omega_p^2,\ k^2 << k_D^2). \tag{10.5}$$

Next, we observe that for $\omega^2 >> (2k^2/k_D^2)\ \omega_p^2$,

$$A(\vec{k},\omega) \simeq [1 - \alpha_0'(\vec{k},\omega)\ R'(\vec{k},\omega)]$$

$$X\ [\varepsilon_0(\vec{k},\omega) + \alpha_0'(\vec{k},\omega)\ u(\vec{k},\omega) + i\alpha_0'(\vec{k},\omega)\ v^*(\vec{k},\omega)]$$

$$\simeq [1 - \alpha_0'(\vec{k},\omega)\ R'(\vec{k},\omega)]$$

$$X\ [\varepsilon_0'(\vec{k},\omega) + \alpha_0'(\vec{k},\omega)\ u(\vec{k},\omega) + i\alpha_0'(\vec{k},\omega)\ v^*(\vec{k},\omega)] \ , \tag{10.6}$$

since at such frequencies

$$\alpha_0''(\vec{k},\omega) = (\pi/2)^{1/2} \left(\frac{\omega}{\omega_p} \frac{k_D^3}{k^3}\right) \exp\left(- \frac{\omega^2}{2\omega_p^2} \frac{k_D^2}{k^2}\right) << |\alpha_0'(\vec{k},\omega)| \ .$$

Consequently,

$$\varepsilon(\vec{k},\omega) \simeq \frac{\varepsilon_0'(\vec{k},\omega)\ \Gamma(\vec{k},\omega)}{1 - \alpha_0'(\vec{k},\omega)\ R'(\vec{k},\omega)} \quad , \tag{10.7}$$

where

$$\Gamma(\vec{k},\omega) = \frac{\varepsilon_0'(\vec{k},\omega) + \alpha_0'(\vec{k},\omega)[u'(\vec{k},\omega) + v''(\vec{k},\omega)]}{\varepsilon_0'(\vec{k},\omega)+\alpha_0'(\vec{k},\omega)[u'(\vec{k},\omega)+v''(\vec{k},\omega)] - i\alpha_0'(\vec{k},\omega)[u''(\vec{k},\omega)+v'(\vec{k},\omega)]} \tag{10.8}$$

primed and double primed symbols referring to real and imaginary parts, respectively. As a comment en route, note that Eqs. (10.7) and (10.8) are valid for

$$\omega^2 \gg (2k^2/k_D^2)\ \omega_p^2 \qquad \text{(necessary and sufficient) if } k^2 > (1/2)k_D^2$$

or

$$\omega^2 \gg \omega_p^2 \qquad \text{(sufficient)} \qquad \text{if } k^2 < (1/2)k_D^2$$

Moreover, Eqs. (10.7) and (10.8) are seen to be suitable for the study of the electron liquid's collective mode behavior when $\omega^2 \simeq \omega_p^2$, $k^2 \ll k_D^2$. However, for frequencies near the plasma frequency, care must be exercised not to use (10.3) or (10.5) in Eq. (10.7).

Our problem now reduces to an assessment of the asymptotic $\omega\to\infty$ behavior of Γ and, in particular, of its real part

$$\Gamma'(\vec{k},\omega) = \frac{[\varepsilon_0' + \alpha_0'(u' + v'')]^2}{[\varepsilon_0' + \alpha_0'(u' + v'')]^2 + [\alpha_0'(u'' + v')]^2} \ . \tag{10.9}$$

From (9.9),

$$u''(\vec{k},\omega\to\infty) \simeq \frac{\alpha_0(k)}{2\pi N}\,\varepsilon'(\vec{k},\omega\to\infty) \sum_{\vec{p}} \frac{\vec{k}\cdot\vec{p}}{p^2} \int_{-\infty}^{\infty} \frac{d\mu}{\mu}\, \hat{a}_2''(\vec{p},\mu;\ \vec{k}-\vec{p},\omega-\mu)$$

$$+ \frac{\alpha_0(k)}{2\pi N}\,\alpha_1''(\vec{k},\omega\to\infty) \sum_{\vec{p}} \frac{\vec{k}\cdot\vec{p}}{p^2} \int_{-\infty}^{\infty} \frac{d\mu}{\mu}\, \hat{a}_2'(\vec{p},\mu;\ \vec{k}-\vec{p},\omega-\mu)$$

$$\simeq \frac{\alpha_0(k)}{2\pi N} \sum_{\vec{p}} \frac{\vec{k}\cdot\vec{p}}{p^2} \int_{-\infty}^{\infty} \frac{d\mu}{\mu}\, \hat{a}_2''(\vec{p},\mu;\ \vec{k}-\vec{p},\omega)$$

$$+ \frac{\alpha_0(k)}{2\pi N}\,\alpha_1''(\vec{k},\omega\to\infty) \sum_{\vec{p}} \frac{\vec{k}\cdot\vec{p}}{p^2} \int_{-\infty}^{\infty} \frac{d\mu}{\mu}\, \hat{a}_2'(\vec{p},\mu;\ \vec{k}-\vec{p},\omega)$$

$$= \frac{\alpha_0(k)}{2\ N} \sum_{\vec{p}} \frac{\vec{k}\cdot\vec{p}}{p^2}\, \hat{a}_2'(\vec{p},0;\ \vec{k}-\vec{p},\omega)$$

$$- \frac{\alpha_0(k)}{2N} \alpha''_1(\vec{k},\omega\to\infty) \sum_{\vec{p}} \frac{\vec{k}\cdot\vec{p}}{p^2} \hat{a}''_2(\vec{p},0;\vec{k}-\vec{p},\omega) \tag{10.10}$$

in virtue of the Kramers-Kronig formulae for $\hat{a}_2$. From the reality of the external quadratic polarizability tensor and the invariance of its longitudinal projection (with respect to $\vec{p}$, $\vec{k}-\vec{p}$ and $\vec{k}$) under spatial inversion,

$$\hat{a}'_2(\vec{p},0;\vec{k}-\vec{p},\omega) = -\hat{a}'_2(\vec{p},0;\vec{k}-\vec{p},-\omega) \quad , \tag{10.11a}$$

$$\hat{a}''_2(\vec{p},0;\vec{k}-\vec{p},\omega) = \hat{a}''_2(\vec{p},0;\vec{k}-\vec{p},-\omega) \quad . \tag{10.11b}$$

Since the corresponding external conductivity $\hat{\sigma}_2 = -(i\omega/4\pi)\hat{a}_2$ is bounded (no finite electric field perturbation is capable of inducing infinite current) as $\omega\to\infty$, it follows from (10.11a,b) that $\hat{a}'_2(\vec{p},0;\vec{k}-\vec{p},\omega\to\infty) \propto 1/\omega$ and $\hat{a}''_2(\vec{p},0;\vec{k}-\vec{p},\omega\to\infty) \propto 1/\omega^2$ at most.* Moreover, the maximum high frequency value of $\hat{a}'_2(\vec{p},0;\vec{k}-\vec{p},\omega)$ is further reduced by the fact that $\hat{\sigma}_2(\vec{p},0;\vec{k}-\vec{p},\omega)$ satisfies a Kramers-Kronig formula: The implication here is that $\hat{\sigma}_2(\vec{p},0;\vec{k}-\vec{p},\omega)$ itself must then tend to zero as $|\omega|\to\infty$ implying, in turn, that $\hat{a}_2(\vec{p},0;\vec{k}-\vec{p},\omega\to\infty) \propto 1/\omega|\omega|$ at most. This being the case, we see from (10.10) that $u''(\vec{k},\omega\to\infty)$ drops off at least as rapidly as $1/(\omega|\omega|)$. Similarly, from (9.10),

$$v'(\vec{k},\omega\to\infty) \simeq -\frac{\alpha_0(k)}{2N} \sum_{\vec{p}} \frac{\vec{k}\cdot\vec{p}}{p^2} \Big[\hat{a}'_2(\vec{p},0;-\vec{k},-\omega) + \hat{a}'_2(-\vec{k},-\omega;\vec{k}-\vec{p},0) \Big] \propto 1/(\omega|\omega|) \quad \text{at most.}$$

In view of the above estimates for u'' and v', it appears that as $\omega\to\infty$, the denominator term $\alpha_0'^2(u''+v')^2$ in Eq. (10.9) drops off at least as fast as $1/\omega^8$ so that one can take $\Gamma'(\omega\to\infty) = 1$. Then recalling that the Vlasov expression

$$\varepsilon'_0(\vec{k}\omega) \simeq 1 - \frac{\omega_p^2}{\omega^2} - 3\frac{k^2}{k_D^2}\frac{\omega_p^4}{\omega^4} \tag{10.12}$$

is derived under the condition that

---

*Note that the effect of $\hat{a}''_2(\vec{p},0;\vec{k}-\vec{p},\omega\to\infty)$ in (10.10) is further diminished by its attachment to $\alpha''_1(\vec{k},\omega\to\infty)$.

$$\omega^2 >> 3(k^2/k_D^2)\ \omega_p^2 \quad , \tag{10.13}$$

one has from Eqs. (10.3) and (10.7) that

$$\varepsilon'(\vec{k}.\omega\to\infty)\Big|_{GKS} \simeq 1 - \frac{\omega_p^2}{\omega^2} - \frac{\omega_p^4}{\omega^4}\left[3\frac{k^2}{k_D^2} + \frac{1}{N}\sum_{\vec{p}} \chi^2 \cdot [ng(|\vec{k}-\vec{p}|) - ng(p)]\right] \tag{10.14}$$

valid either for $\omega^2 >> 3(k^2/k_D^2)\ \omega_p^2$, $k^2 > k_D^2/3$

or for $\omega^2 >> \omega_p^2$, $k^2 < k_D^2/3$.

At long wavelengths ($k\to0$), Eq. (10.14) becomes [cf. (10.5) and (10.7)]

$$\varepsilon'(\vec{k}\to0,\omega\to\infty)\Big|_{GKS} \simeq 1 - \frac{\omega_p^2}{\omega^2} - 3\frac{k^2}{k_D^2}\frac{\omega_p^4}{\omega^4}\left(1 - \frac{4\gamma I}{45\pi}\right) , \tag{10.15}$$

$$(\omega^2 >> \omega_p^2,\ k^2 << k_D^2)$$

in exact agreement with the conductivity sum rule [Ichimaru, 1977]. Our results (10.14) and (10.15) hold both for weakly coupled ($\gamma<1$) and strongly coupled ($\gamma>1$) electron liquids.

XI. SUMMARY AND CONCLUSIONS

An approximation scheme has been proposed for the calculation of the wave vector- and frequency-dependent dielectric response function. The scheme combines successive perturbation expansions (in $\hat{E}$) of the first BBGKY kinetic equation in the velocity-average-approximation with equilibrium fluctuation-dissipation theorems. The principal result is a hierarchy of coupled GKS polarizability relations; only the first two of these relations have been dealt with in these lectures.

In the zero frequency limit, the first GKS equation linking the linear and quadratic polarizabilities, $\underset{1}{a}_{GKS}$ and $\underset{2}{a}_{GKS}$, is shown to be identical to the second BBGKY static equation linking the pair and triplet correlation functions. The correspondence between the second GKS polarizability equation (involving $\underset{1}{a}_{GKS}$, $\underset{2}{a}_{GKS}$ and $\underset{3}{a}_{GKS}$) and the third BBGKY static equation has yet to be established. Both $\underset{1}{a}_{GKS}$ and $\underset{2}{a}_{GKS}$ exactly satisfy their zero frequency long wavelength compressibility sum rules; the implication here is that their pair and triplet correlation function relatives have correct long range behavior. We have not yet assessed the short wavelength behavior of a_{1GKS} and a_{2GKS}. In this connection, we note that our approximation scheme is to be rendered self-consistent by an appropriate decomposition of a_{3GKS} entirely in terms of a_{1GKS} and a_{2GKS}. The choice of a_{3GKS}, however, will probably have to be made in such a way that the known short wavelength requirements on a_{1GKS} and a_{2GKS} are met while preserving at the same time their excellent long wavelength characteristics.

An expression has been derived for the GKS dielectric response function $\varepsilon(\vec{k}\omega)$. In the high frequency limit ($\omega^2 >> \omega_p^2$) and at long wavelengths ($k^2 << k_D^2$), it is shown that $\varepsilon_{GKS}(k\to 0, \omega\to\infty)$ exactly satisfies its conductivity sum rule.

APPENDIX A

In this Appendix, we show how the nonequilibrium two-point function can be expressed in terms of equilibrium three- and four-point functions. The calculation of the nonequilibrium two-point function first by ensemble averaging in the Eulerian picture, i.e.,

$$<n_{\underset{\sim}{k}-\underset{\sim}{q}} n_{\underset{\sim}{q}}>(t) = \prod_{i=1}^{N} \int d^3 \underset{\sim}{x}_i \int d^3 \underset{\sim}{v}_i \Omega(\Gamma,t)\, n_{\underset{\sim}{k}-\underset{\sim}{q}} n_{\underset{\sim}{q}} \quad , \tag{A1}$$

calls for solving the Liouville equation

$$\frac{\partial\Omega(\Gamma,t)}{\partial t} + i\, L(\Gamma,t)\, \Omega(\Gamma,t) = 0 \quad , \tag{A2}$$

$$L(\Gamma,t) \equiv -i[H(\Gamma,t),\ldots] \quad , \tag{A3}$$

for the time evolution of the nonequilibrium distribution function $\Omega(\Gamma,t)$.

The unperturbed state of the electron liquid in the infinite past is characterized by the macrocanonical distribution function

$$\Omega^{(0)}(\Gamma) = \frac{e^{-\beta H^{(0)}(\Gamma)}}{\int d\Gamma\, e^{-\beta H^{(0)}(\Gamma)}} \quad , \tag{A4}$$

where

$$H^{(0)}(\Gamma) = \sum_i \frac{p_i^2}{2m} + \frac{1}{2} \sum_{i \neq j}\sum \frac{e^2}{|\vec{x}_i - \vec{x}_j|}$$

is the unperturbed Hamiltonian and

$$\int d\Gamma \equiv \prod_{i=1}^{N} \int d^3\vec{x}_i\, d^3\vec{v}_i$$

is a more compact notation. The introduction of a sufficiently weak external potential $\hat{\phi}$ into the system then perturbs $H^{(0)}$ and $L^{(0)}$ by amounts

$$H^{(1)}(\Gamma,t) = -\sum_{i=1}^{N} e\hat{\phi}(\vec{x}_i,t) = -\frac{1}{V}\sum_{\vec{k}} e\hat{\phi}(\vec{k},t)\, n_{-\vec{k}} \quad , \tag{A5}$$

$$L^{(1)}(\Gamma,t) = -i[H^{(1)}(\Gamma,t),\ldots] = \frac{ie}{V}\sum_{\vec{k}} \hat{\phi}(\vec{k},t)[n_{-\vec{k}},\ldots] \quad . \tag{A6}$$

The subsequent perturbation of (A2) results in the formal solution

$$\Omega(\Gamma,t) = \Omega^{(0)}(\Gamma) + \Omega^{(1)}(\Gamma,t) + \Omega^{(2)}(\Gamma,t) + \ldots \tag{A7}$$

where

$$\Omega^{(1)}(\Gamma,t) = -i\int_0^\infty d\tau\; e^{-i\tau L^{(0)}} L^{(1)}(t-\tau)\;\Omega^{(0)} \quad , \tag{A8}$$

$$\Omega^{(2)}(\Gamma,t) = -\int_0^\infty d\tau\int_0^\infty d\tau'\; e^{-i\tau L^{(0)}} L^{(1)}(t-\tau)\; e^{-i\tau' L^{(0)}}$$

$$\cdot\; L^{(1)}(t-\tau-\tau')\;\Omega^{(0)} \quad , \tag{A9}$$

$$\cdot\;\cdot\;\cdot\;\cdot\;\cdot\;\cdot\;\cdot\;\cdot\;\cdot\;\cdot$$

Starting first with the calculation of $\Omega^{(1)}(\Gamma,t)$ and $\langle n_{\vec{k}-\vec{q}}\, n_{\vec{q}}\rangle^{(1)}(t)$, we have from (A6) that

$$L^{(1)}(t-\tau)\;\Omega^{(0)} = (ie/V)\sum_{\vec{k}''}\hat{\phi}(\vec{k}'',t-\tau)[n_{-\vec{k}''},\Omega^{(0)}]$$

$$= (i\beta e\Omega^{(0)}/V)\sum_{\vec{k}''}\hat{\phi}(\vec{k}'',t-\tau)[H^{(0)},n_{-\vec{k}''}]$$

$$= -(i\beta e\Omega^{(0)}/V)\sum_{\vec{k}''}\hat{\vec{E}}(\vec{k}'',t-\tau)\cdot\vec{J}_{-\vec{k}''} \quad , \tag{A10}$$

where

$$\vec{J}_{-\vec{k}''} = \sum_{i=1}^{N}\vec{v}_i e^{i\vec{k}''\cdot\vec{x}_i} \quad .$$

Consequently,

$$\Omega^{(1)}(\Gamma,t) = -\frac{\beta e\Omega^{(0)}}{V}\sum_{\vec{k}''}\int_0^\infty d\tau\;\hat{\vec{E}}(\vec{k}'',t-\tau)\cdot e^{-i\tau L^{(0)}}\vec{J}_{-\vec{k}''} \tag{A11}$$

and

$$\langle n_{\vec{k}-\vec{q}}\, n_{\vec{q}}\rangle^{(1)}(t) = -\frac{\beta e}{V}\sum_{\vec{k}''}\frac{1}{k''}\int_0^\infty d\tau\;\hat{E}(\vec{k}'',t-\tau)\int d\Gamma\Omega^{(0)} n_{\vec{k}-\vec{q}}\, n_{\vec{q}}$$

$$\cdot\ e^{-i\tau L^{(0)}}\vec{k}''\cdot\vec{J}_{-\vec{k}''}\quad . \tag{A12}$$

In order to ultimately express $\langle n_{\vec{k}-\vec{q}}n_{\vec{q}}\rangle^{(1)}(t)$ in terms of equilibrium two-time three-point functions, we next write (A12) as

$$\langle n_{\vec{k}-\vec{q}}(\Gamma(t))n_{q}(\Gamma(t))\rangle^{(1)}(t) \tag{A13}$$

$$= -(\beta e/V)\sum_{\vec{k}''}\frac{1}{k''}\int_0^{\infty}d\tau\ \hat{E}(\vec{k}'',t-\tau)\int d\Gamma(t)\ \Omega^{(0)}(\Gamma(t))n_{\vec{k}-\vec{q}}(\Gamma(t))$$

$$\cdot\ n_{\vec{q}}(\Gamma(t))\ e^{-i\tau L^{(0)}}\vec{k}''\cdot\vec{J}_{-\vec{k}''}(\Gamma(t))\quad . \tag{A13}$$

Now, by assigning the label $\Gamma(0)$ to a particular parcel of phase fluid at $t = 0$, its position in phase space at time t can be written in Lagrangian notation as $\Gamma(t) = \Gamma(\Gamma(0),t)$, so that (A13) becomes

$$= -(\beta e/V)\sum_{\vec{k}''}\frac{1}{k''}\int_0^{\infty}d\tau\ \hat{E}(\vec{k}'',t-\tau)\int d\Gamma(\Gamma(0),t)\ \Omega^{(0)}(\Gamma(0),t)$$

$$\cdot\ n_{\vec{k}-\vec{q}}(\Gamma(0),t)\ n_{\vec{q}}(\Gamma(0),t)\ e^{-i\tau L^{(0)}}\vec{k}''\cdot\vec{J}_{-\vec{k}''}(\Gamma(0),t)\quad . \tag{A14}$$

But

$$d\Gamma(\Gamma(0),t)\ \Omega^{(0)}(\Gamma(0),t) = d\Gamma(\Gamma(0),0)\ \Omega^{(0)}(\Gamma(0),0)\quad . \tag{A15}$$

Moreover, in this Lagrangian picture, while Ω remains stationary, dynamical quantities A like the microscopic charge and current densities evolve according to the equation

$$\frac{\partial A(\Gamma(0),t)}{\partial t} = iLA(\Gamma(0),t)\quad . \tag{A16}$$

In particular, we observe that

$$\vec{J}_{-\vec{k}''}(\Gamma(0),t) = e^{iL^{(0)}t}\vec{J}_{-\vec{k}''}(\Gamma(0),0) \tag{A17}$$

is the solution of the microscopic equation

$$\frac{\partial\vec{J}_{-\vec{k}''}(\Gamma(0),t)}{\partial t} = iL^{(0)}\vec{J}_{-\vec{k}''}(\Gamma(0),t)$$

since, in this picture, the operator L^0 is stationary. Then taking account of (A15) and the time shifting property (A17), Eq. (A14) becomes

$$= -(\beta e/V) \sum_{\vec{k}''} (1/k'') \int_0^\infty d\tau \; \hat{E}(\vec{k}'', t-\tau) \int d\Gamma(\Gamma(0),0) \; \Omega^{(0)}(\Gamma(0),0)$$

$$\cdot \; n_{\vec{k}-\vec{q}}(\Gamma(0),t) \; n_{\vec{q}}(\Gamma(0),t) \; \vec{k}'' \cdot \vec{J}_{-\vec{k}''}(\Gamma(0),t-\tau)$$

$$\equiv -(\beta e/V) \sum_{\vec{k}''} (1/k'') \int_0^\infty d\tau \; \hat{E}(\vec{k}'', t-\tau) \int d\Gamma(0) \; \Omega^{(0)}(0) \; n_{\vec{k}-\vec{q}}(t)$$

$$\cdot \; n_{\vec{q}}(t) \; k'' \cdot \vec{J}_{-\vec{k}''}(t-\tau)$$

$$= -(\beta e/V) \sum_{\vec{k}''} (1/k'') \int_0^\infty d\tau \; \hat{E}(\vec{k}'', t-\tau) \langle n_{\vec{k}-\vec{q}}(t) \; n_{\vec{q}}(t) \; \vec{k}'' \cdot \vec{J}_{-\vec{k}''}(t-\tau) \rangle^{(0)}$$

$$= -(\beta e/V) \sum_{\vec{k}''} (1/k'') \int_0^\infty d\tau \; \hat{E}(\vec{k}'', t-\tau) \langle n_{\vec{k}-\vec{q}}(0) \; n_{\vec{q}}(0) \; \vec{k}'' \cdot \vec{J}_{-\vec{k}''}(-\tau) \rangle^{(0)}$$

$$= (i\beta e/Vk) \Big[\hat{E}(\vec{k}, t) \langle n_{\vec{k}-\vec{q}}(0) \; n_{\vec{q}}(0) \; n_{-\vec{k}}(0) \rangle^{(0)}$$

$$- \frac{\partial}{\partial t} \int_0^\infty d\tau \; \hat{E}(\vec{k}, t-\tau) \langle n_{\vec{k}-\vec{q}}(0) \; n_{\vec{q}}(0) \; n_{-\vec{k}}(-\tau) \rangle^{(0)} \Big] \; , \qquad \text{(A18)}$$

where we have exploited the invariance of the equilibrium correlation under spatial and temporal translation. In conclusion,

$$\langle n_{\vec{k}-\vec{q}} n_{\vec{q}} \rangle^{(1)}(t) = (i\beta e/Vk) \Big[\hat{E}(\vec{k}, t) \langle n_{\vec{k}-\vec{q}} n_{\vec{q}} n_{-\vec{k}} \rangle^{(0)}$$

$$- \frac{\partial}{\partial t} \int_0^\infty d\tau \; \hat{E}(\vec{k}, t-\tau) \langle n_{\vec{k}-\vec{q}}(0) \; n_{\vec{q}}(0) \; n_{-\vec{k}}(-\tau) \rangle^{(0)} \Big] \; . \qquad \text{(A19)}$$

It can be similarly shown that to $O(\hat{E}^2)$,

$$\langle n_{\vec{k}-\vec{q}}\, n_{\vec{q}}\rangle^{(2)}(t)$$

$$= (\beta e^2/2V^2) \sum_{\vec{k}'} \frac{k'_\mu\, k''_\nu}{k'^2 k''^2} \int_{-\infty}^{\infty} d\tau' \int_{-\infty}^{\infty} d\tau''\ \hat{E}_\mu(\vec{k}', t-\tau')\ \hat{E}_\nu(\vec{k}'', t-\tau'')$$

$$\cdot\ \theta(\tau')\ \theta(\tau'')$$

$$\cdot \Bigg[-\beta \frac{\partial^2}{\partial\tau'\,\partial\tau''} \langle n_{-\vec{k}'}(-\tau')\ n_{-\vec{k}''}(-\tau'')\ n_{\vec{k}-\vec{q}}(0)\ n_{\vec{q}}(0)\rangle^{(0)}$$

$$+ \left(\frac{\partial}{\partial\tau''} \langle [n_{-\vec{k}'}(0), n_{-\vec{k}''}(\tau'-\tau'')]\ n_{\vec{k}-\vec{q}}(\tau')\ n_{\vec{q}}(\tau')\rangle^{(0)}\right)$$

$$\cdot\ \theta(\tau'' - \tau')$$

$$+ \left(\frac{\partial}{\partial\tau'} \langle [n_{-\vec{k}''}(0), n_{-\vec{k}'}(\tau''-\tau')]\ n_{\vec{k}-\vec{q}}(\tau'')\ n_{\vec{q}}(\tau'')\rangle^{(0)}\right)$$

$$\cdot\ \theta(\tau' - \tau'') \Bigg], \qquad \text{(A20)}$$

where $\vec{k}'' = \vec{k} - \vec{k}'$ and θ is the unit step function.

We note that for the case of static perturbations (cf. Eq. (3.6)) where $\hat{E}(\vec{k},t) = \hat{E}(\vec{k},t=0)$, Eqs. (A19) and (A20) reduce to (4.3) and (4.4).

APPENDIX B

The following calculation, made for the electron liquid ocp, shows how the current-density-current equilibrium correlation function reduces to the pair correlation function:

$$\left.\langle \vec{J}_{\vec{p}} n_{\vec{k}-\vec{p}} \vec{J}_{-\vec{k}} \rangle^{(0)}\right|_{\vec{p},\vec{k}-\vec{p},\vec{k}\neq 0} = \sum_{i=1}^{N} \sum_{j=1}^{N} \sum_{\ell=1}^{N} \int d\Gamma\, \Omega^{(0)} \vec{v}_i e^{-i\vec{p}\cdot\vec{x}_i} e^{-i(\vec{k}-\vec{p})\cdot\vec{x}_j} \times \vec{v}_\ell\, e^{i\vec{k}\cdot\vec{x}_\ell}$$

$$= \sum_i \sum_j \sum_\ell \int d\Gamma\, e^{-i\vec{p}\cdot\vec{x}_i} e^{-i(\vec{k}-\vec{p})\cdot\vec{x}_j} e^{i\vec{k}\cdot\vec{x}_\ell} \cdot \vec{v}_i\, \Omega^{(0)} \frac{\partial H^{(0)}}{\partial \vec{p}_\ell}$$

$$= -(1/\beta) \sum_i \sum_j \sum_\ell \int d\Gamma\, e^{-i\vec{p}\cdot\vec{x}_i} e^{-i(\vec{k}-\vec{p})\cdot\vec{x}_j} \cdot e^{i\vec{k}\cdot\vec{x}_\ell} \vec{v}_i \frac{\partial \Omega^{(0)}}{\partial \vec{p}_\ell}$$

$$= \overleftrightarrow{I}(1/\beta m) \sum_i \sum_j \int d\Gamma\, \Omega^{(0)} e^{i(\vec{k}-\vec{p})\cdot(\vec{x}_i-\vec{x}_j)}$$

$$= \overleftrightarrow{I}(1/\beta m) \langle n_{\vec{k}-\vec{p}} n_{\vec{p}-\vec{k}} \rangle^{(0)} = \overleftrightarrow{I}(N/\beta m)$$

$$\cdot \{[1 + ng(|\vec{k} - \vec{p}|)] + N\delta_{\vec{k}-\vec{p}}\} \quad , \qquad \text{(B1)}$$

where $\overleftrightarrow{I}$ is the unit tensor and where we have made use of Hamilton's equation $\vec{v}_\ell = \partial H^{(0)}/\partial \vec{p}_\ell$ (ℓ=1,2,...,N) and the form (A4) of the macrocanonical distribution $\Omega^{(0)}$. Similarly

$$\langle J_{\vec{k}-\vec{p}}\, n_{\vec{p}}\, J_{-\vec{k}} \rangle^{(0)} \Big|_{\vec{p},\vec{k}-\vec{p},\vec{k}\neq 0} = \overleftrightarrow{I}(N/\beta m)\{[1 + ng(p)] + N\delta_{\vec{p}}\} \quad . \qquad \text{(B2)}$$

We note that in applying (B1) and (B2) to Eq. (10.2), the terms $\overleftrightarrow{I}(N^2/\beta m)\delta_{\vec{k}-\vec{p}}$ and $\overleftrightarrow{I}(N^2/\beta m)\delta_{\vec{p}}$ should not be included since they are exactly cancelled by the $-\overleftrightarrow{I}(N^2/\beta m)\delta_{\vec{k}-\vec{p}}$ and $-\overleftrightarrow{I}(N^2/\beta m)\delta_{\vec{p}}$ contributions from the static ion background.

REFERENCES

Golden, K. I. and G. Kalman, to be published.

Golden, K. I., G. Kalman and T. Datta, 1975, Phys. Rev. A11, 2147.

Golden, K. I., G. Kalman and M. B. Silevitch, 1972, J. Stat. Phys. 6, 87.

Golden, K. I., G. Kalman and M. B. Silevitch, 1974, Phys. Rev. Lett. 33, 1544.

Ichimaru, S., 1977, Phys. Rev. A15, 744.

Kalman, G., T. Datta and K. I. Golden, 1975, Phys. Rev. A12, 1125.

O'Neil, T. and N. Rostoker, 1965, Phys. Fluids 8, 1109.

Singwi, K. S., A. Sjölander, M. P. Tosi and R. H. Land, 1969, Solid State Commun. 7, 1503.

Singwi, K. S., A. Sjölander, M. P. Tosi and R. H. Land, 1970, Phys. Rev. B1, 1044.

Singwi, K. S., M. P. Tosi, R. H. Land and A. Sjölander, 1968, Phys. Rev. 176, 589.

Totsuji, H. and S. Ichimaru, 1973, Progr. Theor. Phys. 50, 753; 1974, Progr. Theor. Phys. 52, 42.

Vashishta, P. and K. S. Singwi, 1973, Phys. Rev. B6, 875.

DYNAMIC BEHAVIOR OF ELECTRONS IN METALS

K. S. Singwi

Department of Physics and Astronomy
Northwestern University
Evanston, Illinois

TABLE OF CONTENTS

DYNAMIC BEHAVIOR OF ELECTRONS IN METALS

K. S. Singwi

Department of Physics and Astronomy
Northwestern University
Evanston, Illinois

I. INTRODUCTION

In these lectures we shall be chiefly concerned with theoretical aspects of the dynamic behavior of electrons in metals. During the last few years both x-ray and electron inelastic scattering experiments in nearly free electron-like metals have yielded very interesting results on the dynamic form factor $S(q,\omega)$ of the electron liquid. At the present time we do not have a full understanding of $S(q,\omega)$. As regards the behavior of electrons in the static case and in the long wavelength limit, I have discussed it in my Antwerp lectures [Singwi, 1976] given two years ago. This aspect of the problem is relatively simpler and better understood. For the sake of continuity and for those of you who are not very familiar with solid state plasmas, I shall review in my first lecture some of the preliminaries.

II. ELEMENTARY CONSIDERATIONS

The behavior of electrons in a real metal is very complicated. Not only do the electrons interact amongst themselves, they interact with the ionic lattice and with the lattice vibrations. As a first approximation, we shall consider the positive ions to be at rest and smeared out so that the electrons move against a uniform rigid positive background. Such a model of a metal is referred to as the jellium model and has long been of interest to theoretical physicists. This model has some aspect of reality too. The main reason for dealing with the jellium model is that anything else is much more difficult. It is not prudent to start with a complicated model when dealing with a complex many-body problem like the one we have at hand.

At zero temperature, the properties of the paramagnetic electron gas depend only on its density n, which is conveniently expressed in terms of a dimensionless parameter r_s through the relation

$$n = \left(\frac{4\pi}{3} r_s^3 a_B^3 \right)^{-1} , \tag{2.1}$$

where a_B is the Bohr radius. For metallic plasmas $n \sim 10^{23}$ cm^{-3} and r_s lies between 2 (for Aℓ) and 6 (for Cs). In terms of r_s, the ground state energy per electron is [for example, Pines and Nozieres, 1966]

$$\varepsilon = \left(\frac{2.22}{r_s^2} - \frac{0.916}{r_s} + \varepsilon_c \right) \text{Ryd.} \tag{2.2}$$

The first term is the kinetic energy and the second is the exchange energy arising as a result of Pauli principle. The last term is what is called the correlation energy. Actually it is the difference between the true energy ε and the first two terms of Eq. (2.2). The correlation energy can be calculated exactly only in the two extreme limits: one [Gell-Mann and Brueckner, 1957; for example, Pines and Nozieres, 1966] of very high density, i.e. $r_s \to 0$ and the other of very low density [Wigner, 1934; 1938], i.e. $r_s \to \infty$. In the former limit is is the kinetic energy that dominates and the system behaves like a gas of charged fermions, while in the latter limit the coulomb forces dominate the motion and the ground state is expected to be a zero-point lattice. In the intermediate region of r_s which is really of interest in metal physics, an exact calculation of $\varepsilon_c(r_s)$ has not been possible and one has to resort to approximate schemes. Different approximations yield values of $\varepsilon_c(r_s)$ which seem to agree within 10 to 20%. Wigner has given the following formula for $\varepsilon_c(r_s)$ which turns out to be reasonably good for all r_s values:

$$\varepsilon_c(r_s) = - \frac{0.88}{r_s + 7.8} \text{Ryd.} \tag{2.3}$$

A. Screening and Plasma Oscillations

Let us introduce a positive test charge in an electron gas. Electrons would gather around the +e charge producing a non-uniform density around it. In the Thomas-Fermi model the change in density is related to the electrostatic potential ϕ by

$$\frac{\delta n(r)}{n} = \frac{3}{2} \frac{\delta E_F}{E_F} = \frac{3}{2} \frac{\phi(r)\ e}{E_F} \tag{2.4}$$

where E_F is the Fermi energy. Poisson's equation is

$$\vec{\nabla} \cdot \vec{D}(r) \doteq \vec{\nabla} \cdot \vec{E}(r) - 4\pi e\ \delta n(r)$$
$$= \left(\nabla^2 - \frac{6\pi\ ne^2}{E_F}\right) \phi(r) \quad , \tag{2.5}$$

where we have made use of Eq. (2.4). Taking the Fourier transform of Eq. (2.5), and using the definition of the dielectric function

$$D(k) = \varepsilon(k)\ E(k) \quad , \tag{2.6}$$

we get

$$\varepsilon(k) = 1 + \frac{6\pi\ ne^2}{E_F} \frac{1}{k^2}$$

or

$$\varepsilon(k) = 1 + \frac{k_{FT}^2}{k^2} \quad , \tag{2.7}$$

where

$$k_{FT} = \left(\frac{6\pi\ ne^2}{E_F}\right)^{1/2}$$

is the inverse of the Thomas-Fermi screening length. It then follows from Eq. (2.7) that the potential in r-space due to the test charge +e is

$$V(r) = \int \frac{d^3k}{(2\pi)^3} \frac{4\pi e}{k^2 \varepsilon(k)} e^{-i\vec{k}\cdot\vec{r}} = \frac{e}{r} e^{-k_{FT} r} \quad . \tag{2.8}$$

Equation (2.8) says that the field of a point charge decays to zero over a distance of order $1/k_{FT}$. The electrons cluster around the +e charge and screen it.

If we now consider one of the electrons of the medium as a foreign charge, the same considerations would apply except that δn would be negative. So that each electron would create a hole around it. Besides this coulomb hole there would also arise a "hole" because of the Pauli principle where electrons of the same spin repel each other. Thus each electron in an electron gas is surrounded by a correlation hole in which it moves. This correlation hole is dynamic in nature. Our aim is to derive an expression for the dielectric function $\varepsilon(k,\omega)$ which is valid for all wave numbers and frequencies. Equation (2.7) is a very special case of $\varepsilon(k,\omega)$.

The kind of fluctuation of electron density which leads to screening, as discussed above, also leads to the plasma oscillation. When one creases a charge disturbance, the electrons would rush to screen it and in doing so they overshoot their mark and are thus pulled back. This results in a charge density oscillation about the state of charge neutrality. The equation of motion for the longitudinal particle current density $\vec{j}(\vec{r},t)$ is

$$\frac{m}{n} \frac{\partial \vec{j}(\vec{r},t)}{\partial t} = e\vec{E}(\vec{r},t) \quad . \tag{2.9}$$

Using the equation of continuity and Poisson's equation, the Fourier transform of the displaced electron density is

$$n(\omega) = n_e(\omega)\, \omega_p^2/(\omega^2 - \omega_p^2) \quad , \tag{2.10}$$

where $\omega_p = (4\pi\, ne^2/m)^{1/2}$ is the plasma frequency and $n_e(\omega)$ is the Fourier transform of the external charge density. Thus $n(\omega)$ has a resonance at $\omega = \omega_p$.

B. Dielectric Response Function

We have already introduced the notion of screening and in a trivial way obtained an expression for the dielectric function in the limit of long wavelength. Our ultimate aim is to derive an expression for the dielectric function which is valid for all wave vectors and frequencies. It is a useful function since it enters in the description of many observed properties of metals. Here I have particularly in mind the dynamic form factor $S(\vec{q},\omega)$ which is measured experimentally through inelastic x-ray and electron scattering and which is directly related to $\mathrm{Im}[1/\varepsilon(\vec{q},\omega)]$.

The dielectric function is defined through the relation

$$V_t(\vec{q},\omega) = V_{ext}(\vec{q},\omega)/\varepsilon(\vec{q},\omega) \quad , \tag{2.11}$$

where V_t is the total potential felt by a test charge, i.e.,

$$V_t(\vec{q},\omega) = V_{ext}(\vec{q},\omega) + \frac{4\pi e^2}{q^2} \langle\rho(\vec{q},\omega)\rangle \quad , \tag{2.12}$$

and where $\langle\rho(\vec{q},\omega)\rangle$ is the induced density due to the external potential $V_{ext}(\vec{q},\omega)$, given by

$$\langle\rho(\vec{q},\omega)\rangle = \chi(\vec{q},\omega)\ V_{ext}(\vec{q},\omega) \quad . \tag{2.13}$$

From the above equations, it follows that

$$\frac{1}{\varepsilon(\vec{q},\omega)} = 1 + \frac{4\pi\ e^2}{q^2}\ \chi(\vec{q},\omega) \quad , \tag{2.14}$$

χ is called the density-density response function.

In the Born approximation, the scattering cross section is related [for example, Pines and Nozieres, 1966] to the function $S(\vec{q},\omega)$ defined by

$$S(\vec{q},\omega) = \sum_n |(\rho_q^+)_{no}|^2\ \delta(\omega - \omega_{no}) \quad , \tag{2.15}$$

where $(\rho_q^+)_{no}$ is the matrix element of the density fluctuation ρ_q^+ between the ground state $|0\rangle$ and the excited state $|n\rangle$ of the system and $\omega_{no} = E_n - E_o$. Here $\vec{q}$ and ω are, respectively, the momentum and energy transfer to the system by the probe. $S(\vec{q},\omega)$ gives information about the excitation spectrum of density fluctuations. It is real and positive and vanishes for negative values of ω at zero temperature.

$S(\vec{q},\omega)$ is closely related by $\chi(\vec{q},\omega)$. In the linear response theory, the latter is given by [for example, Pines and Nozieres, 1966]

$$\chi(\vec{q},\omega) = \sum_n |(\rho_q^+)_{no}|^2 \left(\frac{1}{\omega - \omega_{no} + i\eta} - \frac{1}{\omega + \omega_{no} + i\eta}\right) . \tag{2.16}$$

From Eqs. (2.14), (2.15) and (2.16) it follows that

$$\text{Im}\ \frac{1}{\varepsilon(\vec{q},\omega)} = -\frac{4\pi^2\ e^2}{q^2}\ S(\vec{q},\omega) \quad . \tag{2.17}$$

Hence

$$S(\vec{q}) = \frac{1}{n}\int S(\vec{q},\omega)\ d\omega = -\frac{q^2}{4\pi^2\ ne^2}\int_o^\infty \text{Im}\left(\frac{1}{\varepsilon(\vec{q},\omega)}\right) d\omega \tag{2.18}$$

The above equation is referred to as the quantum mechanical fluctuation-dissipation theorem. Knowing $\varepsilon(\vec{q},\omega)$, we can calculate $S(\vec{q})$ from Eq. (2.18) and hence the pair correlation function g(r).

Some of the important sum rules are:

(i) The compressibility sum rule,

$$\lim q \to 0 \ \varepsilon(q,0) = 1 + K \frac{k_{FT}^2}{q^2} \quad , \tag{2.19}$$

where K is the ratio of the compressibility of the interacting to that of the non-interacting electron gas.

(ii) The f-sum rule,

$$\int_0^\infty d\omega \ \omega \ S(\vec{q},\omega) = \frac{nq^2}{2m} \quad . \tag{2.20}$$

(iii) The third moment sum rule [Puff, 1965]

$$\begin{aligned} \int_0^\infty d\omega \ \omega^3 \ S(\vec{q},\omega) &= \frac{nq^2}{2m} \left[\left(\frac{q^2}{2m}\right)^2 + 4\left(\frac{q^2}{2m}\right) T_{KE} \right] \\ &+ \left(\frac{nq^2}{m}\right)^2 \frac{2\pi e^2}{q^2} \\ &+ \frac{n}{2m^2} \sum_{\vec{k}} (\vec{q}\cdot\vec{k})^2 \frac{4\pi e^2}{k^2} [S(|\vec{q}-\vec{k}|-S(\vec{k})] \quad , \end{aligned} \tag{2.21}$$

where T_{KE} is the kinetic energy per particle in the interacting system. A good dielectric function should satisfy the above sum rules.

III. EQUATION OF MOTION APPROACH

Instead of following the conventional diagrammatic technique to calculate the density response function $\chi(\vec{q},\omega)$, I shall adopt the equation of motion approach for two main reasons: (i) In the former approach even in the simplest approximation such as the RPA, one needs to sum an infinite set of diagrams and it is not obvious as to which are the most relevant diagrams to sum as one tries to go beyond the RPA. (ii) In the latter it is easier to make contact with the phenomenological theories which have had reasonable success. Besides, this approach has the "flavor" of classical fluid dynamics and is, therefore, familiar to many physicists. Since we are dealing with the problem of an electron gas which is quantum mechanical in nature, we start with the Wigner distribution functions which are the counterparts of the classical phase-space distribution functions.

A. Wigner Distribution Functions

[For example, Balescu, 1975]. The one-particle Wigner function in coordinate space is defined as

$$f^{(1)}_{\vec{P}\sigma}(\vec{R},t) = \int d^3r\, e^{-i\vec{p}\cdot\vec{r}} \langle\psi^+_\sigma(R - \vec{r}/2;\ t)\ \psi_\sigma(R + \vec{r}/2;\ t)\rangle \quad , \tag{3.1}$$

where $\psi^+_\sigma(\vec{r},t)$ and $\psi_\sigma(\vec{r},t)$ are, respectively, the creation and annihilation operators for an electron of spin σ at the space-time point $(\vec{r},t)$, and the bracket $\langle\ \rangle$ denotes the expectation value in the ground state of the many particle system. Note the following properties of $f^{(1)}$:

$$\sum_{\vec{P}} f^{(1)}_{\vec{P}}(\vec{R},t) = \langle\psi^+_\sigma(\vec{R},t)\ \psi_\delta(\vec{R},t)\rangle = \langle n_\sigma(\vec{R},t)\rangle \quad . \tag{3.2}$$

$$\int d^3R\ f^{(1)}_{\vec{P}\sigma}(\vec{R},t) = \langle\psi^+_\sigma(\vec{p},t)\ \psi_\sigma(\vec{p},t)\rangle \quad . \tag{3.3}$$

The Fourier transform of $f^{(1)}$ is defined as:

$$f^{(1)}_{\vec{P}\sigma}(\vec{q},t) = \int d^3R\ e^{-i\vec{q}\cdot\vec{R}}\ f^{(1)}_{\vec{P}\sigma}(\vec{R},t)$$

$$= \langle\psi^+_\sigma(\vec{p} - \frac{\vec{q}}{2}\ ,\ t)\ \psi_\delta(\vec{p} + \frac{\vec{v}}{2}\ ,\ t)\rangle \tag{3.4}$$

Similarly, we define the 2-particle Wigner function as

$$f^{(2)}_{\vec{P}\sigma,P'\sigma'}(\vec{R},\vec{R};\ t) + f^{(1)}_{\vec{P}\sigma}(\vec{R},t)\ f^{(1)}_{\vec{P}'\sigma'}(\vec{R}',t)$$

$$= \int d^3r \int d^3r\quad {-i\vec{p}\cdot\vec{r}-i\vec{p}'\cdot\vec{r}'}\ \langle\psi^+_\sigma(\vec{R} - \frac{\vec{r}}{2}\ ,\ t)$$

$$\cdot\ \psi^+_{\sigma'}(\vec{R}' - \frac{\vec{r}'}{2}\ ,\ t)\ \psi_{\sigma'}(\vec{R}' + \frac{\vec{r}'}{2}\ ,\ t)\ \psi_\sigma(\vec{R} + \frac{\vec{r}}{2}\ ,\ t)\rangle \quad . \tag{3.5}$$

And the corresponding Fourier transform is defined as

$$f^{(2)}_{\vec{P}\sigma,\vec{P}'\sigma'}(\vec{q},\vec{q}',t) + f^{(1)}_{\vec{P}\sigma}(\vec{q},t)\, f^{(1)}_{\vec{p}'\sigma'}(\vec{q}',t)$$

$$= \langle \psi^{+}_{\vec{p}-(q/2)\sigma}(+)\, \psi^{+}_{\vec{p}'-(q'/2)\sigma'}(t)\, \psi_{\vec{p}'+(q'/2)\sigma'}(t)$$

$$\cdot\, \psi_{\vec{p}+(q/2)\sigma}(t)\rangle \qquad (3.6)$$

Henceforth, we shall be using a^+ and a for the creation and annihilation operators, respectively, instead of ψ^+ and ψ.

The Hamiltonian of a system of N electrons on a uniform positive background and perturbed by an infinitessimal weak external field is:

$$H = \frac{\hbar^2}{2m} \sum_{\vec{k}\sigma} k^2\, a^{+}_{\vec{k}\sigma}\, a_{\vec{k}\sigma} + \frac{1}{2V} \sum_{\vec{q}} v(q) \sum_{\vec{k}\sigma} \sum_{\vec{k}'\sigma'} a^{+}_{\vec{k}-(1/2)\vec{q}\sigma}$$

$$a^{+}_{\vec{k}'+(1/2)\vec{q}\sigma'}\, a_{\vec{k}'-(\vec{q}/2)\sigma'}\, a_{\vec{k}+(1/2)\vec{q}\sigma} + \frac{1}{V} \sum_{\vec{q}} \phi^{ext}(-\vec{q},t)$$

$$\sum_{\vec{k}\sigma} a^{+}_{\vec{k}-(1/2)\vec{q}\sigma}\, a_{\vec{k}+(1/2)\vec{q}\sigma} \quad , \qquad (3.7)$$

where

$$V(q) = \frac{4\pi e^2}{q^2} \qquad \text{if } q \neq 0$$

$$= 0 \qquad \text{if } q = 0 \qquad (3.8)$$

In the presence of the external field the one-particle Wigner function is perturbed and we write it as

$$f^{(1)}_{\vec{k}\sigma}(\vec{q},t) = \langle a^{+}_{\vec{k}-(1/2)\vec{q}\sigma}(t)\, a_{\vec{k}+(1/2)\vec{q}\sigma}(t)\rangle$$

$$= \delta^{o}_{\vec{q}}\, n_{\vec{k}\sigma} + \bar{f}^{(1)}_{\vec{k}\sigma}(\vec{q},t) \quad , \qquad (3.9)$$

where the first term on the right-hand-side corresponds to the equilibrium situation and the second term denotes deviation from equilibrium. The operator equation of motion is

$$i\hbar \frac{d}{dt}\left[a^{+}_{\vec{k}-(1/2)\vec{q}\sigma}(t)\; a_{\vec{k}+(1/2)\vec{q}\sigma}(t)\right]$$

$$= \left[a^{+}_{\vec{k}-(1/2)\vec{q}\sigma}(t)\; a_{\vec{k}+(1/2)\vec{q},\sigma}(t),\; H\right] \tag{3.10}$$

Using the usual commutation rules for a^+ and a, the commutator in Eq. (3.10) can be straightforwardly evaluated and remembering that the deviation from equilibrium $\bar{f}^{(1)}$ is small, we get the following equation [Niklasson, 1974] for $\bar{f}^{(1)}(\vec{q},\omega)$

$$\left(\hbar\omega - \frac{\hbar^2}{m}\vec{k}\cdot\vec{q}\right)\bar{f}^{(1)}_{\vec{k}\sigma}(\vec{q},\omega) = \frac{1}{V}\left(n_{\vec{k}-(q/2)\sigma} - n_{\vec{k}+(\vec{q}/2)\sigma}\right)$$

$$\cdot\ [\phi^{ext}(\vec{q},\omega) + v(q)\,\bar{n}(\vec{q},\omega)] + \frac{1}{V}\sum_{\vec{q}'} v(q') \sum_{\vec{k}'\sigma'}$$

$$\cdot\ \bar{f}^{(2)}_{\vec{k}-(\vec{q}'/2)\sigma,k'\sigma'}(\vec{q}-\vec{q}',\vec{q}';\omega)$$

$$\cdot\,\bar{f}^{(2)}_{\vec{k}+(\vec{q}'/2)\sigma,k'\sigma'}(\vec{q}-\vec{q}',\vec{q}',\omega)\quad, \tag{3.11}$$

where $\bar{f}^{(2)}$ is the deviation from equilibrium of the two-particle Wigner distribution function. Equation (3.11) is exact and is, obviously, the first in the hierarchy of an infinite set of equations. Niklasson [1974] was able to go one step further, i.e. he was able to write the equation for $\bar{f}^{(2)}$ which involves 3-particle correlation function. One would like to close this hierarchy by some suitable approximation which hopefully would contain most of the important physics in it. Attempts in this direction are now being made but so far not much success has been achieved. We shall return to this question later.

Notice that the first term on the right-hand-side of Eq. (3.11) corresponds to the Hartree mean field and all the effects of exchange and coulomb correlations are contained in the last term via the induced change in the correlated part of the two-particle distribution function. If we neglect this term entirely, we get the standard RPA result, i.e.,

$$\varepsilon(\vec{q},\omega) = 1 - v(q)\,\chi^{o}(\vec{q},\omega) \quad , \qquad (3.12)$$

where $\chi^{o}(\vec{q},\omega)$ is the Lindhard function. Thus within the RPA, the electrons respond as free particles to the Hartree field. We mentioned earlier while discussing screening that each electron in an electron gas is surrounded by a correlation hole which arises because of two-particle correlations. This effect is obviously not present in RPA.

B. Mean Field Theories

Theories which go beyond RPA attempt to take into account the presence of the correlation hole around each electron in a semi-phenomenological way by modifying the Hartree field. Equation (3.11) is approximated by

$$\overline{n}(\vec{q},\omega) = \chi^{o}(\vec{q},\omega)[\phi^{ext}(\vec{q},\omega) + v(q)[1 - G(\vec{q},\omega)]\,\overline{n}(\vec{q},\omega)] \quad , \qquad (3.13)$$

where $G(\vec{q},\omega)$ is an unknown function representing all the complications arising from the last term of Eq. (3.11). Obviously, it is a gross simplification. In general it is a complex function. Besides, it is even doubtful that Eq. (3.11) can be written in the above simple form. Interaction between particles should also lead to a modification of $\chi^{o}(q,\omega)$. In the static case, Singwi et al [1968] have given the following expression for G(q).

$$G(q) = -\frac{1}{n}\int \frac{d\vec{q}\,'}{(2\pi)^3}\frac{\vec{q}\cdot\vec{q}\,'}{q'^2}\,[S(|\vec{q} - \vec{q}\,'|) - 1] \quad , \qquad (3.14)$$

which leads to a simple mean-field type expression for the dielectric function

$$\varepsilon(q,\omega) = 1 - \frac{v(q)\,\chi^{o}(q,\omega)}{1 + v(q)\,G(q)\,\chi^{o}(q,\omega)} \qquad (3.15)$$

A new and interesting feature of this theory is that it is self-consistent. It was later modified [Vashishta and Singwi, 1972] to take care of the compressibility sum rule. In its modified verison the theory has been fairly successful in its use in the calculation of phonon spectra and correlation energy. For a detailed discussion of this approach as well as of other authors, we refer to the review article by Singwi and Tosi [to appear, 1978].

IV. DYNAMICAL BEHAVIOR OF THE ELECTRON LIQUID

During recent years both inelastic electron and x-ray scattering experiments have been performed to measure the energy loss spectra over a wide range of momentum transfer. I shall summarize here some of the more important experimental results in metals:

(a) The plasmon dispersion relation in the small wave vector region $q << q_c$ (q_c being the critical wave vector at which the plasmon joins the particle hole continuum) is well-represented by

$$\omega_p(q) = \omega_p(0) + \alpha\hbar q^2/m \quad , \tag{4.1}$$

where the coefficient α for $A\ell$ ranges from 0.38 ± 0.02 (Batson et al [1976]) to 0.42 (Gibbons et al [1976]). Höhberger et al [1975]) gave $\alpha = 0.401$ whereas Zacharias' [1975] value is 0.40.

(b) For $q \gtrsim q_c$ (q_c for $A\ell$ is $0.8\ q_F$), the line shape is strongly broadened and is asymmetric. The more recent electron scattering results of Batson et al [1976] are shown in Figure 1. In this region Zacharias [1975] sees an almost dispersionless peak, which Batson et al believe to be due to multiple scattering.

(c) In the region $q \gtrsim 1.5\ q_F$ the electron scattering experiments are affected by poor energy resolution. In this region, up to $q \simeq 2q_F$, x-ray scattering experiments [Platzman and Eisenberger, 1974] yield interesting and unexpected results. The excitation spectrum has a double peak structure or a shoulder on the high energy side of the main peak, as shown in Figure 2. For $q < 2q_F$ the latter has larger strength than the former and at $q \simeq 2q_F$ there occurs a switching over of the two strengths. This peak is reported to show almost no dispersion in the entire region $q_c < q < 2q_F$. These features have been observed to be common to such diverse systems as $A\ell$, Be and graphite, thus suggesting that they are an electron gas property rather than a one-electron band structure effect. There seems to be some discrepancy between the electron scattering results of Batson et al [1976] and the x-ray scattering results of Platzman and Eisenberger [1974] in the region where they overlap.

(d) For $q > 2q_F$, the x-ray scattering experiment shows a single broad peak corresponding to free particle excitations.

Can we understand these features on the basis of the approximate theories of the dielectric function? First of all, the mean field type of theories for the dielectric response function give the following expression [Singwi and Tosi, to appear, 1978] for the plasmon dispersion,

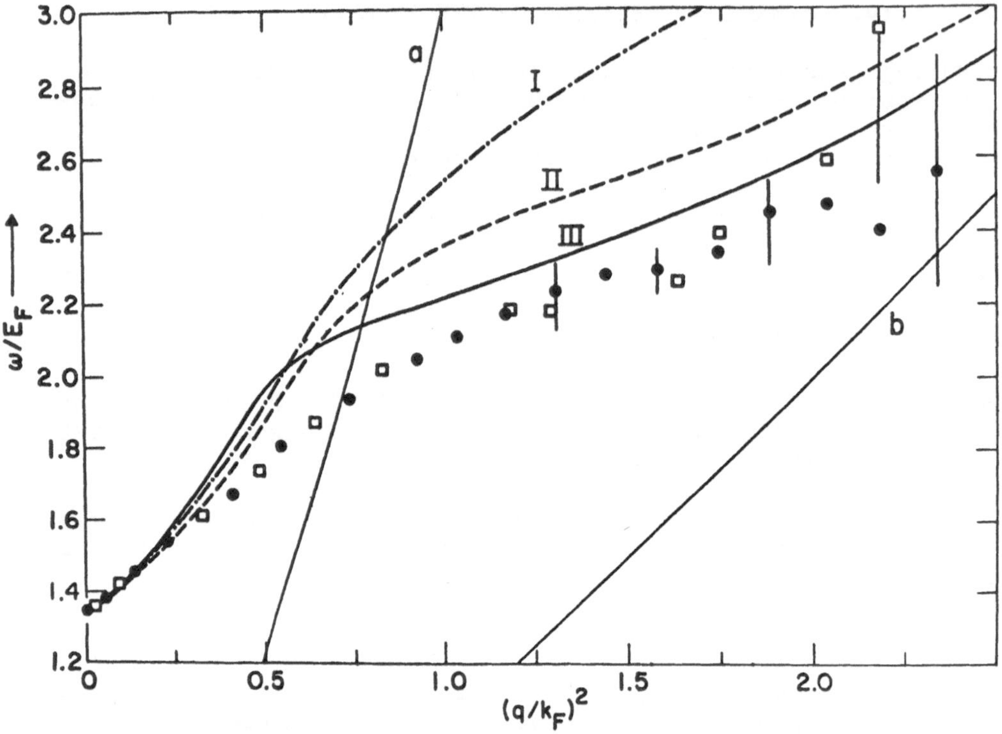

Figure 1. Position of the peak in $S(q,\omega)$ of Aℓ vs. the square of the wave number. The experimental points are due to P. E. Batson, C. H. Chen and J. Silcox, Phys. Rev. Lett. 37, 937 (1976). The curves marked with Roman numerals represent various theoretical results, as discussed in the text. Curve a is the edge of the particle-hole continuum, and curve b is the free particle dispersion curve $\omega = \hbar q^2/2m$. From A. K. Gupta and K. S. Singwi (to be published).

$$\omega_p^2(q) = \omega_p^2 + q^2 \left\{ \frac{6}{5}\frac{E_F}{m} - \omega_p^2 \lim_{q\to 0} [G(q)/q^2] \right\} + \ldots \qquad (4.2)$$

For $r_s = 2$ corresponding to the density of Aℓ, Eq. (3.17) yields $\alpha_{RPA} = 0.451$ in the RPA and $\alpha_{VS} = 0.363$ in the Vashishta-Singwi scheme [Vashishta and Singwi, 1972], against a value of 0.429 from the third moment sum rule. The experimental values fall in the range 0.38-0.42. One concludes that a frequency dependent G is needed for quantitative accuracy.

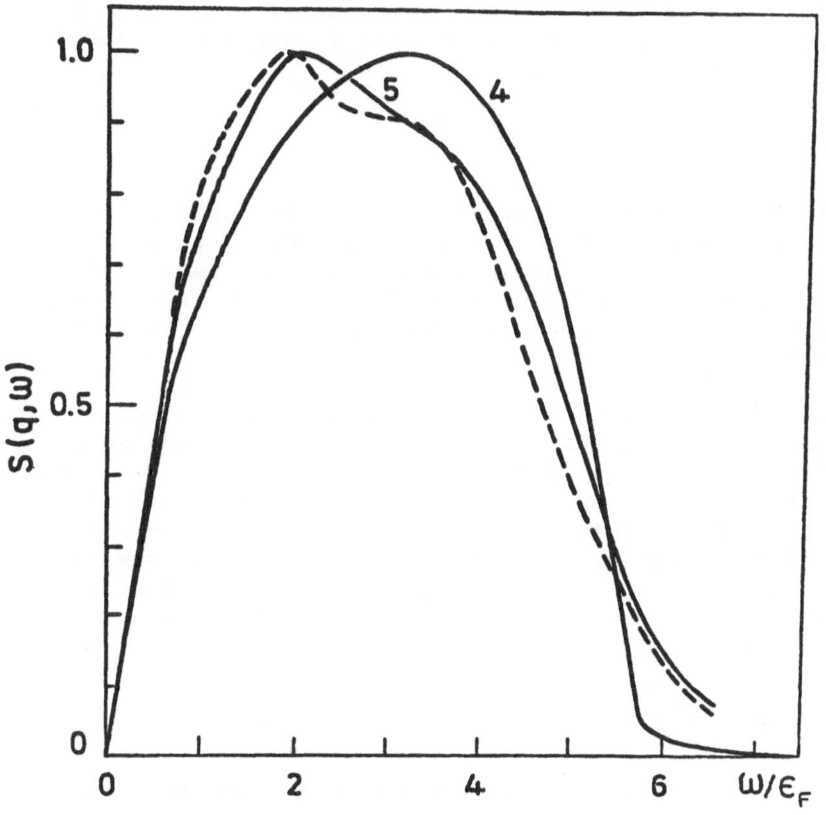

Figure 2. $S(q,\omega)$ of $A\ell$ vs. frequency at $q = 1.6\ q_F$. The dashed curve reports the x-ray scattering results of P. M. Platzman and P. Eisenberger, Phys. Rev. Lett. 33, 152 (1974). Curve 5 gives the theoretical results of G. Mukhopadhyay et al, Phys. Rev. Lett. 34, 950 (1975). Curve 4 is a typical Mori-formalism result. From G. Mukhopadhyay and A. Sjölander, Phys. Rev. (in press.

An ω-dependent local field factor enters the polarization potential theory of Pines [1966]. In this theory the density-density response function is given by

$$\chi(q,\omega) = \frac{\chi_{sc}(q,\omega)}{1 - \left[\psi_1(q) + \frac{\omega^2}{q^2}\psi_2(q)\right]\chi_{sc}(q,\omega)} \qquad (4.3)$$

where $\chi_{sc}(q,\omega)$ is the response to the external field plus the induced polarization potentials. ψ_1 and ψ_2 are two unknown functions. The back-flow term

$$\frac{\omega^2}{q^2}\,\psi_2(q)$$

is crucial in the theory of density fluctuation spectrum of both He^4 and He^3 and leads to mass renormalization. For the electron liquid the back-flow term should not be all that important. We shall now proceed to determine ψ_1 and ψ_2 and then calculate the plasmon dispersion. The analysis given here is due to Gupta and Singwi [to be published].

Equation (4.3) should be considered as a phenomenological extension of the Landau theory [for example, Pines and Nozieres, 1966] of Fermi liquids for finite values of q and ω. In making contact with the Landau theory, one finds that $\chi_{sc}(q,\omega)$ should reduce in the Landau limit to the polarizability of an ideal Fermi gas of particles with effective mass m*. This suggests that for a finite q one should take a q-dependent effective mass. By requiring that the f-sum rule be satisfied by χ given in Eq. (4.3), one finds

$$\frac{m^*(q)}{m} = 1 + \frac{n\psi_2(q)}{m} \quad , \tag{4.4}$$

m is the bare electron mass. The third moment sum rule is satisfied if

$$\mu_3 = \mu_3^{sc}\left(\frac{m^*(q)}{m}\right)^2 + \left(\frac{nq^2}{m}\right)^2 \psi_1(q) \quad , \tag{4.5}$$

where μ_3 is the <u>exact</u> third moment of $\chi(q,\omega)$ and μ_3^{sc} is the third moment of $\chi_{sc}(q,\omega)$. As a first approximation $\psi_1(q)$ can be taken to be the same as given by Vashishta and Singwi [1972], at least if $m^*(q)/m$ is close to unity, i.e.

$$\psi_1(q) = \frac{4\pi e^2}{q^2}\,[1 - G_{VS}(q)] \tag{4.6}$$

Equation (4.5) then leads to

$$\frac{1}{m^*(q)} = \frac{1}{m} = \left\{\frac{2q^2}{m}\langle T_c\rangle + \omega_p^2[G_{VS}(q) - G'(q)]\right\}\left[\frac{q^4}{4m} + \frac{3q^2}{5}\frac{k_F^2}{m}\right] \tag{4.7}$$

where $\langle T_c\rangle$ is the correlation kinetic energy per particle and

$$G'(q) = -\frac{1}{4\pi^2 n}\int_0^\infty dk[S(k) - 1]\, k^2 \left[\frac{5}{6} - \frac{k^2}{2q^2} + \frac{(k^2 - q^2)^2}{4kq^2} \cdot \ell n \left|\frac{k+q}{k-q}\right|\right] \quad (4.8)$$

From a knowledge of S(k) from VS theory, or for that matter from any other theory which gives a physically reasonable pair distribution function and $\langle T_c \rangle$ from the work of Vaishya and Gupta [1973], we can calculate $m^*(q)/m$ from Eq. (4.7). In Figure 3 we have given the results for $m^*(q)/m$ for $r_s = 2$. It is seen that $m^*(q)/m$ is quite close to unity and rises rather gently with q. The value of the plasmon dispersion parameter α as calculated on this basis for Aℓ is 0.426 which is in good agreement with the observed value (0.38 - 0.42). In Figure 1 curve III gives the results for the peak position in $S(q,\omega)$ over a large range of q values. In the same figure, curve I is calculated from the RPA and curve II from the dielectric function of Vashishta and Singwi [1972]. It will be noticed that the inclusion of the back-flow term in $\chi(q,\omega)$ brings the peak position closer to experiment at least in the region of the particle-hole continuum. Although curve III is a considerable improvement over the earlier theories, the agreement with experiment is still not satisfactory, especially for q around q_c. The cause of this discrepancy is to be sought in our neglect of multi-pair excitations, which give rise to plasmon damping for $q < q_c$. Multi-pair excitations, in the absence of a first principle theory, can be introduced phenomenologically, thus increasing the number of parameters in the theory. An attempt in this direction is worthwhile pending further developments.

In the large q-region, x-ray scattering experiments have revealed a double-peak structure in $S(q,\omega)$. Assuming that this structure is a genuine jellium property, none of the existing theories yields such a structure. Rather, one finds a single broad peak in this region of q, although the peak position shifts by a few eV below the RPA peak with the introduction of a local field correction G(q). A further downward shift by several eV, which brings the peak in approximate agreement with experiment, has been obtained by introducing self-energy effects [Mukhopadhyay, Kalia and Singwi, 1975] in the single pair spectrum. As shown in Figure 2, one obtains a shoulder on the high energy side of the peak, which, however, could be washed out when convoluted with an experimental resolution function of width 5 eV. One very serious difficulty with the theory [Mukhopadhyay, Kalia and Singwi, 1975] is that it violates the equation of continuity. This problem has also been reexamined [Mukhopadhyay and Sjölander, in press; Jindal, Singh and Pathak, 1977] more recently within the framework of Mori formalism without much success.

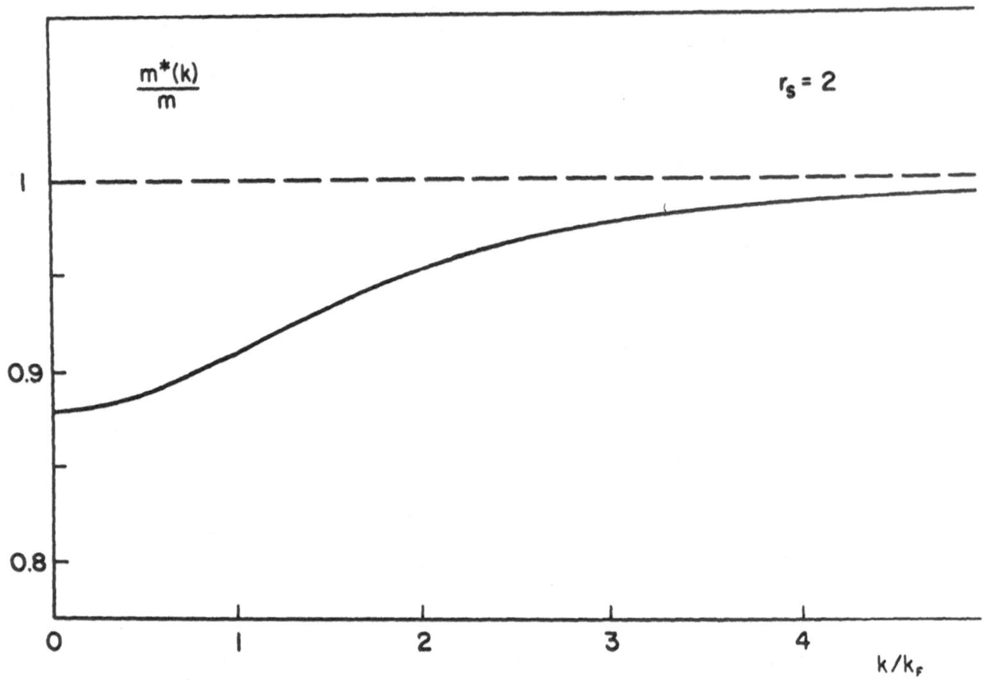

Figure 3. Effective mass $m^*(q)/m$ vs. q for $r_s = 2$. From A. K. Gupta and K. S. Singwi (to be published).

Let me conclude by saying that our understanding of the dynamical behavior of electrons in the electron liquid is far from complete. My own view is that progress would be made by a careful examination of the equation of motion [Niklasson, 1974] of the non-equilibrium two-particle Wigner distribution function and introducing a physically reasonable approximation which would terminate the hierarchy at this level. We are continuing attempts in this direction.

ACKNOWLEDGEMENT

My thanks are due to P. K. Aravind for going through the manuscript.

REFERENCES

Batson, P. E., C. H. Chen and J. Silcox, 1976, Phys. Rev. Lett. 37, 937.

Gell-Mann, M. and K. A. Brueckner, 1957, Phys. Rev. 106, 364.

Gibbons, P. C., S. E. Schnatterley, J. J. Ritsko, and J. R. Fields, 1976, Phys. Rev. B13, 2451.

Gupta, A. K. and K. S. Singwi, to be published.

Höhberger, J., A. Otto and E. Petri, 1975, Solid State Commun. 16, 175.

Jindal, V. K., H. B. Singh and K. N. Pathak, 1977, Phys. Rev. B15, 252.

Mukhopadhyay, G., R. K. Kalia and K. S. Singwi, 1975, Phys. Rev. Lett. 34, 950.

Mukhopadhyay, G. and A. Sjölander, in press, Phys. Rev.

Niklasson, G., 1974, Phys. Rev. B10, 3052.

Pines, D., 1966, in Quantum Fluids, edited by D. F. Brewer, North Holland, Amsterdam.

Platzman, P. M. and P. Eisenberger, 1974, Phys. Rev. Lett. 33, 152.

Puff, R. D., 1965, Phys. Rev. 137, A406.

Singwi, K. S., M. P. Tosi, R. H. Land and A. Sjölander, 1968, Phys. Rev. 176, 589.

Singwi, K. S., 1976, in Linear and Nonlinear Electron Transport in Solids (NATO Advanced Study Institute series), edited by J. T. Devreese and V. E. van Doren, Plenum Press, New York.

Singwi, K. S. and M. P. Tosi, 1978, to appear in Solid State Physics, edited by H. Ehrenreich, F. Seitz and D. Turnbull, Academic Press, New York.

Vaishya, J. S. and A. K. Gupta, 1973, Phys. Rev. B7, 4300.

Vashishta, P. and K. S. Singwi, 1972, Phys. Rev. B6, 875.

Wigner, E. P., 1934, Phys. Rev. 46, 1002.

Wigner, E. P., 1938, Trans. Faraday Soc. 34, 678.

Zacharias, P., 1975, J. Phys. F5, 645.

See, for example, Balescu, R., 1975, Equilibrium and Nonequilibrium Statistical Mechanics, Wiley, New York.

See, for example, Pines, D. and P. Nozières, 1966, The Theory of Quantum Liquids, W. A. Benjamin, New York.

MICROSCOPIC THEORY OF TIME DEPENDENT FLUCTUATIONS IN COULOMB SYSTEMS

Marc Baus*

Chimie-Physique II**, C.P. 231
Université Libre de Bruxelles
B-1050 Bruxelles, Belgique

---

*Chercheur Qualifié du F.N.R.S.

**Association Euratom-État Belge.

TABLE OF CONTENTS

MICROSCOPIC THEORY OF TIME DEPENDENT FLUCTUATIONS IN COULOMB SYSTEMS

Marc Baus

Chimie-Physique II, C.P. 231
Université Libre de Bruxelles
B-1050 Bruxelles, Belgique

I. INTRODUCTION

In these two chapters we would like to present an overview of a number of results we have obtained in recent years for some of the simplest models of Coulomb systems.

If in a system of particles at least part of the interaction potential is coulombic then, as a consequence of the long range of the Coulomb potential, this system is bound to exhibit properties which are also macroscopically different from those of uncharged particle systems. Such systems of charged particles or, briefly, Coulomb systems form part of real matter as plasmas, molten salts or any other more sophisticated ionic fluid. Many of the other lecturers will describe the relevance of such charged fluids in situations ranging from astrophysics to fusion devices. Here we will only be concerned with the simplest models of such coulomb systems: (1) the one component plasma (OCP) consisting of one species of mobile charged particles immersed in an inert neutralizing background, (2) the two component plasma (TCP) consisting of a globally neutral mixture of two oppositely charged mobile species and, (3) the binary ionic mixture (BIM) consisting of a mixture of two mobile ion species of equal charge sign immersed in an inert background of opposite sign. We will moreover neglect all electromagnetic interactions other than coulombic and describe the system classically. Some care has therefore to be exercised as such model systems can clearly not always be directly compared with real matter situations. These models do, however, deserve a study of their own as they do fully exhibit the "coulomb problem" and, for instance, are directly accessible in computer experiments. We will further limit the scope of our results by considering only the space-time correlation functions of the equilibrium fluctuations in these coulomb systems. In the limit of long-wavelengths the correlation functions of these coulomb systems will manifest macroscopic properties drastically different from those exhibited by neutral particle systems. However, and this will be the main point of this Introduction, these results can be obtained for charged fluids with at least as much generality and rigor as for ordinary fluids. Hence there is no need to rush from the start into some approximation scheme in which case one easily loses track of the general properties of fluids. In order to emphasize this we will devote the first lecture to the exact results one can obtain for coulomb systems and leave the approximate kinetic theoretic results for the second lecture.

Except for the results related to the binary ionic mixtures all of the material presented here is available in published or to be published form and we will thus skip most of the algebraic work and present the material with a more physical emphasis.

II. GENERAL LONG WAVELENGTH RESULTS FOR COULOMB SYSTEMS

A. The Microscopic Method

As stated in Chapter I, our main concern will be with the equilibrium fluctuations and more precisely with the equilibrium fluctuations of the conserved variables. The space-time correlation functions of these fluctuations will be seen to reflect the properties of all thermodynamic and transport quantities ordinarily accessible to the experimentalist. In order to obtain these space-time correlation functions it is convenient to consider the phase-space density $f(\vec{r},\vec{p};\, t)$:

$$f(\vec{r},\vec{p};\, t) = \sum_{j=1}^{N} \delta[\vec{r} - \vec{x}_j(t)]\; \delta[\vec{p} - \vec{p}_j(t)] \qquad (2.1)$$

which describes the density of particles at time t at the point $(\vec{r},\vec{p})$ of the one-particle phase space[1] due to the N particles whose natural trajectories are described by $[\vec{x}_j(t), \vec{p}_j(t)]$. The interest of $f(\vec{r},\vec{p};\, t)$ stems from the fact that it is a dynamical variable, and as such obeys a Liouville equation[2], and at the same time its ensemble average [3] is proportional to the ordinary one-particle distribution function which is the object studied by kinetic theory. Moreover the conserved variables[4] can be obtained by taking the momentum moments of $f(\vec{r},\vec{p};\, t)$, for instance the local microscopic number density $n(\vec{r},t)$ is obtained as:

$$n(\vec{r},t) = \int d\vec{p}\; f(\vec{r},\vec{p};\, t) \equiv \sum_{j=1}^{N} \delta[\vec{r} - \vec{x}_j(t)] \tag{2.2}$$

whereas its fluctuation would read:

$$\delta n(\vec{r},t) = \int d\vec{p}\; \delta f[\vec{r},\vec{p};\, t) \equiv n(\vec{r},t) - n \tag{2.3}$$

with

$$\delta f(\vec{r},\vec{p};\, t) \equiv f(\vec{r},\vec{p};\, t) - \langle f(\vec{r},\vec{p};\, t)\rangle \tag{2.4}$$

where in general $\langle A\rangle$ denotes the equilibrium average[5] of A. The central quantity containing all the useful information about the statics and dynamics of the equilibrium fluctuations is the <u>fluctuation spectrum</u> of f. This quantity is defined as the two-point correlation function of the fluctuations δf:

$$S(\vec{r},\vec{p},t\,|\,\vec{r}',\vec{p}',t') = \langle \delta f(\vec{r},\vec{p};\, t)\; \delta f(\vec{r}',\vec{p}';\, t)\rangle \tag{2.5}$$

which because of the <u>invariance</u> of the equilibrium ensemble for space-time translations can be rewritten as:

$$S(\vec{r},\vec{p},t\,|\,\vec{r}',\vec{p}',t') \equiv S(\vec{r} - \vec{r}',\; t - t';\; \vec{p},\; \vec{p}') \tag{2.6}$$

and hence we can Fourier-Laplace transform S as:

$$S(\vec{k}z;\; \vec{p}\vec{p}') = \int d\vec{r} \int_0^{\infty} dt\; e^{-i\vec{k}\cdot\vec{r}+izt}\; S(\vec{r},t;\; \vec{p}\vec{p}'); \quad \mathrm{Im}z > 0 \tag{2.7}$$

We now observe that as $f(\vec{r},\vec{p};\, t)$ obeys the Liouville equation, we can apply Mori's projection operator method to it as first done by Ackasu and Duderstadt [1969]. The intermediate algebra is available in the literature and we will skip it here[6]. This algebra culminates into an exact <u>kinetic equation</u> for S:

$$zS(kz;\; \vec{p}\vec{p}') - \int dp''\; \Sigma(\vec{k}z;\; \vec{p}\vec{p}'')\; S(\vec{k}z;\; \vec{p}''\vec{p}') = iS(\vec{k},t{=}0;\; \vec{p}\vec{p}') \tag{2.8}$$

where Σ is the so-called memory function which splits naturally into three parts, $\Sigma = \Sigma^0 + \Sigma^s + \Sigma^c$, the free-flow term Σ^0:

$$\Sigma^0(\vec{k}z;\; \vec{p}\vec{p}') = \vec{k}\cdot\vec{v}\;\delta(\vec{p} - \vec{p}')\;; \quad \vec{p} = m\vec{v} \tag{2.9}$$

the self-consistent field term:

$$\Sigma^s(\vec{k}z;\ \vec{p}\vec{p}\,') = -\vec{k}\cdot\vec{v}\ \psi(\vec{p})\ c(\vec{k}) \tag{2.10}$$

and the genuine collision term Σ^c:

$$\Sigma^c(\vec{k}z;\ \vec{p}\vec{p}\,') = [\delta f(\vec{k}p)|LQ(z - QLQ)^{-1}\ QL|\delta f(\vec{k}p')][n\psi(\vec{p}\,')]^{-1} \tag{2.11}$$

which can also be interpreted as a nonlocal collision operator[7].

The kinetic equation (2.8) is the first step of the present method and as such it is worthwhile to try to understand better its physical contents. Let us make the following remarks:

(1) So far no approximations have been made and hence this kinetic equation is exact. What it does is to separate the one-particle dynamics described by S from the more than one particle dynamics lumped together into the collision term Σ^c. Such exact equations can be very deceptive because of the difficulty to extract something useful out of them. In the present case I would like, however, to call attention to a number of particular features of Eq. (2.8).

(2) What do we need to know before studying the dynamics of our system? The first thing is clearly the initial condition of S:

$$S(\vec{k},t=0;\ \vec{p}\vec{p}\,') = n\psi(\vec{p})[\delta(\vec{p} - \vec{p}\,') + \psi(\vec{p}\,')\ h(k)] \tag{2.12}$$

where n is the average number density,

$$\psi(\vec{p}) = \left(\frac{\beta}{2\pi m}\right)^{3/2} \exp - \frac{\beta p^2}{2m}$$

the Maxwellian and h(k) the Fourier transform of the equilibrium binary correlation function. Integrating Eq. (2.12) over p and p' we obtain the so-called static structure factor S(k) describing the equilibrium density fluctuations:

$$S(k) = \int d\vec{p}\ d\vec{p}\,'\ S(k,\ t=0;\ \vec{p}\vec{p}\,') = n[1 + h(k)] = n[1 - c(k)]^{-1} \tag{2.13}$$

where we have also introduced the Ornstein-Zernike direct correlation function c(k) appearing also in Eq. (2.10). As S(k), h(k) and c(k) are about the only equilibrium properties which one has reasonable information about, we are not yet in too bad shape.

(3) If we compare the kinetic Eq. (2.8) with some standard results we immediately notice that there is no inhomogeneous term here whereas the self-consistent potential is not just the bare potential but has been renormalized into the direct correlation

function $c(k)$ which reduces to $-n\beta V(k)$ only for weak-coupling. These two features are in fact closely related.

(4) Indeed, this relation can be understood if we remember that a kinetic theory for the equilibrium correlations as described by S, is closely related to a <u>linearized</u> kinetic theory for non-equilibrium distribution functions. Let us recall therefore that if at t = 0 the system was in equilibrium with a hamiltonian perturbation of the form

$$\delta H = -\frac{1}{\beta}\int d\vec{r}\, d\vec{p}\, \delta u(\vec{r}\vec{p})\, f(\vec{r},\vec{p};\, t=0)$$

then its relaxation in the absence of δH can be described as:

$$\langle f(\vec{r}\vec{p}t)\rangle_{n.eq.} = \langle f(\vec{r}\vec{p}t)\rangle_{eq.} + \int d\vec{r}\,'\, d\vec{p}\,'\, S(\vec{r} - \vec{r}\,',\, t;\, \vec{p}\vec{p}\,') \cdot \delta u(\vec{r}\,',\vec{p}\,') + O[(\delta u)^2] \tag{2.14}$$

from which we get:

$$S(\vec{r} - \vec{r}\,', t;\, \vec{p}\vec{p}\,') = \left.\frac{\delta\langle f(\vec{r},\vec{p},t)\rangle_{n.eq.}}{\delta u(\vec{r}\,',\vec{p}\,')}\right|_{\delta u=0} \tag{2.15}$$

and hence the equation obeyed by S is the linearized version of the equation obeyed by the non-equilibrium one-particle distribution function $\langle f\rangle_{n.eq.}$. The latter equation always contains an inhomogeneous term depending on the initial correlations and at the same time a bare self-consistent field term. One usually neglects the effect of the initial correlations. In the present case, however, the initial correlations contain <u>equilibrium</u> correlations part of which decays only slowly contributing to the transport regime and providing a bath for the non-equilibrium part of the distribution. In view of this, what has been performed in Eq. (2.8) is a renormalization of the average force field which takes into account these equilibrium correlations $[V(k) \to c(k)]$ and eliminates the inhomogeneous term completely. This constitutes a major advantage of Eq. (2.8).

(5) Finally, the collision term, Eq. (2.11), as it stands looks rather awful. In the next chapter, we will indicate how it can be re-expressed in terms of a contracted four-point correlation function which when approximated in terms of the two-point function S yields a very interesting closed, approximate, non-linear, kinetic equation for S. This equation will be the basis for the approximate results to be discussed in the next lecture. For the exact results we will be concerned with here, the present form, Eq. (2.11), of Σ^c is in fact convenient because the presence of the Liouville operator L in it yields us an easy access to the microscopic conservation laws which represent an important bit of information.

We are now ready to take the next step in the present method. This step is based on the observation that the kinetic equation for S contains still too much information which is not readily accessible to any experiment whatsoever. The quantities which are really of direct physical interest are the hydrodynamic correlation functions which are obtained from S by taking its momentum space moments corresponding to the hydrodynamical variables; the local number density, the momentum density and the energy density* is readily accessible. As the most interesting hydrodynamic correlation function is the density-density correlation function we will not discuss here the necessary modifications in order to treat the total energy density [Jhon and Forster, 1975]. These hydrodynamic correlation functions can be most conveniently written in scalar product form:

$$G_{ij}(\vec{k}z) \equiv \int d\vec{p}\, d\vec{p}\,' \frac{u_i(p)}{a_i} S(\vec{k}z;\, \vec{p}\vec{p}\,') \frac{u_j(\vec{p}\,')}{a_j}$$

$$\equiv \langle i|S(\vec{k}z)(n\psi)^{-1}|j\rangle \qquad (2.16)$$

where the different hydrodynamical states are labelled as follows:

$$i = n \;;\; u_n = 1 \;;\; a_n^2 = n \qquad (2.17a)$$

$$i = g_i \;;\; u_{g_i} = p_i \;;\; a_{g_i}^2 = nm\beta^{-1} \;;$$

$$i = (1,2,3) = (\ell,t_1,t_2) = (\ell,\perp) \qquad (2.17b)$$

$$i = \varepsilon \;;\; u_\varepsilon = \frac{\vec{p}^{\,2}}{2m} - \frac{3}{2}\beta^{-1} \;;\; a_\varepsilon^2 = \frac{3}{2} n\beta^{-1} \qquad (2.17c)$$

the a_i being normalization constants such that:

$$\langle i|j\rangle \equiv \int d\vec{p}\, \frac{u_i(\vec{p})}{a_i} n\psi(p) \frac{u_j(\vec{p})}{a_j} = \delta_{ij} \qquad (2.18)$$

Using once more the Mori algebra the kinetic Eq. (2.8) is readily transformed [(6)] into an algebraic set of equations for the hydrodynamic correlations function G_{ij}:

$$\sum_{j=1}^{5} [z\delta_{ij} - \Omega_{ij}(\vec{k}z)]\, G_{ji'}(\vec{k}z) = iG_{ii'}(\vec{k},t=0) \qquad (2.19)$$

The solutions of this algebraic system can also be written down by means of Kramer's determinants as:

* Notice that in the one-particle space we are working with only the kinetic energy density.

$$G_{ij}(\vec{k}z) = \Delta_{ij}(\vec{k}z)/\Delta(\vec{k}z) \tag{2.20}$$

where:

$$\Delta(\vec{k}z) = \det|zI - \Omega(\vec{k}z)| \tag{2.21}$$

while Δ_{ij} is the co-factor of G_{ij} in the system of Eqs. (2.19). Equation (2.19) is the second main step in the present method. It provides us with an exact explicit expression, Eq. (2.20), for the hydrodynamic correlation functions in terms of the initial condition:

$$G_{ij}(\vec{k}, t = 0) \equiv \delta_{ij}[1 + \delta_{in} h(k)] \tag{2.22}$$

controlled once more by the binary equilibrium correlation function h(k), and the hydrodynamic transport matrix:

$$\Omega_{ij}(\vec{k}z) = \langle i|\Sigma(\vec{k}z)|j\rangle + \langle i|\Sigma(kz)\,\overline{Q}[z - \overline{Q}\Sigma(kz)\,\overline{Q}]^{-1}\,\overline{Q}\Sigma(kz)|j\rangle \tag{2.23}$$

controlled by the memory function $\Sigma^{(8)}$. A joint use of the system's microscopic conservation laws with the system's rotational invariance properties simplifies life a lot as one can show on this basis[6] that the only non-vanishing matrix elements of the transport matrix Ω_{ij} are those listed here:

$$\Omega_{n\ell}(\vec{k}z) = kv_o \tag{2.24a}$$

$$\Omega_{\ell n}(kz) = kv_o[1 - c(k)];\ \Omega_{\ell\ell}(kz) = -ik^2D_\ell(kz);\ \Omega_{\ell\varepsilon}(kz) = kD_{\ell\varepsilon}(kz) \tag{2.24b}$$

$$\Omega_{\varepsilon\ell}(kz) = kD_{\varepsilon\ell}(kz);\ \Omega_{\varepsilon\varepsilon}(kz) = -ik^2D_\varepsilon(kz) + zB_\varepsilon(z) \tag{2.24c}$$

$$\Omega_{t_it_i}(kz) = -k^2D_\perp(kz) \tag{2.24d}$$

where $v_o^2 = (m\beta)^{-1}$, while the i-factors have been introduced for later convenience where k-factors have been pulled out by using the microscopic conservation laws[6]. Finally, a closer inspection reveals that the self-consistent field term Σ^s of Σ drops out from all Ω_{ij} matrix elements (2.23) except $\Omega_{\ell n}$ displayed in Eq. (2.24b). This is a very lucky feature as Σ^s is very sensitive to the nature of the potential. This is, however, as far as one can go from first principles. To proceed we have to make two assumptions which are generally believed to be true but whose precise status depends on one's requirements for rigor. First, we will assume that the functions D_ℓ, $D_{\ell\varepsilon}$, $D_{\varepsilon\ell}$, D_ε, B_ε and $D_\perp$ as defined by Eq. (2.24) exist as functions of $\vec{k}$ and z when $k \to 0$ and/or $z \to 0$. This can be boiled down to a condition of weak analyticity on $\Sigma^c(\vec{k}z)$. A direct proof of this condition is difficult but if this were not a valid assumption we would have to abandon most of the standard kinetic theoretical results which in turn would be rather surprising. Finally, we have to assume something about the statics. Here it is generally believed that the following relation holds for the direct correlation function c(r) of a fluid of particles interacting through the potential V(r):

$$\lim_{r\to\infty} c(r) = -n\beta V(r) \tag{2.25}$$

This relation states that for large distances ($r \to \infty$) the fluid is always weakly coupled because the right hand side of Eq. (2.25) is also the weak coupling limit of c(r). A direct proof of Eq. (2.25) appears to be underway[9]. For the case of the OCP, Eq. (2.25) is moreover very well verified by computer simulations [Hansen, 1973]. For the OCP we translate Eq. (2.25) into Fourier language as:

$$c(k) = \frac{-k_D^2}{k^2} + \hat{c}(k) \quad ; \quad k_D^2 = 4\pi\, e^2\, n\beta \tag{2.26}$$

where $\hat{c}(k)$ is some unknown function, which vanishes with the coupling $\lambda = k_D^3/n$ and is regular at $k = 0$. It is also convenient to include all non-coulombic forces, whenever present, into $\hat{c}(k)$. The quantity of interest reads then:

$$1 - c(k) = \frac{k_D^2}{k^2} + \frac{k_D^2}{k_s^2} + 0(k^2) \tag{2.27}$$

in agreement with the alternative definition using the static dielectric function $\varepsilon(k,o)$:

$$S(k) = n\,\frac{k^2}{k_D^2}\left[1 - \frac{1}{\varepsilon(k,o)}\right] \tag{2.28}$$

and the definition of a screening wavevector k_s through [Pines and Nozieres, 1966]:

$$\lim_{k\to o} \varepsilon(k,o) = 1 + \frac{k_s^2}{k^2} \tag{2.29}$$

Sometimes one also introduces the isothermal compressibility, χ_T, and the isothermal sound speed, c, through the subsidiary definitions:

$$\frac{k_s^2}{k_D^2} = \frac{v_o^2}{c^2} = \frac{\chi_T}{\chi_T^o} \tag{2.30}$$

v_o and $\chi_T^o = \beta/n$ being the perfect gas ($\lambda = 0$) values of c and χ_T respectively. The physical interpretation of χ_T and c has, however, to be handled with care[10].

Equations (2.8), (2.19) and (2.27) constitute the three main steps in the present method. Notice that Eq. (2.27) contains the so-called Stillinger-Lovet condition[9]:

$$S(k) = n[1 - c(k)]^{-1} = \frac{nk^2}{k_D^2} - \frac{nk^4}{k_D^4} \cdot \frac{k_D^2}{k_s^2} + O(k^6) \qquad (2.31)$$

whereas Eq. (2.26) indicates that both the potential V(k) and c(k) are singular for small k whereas they are regular in the absence of Coulomb forces. Now, as it is precisely in the long wavelength limit ($k \to 0$) that we expect this microscopic theory to reveal the system's macroscopic properties we immediately see that Coulomb systems are bound by Eqs. (2.25)-(2.27) to exhibit macroscopic properties which will be profoundly different from those of uncharged particle systems under similar conditions of temperature and density.

B. The One Component Plasma (OCP)

Now that we have sketched the main points of the method we can inquire for the results. Here we will focus our attention on the density-density correlation function[11] which, in the long-wavelength long-time limit, is known to reveal all the macroscopic information obtainable from fluid systems. As an hors-d'oeuvre let us illustrate, however, some points of the algebra on the simpler case of the transverse momentum correlation function

$$G_{t_i t_i}(\vec{k}z) \quad .$$

Because of rotational invariance the transverse (t_i) and longitudinal (n, ℓ, ε) states cannot couple and the dispersion equation (2.21) factorizes as:

$$\Delta(kz) \equiv \Delta_\perp(kz) \cdot \Delta_T(kz) = 0 \qquad (2.32)$$

separating hereby the transverse[12] and longitudinal modes of the system[6]. Moreover, as already clear from Eq. (2.24d), the different transverse directions are equivalent, hence, the transverse modes are degenerate and very simply obtained by solving their dispersion relation:

$$z + ik^2 D_\perp(k,z) = 0 \qquad (2.33)$$

for instance for z as a function of k, $z = z(k)$. From Eq. (2.33) we immediately see that the only transverse hydrodynamic mode, i.e. a mode such that $z(k = 0) = 0$, is:

$$z = -i\, k^2\, D_\perp(\vec{k} = \vec{0},\ z = 0) + O(k^4) \tag{2.34}$$

which is readily identified as the (twice degenerate) shear mode of shear viscosity $\eta = n\, m\, D_\perp(\vec{0},0)$. [In Eq. (2.34) and elsewhere in this text, $O(k^4)$ indicates that a small-k expansion has been used but not that the next term of this expansion is precisely of order k^4.] The contribution of this shear mode to the transverse momentum correlation function

$$G_{t_i t_i} \equiv G_\perp$$

is all that survives in the hydrodynamical limit of large t and small k (we use a superscript H in order to indicate that only the dominant term in this limit has been retained):

$$\lim_{\substack{k\to 0\\ t\to\infty}} G_\perp(\vec{k},t) \equiv G^H_\perp(\vec{k},t) = \exp - \frac{k^2 \eta t}{nm} \tag{2.35}$$

which is a well-known result of hydrodynamic fluctuation theory [Landau and Lifshitz, 1959]. Hence, we have been able to come all the way down from the microscopic level to the macroscopic description. Because of our restriction to the (longitudinal) Coulomb forces no explicit Coulomb effects did show up in $G_\perp(\vec{k},t)$. In the next lecture we will indicate, however, that an approximate but more detailed study of η does reveal some interesting Coulomb effects. Let us return now to the main object of interest here, the density-density correlation function G_{nn}. We should stress from the beginning that in the case of the OCP, because of electroneutrality, the fluctuations in number density and charge density are proportional to each other and both are thus described by G_{nn}. As the algebra involved is rather elementary(6) we will quote only the result. First, we would like, however, to recall the analogous result for an ordinary uncharged fluid:

$$G^H_{nn}(\vec{k},z) = \frac{\chi_T}{\chi^o_T}\left[\frac{i(1 - c_v/c_p)}{z + i\,k^2(\kappa/nmc_p)} + \sum_{\pm} \frac{i\, c_v/c_p}{\pm z \pm \bar{c}k + ik^2\,\Gamma}\right] \tag{2.36}$$

where c_v/c_p is the specific heat ratio, κ the thermal conductivity, $\bar{c}$ the isentropic sound speed and Γ the sound absorption coefficient. This is the well-known Landau-Placzek formula which was first derived from hydrodynamic fluctuation theory [Landau and Lifshitz, 1959] but which was shown to be a rigorous consequence of the microscopic dynamics in the macroscopic hydrodynamical region of small k and large t by the present method [Forster and Martin, 1970; Forster, 1974] as well as by the method of hydrodynamical modes [Resibois, 1970]. By way of contrast, the equivalent result for the OCP reads:

$$G_{nn}^{H}(\vec{k},z) = \frac{k^2}{k_D^2} \sum_{\pm} \frac{i}{\vec{z} + \omega_p(1 + \frac{1}{2} k^2 \gamma_p) + i\, k^2\, \Gamma_p} \tag{2.37}$$

which hardly reminds one of the Landau-Placzek formula (2.36). There is, however, a close connection between both results which can be understood on the basis of the present microscopic theory. Treating the neutral and charged fluid cases on a par one arrives at the following result which for a change we write in the time language:

$$G_{nn}^{H}(\vec{k},t) = G_{nn}(k)\left\{\left(\frac{\bar{c}k}{\Omega(k)}\right)^2 a(o)\ \exp -k^2 D_T t + \left[1 - \left(\frac{\bar{c}k}{\Omega(k)}\right)^2 a[\Omega(k)]\right]\cos \Omega(k)t\ \exp - \frac{k^2}{2}\Gamma t\right\} \tag{2.38}$$

where $a(z)$ is such that $a(z{=}0) = 1 - c_V/c_p$. The difference in overall amplitude between Eq. (2.36) and Eq. (2.37) is entirely due to the difference in statics. Indeed for the static structure factor $G_{nn}(k)$ we have:

$$\lim_{k\to o} G_{nn}(k) = \frac{1}{(\chi_T^o/\chi_T) + 0(k^2)} \simeq \frac{\chi_T}{\chi_T^o} \tag{2.39}$$

for an uncharged fluid, whereas for the OCP we have (see Eq. (2.27)):

$$\lim_{k\to o} G_{nn}(k) = \frac{1}{(k_D^2/k^2) + (k_D^2/k_s^2) + 0(k^2)} \simeq \frac{k^2}{k_D^2} \tag{2.40}$$

i.e. a vanishingly small static structure factor. As far as the dynamics is concerned both systems exhibit a heat conduction mode, $z = -i\, k^2\, D_T$, of heat conductivity D_T:

$$D_T = \frac{\kappa}{nmc_v}\left[1 + \left(\frac{c_p}{c_v} - 1\right)\frac{\chi_T^o}{\chi_T} G_{nn}(k \to o)\right]^{-1} \tag{2.41}$$

and two oppositely propagating modes

$$z_{\pm} = \pm\, \Omega(k) - \frac{ik^2}{2}\Gamma \quad .$$

Using the result of Eqs. (2.39)-(2.40) in Eq. (2.41) we see that the two heat modes differ by a factor c_v/c_p pointing to a weaker coupling of the energy and density fluctuations in the case of the OCP. For the case of the propagating mode the same Coulomb singularity as reflected in Eq. (2.27) shifts the sound frequency $\Omega(k) = \bar{c}k$ of ordinary fluids into the plasma frequency

$$\Omega(k) = \omega_p(1 + \frac{1}{2} k^2 \gamma_p)$$

and hence the intensity of the heat mode in Eq. (2.38) is also down by a factor of $O(k^2)$ while the plasma modes alone are seen to exhaust the sum rule:

$$\lim_{t\to o} \lim_{k\to o} G_{nn}(k,t) = \lim_{t\to o} \lim_{k\to o} G^H_{nn}(k,t) \quad . \tag{2.42}$$

Finally the damping of the plasma modes can no longer be expressed in terms of transport coefficient as for an ordinary fluid. We will come back to this problem in the next chapter. The Coulomb effects are thus seen to affect profoundly the macroscopic behavior of the density-density correlation function $G_{nn}(\vec{k},t)$.

C. The Two-Component Plasma (TCP) and the Binary Ionic Mixture (BIM)

Before considering some of these features of the OCP in more detail, let us also sketch a few exact long-wavelength results for the other Coulomb systems; the two-component plasma (TCP) and the binary ionic mixture (BIM). In these two component systems we again assume only simple Coulomb forces between the charges of either specie subject moreover to some not further specified short range forces. In the case of the TCP the charges of different species are of opposite sign and hence the TCP offers a simple model for a molten salt or a genuine plasma. In the BIM the charges of different species are, on the contrary, of the same sign and this model can be used to describe for instance a $H^+ - He^{++}$ mixture. Except for the presence of a neutralizing background the BIM is the direct Coulomb analogue of the ordinary liquid mixture.

Many of the various steps of the microscopic theory outlined in Chapter II, Section A for the OCP can be taken over immediately to a system of an arbitrary number of species with arbitrary charge sign relations. There is, however, one important technical point which does appear only for many component systems. This point is related to the fact that the electrical current density is an important variable of Coulomb systems which is not a conserved variable. In the OCP case the charge and mass currents are proportional to each other and hence both are conserved. In the many component case this fact leads to the unpleasant feature that the self-consistent-field term Σ^s produces a lot of singular terms in the system's hydrodynamical space so that it becomes rapidly hopeless to study its long-wavelength behavior. As suggested elsewhere [Baus, 1977] a possible way out to this problem is to use not directly the hydrodynamical variables but instead the so-called multi-fluid variables whose sums correspond to the hydrodynamical variables while their differences correspond to non-conserved relaxation variables including for instance the electrical current. When this is done we can proceed as in Section A except that now all quantities become matrices in species space, for example, Ω_{ij} becomes

$$\Omega_{ij}^{\sigma\sigma'} \equiv \Omega_{i\sigma\ j\sigma'} \quad , \quad \equiv \Omega_{i_\sigma\ j_\sigma} \quad ,$$

σ and σ' running over the various species present. The fact that the multi-fluid variables (i_σ) are not conserved also modifies Eqs. (2.24), for instance now we have:

$$\Omega_{\ell\ell}^{\sigma\sigma'}(\vec{k}z) = -i[\gamma_\ell^{\sigma\sigma'}(z) + k^2 D_\ell^{\sigma\sigma'}(\vec{k}z)] \tag{2.43}$$

instead of

$$\Omega_{\ell\ell}(\vec{k}z) = -i\ k^2\ D\ (\vec{k}z)$$

and similarly for the other transport matrix elements.

As a consequence all singular matrix elements are again concentrated in $\Omega_{\ell n}$ but now there is one for each pair of species σ, σ': $\Omega_{\ell n}^{\sigma\sigma'}$. The algebra remains feasible but nevertheless becomes rather involved even for systems of only two components. As a second consequence of our using the multi-fluid variables we have to treat on a par with the hydrodynamical modes $[z(k = o) = 0]$ a set of relaxation modes $[z(k = o) \neq 0]$. For example, along with the transverse shear mode:

$$z_\perp(k) = -i\ k^2\ D_\perp + O(k^4) \tag{2.44}$$

There now appears a set of interspecies transverse momentum relaxation modes of the form:

$$\overline{z}_\perp(k) = -i(\overline{\gamma}_\perp + k^2\ \overline{D}_\perp + O(k^4)] \tag{2.45}$$

where $\overline{\gamma}_\perp$ is a microscopically defined interspecies relaxation frequency [Baus, 1977] but these correlation functions will suffice to illustrate our purpose here.

Let us recall once more that the algebra gets quickly involved with the number of species present so that we have to exclude the interesting case of the electrolytes which are three component systems. Moreover, even for two component systems we have to distinguish at an early state between the TCP and the BIM. That these two systems do indeed behave rather differently can be seen from a study of their dispersion equation $\Delta(\vec{k}z) = o$ at infinite wavelength ($k = o$). Just as for the case of the OCP we can separate the longitudinal (L) and transverse ($\perp$) modes, $\Delta = \Delta_L \cdot \Delta_\perp$, with for example:

$$\Delta_L(\vec{k}z) = \det|z\,\delta_{ij}\,\delta_{\sigma\sigma'} - \Omega_{ij}^{\sigma\sigma'}(\vec{k}z)|; \quad \begin{matrix}(i,j) = n,\ell,\varepsilon \\ (\sigma,\sigma') = 1,2\end{matrix} \tag{2.46}$$

After some amount of algebra [Baus, 1977] we can rewrite Δ_L as $\Delta_L = \Delta_\varepsilon\,\Delta_{n\ell} + \Delta_{n\ell\varepsilon}$ with Δ_ε and $\Delta_{n\ell}$ defined as in Eq. (2.46) but with i and j restricted to respectively ε and (n,ℓ). The remainder $\Delta_{n\ell\varepsilon}$ describes the coupling between the energy and density fluctuations. At $k = o$, $\Delta_{n\ell G}$ vanishes and we can then moreover separate the longitudinal modes into heat modes ($\Delta_\varepsilon = o$) and density modes ($\Delta_{n\ell} = o$). Concentrating on the latter, their dispersion equation at $k = o$ turns out to be rather simple:

$$\Delta_{n\ell}(k{=}o,z) \equiv z^2(z^2 - \Omega_p^2) + i\,\gamma_\ell(z)\,z(z^2 - \Omega^2) = 0 \tag{2.47}$$

where Ω_p is the mean-field plasma frequency:

$$\Omega_p^2 = \sum_{\sigma=1,2} \omega_\sigma^2 \equiv \sum_{\sigma=1,2} 4\pi\,\rho_{e,\sigma}^2/\rho_{m,\sigma} \tag{2.48}$$

where $\rho_{e,\sigma} = e_\sigma\,n_\sigma$ is the average charge density and $\rho_{m,\sigma} = m_\sigma\,n_\sigma$ the average mass density of specie σ, hence

$$\omega_\sigma^2 = 4\pi\,\varepsilon_\sigma^2\,n_\sigma/m_\sigma \quad .$$

By way of contrast, the second characteristic frequency of Eq. (2.47), Ω can be called a hydrodynamical plasma frequency because it can be expressed in terms of the hydrodynamical variables of total charge

$$\rho_e = \sum_{\sigma=1,2} \rho_{e,\sigma}$$

and total mass

$$\rho_m = \sum_{\sigma=1,2} \rho_{m,\sigma}$$

densities:

$$\Omega^2 = 4\pi\,\rho_e^2/\rho_m \equiv \Omega_p^2 - \Omega_o^2 \quad . \tag{2.49}$$

From Eq. (2.49) we also see that Ω is always smaller than Ω_p. The subsidiary frequency Ω_o is defined as:

$$\Omega_o^2 = 4\pi \, \rho_o^2/\rho_m \quad ; \quad \rho_o^2 = \rho_{m,1} \, \rho_{m,2} \left(\frac{\rho_{e,1}}{\rho_{m,1}} - \frac{\rho_{e,2}}{\rho_{m,2}}\right)^2 \tag{2.50}$$

whereas finally the relaxation frequency γ_ℓ appearing in Eq. (2.47) is given in terms of the transport matrix elements by:

$$-i \, \gamma_\ell(z) \equiv \sum_{\sigma=1,2} \Omega_{\ell\ell}^{\sigma\sigma}(z,k = o) \tag{2.51}$$

The difference in physical behavior between the TCP and BIM is now obvious from Eq. (2.47). Let us first consider the case of the TCP. In this case electroneutrality requires

$$\rho_e = \sum_\sigma \rho_{e,\sigma} = 0$$

and hence from Eq. (2.49) we have $\Omega = 0$. The dispersion equation (2.47) now reduces to:

$$\Delta_{n\ell}(k = o,z) \equiv z^2[z^2 - \Omega_p^2 + i \, \gamma_\ell(z) \, z] = 0 \tag{2.52}$$

Hence we see that in such a system two hydrodynamical modes, $z(k = o) = 0$, will develop besides the already mentioned heat mode and the two transverse shear modes. There will thus be five hydrodynamical modes in a TCP whereas in an ordinary liquid with two conserved densities (binary mixture) one expects in fact six hydrodynamical modes, one for each conserved variable. Equation (2.52) indicates that this would be the case if the charge, and hence Ω_p, was to vanish. As we will indicate later the two hydrodynamical modes which originate from Eq. (2.52) are sound modes. The mode which is missing therefore when $\Omega_p \neq 0$ is the mass diffusion mode which according to Eq. (2.52) is seen to be coupled to the inter-species momentum relaxation mode producing hereby two damped charge relaxation or plasma oscillation modes.

In the case of the BIM, $\rho_e \neq 0$ and balanced by the inert background charge. The mode structure of the BIM will thus be completely different from that of the TCP and resembles in fact more that of the OCP from which it nevertheless also differs. Putting

$$\Omega^2 = \Omega_p^2 - \Omega_o^2$$

(see Eq. (2.49)) into Eq. (2.47) we rewrite it as:

$$\Delta_{n\ell}(k=o,z) = z\left\{(z^2 - \Omega_p^2)[z + i \, \gamma_\ell(z)] + i \, \gamma_\ell(z) \, \Omega_o^2\right\} \tag{2.53}$$

In this case there is therefore no possibility for obtaining sound modes, just as for the case of the OCP except that here we get an additional diffusion mode. Consequently in the BIM there are only four hydrodynamical modes. From Eq. (2.53) we see that for the remaining modes we get a coupling between the plasma modes and the interspecies momentum relaxation mode producing three sets of relaxation modes. In the exceptional case $\Omega_o = 0$, which according to Eq. (2.50) requires $e_1/m_1 = e_2/m_2$ and hence the fluctuations in mass and charge density will be proportional to each other, this coupling vanishes and we get two undamped plasma oscillations and an interspecies momentum relaxation mode. For $\Omega_o \neq 0$ these modes couple together producing a k-independent damping of the plasma modes and a shift in their natural oscillation frequency. The oscillation frequency of the plasma modes becomes now a function of the coupling parameter. Indeed, from Eq. (2.53) we can rewrite the dispersion equation for the plasma modes as:

$$z^2 - \Omega_p^2 + \frac{i\,\gamma_\ell(z)\,\Omega_o^2}{z + i\,\gamma_\ell(z)} = 0 \tag{2.54}$$

which indicates that Ω_p is the oscillation frequency of the plasma modes in the limit of weak coupling ($\gamma_\ell \to 0$) whereas this frequency is lowered (see Eq. (2.49)) to Ω in the limit of strong coupling ($\gamma_\ell \to \infty$). For finite coupling one has to solve Eq. (2.54) and, according to recent molecular dynamics simulations [McDonald, Hansen and Vieillefosse, preprint], the transition from Ω_p to Ω does not appear to be simple.

The situation both for the TCP and the BIM becomes quickly involved when $k \neq 0$ and here we will only comment on the results [Baus, 1977] as these do remain relatively simple but show nevertheless a much richer structure than for ordinary uncharged fluids. Let us recall first that for an ordinary binary mixture of uncharged particles there are four longitudinal hydrodynamical modes, $z_\pm^s$ and $z_\pm$:

$$z_\pm^s = \pm\,\bar{c}k - i\,\frac{k^2}{2}\,\Gamma \tag{2.55a}$$

$$z_\pm = -i\,k^2\,D_\pm \tag{2.55b}$$

$$\bar{z}_\varepsilon = -i(\bar{\gamma}_\varepsilon + k^2\,\bar{D}_\varepsilon) \tag{2.55c}$$

$$\bar{z}_\ell = -i(\bar{\gamma}_\ell + k^2\,\bar{D}_\ell) \tag{2.55d}$$

to which we have added two sets of relaxation modes, $\bar{z}_\varepsilon$ and $\bar{z}_\ell$, as a result of our using six longitudinal multi-fluid variables (three for each specie). The four hydrodynamical modes describe the sound waves ($z_\pm^s$) and the coupled mass and heat diffusion processes

($z_\pm$) whereas the relaxation modes describe the interspecies energy ($\overline{z}_\varepsilon$) and longitudinal momentum ($\overline{z}_\ell$) relaxation processes. In the case of a TCP we have three hydrodynamical modes, $z_\pm^s$ and z_ε, and three sets of relaxation modes, $\overline{z}_\varepsilon$ and $\overline{z}_\pm$, which we list as:

$$z_\pm^s = \pm \overline{c}k - i \frac{k^2}{2} \Gamma' \tag{2.56a}$$

$$z_\varepsilon = -i\, k^2 D_\varepsilon \cdot \frac{c^2}{\overline{c}^2} \tag{2.56b}$$

$$\overline{z}_\varepsilon = -i(\overline{\gamma}_\varepsilon + k^2\, \overline{D}_\varepsilon) \tag{2.56c}$$

$$\overline{z}_\pm = \omega_\pm + \frac{k^2}{2} (\gamma_\pm - i\, \Gamma_\pm) \tag{2.56d}$$

where again $z_\pm^s$ describes sound waves, z_ε heat diffusion, $\overline{z}_\varepsilon$ interspecies heat relaxation and $\overline{z}_\pm$ charge relaxation processes. The Coulomb phenomena have produced, however, a number of remarkable differences with the corresponding processes in uncharged fluids. For instance, the sound absorption rate Γ' has been modified whereas the heat diffusion process (z_ε) has been decoupled from the mass diffusion process which instead is now coupled to the plasma oscillations producing thereby two charge relaxation modes $\overline{z}_\pm$. The heat diffusion process in the TCP is now similar to the one occurring in ordinary binary mixtures; except that here it is always strictly uncoupled from the mass diffusion process. This heat mode differs now by a factor $c^2/\overline{c}^2$ (which is also equal to the specific heat ratio c_v/c_p) from the one occurring in the OCP (cf. Eq. (2.41)). The most important modification with respect to the OCP is, however, the fact that here the plasma oscillations suffer a damping even at k = 0 according to (see Eq. (2.56d)):

$$\omega_\pm = \pm \left[\Omega_p^2 - \left(\frac{\overline{\gamma}_\pm}{2} \right)^2 \right]^{1/2} - \frac{i}{2} \overline{\gamma}_\pm \tag{2.57}$$

where $\overline{\gamma}_\pm$ is defined by the auxiliary relation:

$$\gamma_\ell (z = \omega_\pm) = \overline{\gamma}_\pm \tag{2.58}$$

For weak coupling ($\overline{\gamma}_\pm << \Omega_p$) Eq. (2.57) reduces to:

$$\omega_\pm = \pm\, \Omega_p \left[1 - \frac{1}{8} \left(\frac{\overline{\gamma}_\pm}{\Omega_p} \right)^2 \right] - \frac{i}{2} \overline{\gamma}_\pm$$

i.e., two slightly damped plasma oscillations occurring slightly below Ω_p. For strong coupling ($\overline{\gamma}_\pm >> \Omega_p$) Eq. (2.57) splits into an electrical conductivity mode (ω_+) and an interspecies momentum relaxation mode (ω_-) according to:

$$\omega_\pm = \begin{cases} -i\ \Omega_p^2/\overline{\gamma}_+ \equiv -i\ 4\pi\ \sigma_o \\ \\ -i\ \overline{\gamma}_- \end{cases} \tag{2.59}$$

In the latter case only can one make contact with some recent phenomenological theories [Giaquinta, Parrinello and Tosi, 1976]. A close inspection of the charge relaxation mode $\overline{z}_+(k)$ indicates that it combines the electrical conductivity process described by Eq. (2.59):

$$\lim_{\Omega_p/\gamma_\ell \to o}\ \lim_{k\to o} \overline{z}_+ = -i\ 4\pi\ \sigma_o \tag{2.60}$$

with the charge diffusion process:

$$\lim_{k\to o}\ \lim_{\Omega_p/\gamma_\ell \to 0} \overline{z}_+ = -i\ k^2\ D_1 + O(k^4) \tag{2.61}$$

in such a way that the diffusion constant D_1 is related to the electrical conductivity by an Einstein relation [Martin, 1967]:

$$\frac{4\pi\ \sigma_o}{D_1} = k_s^2 \tag{2.62}$$

where k_s is a screening wave vector analogous to the one defined by Eq. (2.29). Finally, let us also quote the results for the long-wavelength mode-structure of the BIM. Here we will consider only the simpler case where $e_1/e_2 = m_1/m_2$. In this case the longitudinal multi-fluid modes consist of two hydrodynamical modes $z_\pm^D$, two plasma modes $z_\pm^P$ and two sets of interspecies relaxation modes:

$$z_\pm^D = -i\ k^2\ D'_\pm \tag{2.63a}$$

$$z_\pm^P = \pm\ \Omega_p (1 + \frac{1}{2}\ \gamma'_p) - i\ \frac{k^2}{2}\ \Gamma'_p \tag{2.63b}$$

$$\overline{z}_\varepsilon = -i(\overline{\gamma}_\varepsilon + k^2\ \overline{D}_\varepsilon) \tag{2.63c}$$

$$\overline{z}_\ell = -i(\overline{\gamma}_\ell + k^2\ \overline{D}_\ell) \tag{2.63d}$$

This mode-structure is seen to mix some features of the OCP with those of ordinary binary mixtures (see Eq. (2.55)).

The intensity with which these modes appear in various correlation functions is of particular importance for their possible observation in real or computer experiments. Let us restrict our attention to the correlation functions for density fluctuations, $G_{nn}(kt)$, and for charge density fluctuations, $G_{\rho\rho}(kt)$. For small k and large enough t only the multi-fluid modes survive and hence we can write ($i = n$ or ρ):

$$G_{ii}(kt) \simeq G_{ii}(k) \sum_j a_i^j(k) \exp -i\, z_j(k)t \tag{2.64}$$

where $G_{ii}(k) \equiv G_{ii}(k,t{=}o)$ is the static correlation function while j runs over the surviving modes only. The overall amplitude for $G_{nn}(kt)$ and $G_{\rho\rho}(kt)$ always differs as for small k we have:

$$\lim_{k\to o} G_{nn}(k) = n \frac{\chi_T}{\chi_T^o} \quad ; \quad \lim_{k\to o} G_{\rho\rho}(k) = \frac{k^2}{4\pi\beta} \tag{2.65}$$

where $n = n_1 + n_2$ is the total density, χ_T the isothermal compressibility of the system and χ_T^o its ideal gas value. Both for the TCP and the BIM, $G_{\rho\rho}(kt)$ is further seen to be dominated by the plasma modes as all other modes have a strength $a_\rho^j(k)$ which vanishes with k. The corresponding dynamic structure factor will thus exhibit a two-peak structure whose further characteristics are determined by Eq. (2.56d) and Eq. (2.63b) for respectively the TCP and the BIM. On the contrary, for the TCP only the three hydrodynamical modes (Eqs. (2.56a-b)) contribute a finite amount to $G_{nn}(kt)$ for small k, just as would be the case for the four hydrodynamical modes (Eqs. (2.55a-b)) of a neutral binary mixture, whereas both the plasma modes and the two hydrodynamical modes (Eqs. (2.63a-b)) contribute to the density fluctuations $G_{nn}(kt)$ of a BIM. In all cases the dominating modes do exhaust the sum rule,

$$\lim_{k\to o} \sum_j a_i^j(k) = 1 \quad ,$$

indicating that if we were to trace back in time the long-time approximation to $G_{ii}(k,t)$, $i = n$ or ρ, we would recover the small k portion of the exact initial condition[13]. This property implies that, at least for small k, the multi-fluid modes do exhaust the long-time behavior of the correlation functions of the fluctuations of the conserved variables. The multi-fluid modes hereby acquire a particularly important status which fully justifies the emphasis we have put on them.

We have now exhausted our possibilities for making general statements about the long-wavelength behavior of Coulomb systems. From here on, further progress can be made only on the basis of appropriate approximation schemes reflecting the particular features of the problem at hand. A few examples will be discussed in our next chapter. Before doing so, it may be useful to summarize what we have achieved up to now. Starting from first principles we have set up exact expressions for the space-time correlation functions of the equilibrium fluctuations of the conserved variables, or, in the two component case, the multi-fluid variables of a series of Coulomb systems (OCP, TCP and BIM). The long-wavelength modes which built up these correlation functions have been analyzed together with the strength with which these modes appear in a given correlation function. All the quantities have been given exact expressions in terms of the binary equilibrium correlations and various matrix elements of the collision operator or memory function of the system. These expressions are therefore valid for all values of the various coupling constants involved as long as the system remains in its fluid phase. The main limitation of these results is their confinement to the long-wavelength region. They constitute the Coulomb analog of the well known Landau-Placzek result for uncharged fluids. These epxressions also display information about most of the system's thermodynamic and transport properties. As such, they provide valuable qualitative information for analyzing the correlation functions obtained from (mainly) computer experiments. At present, wherever possible, qualitative agreement with these experiments has been achieved. These experiments do, however, provide us also with quantitative information and with information about shorter wavelength regimes. Analysis of the latter results does require further approximations to which we now turn our attention.

III. APPROXIMATE KINETIC THEORY OF THE OCP

A. The Approximation Method

In order to extract some quantitative information from the preceding scheme we have to analyze how a given quantity, for instance a transport coefficient, varies with respect to the thermodynamic parameters of the system. When suitably non-dimensionalized all previously introduced quantities of the OCP are seen to depend on the thermodynamic state only through the plasma coupling parameter[14] $\lambda = k_D^3/n$ which measures the inverse of the number of particles in a Debye cube. This λ-dependence stems from the equilibrium binary correlations, say $S(k)$, and from the memory function $\Sigma^c(\vec{k}z)$. We will always adopt the viewpoint that we can leave the equilibrium problem to the specialists and borrow this equilibrium information from the literature and hence take it as input in the evaluation of transport properties. The central quantity to approximate becomes then the memory function or collision operator $\Sigma^c(\vec{k}z)$. The form given to Σ^c in Eq. (2.11) was convenient in our first chapter where we explored the consequences of the conservation laws. Such an N-body construct is, however, inconvenient for introducing approximations. Let us start therefore from the alternative and equivalent [Mazenko, 1974; Lindenfeld, 1977; Gross, this volume] form of Σ^c expressed in terms of a (contracted) four-point correlation function C:

$$i\ \Sigma^c(1,2;\ t)\ n\psi(2) = \int d1'\ d2'\ L_I(11')\ L_I\ C(11';\ 22';\ t) \tag{3.1}$$

where 1 stands for $(\vec{x}_1,\vec{p}_1)$, etc., ψ for the Maxwellian and $L_I(12)$ for the interaction operator:

$$L_I(12) = -\frac{\partial}{\partial\vec{x}_1} V(|\vec{x}_1 - \vec{x}_2|) \cdot \left(\frac{\partial}{\partial\vec{p}_1} - \frac{\partial}{\partial\vec{p}_2}\right) \tag{3.2}$$

associated with the potential V(r). As a consequence of the cluster properties of C we can split it in a so-called disconnected part, C_D, and a connected part, C_C, according to $C = C_D + C_C$ and:

$$C_D(11';\ 22';\ t) = S(12;\ t)\ S(1'2';\ t) + S(12';\ t)\ S(1'2;\ t) \tag{3.3}$$

where S(12; t) is the two-point function defined by Eq. (2.5). This separation of C into two parts has a physically obvious meaning. Indeed the contribution from C_D to Σ^c via Eq. (3.1) describes a collision process in which the colliding particles propagate through the medium, independently from one another, but with their exact one particle propagator. The contribution from C_C to Σ^c describes on the contrary a collision process in which the colliding particles stay close together during the collision. The contribution from C_C to Σ^c is both very complicated and of higher order in the plasma parameter. In the following we will always neglect it. This is sometimes referred to as the disconnected approximation. In weakly coupled situations this approximation is very good and includes for instance the well-known Balescu-Guernsey-Lenard (BGL) result [Baus and Wallenborn, 1977]. For strongly coupled situations this approximation ($C \simeq C_D$) has, however, an important drawback. Indeed when traced back to t = 0 the disconnected approximation of Σ^c clearly leads to an incorrect value of $\Sigma^c(t = 0)$. The exact value of $\Sigma^c(t = 0)$ is in fact known [Linnebur and Duderstadt, 1973] and expressible in terms of two-body static correlation functions. As we have argued elsewhere [Wallenborn and Baus, 1977] this piece of information should not be thrown away so easily especially as our leitmotif will be that, due to the occurrence of rapid plasma oscillations at later times, the short-time dynamics of charged particle systems acquires a much stronger emphasis than we are used to from ordinary fluids. Similarly, due to the long range of the Coulomb potential the use of exact static properties becomes of overwhelming importance. In order to re-incorporate the close collisions into the disconnected approximation, at least at t = 0, we will use a renormalized version of the present theory. Such a renormalization scheme was first worked out by Mazenko [1974]. Elsewhere we have adopted his scheme so that it yields back the exact initial condition, $\Sigma^c(t = 0)$, even in the disconnected approximation [Wallenborn and Baus, 1977]. That the physical approximations underlying this rather novel strategy appear to be correct is, in our opinion, borne out by the fact that even with a very poor description of the dynamics of the collision process we have obtained values of the shear viscosity ranging over more than three orders of magnitude of the plasma parameter Γ which agree astonishingly well with the results from computer simulations. We will come back to these results in Section C. To close this introduction to the approximation methods let us quote another result which shows that even for weak coupling our approximation scheme can improve some known results. For weak coupling we can approximate S in the right hand side of Eq. (3.3) by the solution of the collisionless or Vlasov approximation to Eq. (2.8). Taking also the local (k = 0) and Markovian (z = 0) limit of $\Sigma^c(\vec{k}z)$ we obtain from our renormalized scheme:

$$i\ \Sigma^c(k = 0), z = 0;\ \vec{p}_1\vec{p}_2)\ n\psi(\vec{p}_2) = \frac{n}{\beta}\int \frac{d\vec{\ell}}{8\pi^3}\,\vec{\ell}\cdot\vec{\partial}_1\,\vec{\ell}\cdot\vec{\partial}_2$$

$$\cdot\ \frac{V(\ell)\ c(\ell)}{|\xi_{\vec{\ell}}(\vec{\ell}\cdot\vec{v}_1)|^2}$$

$$\cdot\ \{\pi\delta(\vec{\ell}\cdot\vec{v}_1 - \vec{\ell}\cdot\vec{v}_2)\ \psi(\vec{p}_1)\ \psi(\vec{p}_2)$$

$$-\ \delta(\vec{p}_1 - \vec{p}_2)\ \psi(\vec{p}_2)\int d\vec{p}_3\ \pi\delta(\vec{\ell}\cdot\vec{v}_1 - \vec{\ell}\cdot\vec{v}_3)\psi(\vec{p}_3)\} \tag{3.4}$$

with a dielectric constant given by:

$$\xi_{\vec{k}}(z) = 1 - \frac{c(k)}{\beta}\int dp\ \frac{\vec{k}\cdot\vec{\partial}\psi(\vec{p})}{z - \vec{k}\cdot\vec{v}} \tag{3.5}$$

If we also approximate the statics by their weak coupling value, i.e. use $c(k) \simeq -n\beta V(k)$ in Eqs. (3.4)-(3.5), then Eq. (3.4) reduces exactly to the linearized BGL expression. As such, Eq. (3.4) has, however, the appealing feature to be divergence-less. Indeed, from the computer results [Hansen, 1973] we know that $c(r = 0)$ is finite and hence $c(k)$ has to vanish faster than k^{-3} for large k. This property of $c(k)$ guarantees then that the large-k divergence familiar from the BGL result is suppressed by Eq. (2.4). Hence no ad hoc cut-off parameters have to be introduced into the theory. After this short introduction to how one can introduce approximations within this general scheme we will illustrate our purpose with two examples.

B. The Plasma Mode

As already pointed out in Section B of Chapter II one of the most striking features of the OCP is that as a consequence of the Coulomb interactions the sound modes of ordinary fluids are shifted into high-frequency plasma oscillations. In the long wavelength limit these plasma modes are weakly damped and were shown in Chapter II, Section B to take the general form:

$$z_{\pm}(k) = \pm \omega_p \left(1 + \frac{1}{2} \frac{k^2}{k_D^2} \gamma_p\right) - \frac{i}{2} \frac{k^2}{k_D^2} \omega_p \Gamma_p \tag{3.6}$$

where ω_p is the plasma frequency ($\omega_p^2 = 4\pi e^2 n/m$), k_D the Debye wave-vector ($k_D^2 = 4\pi e^2 n\beta$), while γ_p and Γ_p are dimensionless quantities which solely depend on the coupling constant $\lambda = k_D^3/n$ and are a measure of the dispersion and absorption, respectively, of the plasma waves. One of the most remarkable results of Section B of Chapter II was to provide us with an exact expression of γ_p and Γ_p. This expression reads explicitly:

$$\gamma_p = \frac{k_D^2}{k_s^2} + \frac{k_D^2}{\omega p} \operatorname{Re} D(k = 0, \omega_p) \tag{3.7a}$$

$$\Gamma_p = - \frac{k_D^2}{\omega_p} \operatorname{Im} D(k = 0, \omega_p) \tag{3.7b}$$

where k_s, the screening wavevector, was defined in Eq. (2.27) whereas $\Omega(kz) \equiv k^2 D(kz)$ is defined in terms of the transport matrix elements of Eq. (2.24) by:

$$\Omega(kz) = \Omega_{\ell\ell}(kz) + \Omega_{\ell\varepsilon}[kz)(z - \Omega_{\varepsilon\varepsilon}(kz)]^{-1}\Omega_{\varepsilon\ell}(kz) \equiv k^2 D(kz) \quad . \tag{3.8}$$

The most important point here is that as displayed by Eq. (3.7) a knowledge of γ_p and Γ_p requires the knowledge of the transport matrix Ω_{ij} and hence of the collision operator Σ^c at the finite frequency $z = \omega_p$. For this reason, and contrary to what happens for uncharged fluids, the plasma wave absorption coefficient, Γ_p, for instance, cannot be expressed in terms of ordinary (zero-frequency) transport coefficients.

Let us now analyze the important modifications undergone by the plasma mode [Baus, 1977] when one varies the coupling parameter

$\lambda = k_D^3/n$. At zero coupling, $\lambda = 0$, one finds back as expected the well-known Vlasov mean-field or collisionless results:

$$\gamma_p(\lambda = 0) = 3 \tag{3.9a}$$

$$\Gamma_p(\lambda = 0) = \left(\frac{\pi}{2}\right)^{1/2} \left(\frac{k_D}{k}\right)^5 \exp\left[-\frac{1}{2}\left(\frac{k_D}{k}\right)^2 - \frac{3}{2}\right] \simeq 0 \tag{3.9b}$$

In this case the damping is not of order k^2 as indicated in Eq. (3.6) but exponentially small as recalled in Eq. (3.9b). This result stems, however, from the unphysical approximation $\lambda = 0$. As soon as $\lambda \neq 0$, whatever small, the conservation laws force the damping to be of order k^2 as in Eq. (3.6). It is the limiting process $\lambda \to 0$ which introduces the Landau singularity leading to the Landau damping of Eq. (3.9b). In other words, the limiting processes, $\lambda \to 0$ and $k \to 0$, do not commute. As a consequence of this non-uniformity in (k,λ)-space it makes little sense to superpose, as is often done in the literature, the Landau and collisional damping.

For small but finite coupling, $0 \neq \lambda << 1$, a simple BGL-like theory can be used in order to compute γ_p and Γ_p from Eq. (3.7). We would like to stress once more, however, that the BGL-theory has to be extended to finite frequencies as a knowledge of the collision operator at $z = \omega_p$ is required for the evaluation of the right hand side of Eq. (3.7). This can be done [Baus, 1977] and results in:

$$\lim_{\lambda \to 0} \omega_p = 3 + 0.35 \frac{\lambda}{4\pi} \tag{3.10a}$$

$$\lim_{\lambda \to 0} \Gamma_p = \frac{2}{15\pi^{3/2}} \lambda(\ln \lambda^{-1} - 0.37) \tag{3.10b}$$

where, as usual in the weak-coupling theory, a large k cut-off uncertainty persists in Eq. (3.10b). The result of Eq. (3.10b) is in agreement with the previously found theoretical results [Dubois and Gilinsky, 1964] whereas Eq. (3.10a) agrees at least qualitatively with the computer findings [Hansen, McDonald and Pollock, 1975] indicating a positive value for $\gamma_p - 3$ for weak coupling. This latter result cannot be obtained from a Markovian theory, i.e. by evaluating the collision operator at $z = 0$ instead of $z = \omega_p$, nor from a high-frequency expansion (sum-rules) except if terms of higher order in λ are introduced [Ichimaru, Totsuji, Tange and Pines, 1975].

For intermediate λ-values no results are as yet known. For large λ-values a further bit of information is available. Indeed for strong coupling we expect the kinetic equation (2.8) to become collision dominated and hence we expect hydrodynamic concepts to show up. It is worthwhile to point out here that for the OCP the relation with respect to hydrodynamics is profoundly different from that of ordinary fluids. Indeed, we have $\Sigma = \Sigma^o + \Sigma^s + \Sigma^c$, where the free-flow term Σ^o is always small ($\sim O(k)$) and the collision term Σ^c always finite in the long-wavelength region $k \to 0$. Hence, for an ordinary fluid where the self-consistent field term Σ^s is also small ($\sim O(k)$) for small k, the kinetic equation will always be dominated by the collision term in the macroscopic region of vanishing k. This is also the reason why, as discussed in Chapter II, Section B, hydrodynamics emerges automatically in the long-wavelength limit. This, however, is not the case for the OCP because here Σ^s is singular ($\sim O(k^{-1})$) for small k and therefore Σ^s can compete with the collision term. If we estimate Σ^s by $\omega_p \cdot k_D/k$ and Σ^c by a collision frequency ω_c, then a collision dominated situation is seen to require

$$\omega_p \frac{k_D}{k} << \omega_c$$

or:

$$\frac{k_D}{k} << \frac{\omega_c}{\omega_p} \tag{3.11}$$

and as here $k << k_D$ we need $\omega_p << \omega_c$ and hence a strongly coupled situation[15]. A more careful analysis reveals [Baus, 1977] that we can expect:

$$D_{ij}(\vec{0}, \omega_p) = D_{ij}(\vec{0}, 0) + O(\omega_p/\omega_c) \tag{3.12}$$

to hold for the various elements of the transport matrix (2.24) involved in Eq. (3.8). If this is the case we obtain for large λ from Eqs. (3.12) and (3.7):

$$\gamma_p \simeq \frac{c_p}{c_v} \cdot \frac{\chi_T^o}{\chi_T} \quad ; \qquad \lambda >> 1 \tag{3.13a}$$

$$\Gamma_p \simeq \frac{k_D^2}{p} \cdot \frac{\phi_M}{nm} \quad ; \qquad \lambda >> 1 \tag{3.13b}$$

where apart from the quantities already introduced in Chapter II, Section B,

$$\phi_M = \frac{4}{3}\eta + \xi_M$$

is the Markovian part of the longitudinal viscosity

$$\phi = \frac{4}{3}\eta + \xi \quad ,$$

composed of the shear (η) and bulk (ξ) viscosities. The restriction of ξ to ξ_M is a rather technical point [Baus, 1977] which should not bother us here. What is important is that, as announced, in this region only can γ_p and Γ_p be expressed in terms of macroscopic, thermodynamic and transport quantities, respectively. The difference with the sound modes of ordinary fluids, where this is always the case, can be entirely ascribed to the fact that a microscopic description of the plasma modes, for instance Eq. (3.7), requires the collision operator at the finite frequency $z = \omega_p$. Finally, we should also observe that Eq. (3.13a) is in agreement with the computer simulations which indicate that for large λ both γ_p and χ_T become negative [Hansen, McDonald and Pollock, 1975]. It would clearly be of interest to have also a quantitative prediction for the transition from the weak, Eq. (3.10), to the strong, Eq. (3.13), coupling region. Such a quantitative transition has been obtained recently for a somewhat simpler problem which we would like to consider now.

C. The Shear Viscosity

The computation of the transport coefficients of the OCP has been currently restricted to the weakly coupled domain ($\lambda \ll 1$). Recently, computer experiments on the OCP have provided us with a number of data for the shear viscosity which belong mainly to the strongly coupled domain [Vieillefosse and Hansen, 1975]. In order to tackle this problem theoretically novel kinetic methods going beyond the BGL theory are necessary. Elsewhere [Wallenborn and Baus, 1977] we have proposed such a kinetic theory. The basic idea there was to emphasize strongly the static and short-time description of the collision term $\Sigma^c(k,t)$. Following the method sketched in Section A, Chapter III we have adopted a renormalized version of the disconnected approximation. Explicitly, our approximation of Σ^c reads:

$$\Sigma^c(\vec{k},t;\ \vec{p}_1\vec{p}_2)\ n\psi(\vec{p}_2) \simeq \frac{i}{2n\beta}\int\frac{d\vec{\ell}}{8\pi^3}\int d\vec{p}_3\ d\vec{p}_4\ \{\vec{\ell}\cdot\vec{\partial}_1,\ \vec{\ell}\cdot\vec{\partial}_2$$

$$\cdot\ c(\ell)\ V(\ell)\ S(\vec{k}-\vec{\ell},t;\ \vec{p}_1\vec{p}_2)\ S(\vec{\ell},t;\ \vec{p}_3\vec{p}_4)$$

$$+\ \vec{\ell}\cdot\vec{\partial}_1(\vec{k}-\vec{\ell})\cdot\vec{\partial}_2\ c(\ell)\ V(k-\ell)$$

$$\cdot\ S(\vec{k}-\vec{\ell},t;\ \vec{p}_1\vec{p}_3)\ S(\vec{\ell}_1 t;\ \vec{p}_4\vec{p}_2)$$

$$+ \vec{k} \cdot \vec{\partial}_1 \, \vec{\ell} \cdot \vec{\partial}_2 \, a(\vec{\ell}, \vec{k} - \vec{\ell}) \, V(\ell)$$

$$\cdot \, S(\vec{k} - \vec{\ell}, t; \vec{p}_3 \vec{p}_2) \, S(\vec{\ell}, t) \, \psi(\vec{p}_1) \, \psi(\vec{p}_4)$$

$$+ (1 \leftrightarrow 2)\} \tag{3.14}$$

where $S(\vec{\ell},t) = \int d\vec{p}\, d\vec{p}'\, S(\vec{\ell},t;\vec{p}\vec{p}')$ while the explicit expression of $a(\vec{\ell}, \vec{k} - \vec{\ell})$ will not be needed here(16) as the third term in the right hand side of Eq. (3.14) does not contribute to the shear viscosity η. This expression of Σ^c can be shown to possess the necessary symmetry relations to lead to appropriate conservation laws and to be such that $\Sigma^c(\vec{k},t=0)$ takes on its exact value. Moreover, Eq. (3.14) generalizes the linearized BGL expression to finite wavelength, finite frequency and finite coupling. Once this choice of Σ^c is made we can compute η from Eqs. (2.33) and (2.23) or explicitly from:

$$\eta = \eta_{dir} + \eta_{ind} \tag{3.15a}$$

$$\eta_{dir} = nm \lim_{k \to o} \frac{1}{k^2} \langle \perp | i \, \Sigma^c(\vec{k}, z = 0) | \perp \rangle \tag{3.15b}$$

$$\eta_{ind} = nm \lim_{k \to o} \frac{1}{k^2} \langle \perp | [\Sigma^o + \Sigma^c(k,o)] \, \overline{Q} [i \, \overline{Q} \, \Sigma^c(\vec{0},0) \, \overline{Q}]^{-1}$$

$$\cdot \, \overline{Q} [\Sigma^o + \Sigma^c(\vec{k},o)] | \perp \rangle \quad . \tag{3.15c}$$

Computing the inverse of Σ^c appearing in Eq. (3.15c) with the aid of a one Sonine polynomial approxiation one obtains from Eqs. (3.14)-(3.15):

$$\eta^* \equiv (\eta_{dir} + \eta_{ind})/\eta_o = \lambda \, I_1(\lambda) + [1 + \lambda \, I_2(\lambda)]^2/\lambda \, I_3(\lambda) \tag{3.16}$$

where η^* is a dimensionless viscosity and $\eta_o = nm\omega_p k_o^{-2}$. The unit of length k_o^{-1} depends in general on whether one uses λ or Γ as coupling constant(14). When using λ the choice $k_o = k_D$ is convenient whereas

$$k_o = a^{-1} \equiv \left(\frac{4\pi}{3} n \right)^{1/3}$$

is well adapted to Γ. Notice also that $ak_D = (3\Gamma)^{1/2}$. Let us illustrate the method further with the aid of one of the three functions, say I_3, which appear in Eq. (3.16). These functions are given by wavevector and time integrals of static and dynamic correlation functions. For instance, for I_3 we have:

$$I_3 = -2\left(\frac{k_o}{k_D}\right) k_o^{-3}\ \omega_p^{-1} \int \frac{d\ell}{8\pi^3} \int_o^{\infty} dt\ c(\ell)(\hat{\vec{\ell}} \cdot \hat{\vec{k}})^2\ G_{nn}(\ell,t)\ G_{\perp}(\ell,t) \tag{3.17}$$

The evaluation of I_3 thus splits into a static and dynamic problem. It is the dynamics which determines the precise way in which the integrand of Eq. (3.17) drops from its exact initial value to zero for large t. We argue that this integrand drops sufficiently quickly so that we can approximate $G_{nn}(\ell,t)$ and $G_{\perp}(\ell,t)$ in Eq. (3.17) by their short time approximation. This then leaves us with a static problem, as now I_3 becomes:

$$I_3 \simeq \frac{1}{10\pi^{3/2}} \int_o^{\infty} dx\ x[1 - \delta(x)] \quad ; \quad x = k/k_o \tag{3.18}$$

To compensate for the poor approximation of the dynamics we now use the exact values of the static form factor S(k) in Eq. (3.18). These values of S(k) are borrowed from the literature. For $0.1 \leq \Gamma \leq 2$ we have used the values of S(k) obtained by numerical integration of the HNC equations. For $2 \leq \Gamma \leq 160$ we have used the Monte Carlo data for $ka \geq 1$ and for $ka \leq 1$ we have used Eq. (2.27) with (see Eq. (2.30) the compressibility computed from the equation of state(17). For large Γ, enough static data are available from which we did obtain the following fit:

$$\frac{\eta}{n\ m\ \omega_p\ a^2} = \frac{\lambda}{60\pi^{3/2}} \cdot \frac{1}{3\Gamma} + \left[1 + \frac{\lambda}{60\pi^2}\ (0.49 - 2.23\Gamma^{-1/3})\right]^2 \Big/ \frac{\lambda}{10\pi^{3/2}} \cdot (2.41\ \Gamma^{1/9}) \tag{3.19}$$

which reproduces the η^* values to within a few percent(18) for $2 \leq \Gamma \leq 160$. Our η^* values compare favorably with the results obtained by Vieillefosse and Hansen [1975] from a generalized hydrodynamics approximation as well as with the value of η^* deduced from an MD experiment [Hansen, McDonald and Pollock, 1975] at $\Gamma = 152.4$. Recently, direct MD computation [Bernue, Vieillefosse and Hansen, 1977] of the Green-Kubo formula of η has produced values of η which compare astonishingly well with our predictions. The ratio of η/η_{MD} yielding 0.97, 0.92 and 1.21 at $\Gamma = 1$, 10 and 100 respectively. For weak coupling our data also indicates a smooth transition to the BGL value [Braun, 1967]:

$$\frac{\eta_{BGL}}{n\, m\, \omega_p\, a^2} = \frac{1}{3\Gamma}\frac{5\pi^{1/2}}{2\varepsilon}\,(\ln \varepsilon^{-1} + 0.346)^{-1} \quad ; \quad \varepsilon = \frac{\lambda}{4\pi} << 1 \tag{3.20}$$

For instance at $\Gamma = 0.1$ we find $\eta/\eta_{BGL} = 1.04$. Hence this relatively simple theory provides good values of η throughout the whole fluid phase of the OCP!

ENDNOTES

(1) When various mobile species are present a species label has to be added but for expository purposes we will omit this complication here.

(2) The variable $f(\vec{r},\vec{p};\, t)$ also obeys the so-called Klimontovich equation but this fact will not be used here.

(3) The average of $f(\vec{r},\vec{p};\, t)$ over the initial phase $\Gamma = \{\vec{x}_j(t = o),\ \vec{p}_j(t = o)\}$ is N times the ordinary one distribution function.

(4) With the exception of the energy because the potential energy is a two body quantity. Hence some care will have to be exercised subsequently in order to treat the energy conservation correctly.

(5) $<A> = \int d\Gamma\, A(\Gamma)\, \rho(\Gamma)$ where Γ denotes the initial phase(3) and $\rho(\Gamma)$ the canonical equilibrium distribution.

(6) In the particular notation which will be used here this algebra can be found in M. Baus: Physica 79A, 377 [1975] where a special effort towards clarity was made!

(7) The details of the algebra leading to Eq. (2.11) can be found elsewhere [Baus, 1975]. Here it suffices to recall that in Eq. (2.11) L denotes the Liouville operator,

$$\partial_t\, f(\vec{r},\vec{p};\, t) = i\, L\, f(\vec{r},\, \vec{p};\, t) \quad ,$$

and $P = 1 - Q$ is Mori's projection operator onto

$$\delta f(\vec{k}\vec{p}) \equiv \delta f(\vec{k}\vec{p};\ t = o).$$

(8) In Eq. (2.23) the presence of $\overline{Q} = 1 - \overline{P}$ with

$$\overline{P} = \sum_{j=1}^{5} |j><j|$$

insures that $\Omega_{ij}(kz)$ is regular at k = o and z = o as detailed elsewhere [Baus, 1975].

(9) For a recent review see the article by G. Stell in "Statistical Mechanics" part A, edited by B. J. Berne, Plenum Press, New York (1977). Notice that this relation is also verified by all known approximation schemes (DH, PY, HNC). See for instance "Theory of Simple Liquids" by J. P. Hansen and I. R. McDonald, Academic Press, London (1976).

(10) For instance, for the OCP, χ_T can become negative [Hansen, 1973].

(11) Transforming $G_{nn}(\vec{k}z)$ back to $\vec{r}$ - t space one obtains the Van Hove function $G_{nn}(\vec{r},t)$ whose Fourier transform, $\hat{G}_{nn}(\vec{k},\omega)$, is the dynamic structure factor.

(12) With respect to the wave vector $\vec{k}$.

(13) See Eq. (2.42) for the analogous property of the OCP.

(14) Related parameters are $\varepsilon = \lambda/4\pi$ and

$$\Gamma = \frac{1}{3}\left(\frac{3\lambda}{4\pi}\right)^{2/3} = e^2 \beta a^{-1}$$

with

$$a^{-1} = \left(\frac{4\pi}{3} n\right)^{1/3} .$$

(15) We can roughly estimate ω_c as $\lambda\omega_p$ for small λ while we still expect ω_c to increase with λ for large λ.

(16) It may suffice to say that it depends in a complicated way on the double and triple equilibrium correlation functions [Wallenborn and Baus, 1977].

(17) These static data have been analyzed in the lectures of J. P. Hansen and H. DeWitt.

(18) These fits are not valid for small Γ (<2) nor do they provide accurate values of the separate integrals I_1, I_2 and I_3 of Eq. (3.16) in regions where the latter contribute little to η. Unfortunately, the latter point was not clearly stated in Wallenborn and Baus [1977]. We have introduced this fit mainly to enable us to localize the minimum of $\eta^*(\Gamma)$. Such a fit has, however, a wider interest.

REFERENCES

Ackasu, A. Z. and J. J. Duderstadt, 1969, Phys. Rev. 188, 479.

Baus, M., 1977, three papers to appear, Phys. A.

Baus, M., 1977, Phys. Rev. A15, 790.

Baus, M. and J. Wallenborn, 1977, J. Stat. Phys. 16, 91.

Bernu, B., P. Vieillefosse and J. P. Hansen, 1977, submitted to Phys. Lett. A.

Braun, E., 1967, Phys. Fluids 10, 731.

Dubois, D. F. and V. Gilinsky, 1964, Phys. Rev. 135, A1519.

Forster, D., 1974, Phys. Rev. A9, 943.

Forster, D. and Martin, P. C., 1970, Phys. Rev. A2, 1575.

Giquinta, P. V., P. Parrinello and M. P. Tosi, 1976, Phys. Chem. Liq. 5, 305.

Gross, E. P., this Volume.

Hansen, J. P., 1973, Phys. Rev. A6, 3096.

Hansen, J. P., I. R. McDonald and E. L. Pollock, 1975, Phys. Rev. A11, 1025.

Ichimaru, S., H. Totsuji, T. Tange and D. Pines, 1975, Progr. Theor. Phys. 54, 1077.

Jhon, M. S. and D. Forster, 1975, Phys. Rev. A12, 254.

Landau, L. D. and E. M. Lifshitz, 1959, Statistical Physics, Pergamon Press, Oxford.

Lindenfeld, M., 1977, Phys. Rev. A15, 1801.

Linnebur, E. J. and J. J. Duderstadt, 1973, Phys. Fluids 16, 665.

Martin, P. C., 1967, Phys. Rev. 161, 143.

Mazenko, G. F., 1974, Phys. Rev. A9, 360.

McDonald, I. R., J. P. Hansen and P. Vieillefosse, preprint.

Pines, D. and Ph. Nozières, 1966, The Theory of Quantum Liquids, Vol. I, W. A. Benjamin, New York.

Resibois, P., 1970, Physica 49, 591.

Vieillefosse, P. and J. P. Hansen, 1975, Phys. Rev. A12, 1106.

Wallenborn, J. and M. Baus, submitted to Phys. Rev. A.

Wallenborn, J. and M. Baus, 1977, Phys. Lett. 61A, 35.

NODAL EXPANSION FOR STRONGLY COUPLED CLASSICAL PLASMAS

C. Deutsch

Laboratoire de Physique des Plasmas
Université Paris XI
Bâtiment 212
91405 ORSAY CEDEX, France

TABLE OF CONTENTS

NODAL EXPANSION FOR STRONGLY COUPLED CLASSICAL PLASMAS

C. Deutsch

Laboratoire de Physique des Plasmas, Université Paris XI

Bâtiment 212, 91405 ORSAY CEDEX, France

I. INTRODUCTION

Perturbative expansions provide the basis of many of the most powerful results as well as the ground of many qualitative powerful pictures. In the equilibrium theory of fluids, the paridigm of perturbation methods is the nodal expansion of the statistical (correlation functions, structure factors, etc.) and the thermodynamic quantities with respect to a small parameters, which implies the exact knowledge of the unperturbed state, and also the existence of a well-defined extrapolating procedure (through resummations for instance) to arbitrarily large values of the expansion parameter. In the case of fully ionized classical Coulomb gases, it is a well-known fact that the "smallness" parameter is uniquely defined [Montroll and Ward, 1958] by the plasma parameter

$$\Lambda = \frac{|\text{binary Coulomb energy at screening length } \lambda_D|}{k_B T}$$

while the reference state is the perfect gas. The nodal expansion is then nothing but that the required adaptation [Salpeter, 1958] of the Mayer density expansion to the case of long-ranged interactions endowed with a well defined Fourier transform in k space.

Such techniques provide one of the three fundamental theoretical tools building up our present understanding of the statistical mechanics of strongly coupled classical plasmas. The other two being the Hypernetted Chain (HNC) approximation of the pair corelations, and the Molecular-Dynamics (MD) numerical simulations of the structure factor and thermodynamics of a few hundred particles enclosed in a box with the aid of computers. It should be appreciated that these different methods work best when used in an interdependent fashion. For instance, the nodal expansion provides the theoretical background used in the numerical manipulations of the HNC scheme [Deutsch, Furutani and Gombert, 1976], while the latter is required [Galam and Hansen, 1976] to complement the small k information demanded by the MD calculations. Moreover, the nodal expansion retains an intrinsic interest when it is the only available technique for an important quantity such as the short range limit of the binary correlation function. Also, the systematic expansion of the equilibrium properties with respect to Λ allows for a smooth albeit systematic building up of the strong correlations when starting from the dilute (Debye) first-order treatment. Up to now, we had in mind the classical three-dimensional plasma. However, we shall take advantage of the peculiar form of the Coulomb interaction, proportional to k^{-2} in k space at all dimensionality $\nu = 2 + \varepsilon$, to generalize the nodal expansion to any ν, and parametrize the equilibrium quantities with respect to ε. The procedure will allow for a very economical and transparent comparison of the high-temperature expansions for different ν values. Another important extension concerns the inclusion of diffraction corrections, always present in a high-temperature plasma when

$$\frac{\hbar}{\sqrt{2\, m_e\, k_B T}} \geq \frac{e^2}{k_B T} \quad ,$$

i.e. $k_B T \geq 1$ Ry for $\nu = 3$.

This generalization is achieved through temperature-dependent effective interactions replacing the bare Coulomb interaction.

II. ONE COMPONENT PLASMAS (OCP) IN ν DIMENSIONS

A. Introduction

To introduce the plasma parameter nodal expansion in the best conditions, let us start from the one-component classical plasma model, an obvious idealization of the realistic plasma with one component smeared out into a continuous rigid neutralizing background. This is a non-zero-temperature extension of the standard jellium (T = 0) model used by the solid-state theorists.

Here, we consider the space dimensionality as a continuous running parameter, so the Coulomb interaction

$$\phi^{(\nu)}(r) = \begin{cases} (\nu - 2)^{-1}\, |r|^{2-\nu}, & \nu \neq 2 \\ \ln |r|^{-1}, & \nu = 2 \end{cases} \quad , \tag{2.1}$$

is the solution of the Poisson equation

$$\Delta \, \phi^{(\nu)} \, (r) = - |\nu-2| \, S_\nu \delta_\nu \, (r), \quad S_\nu = \frac{2\pi^{\nu/2}}{\Gamma(\frac{\nu}{2})} \tag{2.2}$$

The total Coulomb energy of N particles with unit charge e, in the presence of an inert and homogeneous neutralizing background reads ($r_{ij} = |\vec{r}_i - \vec{r}_j|$)

$$W^{(\nu)}(N) = \frac{e^2}{2} \sum_{i \neq j} \phi^{(\nu)}(r_{ij}) - \rho e^2 \sum_{j=1}^{N} \int d^\nu \vec{r} \phi^{(\nu)}(|\vec{r} - \vec{r}_{ij}|)$$

$$+ \frac{\rho^2 e^2}{2} \iint d^\nu \vec{r} \, d^\nu \vec{r}' \, \phi^{(\nu)}(|\vec{r} - \vec{r}'|) \ldots$$

$$= \frac{e^2}{2} \sum_{i \neq j}^{N} \phi^{(\nu)}(r_{ij}) + \frac{N e^2}{2R^{\nu-2}} \sum_{i=1}^{N} \left(\frac{|r_i|}{R}\right)^2 - \tilde{B}_\nu(\nu,R,N) \tag{2.3}$$

where $V = \frac{S_\nu}{\nu} R^\nu$ and

$$\tilde{B}_\nu(\nu,R,N) = \left[\frac{\nu - 1}{\nu + 2} + \phi^{(\nu)}(R) \, R^{\nu-2}\right] \frac{e^2 N^2}{2R^{\nu-2}} \tag{2.4}$$

denotes the background self-energy. Equation (2.3) makes clear that $\nu = 2$ is a landmark with respect to the Coulomb tail behavior. For $\nu > 2$, it is a decreasing function of r, while it increases when $\nu < 2$.

B. ν-Dimensional Coulomb Interaction

The compact expression (2.3) may be immediately extended to any real ν through Eq. (2.1), in agreement with the Fourier transform

$$\phi^{(\nu)}(k) = - \frac{S_\nu}{k^2} \, , \text{ all } \nu, \tag{2.5}$$

of the Poisson Eq. (2.2). In order to work out also the nodal expansion with a continuous ν, we need appropriate Fourier relationships between the two forms of the Coulomb interaction. This may be obtained from the straightforward extension of the volume integral

$$\int_0^R r^{\nu-1} \, dr \prod_{i=1}^{\nu-1} \int_0^\pi \sin^{\nu-i-1} \phi_i \, d\phi_i = \frac{2\pi^{\nu/2}}{\nu\Gamma(\nu/2)R^\nu} = \frac{S_\nu}{\nu} R^\nu \tag{2.6}$$

and the Wilson-like quadratures [Deutsch, 1976; 1977] $[K_\nu = S_\nu (2\pi)^{-\nu}]$

$$(2\pi)^{-\nu} \int d^\nu \vec{k}\, f(\vec{k} \cdot \vec{k}_1) = \frac{K_{\nu-1}}{2\pi} \int_0^\infty dk \int_0^\pi d\phi\, k^{\nu-1} (\sin\phi)^{\nu-2}$$

$$\cdot\, f(k^2, k_1 \cdot k \cos\phi) \qquad (2.7)$$

As in $\nu = 3$, $\phi^{(\nu)}(k)$ is obtained through a regular quadrature for $\varepsilon > 0$, while the inverse transform is only meaningful in the usual Tauberian limit

$$\lim_{a\to o} \int_0^\infty dr\, e^{-ar}\, r^{\varepsilon/2}\, J_{\frac{\varepsilon}{2}}(kr) \simeq \frac{2^{\varepsilon/2}}{k^{1+(\varepsilon/2)}} \Gamma\left(\frac{1+\varepsilon}{2}\right), \quad \varepsilon > 0$$

these isometries may be pushed down [Deutsch, 1976; 1977] to $\varepsilon < 0$, by imposing $C_v > 0$.

C. Pair Correlation Function

(1) Basic formalism and first-order. We are thus allowed to develop the standard [Salpeter, 1958; Deutsch, Furutani and Gombert, 1976; Deutsch, 1976; 1977] high-temperature formalism based on the perturbative analysis of the pair correlation function

$$g_2(r) = \exp\,[\beta\, W_2(r)] \quad , \quad \beta = (k_B T)^{-1} \qquad (2.8)$$

with respect to the dimensionless plasma parameter $\left(\lambda_D^2 = \frac{k_B T}{S_\nu\, \rho e^2}\right)$

$$\Lambda_\varepsilon = \frac{e^2}{k_B T\, \lambda_D^\varepsilon}$$

in terms of the potential of average force

$$W_2(r) = -u(r) + \sum_{k=1}^\infty \beta_k(r)\, \rho^k \quad , \quad u(r) = e^2\, \phi^{(\nu)}(r) \ , \qquad (2.9)$$

and the simple 12-reducible cluster integrals

$$\beta_k(r) = \frac{1}{k!} \int \cdots \int d^\nu \vec{r}_3 \ldots d^\nu \vec{r}_{k+2}\, \Sigma^{(k)} \prod_{i<j} f_{ij}, \ \ldots \qquad (2.10)$$

$\Sigma^{(k)}$ denotes the summation over all possible 12-irreducible cluster diagrams that can be obtained from the root points 1 and 2,k refers to the number of nodal points. $f_{ij} = \exp\,[-\beta u(r_{ij})] - 1$ is the Mayer function. Equation (2.10) is valid in the $N, V \to \infty$ limit for finite k.

The high-temperature assumption is introduced through

$\beta e^2|\phi^{(\nu)}(r_{ij})| << 1$ with $r_{ij} \sim \rho^{-1/\nu}$, and without any further separate restriction on the number density ρ. Thus we hope to find a small parameter in terms of which the cluster expansion may be constructed with [Salpeter, 1958]

$$u(r_{ij}) > M >> 0 \quad , \quad r_{ij} < r_M, \ldots \tag{2.11a}$$

$$U(r_{ij}) \sim \varepsilon' \quad , \quad r_M < r < \lambda', \ldots \tag{2.11b}$$

$$U(r_{ij}) \text{ decreases faster than } r^{-\nu} \ , \quad r > \lambda', \ldots \tag{2.11c}$$

and $r_M/\lambda' << \varepsilon' << 1$. f_{ij} is then approximated by -1 in Eq. (2.11a), if order ε' in Eq. (2.11b), and negligible in Eq. (2.11c) with

$$\int d^\nu \vec{r}\, u(r) \sim \varepsilon' \lambda'^\nu \quad .$$

Now, only cases where the range of the potential is long compared to $\rho^{-\nu^{-1}}$ will be considered. Each β_k contains k field points and ℓ lines. The order of magnitude is given by

$$\varepsilon'^{\ell}(\rho\lambda'^{\nu})^k = \varepsilon'^{\ell-k}(\rho\varepsilon' \lambda'^{\nu})^k \quad .$$

Although ε' is by definition small, $\rho\varepsilon' \lambda'^\nu$ may be large for sufficiently large λ'. It is therefore useful to regroup the cluster expansion terms for $w_2(r)$ according to $\ell - k$. The only dimensionless parameter in the problem being $\beta e^2 \phi^{(\nu)}(r)$, one has to put $\varepsilon' = \Lambda_\varepsilon$. In order to get a realistic result free from the harmonic symmetry-breaking term in Eq. (2.3), we first restrict to $\varepsilon > 0$. So, the first-order ($\ell - k = 1$) contribution to $W_2(r)$ is the expected Debye chain

$$\delta(r) = f(r) + \sum_{n=1}^{\infty} \rho^n \int \ldots \int d^\nu \vec{r}_3 \ldots d^\nu \vec{r}_{n+2}\, f(r_{13}) \ldots f(r_{n+2,2}) \tag{2.12}$$

The introduction of $f_{ij} \simeq -\beta U(r_{ij})$ in the above leads to $[\tilde{V}(k) = -\beta e^2 S_\nu k^{-2}]$

$$C_\nu(r) \equiv \delta(r) = (2\pi)^{-\nu} \int d^\nu \vec{k}\, e^{ikr\cos\theta} \frac{\tilde{V}(k)}{1 - \rho\tilde{V}(k)}$$

$$= - \frac{\Lambda_\varepsilon}{2^{\varepsilon/2}\,\Gamma(1 + \frac{\varepsilon}{2})} \cdot \frac{K_{\varepsilon/2}(r/\lambda_D)}{(r/\lambda_D)^{\varepsilon/2}} \tag{2.13}$$

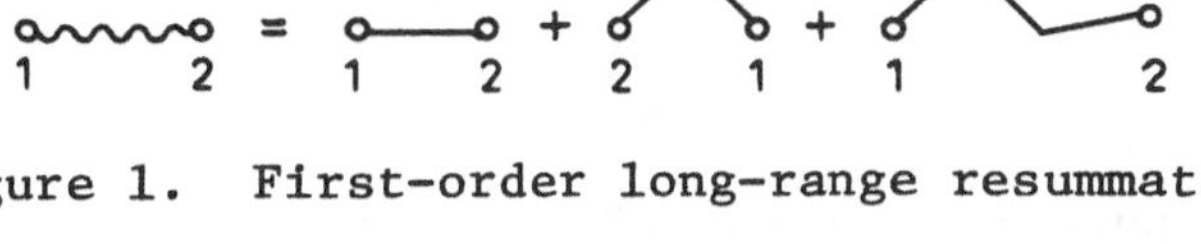

Figure 1. First-order long-range resummation.

giving back at once the well-known results

$$C_3(r) = -\frac{\beta e^2}{r} \exp(-r/\lambda_D) \quad , \quad C_2(r) = -\beta e^2 K_o(r/\lambda_D)$$

$$C_1(r) = -\beta e^2 \lambda_D \exp(-r/\lambda_D) \quad , \quad \ldots \qquad (2.14$$

$K_{\varepsilon/2}(x)$ is the second kind modified Bessel function, with the expected parametrized dimensionality dependence, allowing the analytic continuation to $\varepsilon < 0$ for truncated OCP models retaining only the particle-particle interaction in Eq. (2.3).

(2) Second- and higher order ($\ell - k \geq 2$). The analysis may be pursued further to high order $n = \ell - k \geq 2$. Restricting first to $n = 2$, we have to pay attention to the simplest 2-bubble made of two Debye lines curved between the root points, i.e.

$$(2a) = \frac{\Lambda_\varepsilon^2}{2!} \frac{K^2_{\varepsilon/2}(r)}{2^\varepsilon \Gamma^2\left(1 + \frac{\varepsilon}{2}\right)} \quad , \quad r \text{ in units of } \lambda_D \quad , \qquad (2.15)$$

The next two graphs are equal to the convolution of (2a) with the single Debye line (Eq. 2.13), while the last two-legged nodal graph is a convolution of (2a) with two single Debye lines located symmetrically, so that

$$(2bc,2d) = (-,+) \frac{\Lambda_\varepsilon^2}{2!}\left(S_{\nu-1} \sqrt{\pi} \frac{\Gamma[(\varepsilon+1)/2]}{\Gamma(1+\varepsilon/2)}\right)^{3/4} \cdot \frac{1}{(2\pi)^\nu r^{\varepsilon/2}}$$

$$\cdot \int_o^\infty \frac{dk\, k\, J_{\varepsilon/2}(kr)}{(k^2+1)^{1,2}} \cdot \int_o^\infty du\, u^{1-\varepsilon/2} K^2_{\varepsilon/2}(u) \, J_{\frac{\varepsilon}{2}}(u)$$

$$(2.16)$$

The total second-order correction to $W_2(r)$ is (2a) + (2bc) + (2d). Equations (2.13) and (2.16) bring into light the central role played by the $\nu = 2$ OCP in this high-temperature analysis. The Debye screening process is the same at all dimensionality. The $\nu = 3$ second order results may be immediately recovered from Eqs.

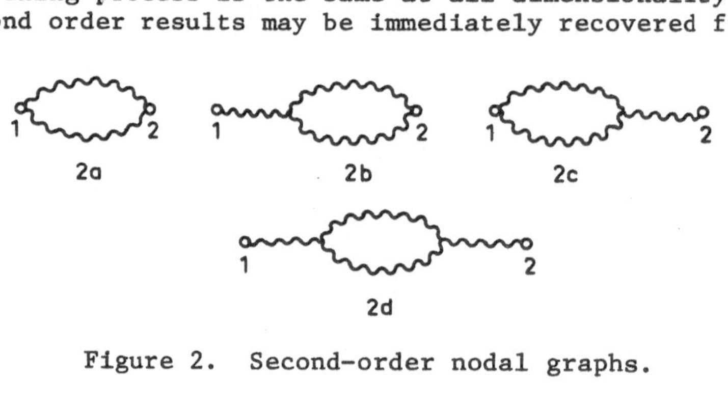

Figure 2. Second-order nodal graphs.

(2.15) and (2.16). More generally, any higher order n convolution graph with $0 \leq m \leq n$ Debye lines and $1 \leq p \leq n-1$ bubbles may be explained as (r in number of λ_D)

$$\left(\frac{\Gamma(1+\varepsilon/2)}{2\sqrt{\pi}}\right)^p r^{-\varepsilon/2} \int_0^\infty \frac{dk\, k^{1+\varepsilon/2} J_{\frac{\varepsilon}{2}}(kr)}{(k^2+1)^m}$$

$$\cdot\ {}_2F_1\left(1, 1-\frac{\varepsilon}{2}\,;\,\frac{3}{2}\,;\,-\frac{k^2}{4}\right)^p$$

$$\underset{r\to\infty}{\simeq} \left(\frac{\Gamma(1-\varepsilon/2)}{2\sqrt{\pi}}\right)^p \frac{r^{m-1}}{2^{m-1}\,\Gamma(m)} \left(\frac{\pi}{2r}\right)^{1/2} e^{-r}, \ \ldots \tag{2.17}$$

The longest convolution chain with m = n and p = n − 1 displays the characteristic nearly ν-independent behavior

$$\left(\frac{\Gamma(1-\varepsilon/2)}{2\sqrt{\pi}}\right)^{n-1} \sqrt{\pi}\, \frac{r^{n-\frac{3}{2}-\frac{\varepsilon}{2}}\, e^{-r}}{2^{n-3/2}\,\Gamma(n)}, \quad r \to \infty \tag{2.18}$$

Therefore the HNC approximation for $g_2(r)$ detailed below for the three dimensional case is expected to hold for any $\nu < 3$. The short range $g_2(r)$ behavior is obtained from the resummation to all orders of the ladder graphs, i.e. the n-bubbles built upon n Debye lines, in the exponential series

$$W_2(r) \underset{r\to o}{\sim} -C_\nu(r) + \frac{C_\nu^2(r)}{2!} + \frac{C_\nu^3(r)}{3!} = \exp\,[-C_\nu(r)] - 1\ , \ \ldots \tag{2.19}$$

So, we get

$$g_2(r) \underset{r\to o}{\sim} \exp\,[-C_\nu(r)] \simeq \begin{cases} \exp\left(-\frac{\Lambda_\varepsilon}{r}\right), & \varepsilon > 0 \\ \exp\left(-\frac{\Lambda}{\varepsilon}\right), & \varepsilon < 0 \\ r^\Lambda, & \varepsilon = 0 \end{cases} \tag{2.20}$$

D. Thermodynamics

Once $g_2(r)$ is known the canonical quantities are immediately obtained from the virial expressions. For the first orders contributions, we can use $g_2(r) \sim 1 + W_2(r)$. This way, one gets

($\rho = N_-/V$, N_- = number of negative pointlike particles).

$$\frac{P_\nu}{k_B T} = \rho\left(1 - \frac{\Lambda_\varepsilon\, \Gamma(1 - \varepsilon/2)}{2^{1+\varepsilon}\, \nu\, \Gamma(1 + \frac{\varepsilon}{2})}\right) \tag{2.21}$$

taking into account the contributions of the unscreened positive background, Eq. (2.21) reproduces the well-known integer equation of state

$$\frac{P_3}{k_B T} = \rho(1 - \frac{\Lambda}{6}) \quad , \quad \frac{P_2}{k_B T} = \rho(1 - \frac{\Lambda}{4})$$

$$P_1 = k_B T\ (1 - \frac{\Lambda}{2}) \tag{2.22}$$

with the $\nu = 1$ result explained in a form equivalent to the two-components Prager result [Prager, 1963]

$$P_1 = 2\rho k_B T - \frac{e}{2}\sqrt{\rho k_B T} \quad ,$$

obtained independently from the Poisson-Boltzmann equation.

The corresponding internal energy reads

$$\frac{E_\nu}{\frac{\nu}{2} N\, k_B T} = 1 - \frac{\Lambda_\varepsilon}{2^\varepsilon\, \nu} \cdot \frac{\Gamma(1 - \varepsilon/2)}{\Gamma(1 + \varepsilon/2)} \quad , \ldots \tag{2.23}$$

The present

$$\frac{E_2}{N\, k_B T} = 1 - \frac{\Lambda}{2}$$

differs from the $\ell n|\frac{r}{R}|$ quantity

$$\frac{E_2}{N\, k_B T} = 1 + \frac{\Lambda}{2}\left[1 - \gamma \ell n \left(\frac{\Lambda_D}{2R}\right)\right]$$

previously obtained [Deutsch and Lavaud, 1974]. This is due to the different $\lim_{r\to\infty} \phi^{(\nu)}(r)$ behavior.

Other first-order thermodynamic quantities are then easily derived, such as

$$\frac{\beta\, F^{exc}}{N} = - \frac{\Lambda_\varepsilon\, \Gamma(1 - \varepsilon/2)}{2^\varepsilon\, \nu\, \Gamma(1 + \varepsilon/2)} \tag{2.24}$$

$$\frac{C_v}{N} = \frac{\nu k_B}{2} + \frac{\Lambda_\varepsilon\, |\varepsilon|\, \Gamma(1 - \varepsilon/2)}{2\nu\, \Gamma(1 + \varepsilon/2)} \tag{2.25}$$

and

$$\frac{S^{exc}}{N} = - \frac{\Lambda_\varepsilon \, |\varepsilon| \, \Gamma(1 - \varepsilon/2)}{2^{1+\varepsilon} \, \nu \, \Gamma(1 + \varepsilon/2)} \tag{2.26}$$

with $S^{exc} = -C_v^{exc}$ when $\nu = 2$. These results extend to any ν the usual OCP polarization picture. C_v^{exc} is positive for all ε provided the Coulomb Fourier transform is taken as

$$- \frac{S_\nu \, |\varepsilon|}{k^2} \quad .$$

Higher order Λ_ε corrections may be obtained from the corresponding $W_2(r)$ expansion. For instance, the corresponding free energy is

$$\beta F^{exc} = - \frac{N^2}{2V} \frac{S_\nu}{S_D^\nu} \int_o^\infty dx \; x^{\nu-1} \sum_{m=3}^{\infty} \left(- \frac{C_\nu(x)}{m!} \right)^m \tag{2.27}$$

while P_ν has to be added

$$\left(1 - \rho \frac{\partial}{\partial \rho}\right) \rho^p \prod_{i=1}^{p} \int d^\nu \vec{r}_i \prod_{k=1}^{p} \left[-C_\nu \left(a r_{ij} \, \rho^{1/\nu} \right) \right] \simeq \frac{\varepsilon}{4 r^{\varepsilon/2}} \tag{2.28}$$

a vanishing contribution for $\nu = 2$.

E. Diffraction Corrections

The nodal expansion starting from the high-temperature regime (small Λ values), we are naturally led to pay attention to the situation described by the inequalities

$$\frac{e^2}{k_B T} \leq \bar{\lambda} \leq \lambda_D \tag{2.29}$$

with the point charges replaced by wave-packets.

The inclusion of the diffraction corrections in the nodal classical expansion requires a classical modelization, which is obtained by approximating [Kelbg, 1965; Gombert and Deutsch, 1974; Deutsch and Gombert, 1976] the two-body high-temperature quantum Slater sum with the classical Gibbs expression through the ansatz

$$\exp\left[-\beta(H_o + H_1)\right] = \exp(-\beta H_1) \exp(-\beta H_o) \, G \tag{2.30}$$

with

$$H_o = \sum_{i=1}^{N} \frac{p_i^2}{2m} \quad , \quad H_1 = \sum_{1\leq i\leq j\leq N} e^2 |r_{ij}|^{-\varepsilon} \tag{2.31}$$

G measures the non-commutativity of H_o and H_1 in the small β range. It is given as a solution of the Bloch-like equation

$$\frac{dG}{d\beta} = -\exp(\beta H_o) \left[\beta [H_1, H_o] + \frac{\beta^2}{2!} \left[H_1, [H_1, H_o] \right] \right] \exp(-\beta H_o) \, G \tag{2.32}$$

The N-body canonical partition function is thus explained in terms of the temperature-dependent effective interaction

$$W_\nu(r) = (2\pi)^{-\nu} \int_{\vec{k}} \phi^{(\nu)}(k) \left[\int_o^1 \exp\left(- \frac{\beta \hbar^2 k^2 \alpha(1-\alpha)}{m} \right) d\alpha \right] e^{i\vec{k}\cdot\vec{r}} \, d\vec{k} \tag{2.33}$$

when specialized to $\nu = 3$, it becomes

$$W_3(r) = r^{-1} \left[1 - \exp\left(- \frac{r^2}{\lambda\!\!\!^-{}^2} \right) \right] + \frac{\sqrt{\pi}}{\lambda\!\!\!^-} \,\mathrm{Erf}\left(\frac{r}{\lambda\!\!\!^-} \right)$$

$$\simeq r^{-1} (1 - e^{-Cr}) \quad , \quad C \sim \lambda\!\!\!^-{}^{-1} \tag{2.34}$$

Therefore, the first order resummation (Eq. 2.14) is changed to

$$\frac{e^2}{r \left(1 - \frac{4}{C^2 \lambda_D^2} \right)^{1/2}} \cdot \left[e^{-\alpha_1 r} - e^{-\alpha_2 r} \right] \tag{2.35}$$

where

$$\alpha_{1,2}^2 = \frac{C^2}{2} \left[1 \mp \left(1 + \frac{4}{C^2 \lambda_D^2} \right)^{1/2} \right] \quad , \quad C\lambda_D > 2$$

yielding Eq. (2.14) in the $T \to \infty$ limit. The effective interaction is locally summable for $r \simeq 0$, so the nodal expansion remains order-by-order well-defined for all n. This remarkable property paves the way to a new class of OCP models free of the Meeron-like resummation [Furutani and Deutsch, 1977] of n-ladders. At the Thermodynamic level the classical $\hbar \to 0$ limit appears in a much more sophisticated [Gombert and Deutsch, to be published] way. For instance, the most diverging (for F for instance) graphs, non-analytic in $\hbar$ should first be summed up to infinity. Nonetheless the few available "exact" results [DeWitt, 1966] for the free energy are reproduced up to all orders by the classical effective interaction techniques through a double expansion in Λ_ε and $\hbar\omega_p/k_BT$, ω_p being the plasma frequency.

III. THREE DIMENSIONAL OCP

In view of the practical importance of the $\nu = 3$ OCP, we detail here a little further some interesting applications for this case.

A. Short Range Limit

The above $\hbar$-extension of the nodal expansion is well-suited to investigate $\lim_{r\to o} g_2(r)$ through a resummation followed by the $\hbar \to 0$ limit. The $g_2(r)$ short range behavior is monitored by the n-ladder graphs ($\ell = n$, $k = 0$) followed by the same structures decorated with one ($\ell = n + 1$, $k = 1$) and two ($\ell = n + 2$, $k = 2$) Debye lines. For instance, we get

$$\frac{(\text{n-ladder with one Debye line})_{r=0}}{(\text{n-ladder})_{r=0}} \leq \frac{\ell n\, \alpha_2}{\alpha_2^2} \underset{h\to o}{\to} 0 \tag{3.1}$$

showing how the first class of graphs diverges in the classical limit. So, the most important contribution in the $r \to 0$ range is ($\Lambda_1 \equiv \Lambda$)

$$\sum_{n=1}^{\infty} \frac{(-\Lambda)^n}{n!} \left[\frac{e^{-\alpha_1 r} - e^{-\alpha_2 r}}{r}\right]^n = \exp\left[-\Lambda\left[e^{-\alpha_1 r} - e^{-\alpha_2 r}\right]r^{-1}\right] - 1 \tag{3.2}$$

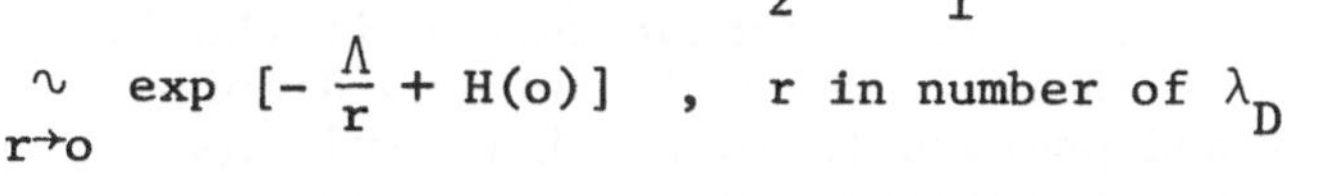

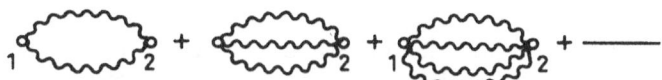

Figure 3. n-Ladders sum.

yielding the classical expression ($\alpha_2 >> \alpha_1 \sim 1$)

$$g_2(r) \underset{r\to o}{\sim} \exp\left[-\frac{\Lambda}{r} + H(o)\right] \quad , \quad r \text{ in number of } \lambda_D \tag{3.3}$$

already obtained by Cooper and DeWitt [1972] from an extrapolation of the HNC equation. This limit is reached with $\alpha_2 \to \infty$, $\alpha_1 \to 1$ followed by $r \to 0$. Had we retained some non-negligible diffraction effects, we should perform $r \to 0$ first, with the result

$$g_2(r) \underset{r\to o}{\sim} \exp\left[-\frac{\beta e^2}{\lambda\!\!\!^{-}} + H(0)\right] \tag{3.4}$$

different from Eq. (3.3). In both cases, H(0) is well approximated by the $r \to 0$ limit of the sum of two series of decorated ladders

$$H(0) = S(1) + S(2) + \alpha_1 \lambda_D \Lambda \tag{3.5}$$

where

$$S(n) = \lim_{r \to o} \frac{2}{\pi r} \int_o^\infty dk \sin kr \left(\frac{1}{k^2 + \alpha_1^2} - \frac{1}{k^2 + \alpha_2^2} \right)^n$$

$$\cdot \int_o^\infty du\, u \sin ku \exp \left[-\Lambda \left(e^{-\alpha_1 U} - e^{-\alpha_2 U} \right) U^{-1} - 1 \right.$$

$$\left. + \Lambda \left(\frac{e^{-\alpha_1 U} - e^{-\alpha_2 U}}{U} \right) \right] \tag{3.6}$$

It is nothing but the constant of the Widom quadratic polynomial in

$$g_2(r) \sim e^{-\frac{\beta e^2}{r} + \text{Polynomial } (r^2)}, \qquad r \to 0 \tag{3.7}$$

Therefore H(0) accounts for the important lowering of the Coulomb barrier in dense plasmas sustaining nuclear reactions.

B. Long Range Limit

Within every order n of the nodal expansion, the slowest decaying chains built upon n-1 2-ladders with n Debye lines intertwined in between, altogether with the corresponding chains with $0 \leq c \leq n - 1$ Debye lines constitute the backbone [Salpeter, 1958] of the $g_2(r)$ infinite limit. It is known that the generic nonconvolution graphs (including the bridge ones) may be included in a straightforward extension of the HNC long-range resummation. The key remark allowing for this important result is the faster than Debye $r \to \infty$ decay of the nonconvolution graphs demonstrated by Deutsch, Furutani and Gombert [1976]. This explains that the asymptotic potential of average force could read

$$W_2(r) \underset{r \to \infty}{\sim} -\frac{\Lambda e^{-r}}{r} + \frac{2}{\pi r} \int_o^\infty dk\, k \sin kr \left(\frac{\frac{k^2}{k^2+1} G_1(k)}{1 + k^2 - \frac{G'(k)}{\Lambda}} - G_2(k) \right) \tag{3.8}$$

with $G_2(k) \sim G_1(k) = G'(k)$ in the $k \to 0$ limit.

$G'(k)$ denotes the sume up to infinity of ladder graphs + suitable combination of bridge graphs. Expression (3.8) thus explains the striking success of the usual HNC numerical procedure [Springer, Pokrant and Stevens, 1973] based upon the iteration of the following set of phenomenological equations

$$W_2(r) = -\frac{\Lambda e^{-r}}{r} + S(r) \tag{3.9}$$

$$g_2(r) - 1 = T(r) + S(r) \tag{3.10}$$

$$T(r) = G(r) - \frac{\Lambda e^{-r}}{r} \tag{3.11}$$

$$S(k) = \frac{k^2\, G(k)\, (k^2+1)^{-1}}{1 + k^2 - \frac{G(k)}{\Lambda}} \tag{3.12}$$

initiated with $S(r) = 0$, while it allows for systematic improvements through the inclusion of nonconvolution graphs.

C. Onset of Short Range Order

From the general asymptotic behavior (3.8), it is clear that the $r \to \infty$ limit of $g_2(r)$ is monitored by the closest poles of the integrand to the real k-axis, in the complex k-plane. Actually one can see [DelRio and DeWitt, 1969; Furutani and Deutsch, 1977] that when Λ increases the roots of the denominator lie on the positive imaginary k-axis. They start respectively from i and 2i until they merge when $\Lambda_C = 4.225$. For $\Lambda \geq \Lambda_C$, the $W_2(r)$ exponential decay is modulated by a cosine factor announcing a non-negligible short-range structure steadily increasing with Λ until the appearance of long-range order with the fluid-solid transition. Up to now, only the infinite n-ladders sum has been used in the calculations [Furutani and Deutsch, 1977]. A more quantitative result should be expected from the inclusion of bridge graphs.

IV. TWO COMPONENT PLASMAS

A well-known long-standing paradign in the classical perturbation theory of equilibrium thermal properties of systems with short as well as long-range two-body forces is its limitation to purely repulsive interactions, in order to prevent a possible collapse due to a lack of stability arising from the strong attractive interactions evidenced by the appearance of bound states. This point of view is at the very basis of the one-component plasma philosophy illustrated in the previous sections. As a consequence, a widespread popular belief tends to relegate systems with attractive forces to a quantum mechanical framework exclusively. With explicit thermodynamical results explained only in the $T \to 0$ limit. Our purpose is to show that such a point of view is far too restrictive. The above noticed limitations, to a large extent, may be removed, provided one puts the emphasis on a high enough mean kinetic energy regime, in order to allow for a complete compensation of the bound states by the scattering states through the Levinson theorem [Dashen and Rajaraman, 1974]. The system we have in mind is the real two-component nonrelativistic electron proton plasma, i.e. real matter in bulk taken in the high temperature regime $k_B T > 1$ Ry. The corresponding two-body sum-over-states for unlike charges may be estimated phenomenologically on sound heuristic basis with the aid of the screened Debye [Grandjouan and Deutsch, 1975] interaction $-e^2 \exp(-r/\lambda_D)/r$, and it is known that the corresponding bound states sum is negligible. So, the bound states formation do not plague any more through the appearance of uncontrolled infinite quantities, the use of the perturbative nodal expansion with respect to the plasma parameter Λ. Starting from the above OCP nodal expansion, we are allowed to extend it in a straightforward way to the present situation. In so doing, we are led to parallel to some extent [Murphy, 1968] a situation already used for charged hard systems, used in the theory of symmetric electrolytes. However, in contradistinction to this particular situation, we are not restricted to a second-order in Λ_ε nodal expansion of the thermal properties, because the short-range behavior of the proton-electron interaction is well taken into account in the high temperature regime through the soft pseudopotential

$$W_{e-i}(r) = -\frac{e^2}{r}\left(1 - e^{-Cr}\right) \quad , \quad C = \frac{1}{\lambda\!\!\!^{-}_{ei}} \quad , \quad \lambda\!\!\!^{-}_{ei} = \frac{\hbar}{\sqrt{2m_{ei}\, k_B T}} \tag{4.1}$$

The above OCP nodal rules have to be modified as follows:

(a) λ_D^2 becomes $k_B T/S_\nu \rho e^2 (C_1 Z_1^2 + C_2 Z_2^2)$, where $C_i = N_i/N$ so that each field point gets an extra factor $(C_1 Z_1^2 + C_2 Z_2^2)^{-1}$.

(b) The interaction between particles i and j is proportional to $Z_i Z_j$, so that each field point is again factorized out with

$$C_1 Z_1^{M_k} + C_2 Z_2^{M_k} \quad ,$$

where M_k denotes the number of lines merging into the field point k. The root points have to be given a factor

$$Z_1^{M_1} \; Z_2^{M_2} \quad .$$

(c) The Debye lines $(k^2 + 1)^{-1}$, in momentum space, corresponding to the bare Coulomb interaction $r^{-\varepsilon}$ have to be replaced by a normalized sum of the three modified Debye lines attached to the direct interaction ion-ion, ion-electron, electron-electron respectively. This is easily performed once we notice that the Debye interaction resumming the long-range behavior of the effective interaction (4.1) may be written as $(C\lambda_D > 2)$:

$$\frac{e^2\left(e^{-\alpha_1 r} - e^{-\alpha_2 r}\right)}{r\sqrt{1 - 4C^2/\lambda_D^2}} \quad , \quad \frac{1}{k^2 + \alpha_1^2} - \frac{1}{k^2 + \alpha_2^2} \tag{4.2}$$

in r-space and k-space respectively.

(d) Putting together the above rules for the field points, we see that each nodal graph with k field points has to be given a factor

$$\left(Z_1^{M_k - 1} - Z_2^{M_k - 1}\right)/(Z_1 - Z_2)$$

when the overall neutrality condition $C_1 Z_1 + C_2 Z_2 = 0$ is taken into account. An immediate by-product of this condition is the vanishing to all order of the longest chains building up the long range behavior of the OCP $g_2(r)$ in the HNC approximation. They are replaced by much more rapidly decaying chains built upon 3-ladders and other more compact nonconvolution graphs with Debye lines intertwined in between. Also, the number of graphs at a given order is drastically reduced. Again the asymptotic behavior of $W_2(r)$ may be systematically resummed yielding a kind of HNC-like expression. However, its analyticity properties in the complex k-plane are completely different from the genuine HNC one. For instance, there is no longer onset of short-range order for a symmetric system when the plasma parameter is increasing, while the two components are considered as Boltzmann classical particles.

A. $\nu \leq 2$ Nodal Expansion

Let us first pay attention to the lower dimensionalities where the above rule (c) is the simpler to apply. The three distinct first-order corrections to the potential of average force are given as $(i,j = e,i)$

$$C_{ij}^{(1)}(r) = -\frac{Z_i Z_j \Lambda_\varepsilon}{2^{\varepsilon/2}\,\Gamma(1 + \varepsilon/2)} \frac{K_{\varepsilon/2}(r/\lambda_D)}{(r/\lambda_D)^{\varepsilon/2}} \quad , \quad Z_{i,j} = \pm 1 \tag{4.3}$$

an obvious extension of the OCP result (2.13).

Higher order corrections start with $\ell - k = 2$. However, as shown on Figure 4 rule (d) drastically reduces the number of available Mayer-Salpeter nodal graphs.

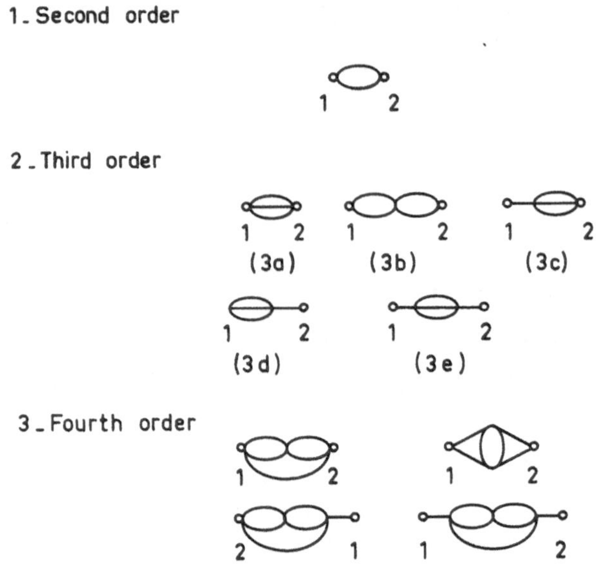

Figure 4. High order TCP nodal graphs. The solid line denotes a Debye interaction.

A similar result holds if one works out the Debye screening process in a field theoretical framework. In the present case, we therefore obtain

$$C_{ij}^{(2)}(r) = \frac{Z_i^2\, Z_j^2\, \Lambda_\varepsilon^2\, K_{\varepsilon/2}^2\,(r/\lambda_D)}{2^{\varepsilon+1}\Gamma(1+\varepsilon/2)^2 (r/\lambda_D)^\varepsilon} \tag{4.4}$$

The first third-order graph (3a) of Figure 4 is

$$C_{ij}^{(3a)}(r) = -\frac{Z_i Z_j\, \Lambda_\varepsilon^3\, K_{\varepsilon/2}^3\,(r/\lambda_D)}{3!\; 2^{3\varepsilon/2}\;\Gamma(1+\varepsilon/2)^3\;(r/\lambda_D)^{3\varepsilon/2}} \tag{4.5}$$

The second (3b) is a convolution product of two 2-bubbles explained by ($Z_i^2 = Z_j^2 = 1$)

$$C_{ij}^{(3b)}(r) = \frac{K_{\nu-1}\,\Lambda_\varepsilon^3}{2\pi}\int d^\nu \vec{k}\; e^{i\vec{k}\cdot\vec{r}}\; G(k)^2 \tag{4.6}$$

where

$$G_2(k) = \frac{K_{\nu-1}}{(2\pi)^{1-\nu}\, 2^{\varepsilon+1}\, \Gamma(1+\varepsilon/2)^2} \int \frac{d^\nu \vec{r}\, e^{i\vec{k}\cdot\vec{r}}\, K^2_{\varepsilon/2}(r)}{r^\varepsilon}$$

$$= \frac{S_{\nu-1}}{2^{\varepsilon+3}\, \Gamma(1+\varepsilon/2)^2} \cdot \frac{\Gamma[(1+\varepsilon)/2]\Gamma(1-\varepsilon/2)}{\Gamma(3/2)}$$

$$\cdot\ {}_2F_1\left(1, 1-\frac{\varepsilon}{2}\,;\, \frac{3}{2};\, -\frac{k^2}{4}\right)^2 \qquad (4.7)$$

in the form

$$C_{ij}^{(3b)}(r) = \frac{K_{\nu-1}\, S^2_{\nu-1}\, \Gamma[(1+\varepsilon)/2]^3\, \Gamma(1+\varepsilon/2)^2\, \Lambda^3_\varepsilon}{2^{3+3\varepsilon/2}\, \Gamma(1+\varepsilon/2)^2\, \pi^{3/2}\, r^{\varepsilon/2}}$$

$$\cdot \int_0^\infty dk\, k^{1+\varepsilon/2}\, J_{\varepsilon/2}(kr)\, {}_2F_1\left(1, 1-\frac{\varepsilon}{2}\,;\, \frac{3}{2}\,;\, \frac{-k^2}{4}\right)^2 \qquad (4.6')$$

suitable for numerical computation with r, in number of λ_D, as a running parameter.

The graphs (3c,d) may be worked out similarly through

$$G_{ij}^{(3c,3d)}(r) = Z_i Z_j \frac{K_{\nu-1}}{2\pi} \Lambda^3_\varepsilon \int d^\nu \vec{k}\, e^{i\vec{k}\cdot\vec{r}}\, G_3(k)\, H(k) \qquad (4.8)$$

with

$$H(k) = -\frac{S_{\nu-1}\, \pi^{1/2}\, \Gamma[(\varepsilon+1)/2]}{\Gamma(1+\varepsilon/2)(k^2+1)} \qquad (4.9)$$

and

$$G_3(k) = -\frac{K_{\nu-1}}{(2\pi)^{1-\nu}\, 2^{3\varepsilon/2}\, \Gamma(1+\varepsilon/2)^3 3!} \left(\frac{2}{k}\right)^{\varepsilon/2}$$

$$\cdot \int_0^\infty dr\, r^{1+\varepsilon/2}\, J_{\varepsilon/2}(kr) \left(\frac{K_{\varepsilon/2}(r)}{r^{\varepsilon/2}}\right)^3 \qquad (4.10)$$

Incidentally, it must be noticed that $G_n(k) < +\infty$, all n. This property is the very basis of the entire classical treatment valid for the finite to all orders OCP and TCP nodal expansions when $\varepsilon < 0$. In view of the intractability of the last quadrature, it is best to explain $G_{ij}^{(3e)}(r)$ in the form

$$G_{ij}^{(3c,3d)}(r) = \frac{Z_i Z_j}{3!} \frac{K_{\nu-1}^2}{(2\pi)^{2-\nu}} \frac{\Lambda_\varepsilon^3 \, S_{\nu-1} \, \pi^{1/2} \, \Gamma[(1+\varepsilon)/2]}{2^{\varepsilon/2} \, \Gamma(1+\varepsilon/2)^4 \, r^{\varepsilon/2}}$$

$$\int_0^\infty \frac{dk \, k \, J_{\varepsilon/2}(kr)}{k^2+1} \int_0^\infty du \, u^{1-\varepsilon} \, K_{\varepsilon/2}^3(u) \, J_{\varepsilon/2}(ku) \tag{4.11}$$

with the aid of

$$\int_0^\infty \frac{dk \, k \, J_{\varepsilon/2}(kr) \, J_{\varepsilon/2}(ku)}{k^2+1} = \begin{cases} I_{\varepsilon/2}(r) \, K_{\varepsilon/2}(u) \; , & r < u \\ I_{\varepsilon/2}(u) \, K_{\varepsilon/2}(r) \; , & r > u \end{cases} \tag{4.12}$$

which allows us to express its r behavior as

$$\sim r^{-\varepsilon/2} \left[\int_0^r du^{1-\varepsilon} \, K_{\varepsilon/2}^3(u) \, I_{\varepsilon/2}(u) \, K_{\varepsilon/2}(r) + \int_0^\infty du \, u^{1-\varepsilon} \, K_{\varepsilon/2}^4(u) \, I_{\varepsilon/2}(r) \right]$$

whence one easily deduces ($\varepsilon > -1$)

$$G_{ij}^{(3c,3d)}(r) \underset{r\to 0}{\sim} \int_0^\infty du \, u^{1-\varepsilon} \, K_{\varepsilon/2}^4(u) \, 2^{-\varepsilon/2} \, \Gamma(1+\varepsilon/2)^{-1} \tag{4.13}$$

$$G_{ij}^{(3c,3d)}(r) \underset{r\to\infty}{\sim} \int_0^\infty du \, u^{1-\varepsilon} \, K_{\varepsilon/2}^3(u) \, I_{\varepsilon/2}(u) \left(\frac{\pi}{2}\right)^{1/2} \frac{e^{-r}}{r^{(1+\varepsilon)/2}}$$

with a nearly Debye-like asymptotic decrease. Actually this last result illustrates the basic trend of the TransFourier-convolution treatment: a given nodal graph may not, irrespective of dimensionality, be simultaneously negligible at r = 0 and r = ∞. The last third-order graph (3e) may be similarly analysed with

$$G_{ij}^{(3e)}(r) = -\frac{Z_i Z_j}{3!}\frac{K^2_{\nu-1}}{(2\pi)^{2-\nu}}\frac{\Lambda_\varepsilon^3}{2^{\varepsilon/2}}\frac{\left[S_{\nu-1}\,\pi^{1/2}\,\Gamma\left(\frac{1+\varepsilon}{2}\right)\right]^2}{\Gamma(1+\varepsilon/2)^5\,r^{\varepsilon/2}}$$

$$\cdot\int_0^\infty \frac{dk\,k\,J_{\varepsilon/2}(kr)}{(k^2+1)^2}\int_0^\infty du\,u^{1-\varepsilon}\,K^3_{\varepsilon/2}(u)\,J_{\varepsilon/2}(ku) \tag{4.14}$$

It must be kept in mind that the present analysis works only for $\varepsilon > -1$, and also that it cannot be straightforwardly pursued to high orders, in view of $K^P_{\varepsilon/2}(u) \sim u^{-P|\varepsilon|/2}$ when $u \to 0$. Actually, short range resummations with the aid of n bubble lines should then be used to all order in Λ_ε, in order to get rid of this difficulty. Nevertheless, the present order-by-order stepwise evaluation of the nodal graphs is expected to remain valid in the $\varepsilon = 0$ limit, on both sides of the cross over dimensionality $\nu = 2$. Also their TCP dependence is concentrated in the overall

$$Z_i^{M_1}\,Z_j^{M_2}$$

factor, because their field points counting factor factorize out trivially to unity (rule c). By extrapolating in a straightforward way the above third-order results, one gets the dominant asymptotic contribution within each order from the alternate chains built up from 3-bubbles and single Debye lines intertwined in between.

Therefore a modified version of the standard OCP hypernetted chain (HNC) approximation, with 3-bubbles replacing 2-bubbles, and lacunary series replacing geometric ones is expected to exhaust $g_2(r)$ for $r \to \infty$. On the other hand, the short range $g_2(r)$ behavior can no longer be taken as monitored by the resummation to all orders of the parallel graphs when $\nu < 2$. Actually, all the nodal structures remain finite and are therefore equivalent in the r = 0 limit, and no simple expression is expected for $g_2(r)$ in this case. Nevertheless, the logarithm divergence of $K_o(z)$ with $z \to 0$ allows to extend immediately to the $\nu = 2$ TCP, the $\nu = 2$ OCP short-range limit in the form

$$\lim_{r\to o} g_2^{ij}(r) \simeq \exp\frac{-z_i z_j\,\Lambda_\varepsilon\,K_{\varepsilon/2}(r)}{2^\varepsilon\Gamma(1+\varepsilon/2)\,r^{\varepsilon/2}} \quad .$$

This latter clearly shows that, while the short-range correlations between like particles behave similarly to their OCP homolog, the a-like ones exhibit the classical divergence.

Finally, one may also wonder about the possible appearance of the short range order when the plasma parameter increases. This point may be well elucidated by explaining the above-mentioned Hypernetted chain approximation through the longest chains resummations

$$\sim Z_i Z_j \left(\frac{2}{r} \right)^{\varepsilon/2} \sum_{n=1,3,5}^{\infty} (-\Lambda_\varepsilon)^n \int_0^\infty dk\, k^{1+\varepsilon/2}$$

$$\cdot \frac{J_{\varepsilon/2}(kr)}{(k^2+1)^n} G_3^{(n-1/2)}(k) \quad , \quad \ldots$$

$$= -\left(\frac{2}{r} \right)^{\varepsilon/2} Z_i Z_j \left\{ \int_0^\infty \frac{dk\, k^{1+\varepsilon/2} J_{\varepsilon/2}(kr)}{(1+k^2) + \Lambda_\varepsilon G_3^{1/2}(k)} \right.$$

$$\left. - \sum_{n=1}^{\infty} \Lambda_\varepsilon^{2n} \int_0^\infty \frac{dk\, k^{1+\varepsilon/2} J_{\varepsilon/2}(kr)\, G_3^{(2n-1)/2}(k)}{(k^2+1)^{2n}} \right\}$$

$$= -Z_i Z_j \left(\frac{2}{r} \right)^{\varepsilon/2} \int_0^\infty dk\, k^{1+\varepsilon/2} J_{\varepsilon/2}(kr)$$

$$\cdot \left\{ \frac{1}{1+k^2+\Lambda_\varepsilon G_3^{1/2}(k)} - \frac{G_3^{1/2}(k)}{(k^2+1)^2 - \Lambda_\varepsilon^2 G_3(k)} \right\} \tag{4.15}$$

A more complete result may be obtained by retaining all the leading chains with 0,1,2,...,n, Debye lines respectively, i.e.

$$W_2^{ij}(r) \simeq Z_i Z_j \sum_{n=3,5,7..} \sum_{p=0}^{n} (-)^p \binom{n}{p} (-\Lambda^{n-1}) \left(\frac{2}{r} \right)^{\varepsilon/2} \Lambda_\varepsilon$$

$$\cdot \int_0^\infty \frac{dk\, k^{1+\varepsilon/2} J_{\varepsilon/2}(kr)\, G_3^{(n-1)/2}(k)}{(k^2+1)^{n-p}}$$

$$- \left(\frac{2}{r} \right)^{\varepsilon/2} \Lambda_\varepsilon^2 Z_i Z_j \int_0^\infty dk\, k^{1+\varepsilon/2} J_{\varepsilon/2}(kr)\, G_3(k) \tag{4.16}$$

where the non-nodal (with respect to the $W_2(r)$ more stringent irreducibility conditions) reducible (3a) has been subtracted out to secure the $k \to \infty$ summability. Performing first the binomial summation within each order yields

$$\Lambda_\varepsilon \left(\frac{2}{r} \right)^{\varepsilon/2} \left\{ \sum_{n=3}^{\infty} \Lambda_\varepsilon^{n-1} \int_0^\infty dk\ k^{1+\varepsilon/2}\ J_{\varepsilon/2}(kr) \left(\frac{k^2}{k^2+1} \right)^n G_3^{(n-1)/2}(k) \right.$$

$$\left. - \sum_{n=2}^{\infty} \Lambda_\varepsilon^{2n-1} \int_0^\infty dk\ k^{1+\varepsilon/2}\ J_{\varepsilon/2}(kr) \left(\frac{k^2}{k^2+1} \right)^{2n} G_3^{(2n-1)/2}(k) \right\} \tag{4.17}$$

for the first term on the righthand side of Eq. (4.16), and summing up as before the geometrics series

$$W_2^{ij}(r) \simeq \left(\frac{2}{r} \right)^{\varepsilon/2} \Lambda_\varepsilon^2\ Z_i Z_j \int_0^\infty dk\ k^{1+\varepsilon/2}\ J_{\varepsilon/2}(kr)$$

$$\cdot \left\{ \Lambda_\varepsilon \frac{k^2}{k^2+1} G_3(k) \cdot \frac{1}{1 + (1/k^2) - \Lambda_\varepsilon G_3^{1/2}(k)} \right.$$

$$\left. - G_3(k) \frac{1}{[1 + (1/k^2)]^2 - \Lambda_\varepsilon^2 G_3(k)} - G_3(k) \right\} \tag{4.18}$$

with

$$\left| \Lambda_\varepsilon \frac{k^2}{k^2+1} G_3^{1/2}(k) \right| \leq 1 \quad ,$$

makes to appear an algebraic extension of the standard HNC long-range approximation, with the bubble function G conveniently restricted to its first term $\Lambda_\varepsilon^3\ G_3(k)$.

The sign $\simeq$ accounts for the obvious but cumbersome ν-dependent overall factors. A complete expression would be obtained by adding the first order Debye term to the above. Nevertheless, Eq. (4.18) is sufficiently well explained to make sure that the onset of short range order could again be obtained from $1 + k^2 - \Lambda_\varepsilon G_3^{1/2}(k) = 0$, fulfilled only for imaginary $k = iK$ values. The corresponding two Λ-dependent zeros $V_1(\Lambda)$ and $V_2(\Lambda)$ are expected to get closer when Λ_ε increases, until they merge at Λ_ε^C, where $g_2(r) \gtrsim 1$ for moderate r values, as shown by the relation

$$K^2 = 1 + \Lambda_\varepsilon K^2 G_3^{1/2}(iK) \tag{4.19}$$

Nevertheless, the negative quantity

$$G_3(iK) \simeq - \int_0^\infty dr\, r^{1-\varepsilon} I_{\varepsilon/2}(Kr)\, K^3_{\varepsilon/2}(r) < 0$$

makes clear that Eq. (4.19) has only one zero, so the $\nu \leq 2$ TCP's do not exhibit the extension of the three-dimensional OCP short-range order (SRO). A result not unexpected on plausible grounds, because we have here as many attractive as repulsive interactions, and nothing more to simulate any hard core effects. Also, in the $\varepsilon = 0$ limit, it does (in a mean field setting) confirm our previous findings about the inexistence of any cooperative (many-body) effects when the critical temperature is approached from above in the two-dimensional TCP. There is no clustering phenomena annonciating in advance a phase transition with the breaking of the long range order.

B. $\nu > 2$ Nodal Expansion

Most of the above discussion is expected to hold for the TCP's nodal expansion with $\nu > 2$, provided due attention is given to rule (c) through a Debye-like interaction taken in the $\nu = 3$ form. Actually, this is only a slight algebraic alteration of the foregoing treatment. Nevertheless, it leads to a much more cumbersome derivation, which explains that we detailed first the basic features of the TCP's nodal expansion on the more symmetric and transparent $\nu < 2$ situation. Following rule (c), the normalized first order Debye line should now be explained as

$$H'(k) = - \frac{S_{\nu - 1}\pi^{1/2}\,\Gamma[(\varepsilon + 1)/2]}{4\Gamma[1 + (\varepsilon/2)]} \left\{ \frac{1}{k^2 + 1} + \frac{2(\alpha_2^2 - \alpha_1^2)}{(k^2 + \alpha_1^2)(k^2 + \alpha_2^2)} + \frac{\alpha_2'^2 - \alpha_1'^2}{(k^2 + \alpha_1'^2)(k^2 + \alpha_2'^2)} \right\} \tag{4.20}$$

extrapolating first the real case $\nu = 3$, where the pair (α_1,α_2) refers to the plus-minus effective interaction, while (α_1',α_2') pertains to the diffraction corrections of the lighter particles ("electrons").

For the sake of simplicity, we shall ignore these latter in the sequel. Taking the corresponding classical limit $\alpha_2 \to \infty$, $\alpha_1 \to 1$, the above becomes

$$H'(k) = -\frac{S_{\nu-1}\,\pi^{1/2}\,\Gamma[(\varepsilon+1)/2]}{2\Gamma[1+(\varepsilon/2)]}\left[\frac{1}{k^2+1}+\frac{\alpha_2^2-\alpha_1^2}{(k^2+\alpha_1^2)(k^2+\alpha_2^2)}\right] \tag{4.9'}$$

in k-space, and

$$c_{ij}^{(1)}(r) = -\frac{Z_i\, Z_j\, \Lambda_\varepsilon}{2^{(\varepsilon/2)+1}\,\Gamma[1+(\varepsilon/2)]\, r^{\varepsilon/2}}$$

$$\cdot\left\{K_{\varepsilon/2}(r) + \alpha_1^{\varepsilon/2} K_{\varepsilon/2}(\alpha_1 r) - \alpha_2^{\varepsilon/2} K_{\varepsilon/2}(\alpha_2 r)\right\} \tag{4.3'}$$

in r-space respectively, with r and $\alpha_{1,2}$ given in λ_D and λ_D^{-1}. In the $T \to \infty$ limit, Eq. (4.3') reduces to Eq. (4.3), as expected. Actually, as far as their derivation is concerned, these expressions are only meaningful for $\nu = 3$. Nevertheless, we find it useful to write them down as ε-dependent expressions, continuing analytically, the integer Debye quantity to real ε ones. Such a procedure may be given further support, through perturbation theory around $\varepsilon = 1$.

V. HIGHER ORDER CORRECTIONS TO THE BOHM DIFFUSION

The diffusion of plasma particles transversally to an arbitrarily strong constant magnetic field is a subject of basic importance in classical plasma physics, in view of its direct relevance to the confinement of sufficiently hot ionized gases able to sustain exo-energetic nuclear yielding thermal neutrons [Taylor and McNamara, 1971]. Up to now, attention has been focused on the stochastic assumptions underlying the many different, although nearly equivalent, derivations of the Bohm-like transverse diffusion coefficient $D_\perp \sim B^{-1}$ in two- and three-dimensions, in agreement with general scaling arguments. With no exceptions, all these treatments make use of a fluid-like picture of the two-component plasma (TCP). The screening length λ_D is then supposed very large, while the dynamics is taken as a functional of the first-order equilibrium properties with respect to the plasma parameter. Here we shall not enter into the details of more sophisticated calculations based on the microscopic Klimontovitch time-dependent one- and two-particle distribution functions, yielding back essentially the same results. Our main purpose is the systematic investigation of the high order corrections in the plasma parameter Λ to $D_\perp^2$, irrespective of the space dimensionality $2 \leq \nu \leq 3$. Therefore the more traditional hydrodynamical approach will prove sufficient for our present needs. In this case, the TCP's equilibrium properties are simply obtained by duplicating those of the standard one-component plasma (OCP) model.

As shown above, such an approximation is valid only when one restricts to the first-order Debye term. The previous investigation of the TCP pair correlation functions has shown that the neutrality condition prevents the extension of the previous treatments based upon the above assumption, to all order in Λ. The importance of the higher corrections to D_I^2 is motivated by several distinct considerations. First, there is the need of a deeper understanding of the fluid-like picture underlying all the derivations of the Bohm result. Otherwise stated, even in a very hot and dilute plasma with a very large λ_D, it should be checked out that the higher order terms are consistently negligible when compared to the first, in order to allow for a coherent use of the perturbation expansion in Λ. Of more direct relevance to experimental Fusion research, we should also mention that the present most successful Tokamak device works in a relatively moderate temperature regime, i.e. $T \sim 1$ Kev instead of the expected 10 Kev. However, the basic need for the present investigation is perhaps arising from the pathological, albeit likely, situations occurring in laser Fusion when huge self-generated magnetic fields (up to 10^6 G) produce hot spots and inhibit very severely the heat penetration from the underdense outside high-temperature plasma toward the dense core of the pellet.

Last, but not least, the strongly magnetized TCP allows for a clear and relatively simple investigation of the time-independent transport quantities. Such a fortunate situation is due to the drastic modelling of the particle dynamics within the framework of the guiding center approximation valid for this case. We should also notice that our continuous parametrization of the equilibrium nodal expansion with respect to the space dimensionality $\nu = 2 + \varepsilon$ gives access to an elegant unified presentation of the hydrodynamic calculation of the Bohm velocity diffusion coefficient.

A. Basic Theory

In this work we address ourselves to the evaluation of the time-independent transverse diffusion coefficient to a very large constant magnetic field B, in a high-temperature fully ionized TCP. The mean kinetic energy per particle is supposed high enough to allow the neglect of any symmetry (Fermi) contribution to the equilibrium properties. The only retained $\hbar \neq 0$ effect is due to the uncertainty principle through the inequality

$$\frac{\hbar}{\sqrt{2m_{ei}\, k_B T}} > \frac{e^2}{k_B T}$$

which means $k_B T > 1$ Ry in three dimensions. These non-zero diffraction corrections usually unnoticed in high-temperature Fusion physics are non-negligible only for small distances comparable to the Bohr radius a_o. So the hydrodynamical modes (convective cells, for instance) supposed to convey most of the particle transport across $\vec{B}$ are not affected by them. Moreover, we shall work with the very plausible assumption that $\vec{B}$ does not change the TCP equilibrium properties. Otherwise stated, the Bohr-Van Leuween theorem applies to the electron wave-packets taken as spinless to simplify matters. Therefore, the transverse velocity coefficient extending the $\nu = 2$ model of charged filaments [Taylor and McNamara, 1971; Montgomery, 1975] aligned with $\vec{B}$ obtains as (c = speed of light)

$$D_\perp = \frac{c^2}{B^2} \int_o^\infty \langle \vec{E}_\perp(o) \cdot \vec{E}_\perp(\tau) \rangle \, d\tau \tag{5.1}$$

in terms of the autocorrelation function of the electric field $\vec{E}(\tau)$ seen by a test charge at time τ, with the guiding center approximation

$$\vec{X}(t) = c \int_o^t \frac{\vec{E}(\tau) \times \vec{B}}{B^2} \, d\tau \tag{5.2}$$

The bracket denotes the usual canonical equilibrium average. To go farther we need a redefinition of the point and cross vector products. This is easily achieved by taking the same scalar product while the vectorial one is interpolated by an infinite matrix arising from the infinite number of vector components in a Euclidian space with real ν. The corresponding integrals should read

$$\int d^\nu \vec{p} \, f(p^2, p \cdot q_1, p \cdot q_2, \ldots)(2\pi)^{-\nu} \quad .$$

Moreover, we take for granted the existence of the discrete sum

$$\vec{E}(\tau) = \sum_{\vec{k}} \vec{E}_{\vec{k}}(\tau) \, e^{i\vec{k}\cdot\vec{X}(\tau)}$$

for all $2 \leq \nu \leq 3$. The problem is one of evaluating the electric field autocorrelation

$$\langle \vec{E}_\perp(o) \cdot \vec{E}_\perp(\tau) \rangle$$

when $\vec{E}(\tau)$ is the electric field seen at time τ by a "test ion" located on the z-axis at $\tau = 0$. $\vec{E}(\tau)$ is related to the Eulerian electric field $\vec{E}(\vec{X},t)$ by $\vec{E}(\tau) = \vec{E}[\vec{X}(\tau),\tau]$ and

$$\vec{E}(\vec{X},t) = \sum_{\vec{k}} \sum_j \frac{4\pi\vec{k}}{ik^2} \frac{e_j}{V} \, e^{i\vec{k}\cdot[\vec{X}-\vec{X}_j(t)]} \equiv \sum_{\vec{k}} \vec{E}_{\vec{k}}(t) \, e^{i\vec{k}\cdot\vec{X}} \tag{5.3}$$

The present Fourier expansion holds for all $0 \leq \varepsilon \leq 1$, because

$$\vec{E}_{\vec{k}}(t) = \int d^{\nu}\vec{x}\; e^{-i\vec{k}\cdot\vec{x}}\; \vec{E}(\vec{x},t)$$

is meaningful for any k. $\vec{X}(\tau)$ is the orbit of the test particle, while $\vec{X}_j(t)$ is the location of the jth charge at time t. In Eq. (5.2) the bracket is explained so that

$$D_{\perp} = \frac{c^2}{D^2} \int_0^{\infty} \sum_{\vec{k}} \langle \vec{E}^*_{\perp\vec{k}}(o) \cdot \vec{E}_{\perp\vec{k}}(\tau)\; e^{i\vec{k}\cdot\vec{X}(\tau)} \rangle\, d\tau \qquad (5.4)$$

To go on farther, one neglects the correlation between the position of the test particle and those of the background plasma (the $\vec{X}(\tau)$). This amounts to saying that

$$|g_2^{ij}(r) - 1| < 1 \quad ,$$

$g_2^{ij}(r)$ denotes any one of the three TCP pair correlation functions, even when higher order corrections are retained. Such an assumption restricts us to plasma parameter values smaller than unity. Since

$$\langle \vec{E}_{\vec{k}_1} \cdot \vec{E}_{\vec{k}_2} \rangle = 0$$

for $\vec{k}_1 \neq -\vec{k}_2$ and a spatially uniform ensemble, we have the statistical factorized out result

$$D_{\perp} = \frac{c^2}{B^2} \int_0^{\infty} \sum_{\vec{k}} \langle \vec{E}^*_{\perp\vec{k}}(o) \cdot \vec{E}_{\perp\vec{k}}(\tau)\rangle \langle e^{i\vec{k}\cdot\vec{X}(\tau)} \rangle\, d\tau \qquad (5.5)$$

evaluated when $\langle \exp[i\vec{k}\cdot\vec{X}(\tau)]$ is known, while the first average reads

$$\langle \vec{E}^*_{\perp\vec{k}}(o) \cdot \vec{E}_{\perp\vec{k}}(\tau)\rangle = \sum_{i,j} \frac{S_{\nu}^2\, e_i e_j}{k^2 V^2} \frac{k_{\perp}^2}{k^2} \langle e^{i\vec{k}\cdot[\vec{X}_i(\tau)-\vec{X}_j(o)]} \rangle \qquad (5.6)$$

where

$$S_{\nu} = \frac{2\pi^{\nu/2}}{\Gamma(\frac{\nu}{2})} \quad .$$

Extending in a straightforward way the $\nu = 2$ and 3 trajectories, we have [Montgomery, 1975]

$$\vec{X}(\tau) = \vec{X}(o) + c\int_o^{\tau} \frac{\vec{E}(\tau') \times \vec{B} d\tau'}{B^2} + \hat{b}\, V_{||}\, t$$

$$+ \int_o^{\tau} d\tau' \int_o^{\tau'} d\tau''\, e_i \frac{\hat{b} \cdot \vec{E}(\tau'')\, \hat{b}}{m_i} \tag{5.7}$$

$\hat{b}$ is the unit vector $||$B. The test charge is taken to be an ion with charge-to-mass ratio e_i/m_i. The initial position $\vec{X}(o)$ may be set equal to zero, while the initial test ion velocity $V_{||}$ is a statistically-distributed quantity assumed to obey the Boltzmann distribution. Evaluating $\langle \exp i\vec{k}\cdot\vec{X}(\tau)\rangle$ is no simple matter. About the best that can be done is the following. Because $\vec{X}(\tau)$ is the portion of a random variable initially localized near $\vec{X}(o) = 0$, and because its probability distribution $P[\vec{X}(\tau),\tau]$ is expected to spread out with time

$$\langle \exp i\vec{k} \cdot \vec{X}(\tau)\rangle = \int d\vec{X}(\tau)\, e^{i\vec{k}\cdot\vec{X}(\tau)}\, P[\vec{X}(\tau),\tau]$$

will damp with increasing τ for any $\tau > 0$. The damping occurs for two reasons: motion "parallel" and "perpendicular" to $\vec{B}$. Only the latter is operative when $\tau = 0$. The damping along $\vec{B}$ will be more extreme than that due to the "perpendicular" motion.

Finally, we need

$$\langle \exp i\vec{k} \cdot [\vec{X}_i(o) - \vec{X}_j(o)]\rangle = \delta(\vec{k}) - \frac{\int d^{\nu}\vec{r}\, g_2^{ij}(r)\, e^{-i\vec{k}\cdot\vec{r}}}{V}$$

$$\equiv \delta(k) - V^{-1}\, S_{ij}(k) \tag{5.8}$$

and $(V_{oj}^2 = k_B T/mj)$

$$\langle \vec{E}_{\perp}(o) \cdot \vec{E}_{\perp}(\tau)\rangle = \sum_{i=j} \sum_{\substack{\vec{k} \\ k_{||}\neq 0}} \frac{S_{\nu}^2\, e_i e_j}{k^2 V^2} \frac{k^2}{k^2}$$

$$\cdot \exp\left(- \frac{k_{||}^2\, V_{oj}^2\, \tau^2}{2}\right) \exp\left(- \frac{k_{||}^2\, V_{ion}^2\, \tau^2}{2}\right)$$

$$+ \sum_{i \neq j} \sum_{\substack{\vec{k} \\ k_{||}=0}} \frac{S_\nu^2 e_i e_j}{k^2 V^2}$$

$$\cdot \exp\left(-\frac{c^2 k_\perp^2}{2B^2} \int_o^\tau d\tau_1 \int_o^\tau d\tau_2 <\vec{E}_\perp(\tau_1) \cdot \vec{E}_\perp(\tau_2)>\right)$$

$$+ \sum_{i \neq j} \sum_{\substack{\vec{k} \\ k_{||}=0}} \frac{S_\nu^2 e_i e_j}{k^2 V^2} \frac{k_\perp^2}{k^2}$$

$$\cdot \exp\left(-\frac{k_{||}^2 V_{oj}^2 \tau^2}{2}\right) \left[\delta(k) - S_{ij}(k) V^{-1}\right]$$

$$\cdot \exp\left(-\frac{k_{||}^2 V_{ion}^2 \tau^2}{2}\right) + \sum_{i \neq j} \sum_{\substack{\vec{k} \\ k_{||}=0}} \frac{S_\nu^2 e_i e_j}{k^2 V^2}$$

$$\cdot \exp\left(-\frac{c^2 k_\perp^2}{2B^2} \int_o^\tau d\tau_1 \int_o^\tau d\tau_2 <\vec{E}_\perp(\tau_1) \cdot \vec{E}_\perp(\tau_2)\right)$$

$$\cdot \left[\delta(k) - S_{ij}(k) V^{-1}\right] \quad , \quad \ldots \qquad (5.9)$$

This is an integral equation for $<\vec{E}_\perp(o) \cdot \vec{E}_\perp(\tau)>$. Performing the i,j summation makes the $\delta(k)$ contribution cancel, altogether with the first two sums thus yielding

$$<\vec{E}\ (o) \cdot \vec{E}\ (\tau)> = \sum_{\substack{\vec{k} \\ k_{||}\neq 0}} \frac{S_\nu^2 e^2 k_\perp^2}{k^4 V^2} \exp\left(-\frac{k_{||}^2 V_{oj}^2 \tau^2}{2}\right)$$

$$\cdot \exp\left(-\frac{k_{||}^2 V_{ion}^2 \tau^2}{2}\right) H(k) + \sum_{\substack{\vec{k} \\ k_{||}=0}} \frac{S_\nu^2 e^2}{k^2 V^2} \qquad (5.10)$$

$$\cdot \exp\left(-\frac{c^2 k_\perp^2}{2B^2} \int_o^\tau d\tau_1 \int_o^\tau d\tau_2 <\vec{E}(\tau_1) \cdot \vec{E}(\tau_2)>\right) H(k)$$

where $H(k) = V^{-1}[-S_{ee}(k) - S_{ii}(k) + 2S_{ei}(k)]$ (5.11)

is the extension of the usual first-order thermal spectrum. Equation (5.10) can be simplified through [Montgomery, 1975]

$$\frac{1}{2} \langle \vec{E}_{\perp}(o) \cdot \vec{E}_{\perp}(\tau) \rangle \equiv Q_{\perp}(\tau)$$

$$R_{\perp}(\tau) \equiv \frac{c^2}{2B^2} \int_o^{\tau} d\tau_1 \int_o^{\tau} d\tau_2\, Q_{\perp}(\tau_2 - \tau_1)$$

$$\frac{d^2 R_{\perp}(\tau)}{d\tau^2} = \frac{c^2\, Q_{\perp}(\tau)}{B^2}$$

in the form

$$2Q_{\perp}(\tau) = \frac{S_{\nu}^2 e^2}{V^2} \sum_{\substack{\vec{k} \\ k_{||} \neq 0}} \frac{k_{\perp}^2}{k^2} H(k)\, e^{-k_{||}^2 V^2 \text{ion}\, \tau^2/2}\, e^{-k_{||}^2 V_{oj}^2 \tau^2/2}$$

$$+ \sum_{\substack{\vec{k} \\ k_{||} = o}} \frac{S_{\nu} e^2}{k^2 V^2} \exp\,[-2k_{\perp}^2 R_{\perp}(\tau)]\, H(k) \qquad (5.12)$$

Defining

$$\varepsilon_b = \frac{S_{\nu}^2 e^2 c^2}{B^2 V^2 k_D^2} ,$$

this becomes, replacing τ by t

$$\frac{1}{2}\left(\frac{d\, R_{\perp}(\infty)}{dt}\right)^2 = \frac{\varepsilon_b}{2} \sum_{\vec{k}_{\perp}} \frac{H(k)}{k_{\perp}^2} \qquad (5.13)$$

But $D_{\perp}/2$ is just $dR_1(\infty)/dt$, so

$$D_{\perp} = 2\varepsilon_b^{1/2} \left[\sum_{\vec{k}_{\perp}} \frac{H(k)}{k_{\perp}^2}\right]^{1/2} + O(\varepsilon_b) \qquad (5.14)$$

To approximate it by an integral for a large volume, we make the replacement

$$\sum_{\vec{k}} \rightarrow \frac{V^{2/\nu}}{2\pi} \int_{k_{min}}^{k_D} k_\perp \, dk_\perp \quad . \tag{5.15}$$

$$k_{min} = \frac{2\pi}{V^{1/\nu}}$$

is the lower limit which results from the finite bose size, while the upper $k_D = \lambda_D^{-1}$ means that we are restricted to the so-called fluid limit [Taylor and McNamara, 1971].

B. Final Calculation of D

The TCP nodal equilibrium properties detailed in Section IV allow us to complete the calculation of the transverse velocity coefficient $D_\perp$ as given by Eq. (5.14). The only remaining task is to explain the $S_{ij}(k)$'s introduced in Eq. (5.8) through the linearized pair correlation function.

$$g_2^{ij}(r) \sim 1 + W_2^{ij}(r) \quad , \qquad \Lambda_\varepsilon \leq 1 \tag{5.16}$$

It is important to notice that the sum $H(k)$ is essentially built upon the more compact and decaying faster-than-Debye nodal graphs.

The convolutions with one and two Debye lines are to be included too, so that the general quantity to be inserted in Eq. (5.10) is

$$\begin{aligned} H(k) &\equiv -S_{ee}(k) - S_{ii}(k) + 2S_{ei}(k) \simeq -W_2^{ii}(k) - W_2^{ee}(k) + 2W_2^{ei}(k) \\ &= 4\left\{H'(k) + \sum_{n=3}^{\infty} \text{Bubble}(k)[1 + H'(k)]^2\right\} \equiv 4H'(k) + H'(k) \end{aligned} \tag{5.17}$$

where Bubble (k) denotes the sum, within a given order n, of the above graphs. The first remarkable result is the absence of second-order contribution. The more important diagrams in the $k \rightarrow 0$ limit are the linear chains built upon 3-bubbles with Debye lines intertwined in between. The first one appears with $n = 5$. $H'(k)$ and Bubble (k) are obtained from the nodal analysis performed in Section IV. However, the most important property which motivates the present work is the negligible contribution left by all the higher order graphs to the expression (5.14) when the integration procedure (5.15) is applied to it, so that

$$\int_{\frac{2\pi}{L}}^{k_D^{-1}} dk_\perp \; H'(k) \; k_\perp \simeq 0 \tag{5.18}$$

in view of the fast decaying properties of the integrand in the $k \to 0$ limit, conveying much of the particle transport across the magnetic field. As a consequence, the usual first-order derivation of the Bohm-like transverse diffusion appears as a very efficient self-consistent procedure eliminating almost completely any contribution from the higher orders in Λ_ε. Moreover, it can be worked continuously for any dimensionality $2 \leq \nu \leq 3$. Therefore, the expression (5.14) for D explained by its first order [Taylor and McNamara, 1971]

$$D = \begin{cases} \sqrt{2}\,\dfrac{ck_BT}{eB}\,\Lambda_o^{1/2}\left[\ell n\left(\dfrac{R}{2\pi\,\lambda_D}\right)\right]^{1/2} & , \quad \nu = 2 \\[2ex] \dfrac{ck_BT}{eB}\left(\dfrac{\Lambda_1}{2}\right)^{1/2}\left(\dfrac{\lambda_D}{R}\,\ell n\,\dfrac{R}{2\pi\,\lambda_D}\right)^{1/2} & , \quad \nu = 3 \end{cases} \tag{5.19}$$

integer quantities appears as remarkably stable with respect to the high order Λ_ε corrections.

REFERENCES

Cooper, M.S. and DeWitt, H.E., Phys. Lett. A40, 391 (1972).

Dashen, R.F. and Rajaraman, R., Phys. Rev. D10, 694 (1974).

DelRio, F., DeWitt, H.E., Phys. Fluids, 12, 791 (1969).

Deutsch, C. and Lavaud, M., Phys. Rev. A9, 2598 (1974).

Deutsch, C., Furutani, Y., and Gombert, M.M., Phys. Rev. A13, 2244 (1976).

Deutsch, C., J. Math. Phys. 17, 1404 (1976) and 18, 1297 (1977).

Deutsch, C., to be published.

DeWitt, H.E., J. Math. Phys. 7, 616 (1966).

Furutani, Y., and Deutsch, C., J. Math. Phys. 18, 292 (1977).

Galam, S. and Hansen, J.P., Phys. Rev. A14, 816 (1976).

Gombert, M.M. and Deutsch, C., Phys. Lett. A47, 473 (1974); Deutsch, C., and Gombert, M.M., J. Math. Phys. 17, 1077 (1976).

Gombert, M.M. and Deutsch, C., J. Physique (Paris) (1978).

Grandjouan, N. and Deutsch, C., Phys. Rev. A11, 522 (1975).

Kelbg, G., Bose S. 70th Birthday Commemoration Volume, Calcutta, p. 100 (1965).

Montgomery, D., Les Houches, 1972 - Plasma Physics, Gordon and Breach (New York) (1975).

Montroll, E.W. and Ward, J.C., Phys. Fluids 1, 55 (1958).

Murphy, T.J., Ph.D. Thesis, The Rockefeller University, New York (1968).

Prager, S., Adv. Chem. Phys. 4, 201 (1963).

Salpeter, E.E., Ann. Phys. (New York) 5, 183 (1958).

Springer, J.F., Pokrant, M.A. and Stevens, F.A., Jr., J. Chem. Phys. 58, 4863 (1973).

Taylor, J.B. and McNamara, B., Phys. Fluids 14, 1492 (1971).

SELECTED TOPICS ON THE EQUILIBRIUM STATISTICAL MECHANICS OF COULOMB SYSTEMS

Ph. Choquard

Laboratoire de Physique Théorique
École Polytechnique Fédérale de Lausanne
P. O. Box 1024
CH 1001 Lausanne, Switzerland

TABLE OF CONTENTS

SELECTED TOPICS ON THE EQUILIBRIUM STATISTICAL MECHANICS OF COULOMB SYSTEMS*

Ph. Choquard

Laboratoire de Physique Theorique, Ecole Polytechnique
Federale de Lausanne, P. O. Box 1024
CH 1001 Lausanne, Switzerland

I. H-STABILITY OF MATTER

A. Introduction

I wish to begin this lecture with a quotation from Elliott Lieb [Lieb, 1976]

> "Thus, while Eq. (6) (the Heisenberg principle) is correct it is a pale reflection of the power of the operator $-\Delta$ to prevent collapse".

The purpose of this lecture is, in effect, to encourage a genuine appreciation of this statement. With this in mind, let me define the system and state the problem.

The system consists of neutral assembly of Q static nuclei of charge $Z|e|$ and of N charged fermions of charge $-|e|$ and number s of spin states. Thus, $N = ZQ$ and with $s = 2$ we have the case of electrons. Choosing as unit of length, a = 1/2 of the Bohr radius, i.e.

$$\hbar^2/2ma^2 = e^2/a$$

(m, electron mass) and as unit of energy $e^2/a = 4Ry$ the Hamiltonian

*Sections II.B, III.C, D, E, and IV.B, C, D of this paper report mostly unpublished work, some of which, incorporated in Sections II.C, IV.C and IV.D, result from stimulating discussions with participants at the Institute. The author is very much indebted to H. J. Raveché for reviewing the manuscript.

of this system with fermions at positions x_i and nuclei at positions R_n is

$$H_N = \sum_i - \Delta_i - V(x_i) + \sum_{i<j} |x_i - x_j|^{-1} + U(Z, \{R_n\})$$

where

$$V(x) = \sum_n Z|x - R_n|^{-1}$$

and

$$U(Z, \{R_n\}) = \sum_{m<n} Z^2|R_m - R_n|^{-1}$$

This Hamiltonian acts on N particle wave functions $\psi(x_1,..x_N;s_1..s_N)$, antisymmetric in the pairs (x_i,s_i), and defined for $x_i \in R^{3N}$ and $s_i \in (1,..s)$. Their norm is given by

$$\langle\psi,\psi\rangle = \sum_{s_i=1}^{s} \int |\psi(x_1..x_N; s_1..s_N)|^2 \, dx_1..dx_N = 1$$

Define

$$T_\psi = N \sum_{s_i=1}^{s} \int |\nabla_1\psi(x_1..x_N; s_1..s_N)|^2 \, dx_1..dx_N \qquad (1.1)$$

to be the kinetic energy given by ψ. Similarly, define

$$\rho_\psi(x) = N \sum_{s_i=1}^{s} \int |\psi(x_1x_2..x_N; s_1..s_N)|^2 \, dx_2..dx_N \quad , \qquad (1.2)$$

to be the single particle density and define

$$\begin{aligned} E_\psi^Q &= \langle\psi,H_N \psi\rangle \\ &= T_\psi - \int \rho_\psi(x) \, V(x) \, dx \\ &+ \langle\psi| \sum_{i<j} |x_i - x_j|^{-1}|\psi\rangle + U(Z, \{R_n\}) \end{aligned} \qquad (1.3)$$

to be the total energy of ψ. Call E_N^Q the ground state energy of the system.

The problem is to show that matter does not collapse, i.e. that the expected value of H_N is bounded below by an extensive quantity, namely the total numbers of particles, independent of the nuclear locations $\{R_n\}$. This is called the H-Stability of the system. Here the nuclear kinetic energy, which adds a positive quantity, can be omitted. In 1967, Dyson and Lenard proved for the first time the existence of an extensive lower bound to E_N^Q. In 1968, Dyson also showed that matter would be unstable if electrons were bosons instead of fermions, thereby establishing Ehrenfest's surmise that without the Pauli exclusion principle matter would collapse. For an assembly of hydrogen atoms, Dyson and Lenard obtained a value of about -10^{14} Ry/particle whereas in 1975, Lieb and Thirring obtained -23 Ry/particle, thirteen orders of magnitude better!

As we will later discuss, the key idea needed to obtain this improved bound involved multiple uses of the Thomas-Fermi (TF) theory. This was first conceived by Leib. In parallel investigations, Lieb and Simon put the TF theory on a firm basis. In order to follow these ideas, let me recall here how the TF energy functional is constructed.

Consider N free fermions with s spins states each in a box of volume Ω. With K_F designating the Fermi momentum, T the kinetic energy and $\rho = N/\Omega$ the particle density, we have, for large N, the well-known relations

$$N = s \frac{\Omega}{(2\pi)^3} \frac{4\pi}{3} K_F^3$$

$$T = s \frac{\Omega}{(2\pi)^3} \frac{4\pi}{5} K_F^5$$

and, in eliminating K_F

$$T/\Omega = \frac{3}{5} {\frac{6\pi^2}{s}}^{2/3} \rho^{5/3} \equiv s^{-2/3} K_c \rho^{5/3}$$

where

$$K_c = \frac{3}{5} (6\pi^2)^{2/3} \simeq 9.116$$

Switch on the electrostatic interaction and imbed Q static nuclei of charge Z in this fluid. The fermion density $\rho = \rho(x)$ becomes a spatially inhomogenous function which must necessarily be non-negative.

Approximate T_ψ by

$$s^{-2/3} K_c \int \rho(x)^{5/3} dx$$

where the domain of integration is R^3, the TF energy functional is then given by

$$\varepsilon(\rho) = s^{-2/3} K_c \int \rho(x)^{5/3} dx - \int V(x) \rho(x) dx$$

$$\frac{1}{2} \int \rho(x) |x - y|^{-1} \rho(y) dx dy + U(Z, \{R_n\}) \qquad (1.4)$$

Then

$$E_N^{TF} = \text{inf.} \left\{ \varepsilon(\rho) : \int \rho(x) dx = N \quad , \quad \rho(x) \geq 0 \right\}$$

is the TF energy for N electrons and the minimizing $\rho(x)$ is supposed to approximate the true ground state electron density and E_N^{TF} approximate the ground state energy. I proceed next with the problem of establishing Lieb-Thirring's lower bound to T_ψ.

The lower bound to T_ψ is intended to: (1) deal with the atomic case N = Q = 1, where Lieb's new uncertainty principle will come into play; (2) extend this uncertainty principle for the many fermion case.

B. The Stability of Atoms

In saying that an atom or ion is stable, we mean that its ground state energy is finite. Consider the Hamiltonian of the hydrogenic ion or atom ($N = Q = 1$, $z \geq 1$), which in our units, is

$$H = -\Delta - Z|x|^{-1} \quad ,$$

where H acts on square integrable functions in R^3. Why is the ground state energy finite, i.e. why

$$\langle\psi, H_\psi\rangle \geq -E_o \langle\psi,\psi\rangle \qquad (1.5)$$

for some finite negative E_o? The standard answer is the Heisenberg uncertainty principle: if

$$T_\psi = \int |\nabla\psi(x)|^2 dx$$

is the kinetic energy, and if

$$\langle x^2\rangle_\psi = \int |x|^2 |\psi(x)|^2 dx$$

then when $\langle\psi,\psi\rangle = 1$

$$T_\psi \langle x^2 \rangle_\psi \geq \frac{9}{4} \tag{1.6}$$

Here, let me again quote Lieb [1976]:

> "The intuition behind applying the Heisenberg uncertainty principle (Eq. 1.6) to the ground state problem (Eq. 1.5) is that if the electron tries to get within a distance R of the nucleus, the kinetic energy T_ψ is at least as large as R^{-2}. Consequently $\langle\psi, H_\psi\rangle \geq R^{-2} - ZR^{-1}$, and this has a minimum $-Z^2/4$ for $R = 2/Z$. The above argument is <u>false</u>! The Heisenberg uncertainty principle says no such thing, despite the endless invocation of the argument. Consider a ψ consisting of two parts, $\psi = \psi_1 + \psi_2$. ψ_1 is a narrow wave packet of radius R centered at the origin with $\int|\psi_1(x)|^2\, dx = 1/2$. ψ_2 is spherically symmetric and has support in a narrow shell of mean radius L and $\int|\psi_2(x)|^2\, dx = 1/2$. If L is large then, roughly, $\int|x|^2|\psi(x)|^2\, dx \sim L^2/2$, whereas $\int|x|^{-1}|\psi(x)|^2\, dx \sim 1/2R$. Thus, from Eq. (1.6) we can conclude <u>only</u> that $T_\psi > 9/2L^2$ and hence that $\langle\psi, H_\psi\rangle \geq 9/2L^2 - Z/2R$. With this wave function, and using <u>only</u> the Heisenberg uncertainty principle, we can make E_o arbitrarily negative by letting $R \to o$".

Thus, a better uncertainty principle is needed which would provide a lower bound to the kinetic energy in terms of some integral of ψ <u>not</u> involving derivatives and which should reflect more accurately the fact that if one compresses a wave function anywhere then the kinetic energy will increase.

Aiming at the $\rho^{5/3}$ dependence of the TF kinetic energy for this better uncertainty principle, Lieb made an astute combination of two inequalities, one due to Sobolev, the other due to Hölder.

Sobolev's inequality states that for a scalar, real $\psi(x)$, continuous in R^3, then, if the following integrals exist,

$$\int |\nabla\psi(x)|^2\, dx \geq K_s \left(\int |\psi(x)|^6\, dx\right)^{1/3}$$

where K_s is the unique minimum of

$$\int|\nabla\psi(x)|^2\, dx \bigg/ \left(\int|\psi(x)|^6\, dx\right)^{1/3} .$$

Rosen has found that the best possible constant is

$$K_s = 3\left(\frac{\pi}{2}\right)^{4/3} \simeq 5.478$$

Application of Sobolev's inequality to the hydrogen atom problem yields, for any ψ

$$\langle\psi, H_\psi\rangle \geq K_s \left(\int \rho(x)^3\, dx\right)^{1/3} - Z\int |x|^{-1}\, \rho_\psi(x)\, dx = h(\rho)$$

and hence, when $\langle\psi,\psi\rangle = 1$

$$\langle\psi, H_\psi\rangle \geq \min\left\{h(\rho) : \rho(x) \geq 0 \int \rho(x)\, dx = 1\right\}$$

This is a soluble exercise of variational calculus involving ρ_ψ alone, which results in

$$E_o \geq -\frac{1}{3} Z^2 \frac{e^2}{a} = -\frac{4}{3} Z^2 R_y$$

which is a very good lower bound indeed, since the exact E_o is, as you are well aware, $-Z^2R_y$.

Next comes Hölder's inequality, a generalization of the triangle inequality which states that

$$\left|\int F(x)\, g(x)\, dx\right| \leq \left(\int |F(x)|^P\, dx\right)^{1/P}\left(\int |g(x)|^q\, dx\right)^{1/q}$$

with $p^{-1} + q^{-1} = 1$ and $p \geq 1$. Take $F = P$, $g = P^{2/3}$, $p = 3$, and $q = 3/2$ and one obtains

$$\left|\int \rho(x)^{5/3}\, dx\right. \leq \left(\int \rho(x)^3\, dx\right)^{1/3}\left(\int \rho(x)\, dx\right)^{2/3}$$

Since we always take $\langle\psi,\psi\rangle = 1$ the two inequalities combine into a new one

$$\int |\nabla\psi(x)|^2\, dx \geq K_s \int \rho_\psi(x)^{5/3}\, dx$$

The constant K_s, which was the best in Sobolev's inequality is not the best here. The best one, call it K_L, is the minimum, proved to exist and to be unique, of

$$\int |\nabla\psi(x)|^2\, dx \Big/ \int \rho_\psi(x)^{5/3}\, dx$$

subject to

$$\int \rho_\psi(x)\, dx = 1 \quad .$$

Solving the problem numerically, Barnes found

$$K_L \simeq 9.578$$

Noticing that K_L is $>K_c$ (= 9.116), where K_c is the constant in Eq. (1.4), we have the somewhat weaker inequality which, however, makes contact with the TF energy functional

$$T_\psi \geq K_c \int \rho_\psi(x)^{5/3}\, dx \tag{1.7}$$

where

$$\int \rho_\psi(x)\, dx = 1 \quad .$$

This is Lieb's remarkable inequality, this better uncertainty principle which fully exploits "the power of the operator $-\Delta$ to prevent collapse" and which should be found in any updated textbook of Quantum Mechanics. Application of this inequality to the hydrogen atom problem results in

$$E_o \geq -3^{1/3}\ \text{Ry}$$

only 8.2% below -4/3.

C. Extension of the Uncertainty Principle to Many Fermions

Can we extend, with the spin factor $s^{-2/3}$ and with ρ_ψ defined by Eq. (1.2), the inequality (Eq. 1.7) to produce on its right-hand-side the TF approximation to the kinetic energy of the N fermions systems and, hence, show that it is a lower bound to the true T_ψ? The answer is "almost" yes, i.e. it is yes but with a constant smaller than K_c, namely by a factor $(4\pi)^{-2/3}$ lately improved to $(4\pi/1.83)^{-2/3}$.

The ideas of the proof can be described as follows: given $\psi(x)$ and hence $\rho_\psi(x)$, consider the N particle Hamiltonian

$$\tilde{H}_N = \sum_{i=1}^{N} h_i$$

with

$$h_i = -\Delta_i + v(x_i)$$

The potential v(x) (≤ 0 and specified below) possesses enough bound states ($\geq$ N/s, a sufficient condition that Lieb can relax) and regularity ($\int |v(x)|^{5/2}\, dx < \infty$),

First find upper and lower bounds to the ground state energy of $\tilde{H}_N$, obtained by filling the N/s lowest energy levels e_n of h, s times each. These bounds will respectively depend on T_ψ, ρ_ψ and v(x).

Then take T_ψ from the upper bound and choose $v(x) = v[\alpha\rho_\psi(x)]$ with α being an adjustable parameter, so that the right-hand-side becomes a homogenous function of $\rho_\psi(x)$ of a certain degree.

Optimize with respect to α and obtain the remarkable Lieb-Thirring inequality

$$T_\psi \geq (4\pi s)^{-2/3} K_c \int \rho_\psi(x)^{5/3}\, dx \quad . \tag{1.8}$$

This result is obtained by starting from

$$E_o = s \sum_{n=1}^{N/s} e_n \equiv s\, \eta_o$$

and by defining η to be the sum over all negative eigenvalues of h. In general, η will be more negative than η_o, so that

$$E_o \geq s\, \eta$$

An upper bound to E_o is then easily provided by

$$<\psi\tilde{H}_N|\psi> = T_\psi + \int v(x)\ \rho_\psi(x)\ dx \geq E_o \tag{1.9}$$

Finding the lower bound to η, hence E_o, is the hard part of the problem. Let me quote Lieb [1976] without reproduction of the proof*:

"If $e_1 \leq e_2 \leq .. \leq 0$ are the negative eigenvalues of h

---

* It implies Young's inequality applied to a suitable integral representation of an upper bound to the number of bound states of h, constructed with the help of Birman-Schwinger operator.

(if any) then

$$|\eta| = \Sigma|e_n| \leq \frac{4\pi}{15\pi^2} \int |v(x)|^{5/2}\, dx \qquad (1.10)$$

we believe the factor 4π does not belong in Eq. (1.10)."

Some insight into this inequality can, however, be gained in calculating the semi-classical approximation η_c to η. This approximation is obtained by saying that a region of volume $(2\pi)^3$ in the six-dimensional space (P,x) can accommodate one eigenstate. Hence, integrating over the set $\theta(h)$ in which $h = P^2 + v(x)$ is negative, we have, with $v(x) = -|v(x)|$

$$|\eta_c| = (2\pi)^{-3} \int_{\theta(h)} dx\, dp\, [P^2 + v(x)]$$

$$= (2\pi)^{-3} \int dx\, 4\pi \int_0^{|v(x)|^{1/2}} P^2\, dp\, [P^2 - |v(x)|]$$

$$= (2\pi)^{-3} \int dx\, 4\pi \left(\frac{1}{5} - \frac{1}{3}\right) |v(x)|^{5/2}$$

$$= -(15\pi^2)^{-1} \int |v(x)|^{5/2}\, dx$$

We notice that apart from the factor 4π we have the right-hand-side of the inequality (Eq. 1.8).

Combining Eqs. (1.8) and (1.10), we have

$$T_\psi - \int \rho_\psi(x)\, |v(x)|\, dx \geq E_o \geq -\frac{4\pi s}{15\pi^2} \int |v(x)|^{5/2}\, dx$$

or

$$T_\psi \geq \int \rho_\psi(x)\, |v(x)|\, dx - \frac{4\pi s}{15\pi^2} \int |v(x)|^{5/2}\, dx$$

Introducing $|v(x)| = \alpha\rho(x)^n$ and choosing $n + 1 = \frac{5}{2}\, n$, i.e. $n = 2/3$, results in

$$T_\psi \geq \left(\alpha - \frac{4s}{15\pi}\, \alpha^{5/2}\right) \int \rho_\psi(x)^{5/3}\, dx$$

which has a maximum for

$$\frac{2s}{3\pi}\alpha^{3/2} = 1$$

Hence,

$$T_\psi \geq \frac{3}{5}\left(\frac{6\pi^2}{4\pi s}\right)^{2/3} \int \rho_\psi(x)^{5/3}\, dx = (4\pi s)^{-2/3} K_c \int \rho_\psi(x)^{5/3}\, dx$$

which is Eq. (1.8).

Having so far dealt with the kinetic part of E_ψ^Q in Eq. (1.3), we consider now its potential part. Since the TF theory comes into play again the relevant results of the work of Lieb and Simon on this subject are quoted (without proof) and the ideas which culminate in the proof of the H-Stability of matter are presented.

D. Some Properties of the TF Theory

Consider the TF energy functional given by Eq. (1.4), then from theorem (1.3) [Lieb, 1976], we have:

> $\varepsilon(\rho)$ has a minimum on the set $\int \rho(x)\, dx = N$ (this property is based on the fact that $\varepsilon(\rho)$ is bounded below)
>
> This minimizing ρ, called ρ_N^{TF} is unique (this property is based on the convexity of $\varepsilon(\rho)$) and satisfies, for the neutral case $Z = NQ$ considered here, the familiar TF equations

$$\frac{5}{3} s^{-2/3} K_c\, \rho_N^{TF}(x)^{2/3} = \phi_N^{TF}(x) \geq 0 \tag{1.11}$$

with

$$\phi_N^{TF}(x) = V(x) - \int |x - y|^{-1} \rho_N^{TF}(y)\, dy \tag{1.12}$$

Equipped with this important existence and uniqueness theorem, we are interested in the TF energy for an isolated, neutral atom of nuclear charge Z. Equations (1.11) and (1.12) are solved numerically. Introducing two scaling parameters $x = \lambda x'$, $\rho = \mu\rho'$, we can require homogeneity of the three contributions to E_Z^{TF}, namely

$$s^{-2/3} K_c\, \lambda^3 \mu^{5/2} \sim Z\lambda^2\mu \sim \lambda^5 \mu^2$$

which yields

$$\lambda = Z^{-1/3} K_c\, s^{-2/3} \quad , \qquad \mu = Z^2 K_c^{-3} s^2 \quad ,$$

that is

$$E_Z^{TF} \sim K_c^{-1} s^{2/3} Z^{7/3}$$

The proportionality constant found numerically is -2.21. Hence

$$E_Z^{TF} = -2.21\, s^{2/3} K_c^{-1} Z^{7/3} \qquad (1.13)$$

This result will turn out to be particularly useful in taking up the next question: What about the E_N^{TF} of the polynuclear case $Q > 1$? It is here that a central theorem, called the no binding-theorem, comes into play. As early as 1955, Sheldon noticed numerically that molecules did not appear to bind in the TF model, e.g. that the energy of a diatomic molecule was less negative than the energy of the two separated atoms. In 1962, Teller proved this to be a general theorem. But it was only recently that a complete and deep understanding was achieved through Lieb and Simon's work. The TF theory is really a large Z theory; it is exact in the $Z \to \infty$ limit and constitutes with the theory of the hydrogen atom two opposite, but rigorous foundations of the many electron problem. For finite Z it describes adequately the bulk of an atom or molecule but it is not precise enough to give binding. Indeed, it should not do so because binding in TF theory would imply that the cores of atoms bind, and this does not happen. Molecular binding is a fine quantum effect: It results indeed from constructive interference between electronic wave functions centered around different nuclei and there are <u>no</u> wave functions in TF theory. In short, this theorem says that E_N^{TF} is bounded below by the sum of Q atomic E_Z^{TF} given by Eq. (1.13) i.e.

$$E_N^{TF} \geq -2.21\, s^{2/3} K_c^{-1} Z^{7/3} Q = -2.21\, s^{2/3} K_c^{-1} Z^{4/3} N$$

and since $\varepsilon(\rho) \geq E_N^{TF}$ we have, lastly

$$\varepsilon(\rho) \geq -2.21\, s^{2/3} K_c^{-1} Z^{4/3} N \qquad (1.14)$$

With this and the T_ψ inequality (1.8), I can now explain Lieb's ingenous way of using the TF theory twice, once for finding a lower bound to the potential part of E_ψ^Q and once for E_ψ^Q itself.

E. H-Stability

Consider an auxiliary charged fluid with $s = 1$, $Z = 1$, of density $n(x)$ with $\int n(x)\, dx = N$ in which the N electrons of locations x_i are embedded and thus interact with this fluid. Assign to this system an energy functional of the TF type but with an arbitrary $K > 0$,

$$\varepsilon(n) = K \int n(\psi)^{5/3}\, dy - \int V_x(y)\, n(\psi)\, dy$$

$$+ \frac{1}{2} \int n(\psi)\,|y - y'|^{-1}\, n(\psi')\, dy\, dy' + \sum_{i<j} |x_i - x_j|^{-1}$$

with

$$V_x(\psi) = \sum_i |x_i - y|^{-1}$$

By virtue of Eq. (1.14)

$$\varepsilon(n) \geq -2.21\, K^{-1}\, N$$

for all x_i. Hence, we obtain the electrostatic inequality

$$\sum_{i<j} |x_i - x_j|^{-1} \geq -K \int n(\psi)^{5/3}\, dy + \int V_x(y)\, n(\psi)\, dy$$

$$- \frac{1}{2} \int n(\psi)\,|y - y'|^{-1}\, n(\psi')\, dy\, dy' - 2.21\, K^{-1}\, N$$

Take the expectation value of this inequality over any N electron normalized antisymmetric wave function, $\psi(x_1 \,..\, x_N;\, s_1 \,..\, s_N)$. On the left-hand-side you obtain the mean electron-electron repulsion and on the right-hand-side the second term becomes

$$\int \rho_\psi(x)\,|x - y|^{-1}\, n(y)\, dx\, dy$$

while the other terms remain unchanged. Choose now $n(x) = \rho_\psi(x)$; the second and third term combine to

$$+ \frac{1}{2} \int \rho_\psi(x)\,|x - y|^{-1}\, \rho_\psi(y)\, dx\, dy$$

Hence, for any $K > 0$ we find

$$\langle\psi\Big|\sum_{i<j} |x_i - x_j|^{-1}\Big|\psi\rangle \geq \frac{1}{2} \int \rho_\psi(x)\,|x - y|^{-1}\, \rho_\psi(y)\, dx\, dy$$

$$- K \int \rho_\psi(x)^{5/3}\, dx - 2.21\, K^{-1}\, N \qquad (1.15)$$

This is the basic inequality used to bound below the true electron-electron repulsion.

Introduce this inequality in E_ψ^Q (Eq. 1.3) and use Eq. (1.8) as a lower bound to T_ψ. Gathering together the terms proportional to $\rho_\psi^{5/3}$ this yields,

$$E_\psi^Q \geq [(4\pi s)^{-2/3} K_c - K] \int \rho_\psi(x)^{5/3} dx - \int V(x) \rho_\psi(x) dx$$

$$+ \frac{1}{2} \int \rho_\psi(x) |x - y|^{-1} \rho_\psi(y) dx dy - 2.21 K^{-1} N$$

$$+ U(Z, \{R_n\}) \tag{1.16}$$

Restrict K, initially defined to be positive but otherwise arbitrary so that

$$(4\pi s)^{-2/3} K_c - K \equiv s^{-2/3} \underline{K} > 0$$

Then, apart from the constant $-2.21 K^{-1}N$, the right-hand-side of Eq. (1.16) is just another TF energy functional $\underline{\varepsilon}(\rho_\psi)$, applied to ρ_ψ but with K_c replaced by $\underline{K}$. By virtue of Eq. (1.14), we have again

$$\underline{\varepsilon}(\rho_\psi) \geq -2.21 s^{2/3} \underline{K}^{-1} Z^{4/3} N$$

Hence

$$E_\psi^Q \geq - 2.21 [(4\pi s^{-2/3} K_c - K)^{-1} Z^{4/3} - K^{-1}]^{-1} N$$

which has a maximum for

$$K = (4\pi s)^{2/3} K_c (1 + Z^{2/3})^{-1}$$

and which yields

$$E_\psi^Q \geq -2.21(4\pi s)^{-2/3} K_c^{-1} (1 + Z^{2/3})^2 N \tag{1.17}$$

Since this true for any ψ it is also true for the ground state. With the better constant $4\pi/1.89$ we have lastly for $s = 2$

$$E_N^Q \geq -1.89(1 + Z^{2/3})^2 N \frac{e^2}{a} = -5.56(1 + Z^{2/3})^2 N \text{ Ry}$$

or -22.24 Ry for hydrogenic matter, a very impressive bound indeed!

Having dealt with the H-stability of matter, I should now take up the problem of the existence of the thermodynamic limit. To establish this, the strategy is to find a finite lower bound estimate to the free energy density $F(N,Q,\Lambda)/\Lambda$ (this is easy once you have proved the H-stability) and a decreasing sequence of upper bound estimates to $F(N_i,Q_i,\Lambda_i)/\Lambda_i$ as $\Lambda_i \to \infty$, for neutral systems (this is the hard part). Now, while the problem of the H-stability was to tame the danger of implosion due to the singular attraction between electrons and nuclei, the problem of the thermodynamic limit is to tame the danger of explosion due to the long range nature of the Coulomb repulsion between like particles. The absence of explosion is due to screening.

In fact, the proof of the existence of the thermodynamic limit has been given by Lieb and Lebowitz in 1969 already. The authors invented a famous geometrical theorem, known as the "cheese-theorem", which tells us how non-overlapping neutral balls of appropriate radii can asymptotically be densely packed in any domain of reasonable shape. Then, static screening is produced by virtue of Newton's theorem which says that non-intersecting neutral balls do not interact.

However, since I have no time to explain this part of the theory I will refer you to Lieb's review article. I cordially invite you to read it anyway, to have an idea of the ingenuity manifested by the authors involved in the proof of the stability of coulombic matter, a fundamental problem of physics.

II. BOUNDS TO THE FREE ENERGY OF THE OCP

A. Introduction

Consider an assembly of N classical particles of charge $+|e|$ and coordinate x_i, in a domain Λ, immersed in a neutralizing background of opposite charge. This is the one-component plasma (OCP), a model of matter which applies to several situations in astrophysics, plasma physics and in the physics of electrolytes.

In the last decade this model has been the subject of extensive numerical and theoretical investigations. The numerous contributions presented during this Institute attest to this fact. The model was comprehensively reviewed by H. DeWitt in his introductory lecture.

In this talk, I wish to restrict myself to the question of bound estimates to the free energy per particle in the three-dimensional OCP. There is no need to advocate the interest for rigorous and accurate upper and lower bounds to the free energy of models of matter for which no exact solution is known.

In the first section we shall discuss the lower bound estimates due to Mermin and to Lieb and Narnhofer. In the second section we present our work on upper bounds estimates and we compare numerically these bounds with DeWitt interpolation formula constructed on the basis of Hansen's data.

B. Lower Bound Estimates

As early as 1968, N. D. Mermim [1968] showed that the structure factor $S(k)$, the internal energy, ε, and Helmholtz free energy, F, per particle of the OCP model were bounded below by their Debye-Huckel values. He showed, furthermore, that these inequalities were true whether the equilibrium state was uniform or crystalline. Mermin's results are

$$S(k) \geq k^2 \ (k_D^2 + k^2)^{-1} \tag{2.1}$$

$$\varepsilon \geq \frac{3}{2} k_B T - \frac{1}{2} e^2 \ k_D \tag{2.1'}$$

and

$$F(\beta,\rho) \geq F_o(\beta,\rho) - \frac{1}{3} e^2 \ k_D \tag{2.1''}$$

where $k_D^2 = 4\pi\beta e^2 \rho$ and $F_o(\beta,\rho)$ is the Helmholtz free energy per particle of an ideal gas at the same temperature and density. The demonstration of Eq. (2.1) and subsequently of Eqs. (2.1') and (2.1") was based on an application of Peirls-Bogoliubov inequality to a judicious choice of functions depending upon the Fourier components of the particle density and upon the potential energy of the system in thermal equilibrium.

No new rigorous results were established until 1975 when Lieb and Narnhofer [1975] proved the existence of the thermodynamic limit of the free energy, internal energy, pressure and entropy for the OCP. They also established the non-equivalence of the grand canonical and canonical ensembles due to the lack of convexity in the density of the free energy. Borrowing an idea due to Onsager they showed in particular that the excess internal energy

$$\varepsilon - \frac{3}{2} k_B T = u$$

was bounded below by the Wigner estimate (see below) and thus, that

$$F(\beta,\rho) \geq F_o(\beta,\rho) - \frac{9}{10} e^2 \ a^{-1} \tag{2.2}$$

where

$$\frac{4\pi}{3} a^3 = \rho_b^{-1} \quad ,$$

ρ_b being the background density = ρ for neutral systems. It is readily noted that, if $k_D a < 2.7$ then Eq. (2.1") is better than Eq. (2.2) and conversely for $k_D a > 2.7$. Since we are primarily interested here in strongly coupled plasmas, I wish to reproduce the Lieb and Narnhofer derivation of Eq. (2.2). Let

$$V = V_{pp} + V_{pb} + V_{bb} \tag{2.3}$$

be the potential energy of the system with

$$V_{pp} = \sum_{i<j} e^2 |x_i - x_j|^{-1}$$

$$V_{pb} = -\sum_i e^2 \rho_b \int_\Lambda dy |x_i - y|^{-1}$$

$$V_{bb} = \frac{1}{2} e^2 \rho_b^2 \int_\Lambda dx\, dy |x - y|^{-1} \tag{2.3'}$$

Add and subtract in Eq. (2.3) the electrostatic energy of a system of N charged balls, immersed in the same background and centered at the positions x_i of the particles, with a charge $+|e|$ spread uniformly over the balls. The radius, α, of the balls is an adjustable parameter. With $e\rho_\alpha(x)$ designating the charge density of the balls, this electrostatic energy, W, is

$$W = \frac{1}{2} e^2 \int_\Lambda dx\, dy [\rho_\alpha(x) - \rho_b] |x - y|^{-1} [\rho_\alpha(\psi) - \rho_b] \tag{2.4}$$

Alternatively, if W_s designates the self-energy of the balls we have also, in a decomposition similar to Eq. (2.3'),

$$W = W_s + W_{pp} + W_{pb} + W_{bb} \tag{2.4'}$$

and $W_{bb} = V_{bb}$. Then

$$\begin{aligned} V = &\ W && \text{(i)} \\ &+ V_{pp} - W_{pp} && \text{(ii)} \\ &+ V_{pb} - W_{pb} && \text{(iii)} \\ &- W_s && \text{(iv)} \end{aligned} \tag{2.5}$$

and we consider these terms individually: (i) is by definition positive and, then, can be omitted for a lower bound estimate; (ii) is also positive. It is a sum of repulsive short range pair potentials which vanish if the balls do not overlap as a consequence of Newton's theorem. More precisely, for a pair of particles separated by the distance $|x| < 2\alpha$, this repulsive potential is

$$\psi(|x|) = e^2 |x|^{-1} - e^2 \left(\frac{3}{4\pi\, \alpha^3}\right)^2 \int_{B(\alpha)} dx'\, dx'' |x + x' - x''|^{-1} \tag{2.6}$$

$$= e^2|x|^{-1} - e^2\left(\frac{3}{4\pi\,\alpha^3}\right)\int_{B(2\alpha)} dy$$

$$\cdot\left(1 - \frac{3}{4\alpha}|y| + \frac{1}{16\alpha^3}|y|^3\;|x+y|^{-1}\right)$$

$$= e^2|x|^{-1} - \frac{e^2}{\alpha}\left(\frac{6}{5} - \frac{1}{2}\frac{|x|^2}{\alpha^2} + \frac{3}{16}\frac{|x|^3}{\alpha^3} - \frac{1}{160}\frac{|x|^5}{\alpha^5}\right) \tag{2.6'}$$

We note that

$$\psi(|x| < 2\alpha) > 0 \quad , \quad \psi(2\alpha) = 0 \quad , \quad \left.\frac{d\psi}{d|x|}\right|_{|x|=2\alpha} = 0$$

and, obviously, that $\psi(|x| > 2\alpha) = 0$, by virtue of Newton's theorem. Hence, (ii) can also be omitted for a lower bound estimate to V. The terms in (iii) yield an extensive, negative, contribution to V. Indeed, for a ball centered at a particle position x and entirely contained in Λ, we have the contribution

$$U_{pb}(x) = -e^2\,\rho_b\int_\Lambda dy\left(|x-y|^{-1} - \frac{3}{4\pi\,\alpha^3}\int_{B(\alpha)} dz\,|x-y+z|^{-1}\right) \tag{2.7}$$

$$= -e^2\,\rho_b\,\frac{2\pi}{5}\,\alpha^2 \tag{2.7'}$$

independent of x as long as $\delta(x,\partial\Lambda) \geq \alpha$, where $\partial\Lambda$ is the boundary of Λ. To prove Eq. (2.7'), we note that, after performing the angular integrations,

$$\int_{B(\alpha)} dz\,|x-y+z|^{-1} = 4\pi\int_0^\alpha |z|^2\,d|z|\left\{\mathrm{Min}\left(\frac{1}{|x-y|}\,,\,\frac{1}{|z|}\right)\right\}$$

Then, when $|x-y| > \alpha$ the term in parenthesis is $|x-y|^{-1}$ and the two terms in Eq. (2.7) cancel. We are thus left with a contribution which has no spatial dependence,

$$u_{pb} = -e^2\,\rho_b\int_{B(\alpha)} dy\left[|y|^{-1} - \frac{3}{\alpha^3}\int |z|^2\,dz\left\{\mathrm{Min}\left(\frac{1}{|y|}\,,\,\frac{1}{|z|}\right)\right\}\right]$$

$$= -e^2\,\rho_p\,\frac{4\pi}{3}\,\alpha^3\left(\frac{3}{2}\,\alpha^{-1} - \frac{6}{5}\,\alpha^{-1}\right)$$

$$= -e^2\,\pi_b\,\frac{2\pi}{5}\,\alpha^2$$

which is Eq. (2.7'). For balls which are not entirely in Λ, the second term in Eq. (2.7) is smaller than before, that is, $u_{pb}(x)$ is less negative than Eq. (2.7'). The same is true if the particles are completely outside of Λ. We have then for the most general case, neutral or not, that

$$V_{pb} - W_{pb} \geq -N\, u_{pb} = -N\, e^2\, \rho_b\, \frac{2\pi}{5}\, \alpha^2$$

Lastly, the self-energy of the balls, W_s, is

$$W_s = \frac{1}{2}\, N\, e^2\, \frac{6}{5\alpha}$$

and then

$$V \geq -N \left(\frac{3}{5\alpha} + \frac{2\pi}{5}\, \alpha^2\, \rho_b\right)$$

The right-hand-side of this inequality is concave in α. There is a best lower bound precisely for $\alpha = a = (3/4\pi\, \rho_b)^{1/3}$ and we reproduce the remarkable lower bound due to Lieb and Narnhofer,

$$V \geq -N \left(\frac{3}{5} + \frac{3}{10}\right) \frac{e^2}{a} = -\frac{9}{10}\, N\, e^2\, a^{-1} \tag{2.8}$$

It is well-known that the right-hand-side of Eq. (2.8) coincides with Wigner's estimate of the binding energy of the OCP. However, the method which leads to this estimate is different. Indeed, Wigner's argument is as follows. A particle is, in the mean, surrounded by a neutralizing background ball of radius a. The energy of this neutral entity is composed of one-half of the ball self-energy and the interaction of the particle with the ball. Neglecting small overlap effects between these neutral entities gives

$$u_W = \frac{1}{2}\, \frac{6}{5}\, \frac{e^2}{a} - e^2\, \frac{3}{4\pi\, \alpha^3} \int_{B(\alpha)} dx\, |x|^{-1}$$

$$= \left(\frac{3}{5} - \frac{3}{2} \right) e^2\, a^{-1} = -\frac{9}{10}\, e^2\, a^{-1}$$

and nicely enough,

$$\frac{3}{5} - \frac{3}{2} = -\frac{3}{5} - \frac{3}{10} \quad .$$

Another point worth mentioning here is that the idea of decomposing V as in Eq. (2.5) is very useful in calculating the binding energy of static Wigner crystals. Consider, for instance, mono-atomic crystals such as the simple cubic, body-centered and face-centered cubic ones. Let $\omega = \rho^{-1} = \rho_b^{-1}$ be the volume of the primitive cells, let $\{R\}$ be the position vectors of the lattice ions measured from a given ion and let $\{k\}$ be the vectors of the corresponding reciprocal lattice. We then find, in the thermodynamic limit that

$$u_o = -\frac{9}{10} e^2 a^{-1} + \frac{1}{2} \sum_{R \neq o} \psi(|R|) + \frac{1}{2} \sum_{k \neq o} \psi(|k|) \tag{2.9}$$

where $\psi(|x|)$ is the short range potential given by Eq. (2.6') and, since $\rho_a(x) - \rho_b$ is periodic over a primitive cell with a smeared out $\rho_a(x)$, the electrostatic energy W of Eq. (2.5)(i) per particle is strictly given by the second sum in Eq. (2.9) with

$$\psi(|k|) = \frac{4\pi e^2}{\omega |k|^2} \frac{g}{(a|k|)^4} \left(\cos(|k|a) - \frac{\sin(|k|a)}{|k|a} \right)^2 \tag{2.9'}$$

The advantages of the decomposition given in Eq. (2.5) for calculating binding energies are that: the thermodynamic limit (Eq. 2.9) can be rigorously established without cutting off the Coulomb potential before taking the thermodynamic limit; for all types of lattices one starts from the same, absolute, Lieb-Narnhofer lower bound; the structure of Eq. (2.9) enables a qualitative understanding of why a b.c.c. lattice can have a lower binding energy than an f.c.c. one. The reason is essentially that the two sums in Eq. (2.9) favor f.c.c. lattices, in x and k space but the second one yields a larger contribution than the first because it is less rapidly convergent. And since the b.c.c. and f.c.c. lattices are reciprocal of one another, the b.c.c. lattice in x space can indeed have a lower energy. Numerical estimates of Eq. (2.9) produce the generally accepted values of $-.880\ e^2/a$, $-.89593\ e^2/a$ and $-.89583\ e^2/a$ for the s.c., b.c. and f.c. cubic lattices. Let us remark finally that the case of poly-atomic crystals can be treated with the same method, provided that the primitive cell be free of dipole moment.

C. Upper Bounds Estimates

We start from the canonical partition function of a neutral OCP in thermal equilibrium at the temperature $T = 1/k_B\beta$ in the volume Λ,

$$Q(\beta,\Lambda,N) = \frac{1}{N!\nu^N} \int_\Lambda dx_1 \ldots dx_N \exp[-\beta V(x)] \tag{2.10}$$

where $V(x)$ is given in Eqs. (2.3) and (2.3') and where $\nu^{1/3}$ is the thermal wavelength.

Our class of lower bounds to Eq. (2.10) is based on an idea of optimal cell decomposition. Consider Λ to be the union of $M = m^3$ with $m\ z_+$, identical and disjoint cells ω_s ($s = 1, \ldots, M$) of shape affine to some primitive cell of the static crystal and of volume $|\omega|$ such that $M|\omega| = |\Lambda|$. If for any $m \geq 1$ we would distribute the N particles in all possible ways in the cells we would just produce an identity, in M, with the right-hand-side of Eq. (2.10). So we consider the simplest limitation on the cell occupation numbers to be zero or one, as in the lattice gas interpretation of Ising systems; but, for this limitation to be physically meaningful, we let M be $\geq N$. If we designate by $c = N/M$ the particle concentration we have, for any given density $\rho = N/\Lambda$,

$$\frac{c}{|\omega|} = \rho \tag{2.11}$$

Given some $M \geq N$, we define the auxiliary partition function,

$$y(\beta,\Lambda,N,M) = \sum_{\{n_s=o,1\}} \prod_{s=1}^{M} \frac{dx_s^{(n_s)}}{\nu^{n_s}} \exp\left[-\beta V(x,\{n_s\})\right] \tag{2.12}$$

$$\sum_s n_s = N$$

with the convention that $dx_s^{(0)} = 1$. The potential energy in Eq. (2.12) is conveniently written as

$$\begin{aligned} V(x,\{n_s\}) &= \frac{1}{2} M\, e^2 \rho_b^2 \int_\omega dx\, dy\, |x - y|^{-1} \\ &- \sum_s e^2 \rho_b \int_\omega dx\, dy\, |x - y|^{-1} \delta(x - x_s)\, n_s \\ &+ \frac{1}{2} \sum_{s \neq t} e^2 \int_\omega dx\, dy [\delta(x - x_s)\, n_s - \rho_b] \\ &\cdot |R_s - R_t + x_s - x_t|^{-1} [\delta(u - y_t)\, n_t - \rho_b] \end{aligned} \tag{2.13}$$

In Eq. (2.13), the vectors R_s ($s = 1, \ldots, M$) indicate the sth cell position and the vector x_s the instantaneous position of a particle in the cell ω_s. The number n_s = zero or one, depending on whether ω_s is occupied or not.

The auxiliary partition function (Eq. 2.12) is manifestly a lower bound to the exact one (Eq. (2.10). We then construct the best lower bound in maximizing Eq. (2.12) with respect to M for any fixed β and ρ, with the constraint that $M \to N$, or $c \to 1$, $\omega \to \rho^{-1}$ in the zero temperature limit where we assume that the ground state configuration is realized by a static b.c.c. lattice. Without the above constraint, we would always obtain $M_{max} = \infty$ thus producing the continuum limit of lattice theories in which case the exact $Q(\beta,\Lambda,N)$ would be reproduced. This is not what we want here: Our objective is to arrive at accurate upper bounds to the free energy of the system in such a way that its crystalline phases might be described. So, with Sup designating the Sup over M for this restricted sequence of cell divisions our model partition function will be written as

$$Q(\beta,\Lambda,N) = \underset{M \geq N}{\text{Sup}} \; y(\beta,\Lambda,N,M) \leq Q(\beta,\Lambda,N) \tag{2.14}$$

It is clear that Eq. (2.14) can be written for other models of matter, and in general, it will be understood that Sup may include appropriate subdivision of M, N and Λ in the case of coexistence of different phases. For neutral systems, for instance, the ground state will again be assumed to be given as the solution of a static solid state physics problem. The restricted Infimum applied to the potential energy of the ground state then tells us that, for densities larger than the density ρ_o at which the static pressure is zero, Λ is made up of N occupied cells of volume $\omega = \rho^{-1}$ and that, for densities smaller than ρ_o the ground state is pictured as a perfect crystal in equilibrium at zero pressure with the vacuum, that is, N cells of volume $\omega_o = \rho_o^{-1}$ are occupied, the remaining $(\Lambda/\omega_o) - N$ being empty. This situation anticipates sublimation which will occur in normal systems at low temperatures but there is no sublimation in the OCP.

We wish now to extract from Eqs. (2.12) and (2.14) easily calculable lower bounds in order to make contact with DeWitt's interpolation formula for the Helmholtz free energy of the OCP in its homogenous phase.

The simplest bound to Eq. (2.12), obtained by using Jensen's inequality, is that given by the Mean-Field Approximation where

$$\langle \delta(x - x_s)\, n_s \rangle_{M.F.A.} = c/\omega = \rho = \rho_b \quad .$$

In this case, the inter-cell interaction drop out of Eq. (2.12) and we find, after elementary calculations, and introducing the plasma parameter $\Gamma = \beta\, e^2\, a^{-1}$ with $(4\pi/3)\, a^3 = \rho^{-1} = \rho_b^{-1}$,

$$y(\beta,\Lambda,N,M) \geq \exp\,[N\, g_I(\beta,\rho,c)]$$

with

$$g_I(\beta,\rho,c) = g_o(\beta,\rho) + \frac{1-c}{c}\,\ell n\,\frac{1}{1-c} - 1 + k_I \Gamma\, c^{2/3} \tag{2.15}$$

where

$$g_o(\beta,\rho) = -\beta F_o(\beta,\rho)$$

and

$$k_I\Gamma c^{2/3} = \frac{1}{c}\beta e^2 \frac{1}{2}\delta \int_{\omega_{w.s.}} dx\, dy\, |x-y|^{-1} = .66\Gamma c^{2/3}$$

The domain of integration is $\omega_{w.s.}$, a Wigner-Switz cell of a b.c.c. lattice of volume $|\omega_{w.s.}| = c/\rho$. The constant k_I has been obtained by numerical integration [Sari, Merlini, Calinon, 1976] and we note that it is 10% larger than 3/5. The bound given by Eq. (2.15) displays two interesting properties: One concerning the second term and one concerning the third one. The second term corresponds to the concentration dependent part of the entropy and we note that as c varies from 0 to 1, this term varies from 0 to -1; there is exactly one Boltzmann constant of entropy change between the two limits. This means that the famous communal entropy problem of the usual cell theory of liquids seems to be soluble in using the optimal cell decomposition method. The third term is an energy term and the point is in its $c^{2/3}$ dependence. In standard lattice statistics (fixed cell volume) we would have here an energy proportional to c and the Sup over c in Eq. (2.15) would have produced a spurious continuous phase transition, with $c_{sup} = 0$ until

$$i_I J = k_I J_c = \frac{1}{2}$$

with $J = r\, a/\omega^{1/3}$ and then a $c_{max} > 0$. On the contrary, Eq. (2.15) gives a $c_{max}(r) \equiv \hat{c}(r)$ for all $r > 0$, a Sup o only at $\Gamma = 0$ and $\hat{c}(r)$ varies from zero to 1 when Γ varies from zero to infinity. For $\Gamma < 1$ in particular, we find

$$\hat{c} \simeq \left(\frac{4}{3} k_I r\right)^3$$

and $g_I \simeq g_o + .029r^3$. In general, we define

$$-\beta F_I(\beta,\rho,r) = g_I[\beta,\rho,\Gamma,\hat{c}(r)] \qquad (2.16)$$

While the bounds Eqs. (2.15) and (2.16) display some instructive properties they are not accurate enough in the strong coupling limit. In this case, we expect that the particle-background interaction (the second term in Eq. (2.13)) will produce an inhomogenous probability density in each cell, which is the same for each cell, namely

$$p_\omega(x) = \langle\delta(x - x_r)\, n_r\rangle$$

with

$$\int_{\omega} p_{\omega}(x)\ dx = c$$

This suggests that we construct another lower bound to Eq. (2.12) with $p_{\omega}(x)$ determined variationally. Since we are interested in the strong coupling limit, $p_{\omega}(x)$ will be sharply peaked, at the center of the cell, and decreases rapidly from the center. Therefore we can easily produce a calculable bound as follows: we limit $p_{\omega}(x)$ to be $\neq 0$ in a ball inscribed in the Wigner-Seitz b.c.c. cell of volume $|\omega|$. The radius of this ball is $r = .89|\omega|^{1/3}$. Inside the ball, $p_{\theta}(x)$ can be simply given, with $x = |\omega|^{1/3} t$ by

$$p_{\theta}(t) = c \frac{\exp\ (-\frac{1}{2}\Gamma\ c^{2/3}\ t^2)}{\int_{\theta} dt\ \exp\ (-\frac{1}{2}\Gamma\ c^{2/3}\ t^2)}$$

Then, by virtue of Newton's theorem, the excess internal energy per particle of the system becomes

$$-k_{II} c^{2/3}\ e^2/a$$

where k_{II} is the b.c.c. value = .89593! Of course we have lost something from the entropy, namely

$$\ell n\ \gamma \simeq -.3903$$

where

$$\gamma = \frac{4\pi}{3} r^3/\omega \equiv \frac{4\pi}{3} t_o^3$$

and, thus, we expect Eq. (2.16) to be better for small Γ. But this lsos is Γ-independent. Hence, we arrive at

$$y \geq \exp\ (N\, g_{II}(\beta,c,\Gamma)$$

where

$$g_{II}(\beta,c,\Gamma) = g_o(\beta,\rho) + \frac{1-c}{c}\ \ell n \frac{1}{1-c} - 1 + \ell n\ \gamma$$

$$+ \ell n\ \frac{3}{t_o^3}\int_o^{t_o} t^2\ dt\ e^{-\frac{1}{2}\Gamma\ c^{2/3}\ t^2} + k_{II}\Gamma\ c^{2/3} \ldots \tag{2.17}$$

Here, as before, g_{II} has a unique maximum $\hat{c}(\Gamma)$ and,

$$-\beta f_{II}(\beta,\rho,\Gamma) = g_{II}[\beta,\rho,\hat{c}(\Gamma),\Gamma] \tag{2.18}$$

Gathering together the lower bounds (Eqs. 2.1" and 2.2), the upper bounds (Eqs. 2.16 and 2.18), and subtracting the perfect gas free energy we have finally that

$$\text{Max}\ (-3^{-1/3}\ \Gamma^{3/2},\ -\frac{9}{10}\ \Gamma) \leq \Delta\beta F(\Gamma) \leq \text{Min}\ [\Delta\beta F_I(\Gamma),\ \Delta\beta F_{\pi}(\Gamma)] \tag{2.19}$$

For $\Delta\beta F(\Gamma)$ itself we take DeWitt interpolation formula which reads

$$\Delta\beta F_{D.W.}(\Gamma) = -2.816 - .50123\ \ell n\ \Gamma + 3.26591\Gamma^{1/4} - .89461\Gamma$$

and we obtain the results shown in Table I.

In the respective columns we have:

Γ, varying from 1 to 153

$A = \text{Max}\ (\beta\Delta F_M,\ \beta\Delta F_{L.N.}) = \text{Max}\ (-3^{1/2}\ \Gamma^{3/2},\ -\frac{9}{10}\ \Gamma)$

$B = \beta\Delta F_{D.W.} - A$

$C = \text{Min}\ (\beta\Delta F_I,\ \beta\Delta F_{II}) - A$

$D = \hat{c}(\Gamma)$

TABLE I

Γ	A	B	C	D
1	-.5774	.1387	.4491	.3216
9	-8.1000	1.7939	3.1150	.9926
17	-15.3000	2.4931	4.0390	.9998
25	-22.5000	3.0141	4.6466	
33	-29.7000	3.4430	5.0954	
41	-36.9000	3.8138	5.4536	
49	-44.1000	4.1442	5.7535	
57	-51.3000	4.4444	6.0129	
65	-58.5000	4.7213	6.2425	
73	-65.7000	4.9792	6.4492	
81	-72.9000	5.2217	6.6377	
89	-80.1000	5.4510	6.8116	
97	-87.3000	5.6692	6.9733	
105	-94.5000	5.8777	7.1247	
113	-101.7000	6.0777	7.2674	
121	-108.9000	6.2702	7.4026	
129	-116.1000	6.4560	7.5312	
137	-123.3000	6.6357	7.6540	
145	-130.5000	6.8101	7.7717	
153	-137.7000	6.9794	7.8849	

We find that $\beta F_{II} \leq \beta F_I$ and $1 \geq \hat{c}(\Gamma) \geq .99$ as of $\Gamma \simeq 10$!

III. EQUILIBRIUM FIELD THEORY OF CLASSICAL COULOMB SYSTEMS

A. Introduction

Consider a neutral assembly of N cations and anions with charge $\pm$ e and with form factors $f_+(x)$, $f_-(x)$ of finite extension, in a finite region of R^3. The local charge density is

$$q(x) = |e|[\rho_+(x) - \rho_-(x)] \tag{3.1}$$

where

$$\rho_s(x) = \sum_j f_s(|x - x_{js}|) \qquad s = +,-$$

and

$$\int f_s(x)\, dx = 1$$

This is a TCP model with soft cores. For the OCP, we would have $\rho_-(x) = \rho_b$, the fixed background density. Let

$$\varepsilon_s = \int f_s(|x|)\,|x - y|^{-1} f_s(|y|)\, dx\, dy$$

be the self-energy of a cation or an anion. In adding and subtracting the self-energy of the two species to the potential energy, $V(\{x_{js}\})$, of the system, we can write

$$V(\{x_{js}\}) = \frac{1}{2}\int q(x)\,|x - y|^{-1} q(y)\, dx\, dy - \frac{1}{2} N(\varepsilon_+ + \varepsilon_-) \tag{3.2}$$

$$\equiv U(q) = \frac{1}{2} N(\varepsilon_+ + \varepsilon_-) \tag{3.2'}$$

Let

$$Q(\beta,\Lambda,N) = \frac{\nu_+^{-N}}{N!}\frac{\nu_-^{-N}}{N!}\int_\Lambda dx_1 \,..\, dx_N\, dy_1 \,..\, dy_N \exp[-\beta V(\{x_{js}\})] \tag{3.3}$$

be the canonical partition function (c.p.f.) at the temperature $T = 1/k_B\beta$ in the volume Λ with $\nu_+^{1/3}$, $\nu_-^{1/3}$ being the thermal wave-lengths of the cation and anion respectively.

The equilibrium statistical mechanics of classical Coulomb systems has a field theoretic formulation in terms of a functional integral representation of the Boltzmann factor exp $[-\beta U(q)]$, for any finite total charge, as an average of

$$\exp [i\ \beta^{1/2} \int q(x)\ \psi(x)\ dx]$$

over a gaussian measure $d\mu(\psi)$ of random variables $\psi(x)$ having mean value zero and covariance

$$v(|x - y|) \equiv \int d\mu(\psi)\ \psi(x)\ \psi(y) \equiv \langle\psi(x)\quad \psi(y)\rangle_o = |x - y|^{-1}$$

What does this mean? Consider the following particular case where this representation can be understood as the infinite volume limit of a situation with periodic boundary conditions for $\psi(x)$ and, a fortiori, for $q(x)$ necessarily neutral over a box Ω. In this case, the Fourier components are such that $q_{-k} = q_k^*$, $\psi_k = \psi_{-k}^*$ because $q(x)$ and $\psi(x)$ are real, and we have

$$U(q) = \frac{1}{2} \int_\Omega q(x)\ v(|x - y|)\ q(y)\ dx\ dy$$

$$= \frac{1}{2} \sum_{k \neq o} \tilde{v}(k)\ q_k\ q_k^* = \sum_{k \neq o}{}' \tilde{v}(k)\ q_k\ q_k^* \tag{3.4}$$

The last sum is over half of the k-space, so that we do not over-count the complex amplitudes, q_k. Then, the basic identity is, for any pair $(k,-k)$,

$$\exp [-\beta\tilde{v}(k)\ q_k\ q_k^* = \int \frac{d|\psi_k|^2\ d(\arg\ \psi_k)}{2\pi\tilde{v}(k)}$$

$$\cdot \exp\left(-\frac{\psi_k\ \psi_k^*}{\tilde{v}(k)} + i\ \beta^{1/2}\ (q_n\ \psi_n + c.c.)\right) \tag{3.5}$$

$$= \int d\mu(\psi_k)\ \exp [i\ \beta^{1/2}\ (q_k\ \psi_k^* + q_k^*\ \psi_k)] \tag{3.5'}$$

To check Eq. (3.5), write

$$q_k = \frac{1}{\sqrt{2}}\ (q_1 + iq_2) \quad , \qquad \psi_k = \frac{1}{\sqrt{2}}\ (\psi_1 + i\ \psi_2) \quad .$$

The covariance of the particular random variable $\psi_k(x) = \psi_k e^{ikx}$ is

$$\langle\psi_k(x)\ \psi_k^*(y)\rangle_o = \int d\mu(\psi_k)\ \psi_k\ \psi_{kx}^*\ e^{ik(x-y)} = \tilde{v}(k)\ e^{ik(x-y)} \tag{3.6}$$

By taking Ω^{-1} times the sum over k and the continuum limit of Eq. (3.6), we have

$$\int \frac{dk}{(2\pi)^3} \langle\psi_k(x)\ \psi_k^*(y)\rangle_o = \langle\psi(x)\ \psi(y)\rangle_o$$

$$= \int d\mu(\psi)\ \psi(x)\ \psi(y) = \int \frac{dk}{(2\pi)^3}\ \tilde{v}(k)\ e^{ik(x-y)} = v(|x-y|)$$

and $v(|x-y|) = |x-y|^{-1}$. In general this representation applies to any interaction potential such that its Fourier transform exists and is positive definite.

In applying Eq. (3.5) to the Boltzmann factor $\exp[-\beta V(\{x_{js}\})]$ with u(q) given in Eq. (3.4), we obtain in the continuum limit

$$\exp[-\beta V(\{x_{js}\})] = \exp[-\beta u(q) + N\frac{1}{2}\beta(\varepsilon_+ + \varepsilon_-)]$$

$$= \int d\mu(\psi)\ \exp\left[i\beta^{1/2}\int_\Lambda q(x)\ \psi(x)\ dx + \frac{1}{2}N\beta(\varepsilon_+ + \varepsilon_-)\right] \tag{3.7}$$

which explains at least in this case, the definition given above. We will come back to this point later.

One of the motivations for using this representation can now be appreciated by observing that the particle coordinates are in q(x) only and this allows us to permute the configuration integrals with the gaussian average. We find

$$Q(\beta,\Lambda,N) = \int d\mu(\psi)\ Q_o(\beta,\Lambda,N,i\psi) \equiv \langle Q_o\rangle_o \tag{3.8}$$

where

$$Q_o(\beta,\Lambda,N,i\psi) = \frac{\nu_+^{-N}}{N!}\frac{\nu_-^{-N}}{N!}\prod_{j,s}\int dx_{js}\ b(x_{js},\psi) \tag{3.9}$$

and,

$$b(x_{js},\psi) = \exp\left[is\beta^{1/2}\int dx\ f_s(x-x_{is})\ \psi(x) + \frac{1}{2}\beta\varepsilon_s\right] \tag{3.9'}$$

The relations at Eqs. (3.8) and (3.9) say that the exact c.p.f. of the neutral system is given by the gaussian average of the c.p.f. of N_+ and N_- independent particles in an imaginary random field $\pm\ i\beta^{-1/2}\ \psi(x)$. Strictly speaking, Eq. (3.8) has been derived for a particular case. We said in the beginning, however, that Eq.

(3.8) was valid for any q(x) even if $N_+ \neq N_-$. In order to understand in physical terms why this is indeed so, we look at Eq. (3.8) from another point of view. Imagine replacing ψ in $d\mu(\psi)$ by $u/\sqrt{t}$, with t being a parameter varying from zero to one. Then Eq. (3.8) can be viewed as the solution at t = 1 of the "diffusion equation"

$$\frac{\partial Q(\,..\psi)}{\partial t} = \frac{1}{2}\int dx\, dy\, v(|x-y|)\, \frac{\delta^2}{\delta\psi(x)\,\delta\psi(y)}\, Q(..\psi)$$

with some initial condition $Q_o(..\psi)$ at t = 0. It is then clear from this interpretation that the "kernel" of this "diffusion equation" is proportional to

$$\exp\left[-\frac{1}{2}\int dx\, dy\, \psi(x)\, v^{-1}(|x-y|)\, \psi(y)\right]$$

where v^{-1}, the inverse of v is such that

$$\int dz\, v^{-1}(|x-z|)\, v(|z-y|) = \delta(x-y)$$

At this point, we can mention another motivation for uisng the representation given by Eq. (3.8). It is that the original problem with long-range interaction, $|x-y|^{-1}$, has been converted into a dual one with short-range interaction involving the inverse of $|x-y|^{-1}$. The proof of the existence of thermodynamic limits [Frohlich and Park, preprint] is thereby considerably simplified in comparison with the direct proofs. Since, for Coulomb interactions,

$$-\frac{1}{4\pi}\Delta|x-y|^{-1} = \delta(x-y)$$

we have that

$$\int dx\, dy\, \psi(x)(|x-y|^{-1})^{-1}\,\psi(y) = -\frac{1}{4\pi}\int dx\, \psi(x)\,\Delta\psi(x)$$

$$= \frac{1}{4\pi}\int (\nabla\psi)^2\, dx - \frac{1}{4\pi}\int \nabla(\psi\nabla\psi)\, dx \tag{3.10}$$

The first term is translationally invariant, whereas the second term is not. This second term, a surface term, was strictly zero in the case considered above, precisely for that reason. This is not so in general, that is for other initial and/or boundary conditions, for the sub-domains $\omega\varepsilon\Lambda$ in the case of partitioning, and a fortiori for systems possessing an excess charge. Yet Eq. (3.8) is always valid, and it is with due consideration of the second term in Eq. (3.10) that the effective interaction between local density fluctuations (screening) will be rigorously derived and that the occurrence of local and global equilibrium inhomogenous charge distributions (anti-screening) can be contemplated. In this lecture, however, we will assume screening and mainly deal with homogenous states.

Returning to Eq. (3.9) we note that

$$\langle b(x_{js},\psi)\rangle_o = \exp\left[-\frac{1}{2}\beta e^2 \int dx\, dy\, F_s(x - x_{js})|x - y|^{-1} \cdot F_s(y - x_{js}) + \frac{1}{2}\beta t_s\right]$$

$$= \exp\left(-\frac{1}{2}\beta t_s + \frac{1}{2}\beta t_s\right) = 1 \quad ,$$

which means that the self-energy, which we had added and subtracted from V(x), disappears again from the gaussian average of the integrands in Q_o, when each is taken individually. Defining the convolution $f \cdot \psi$ by $\psi(f)$, and rescaling

$$\beta^{1/2} e\, \psi(x) = \phi(x)$$

with the new, dimensionless, covariance

$$\langle\phi(x)\, \phi(y)\rangle_o = \beta e^2 |x - y|^{-1}$$

we introduce the standard notation of field theory, due to Wick,

$$b(x_{js},\psi) = \frac{\exp\,[is\, \phi(f_s)]}{\langle\exp\,[is\, \phi(f_s)]\rangle_o} \equiv :e^{is\phi(f_s)}: \tag{3.11}$$

Hence, Eq. (3.8) becomes

$$Q(\beta,\Lambda,N) = \frac{1}{(N!)^2}\int d\mu(\psi)\, \prod_{js} \nu_s^{-1} \int dx_{js}\left(:e^{is\phi(f_s)}:\right) \tag{3.12}$$

As a next step we consider the grand canonical ensemble in two cases, one with strict charge neutrality and the other with charge neutrality in the mean. With z the fugacity, the grand canonical partition function is constructed in the first case by summing $z^{2N}Q(\beta,\Lambda,N)$ over all N with $Q(\beta,\Lambda,0) = 1$, which preserves strict charge neutrality.

The other case is defined by independently summing the series

$$z^{N_+ + N_-} Q(\beta,\Lambda,N_+,N_-)$$

over N_+ and N_- and in requiring charge neutrality in the mean only. By introducing the pressure functional

$$\beta P(z,\phi) \equiv z\left[\nu_+^{-1} : e^{i\phi(f_+)} + \nu_-^{-1} : e^{-i\phi(f_-)}\right] \tag{3.13}$$

we have in this case that,

$$Z(\beta,z,\Lambda) = \int d\mu(\psi)\; e^{\beta\int_\Lambda P(\phi)dx} \tag{3.14}$$

An explicit representation of the first ensemble can be given in writing the Kronecker symbol

$$\delta_{N_+,N_-}$$

as

$$\delta_{N_+,N_-} = \int_o^{2\pi} \frac{d\alpha}{2\pi}\; e^{i\alpha(N_+-N_-)}$$

We find, for the grand partition function (g.p.f.) of the neutral ensemble, that

$$\overline{Z}(\beta,z,\Lambda) = \int_o^{2\pi} \frac{d\alpha}{2\pi} \int d\mu(\psi)\; e^{\beta\int_\Lambda P(\phi+\alpha)dx} \tag{3.15}$$

or in shifting the phase α,

$$\overline{Z}(\beta,z,\Lambda) = \int_o^{2\pi} \frac{d\alpha}{2\pi} \int d\mu(\psi - \alpha)\; e^{\beta\int_\Lambda P(\phi)dx} \tag{3.15'}$$

$$\equiv \int_o^{2\pi} \frac{d\alpha}{2\pi}\; \hat{Z}(\alpha,\beta,z,\Lambda) \tag{3.15''}$$

With the g.p.f. $\hat{Z}(\alpha,\beta,z,\Lambda)$ just defined we can treat other multi-component systems and obtain $Z = \hat{Z}(\alpha = o)$ or $\overline{Z}$ as in Eq. (3.15"). For the two-component systems considered here it is intriguing to observe (an allusion to anti-screening?) that, since $P(\phi + \pi) = -P(\phi)$, we can also write Eq. (3.15) as

$$\overline{Z}(\beta,z,\Lambda) = \int_{-\pi/2}^{+\pi/2} \frac{d\alpha}{2\pi} \int d\mu(\phi)\; \left(e^{+\int_\Lambda \beta P(\phi)dx} + e^{-\int_\Lambda \beta P(\phi)dx} \right)$$

If we consider the particular case of a symmetric TCP where $\nu_+ = \nu_- = \nu$ and $f_+ = f_-$, then Eq. (3.14) becomes

$$Z(\beta,z,\Lambda) = \int d\mu(\phi)\ e^{2z\nu^{-1}\int_\Lambda dx :\cos\phi(f):} \tag{3.16}$$

This is a famous model of euclidian field theory. For classical and quantum systems with distinguishable particles interacting with a long-range potential of positive type, hence, including the model in Eq. (3.16), Frohlich and Park [preprint] have recently proved the existence of the thermodynamic limit for the pressure and for the correlation functions for arbitrary $\beta \geq 0$ and $z \geq 0$. Debye screening, that is, exponential clustering of correlation functions for a dilute symmetric TCP has recently been proved by D. C. Bridges [preprint] for a lattice model. To prove screening, which we do not do in this lecture, it is important to take into account the fact that $d\mu(\phi)$ is not translationaly invariant in ϕ, as shown before, while $P(\phi)$ in Eq. (3.16) is periodic in ϕ with a period 2π and thus can be expanded in each sector $(2k - 1)\pi \leq \phi \leq (2k + 1)\pi$ around the maxima $\phi_{o,k} = 2k\pi$.

As a next model consider the OCP. Here

$$q(x) = e \int dx\ [\rho(x) - \rho_b]$$

and, using the notation given in Eq. (3.11) we have for the c.p.f.

$$Q(\beta,\rho_b,\Lambda,N) = \frac{\nu^{-N}}{N!} \int d\mu(\phi) \prod_j dx_j\ : e^{i\phi(f_j)} : e^{-i\int_\Lambda \rho_b \phi(x) dx} \tag{3.17}$$

For the g.p.f., we find

$$Z(\beta,z,\Lambda) = \int d\mu(\phi)\ e^{\int_\Lambda \beta P(\phi) dx} \tag{3.17'}$$

with

$$\beta P(\phi) = z\nu^{-1} : e^{i\phi(f)} : -i\rho_b \phi \tag{3.18}$$

Finally, let us consider a TCP model with hard core repulsion. Although the field theoretic formulation cannot handle such singular interactions, the following can be done. Let us switch off the electrostatic interaction altogether, then we are left with a system of hard spheres which plays the role of a reference system. Suppose that we possess a reasonable approximation to the Ursell-Mayer expansion of the pressure functional of a hard sphere system in an inhomogenous external potential u(x); say

$$P_{ref}[z\ e^{-\beta u(x)}] \qquad .$$

Switch on the electrostatic interaction again, then the g.p.f. of the model will be given by

$$Z(\beta,z,\Lambda) = \int d\mu(\phi)\, e^{\beta\int_\Lambda P_{ref}(z{:}e^{i\phi}{:})dx}$$

This formulation offers interesting possibilities to investigate the equilibrium properties of a realistic model of electrolytes.

In the next two sections we will concentrate our attention on the OCP.

B. Field-Field and Density-Density Correlation Functions

We wish here to derive for the OCP model and in the grand canonical ensemble a few useful and exact relations between the one- and two-field correlation functions, and the one- and two-particle equilibrium densities. For this purpose, let us add an external field $\phi_e(x)$ to the fluctuating one, $\phi(x)$; the g.p.f., given by Eq. (3.17), can be written in two ways, namely

$$Z(\beta,i\phi_e,\Lambda) = \int d\mu(\phi)\, e^{\int_\Lambda \beta P(\phi+\phi_e)dx} \tag{3.19}$$

and

$$Z(\beta,i\phi_e,\Lambda) = \int d\mu(\phi - \phi_e)\, e^{\int_\Lambda \beta P(\phi)dx} \tag{3.20}$$

We calculate $\delta\, \ell n\, Z/i\, \delta\, \phi_e(x)$ at $\phi_e = 0$ from Eqs. (3.19) and (3.20) and identify the results. Defining for any meaningful functional $A(\phi)$ the exact mean value

$$\langle A(\phi)\rangle \equiv Z^{-1} \int d\mu(\phi)\, A(\phi)\, e^{\int_\Lambda \beta P(\phi)dx}$$

we obtain from Eq. (3.19)

$$\left.\frac{\delta\, \ell n\, Z}{i\, \delta\, \phi_e(x)}\right|_{\phi_e=0} = z\nu^{-1} \langle {:}e^{i\phi(x)}{:}\rangle - \rho_b = \rho(x) - \rho_b$$

Using the nature of the kernel discussed in the preceeding section, we find from Eq. (3.20) that

$$\left.\frac{\delta\, \ell n\, Z}{i\, \delta\, \phi_e(x)}\right|_{\phi_e=0} = -\frac{1}{4\pi\, i\, \beta e^2}\, \Delta\, \langle\phi(x)\rangle$$

Thus,

$$\frac{1}{i} \Delta \langle\phi(x)\rangle = -4\pi \beta e^2[\rho(x) - \rho_b] \tag{3.21}$$

which is the Poisson equation for $-i\langle\phi(x)\rangle$. Since the right-hand-side of Eq. (3.21) is real, this equation is physically meaningful in the homogenous case where $\rho = \rho_b$ and $\langle\phi\rangle = 0$ in the thermodynamic limit, and also in the inhomogenous case if $\langle\phi(x)\rangle$ acquires an imaginary component, say $i\phi(x)$. The same remark applies to the symmetric TCP.

For the remainder we consider homogenous states and proceed with the second functional derivative of ℓn Z with respect to $i\,\phi_e(x)$. The particle-background interaction no longer comes into play and from Eq. (3.19) we obtain

$$\left.\frac{\delta^2 \ell n\, Z}{i^2\, \delta\, \phi_e(x)\; i\; \phi_e(y)}\right|_{\phi_e=0} = z\nu^{-1} \langle :e^{i\phi(x)}:\rangle\, \delta(x-y)$$

$$+ z^2\nu^{-2} \langle :e^{i\phi(x)}::e^{i\phi(y)}:\rangle$$

$$- z^2\nu^{-2} \langle :e^{i\phi(x)}:\rangle\langle :e^{i\phi(x)}:\rangle$$

$$= \rho\, \delta(x-y) + \rho_2(|x-y|) - \rho^2$$

where $\rho_2(|x-y|)$ is the two-particle density. Defining the net or truncated pair correlation function $h(|x-y|)$ through

$$\rho_2(|x-y|) = \rho^2[1 + h(|x-y|)]$$

we obtain the familiar expression

$$\left.\frac{1}{i^2}\frac{\delta^2 \ell n\, Z}{\delta\, \phi_e(x)\; \delta\, \phi_e(y)}\right|_{\phi_e=0} = \rho\, \delta(x-y) + \rho^2\, h(|x-y|) \tag{3.22}$$

On the other hand, we find from Eq. (3.20)

$$\left.\frac{\beta e^2}{i^2}\frac{\delta^2 \ell n\, Z}{\delta\, \phi_e(x)\; \delta\, \phi_e(y)}\right|_{\phi_e=0} = v^{-1}(|x-y|)$$

$$- \int dz\; v^{-1}(|x-z|)\; v^{-1}(|y-z|)$$

$$\cdot \langle\phi(x)\; \phi(y)\rangle \tag{3.22'}$$

Let

$$D_o(|x - y|) = \langle\phi(x)\ \phi(y)\rangle_o = \beta e^2\ v(|x - y|) = \beta e^2 |x - y|^{-1}$$

and

$$D(|x - y|) = \langle\phi(x)\ \phi(y)\rangle$$

be the bare (unperturbed) and perturbed field-field correlation functions. Let $\tilde{D}_o(k)$ and $\tilde{D}(k)$ be their Fourier transform and let $D_o(k) = \rho\ \tilde{D}_o(k)$ and $D(k) = \rho\ \tilde{D}(k)$ be dimensionless Fourier transforms. With these,

$$D_o(k) = \rho\ \frac{4\pi\ \beta e^2}{k^2} = \frac{k_D^2}{k^2}$$

where k_D^{-1} is the Debye length. Fourier transforming Eqs. (3.22) and (3.22') and dividing by $\rho \neq 0$, we obtain the relation

$$D_o^{-1}(k) - D_o^{-2}(k)\ D(k) = 1 + h(k) \tag{3.23}$$

But, $1 + h(k)$ is the static structure factor $S(k)$ and this is known to be related with the static dielectric function $\varepsilon(k)$ through

$$D_o(k)\ S(k) = 1 - \varepsilon^{-1}(k)$$

Hence

$$D(k) = D_o(k)[1 - D_o(k)\ S(k)]$$

that is

$$D(k) = D_o(k)\ \varepsilon^{-1}(k) \tag{3.24}$$

This is an exact and fundamental relation between the equilibrium field-field correlation function and the static dielectric function. It is this direct connection which makes the functional integral method so useful. Our purpose is next to set up the Dyson equation for $D(k)$ so that we may actually calculate $\varepsilon(k)$ either numerically or analytically, in a given, well-defined, approximation.

C. The Dyson Equations for the OCP in the Grand Canonical and Canonical Ensembles

The Dyson equations for interacting fermions or bosons systems are standard material in texts on the many-body theory of condensed matter (see, e.g., Fetter and Walecka [1971]). For classical systems, however, the situation is simpler than in the quantum case and the interested reader can find a self-contained exposé of the necessary background in our monograph on "The Anharmonic Crystal" [1967]. Here we give only a simple sketch of how the implicit, or self-consistent, Dyson equations are obtained in starting from Eq. (3.17'). The problem is to calculate

$$\lim_{\Lambda\to\infty} \Lambda^{-1} \ln Z(\beta,z,\Lambda) = \psi(\beta,z)$$

as efficiently as possible while keeping in mind that there are no exact solutions in three dimensions. In brief, this is achieved as follows: Replace the bare gaussian measure $d\mu(\phi)$ by a trial gaussian measure $d\mu^*(\phi)$ with a covariance D^* to be determined variationally; compensate $\psi(\beta,z)$ for this change of measure and evaluate the effect of $P(\phi)$ as an infinite series beginning with

$$z\nu^{-1}\langle :e^{i\phi}:\rangle_*$$

of irreducible contributions from products of the form

$$\langle P[\phi(x_1)] \ldots P[\phi(x_n)]\rangle_*$$

which are integrated over (n - 1) variables.

The rules for the irreducible contributions are that: From each one of the $n > 1$ vertices $x_1 \ldots x_n$ emerge at least three correlation lines $D^*(|x_i - x_j|)$; no graph constructed in this way can be split in two parts by cutting two lines; a factor

$$z\nu^{-1}\langle :e^{i\phi}:\rangle_*$$

is assigned to each vertex. Combinatorial factors associated with the number of ways that equivalent graphs can be obtained have to be calculated for each case. Call $W(z,D^*)$ the functional obtained in this way.

Defining

$$\langle :e^{i\phi(x)}:\rangle_* = e^{\frac{1}{2}\langle\phi\phi\rangle_o - \frac{1}{2}\langle\phi\phi\rangle_*} = e^{\frac{1}{2}\lambda^*}$$

with

$$\lambda^* = \int \frac{dk}{(2\pi)^3} [D_o(k) - D^*(k)] \tag{3.25}$$

the first two terms of this functional read

$$W(z,D^*) = z\nu^{-1}\, e^{\frac{1}{2}\lambda^*} + \frac{1}{2}(z\nu^{-1})^2\, e^{\lambda^*} \int dx \sum_{n=3}^{\infty} \frac{(-1)^n}{n!} D^*(x)^n + \ldots \tag{3.26}$$

$$= z\nu^{-1}\, e^{\frac{1}{2}\lambda^*} + W_{ir}\,(z,D^*) \tag{3.26'}$$

where the index ir means irreducible. Then, $\psi(z,D^*)$ becomes

$$\psi(z,D^*) = \frac{1}{2}\int \frac{dk}{(2\pi)^3}\,[\ell n\, \tilde{D}^*(k)/\tilde{D}_o(k) + 1 - \tilde{D}(k)/\tilde{D}_o(k)]$$

$$+ z\nu^{-1}\, e^{\frac{1}{2}\lambda^*} + W_{ir}(z,D^*) \tag{3.27}$$

and D^* is determined in such a way that $\delta\Psi/\delta D^* = 0$. This gives

$$\tilde{D}(k)^{-1} - \tilde{D}_o(k)^{-1} + 2(2\pi)^3 \frac{\delta\, W(z,D^*)}{\delta\, D^*(k)} = 0$$

Defining the two-point correlation functions $\tilde{G}(z,k)$ and $\tilde{\Gamma}(z,k)$ through

$$\tilde{G}(z,k) = -2(2\pi)^3 \frac{\delta\, W}{\delta\, D^*(k)} \tag{3.28}$$

and

$$\tilde{\Gamma}(z,k) = \rho(z)^{-2}\, \tilde{G}(z,k) \tag{3.28'}$$

we obtain

$$\tilde{D}(z,k) = \tilde{D}_o(z,k)[1 + \tilde{D}_o(k)\, \tilde{G}(z,k)]^{-1} \tag{3.29}$$

which is the Dyson equation in the grand canonical ensemble.

It is appropriate to mention here that there is a direct relation between $\tilde{G}(z,k)$ and the polarizability $\alpha(k)$. Using

$$\varepsilon(k) = 1 + 4\pi\, \alpha(k)$$

and Eq. (3.24), we have induced that

$$\alpha(k) = \frac{\beta e^2}{k^2}\, \tilde{G}(z,k)$$

For $D^* = D$, $\psi(z,D^*)$ reaches its stationary value supposedly the exact $\psi(z,\beta)$, that is, β times the pressure. Although Eqs. (3.28) and (3.29) can be studied rather systematically, we have to keep in mind that, for the OCP, $\rho = \rho_b$ is the independent variable. This means one should convert the fugacity expansion into a density expansion just as in the Ursell-Mayer theory. Since $\psi(z,D)$ is stationary with respect to D we have that

$$\rho = z\left(\frac{\partial\psi}{\partial z}\right)_\beta = z\left(\frac{\partial\psi}{\partial z}\right)_{\beta,D} = z\left(\frac{\partial W}{\partial z}\right)_0 = \rho_b \qquad (3.30)$$

and the conversion is carried out in making the Legendre transform

$$\chi(\rho,D) = \psi(z,D) - \rho\ \ln z \qquad (3.31)$$

which defines $-\rho$ times the Helmholtz free energy density $\rho F(\beta,\rho)$. In field theory this fugacity to density conversion would be represented by the replacement of

$$z\nu^{-1}:e^{i\phi}:$$

by

$$\rho:e^{i\phi}:$$

in $W_{ir}(z,D)$ where

$$:e^{i\phi}: \equiv \frac{e^{i\phi}}{\langle e^{i\phi}\rangle}$$

Note that the exact expectation value of $e^{i\phi}$, which is ρ, enters in the denominator. The canonical free energy functional becomes

$$\chi(\rho,D^*) = \frac{1}{2}\int \frac{dk}{(2\pi)^3}\left[\ln D^*(k)/D_o(k) + 1 - D^*(k)/D_o(k)\right]$$

$$+ \frac{1}{2}\, e\lambda^* + W_{ir}(\rho,D^*) \qquad (3.32)$$

and we recall that λ^* is given by Eq. (3.25) and that $W_{ir}(\rho,D^*)$ is given in terms of an infinite series of irreducible contributions, the same as in $W_{ir}(z,D^*)$, but from products of the form

$$\langle e:e^{i\phi(x_1)}:e:e^{i\phi(x_2)}: \ldots \rangle \quad .$$

It should be pointed out here that the free energy functional given by Eq. (3.32) could have directly been obtained from the canonical partition function given by Eq. (3.17). In fact, this was originally done [Choquard, 1972], in a heuristic way, in order to study the onset of short range order in the OCP by a self-consistent, asymptotic analysis of the field theoretic analogue of the HNC approximation [Choquard and Sari, 1972]. Now, the stationarity condition applied to Eq. (3.32) yields, using Eq. (3.25),

$$\tilde{D}(k)^{-1} - \tilde{D}_o(k)^{-1} = \rho - 2(2\pi)^3 \frac{\delta W_{ir}(\rho,D)}{\delta \tilde{D}(k)}$$

Defining $\tilde{G}_{ir}(\rho,k)$ and $\tilde{\sigma}(\rho,k)$ through

$$\tilde{G}_{ir}(\rho,k) = -2(2\pi)^3 \frac{\delta W_{ir}}{\delta \tilde{D}(k)}$$

and

$$\tilde{\sigma}(\rho,k) = \rho^{-2} \tilde{G}_{ir}(\rho,k)$$

we obtain, with our convention concerning dimensionless Fourier transform,

$$\Gamma(\rho,k) = 1 + \sigma(\rho,k) \tag{3.33}$$

and

$$D(\rho,k) = D_o(k)[1 + D_o(k)\ \Gamma(\rho,k)]^{-1} \tag{3.33'}$$

this is the Dyson equation for the OCP in the canonical ensemble.

It is instructive to compare the solutions of the Dyson equations in the grand canonical and canonical ensembles. The motivation is that these ensembles are not equivalent as shown by Lieb and Narnhofer. We wish to understand concretely what this inequivalence means and, to begin with, we study these equations in the approximation where W_{ir} is neglected.

Neglecting $W_{ir}(\rho,D)$ in the canonical ensemble defines the familiar Debye-Hückel approximation. It results from Eq. (3.32) by omitting W_{ir} or simply by taking the term in $:\phi^2:$ from

$$\ell n\ \Lambda^{-1} \int_\Lambda dx(:e^{i\phi}:)$$

of Eq. (3.17). In this case we have

$$\tilde{D}(k)^{-1} = \tilde{D}_o(k)^{-1} + \rho = \tilde{D}_o(k) + \frac{k_D^2}{4\pi\ \beta e^2}$$

or

$$D(k) = k_D^2(k^2 + k_D^2)^{-1}$$

and

$$\beta F(\beta,\rho) = \beta F_o = \frac{1}{3}\beta e^2 k_D$$

or

$$\chi(\beta,\rho) = -\beta\rho F(\beta,P) = \rho(1 - \ell n\, \rho\nu) + \frac{1}{3}\rho\,\beta e^2 k_D$$

In the grand canonical ensemble the situation is more subtle. We have ψ of Eq. (3.26) with $W_{ir} = 0$. This defines a self-consistent "Hartree" type of approximation with

$$W(z,D) = z\nu^{-1} e^{\frac{1}{2}\lambda(D)}$$

$$\rho(z,D) = z\nu^{-1} e^{\frac{1}{2}\lambda(D)}$$

and with Eq. (3.25),

$$\tilde{D}(k)^{-1} = \tilde{D}_o(k)^{-1} + z\nu^{-1} e^{\frac{1}{2}\lambda(D)} = \tilde{D}_o(k)^{-1} + \frac{k_D^2(z,\lambda)}{4\pi\,\beta e^2}$$

The self-consistency condition is, in using Eq. (3.25),

$$\lambda = \int \frac{dk}{(2\pi)^3}\frac{4\pi\,\beta e^2}{k^2}\frac{k_D^2(z,\lambda)}{k^2 + k_D^2(z,\lambda)}$$

$$= \beta e^2 k_D(z,\lambda) = \beta e^2(4\pi\,\beta e^2\, z\nu^{-1})^{1/2} e^{\frac{1}{4}\lambda}$$

We observe that $z(\beta,\lambda)$ is a two-valued function of λ with a maximum, z_c, at $\lambda = \lambda_c = 4$, which is

$$z_c = \frac{1}{4\pi}\nu(\beta e^2)^{-3}\, 16e^{-2}$$

The first portion, $z_1(\beta,\lambda)$, is monotonically increasing from zero to z_c as λ varies from zero to λ_c; the second portion, $z_2(\beta,\lambda)$ is monotonically decreasing from z_c to zero as λ varies from λ_c to ∞. There is no solution for $z > z_c$. In this approximation we can easily calculate the stationary value of the functional $\psi(z,D)$ from Eq. (3.25) and we find (note that, for the TCP, this "Hartree" approximation gives a lower bound to the exact $\psi_{TCP}(z,D)$

$$\psi(\beta,z,D) = \beta P(\beta,z) = \rho[\beta,z)(1 - \frac{1}{6}\lambda(\beta,z)]$$

which is also the canonical pressure derived from the Debye-Hückel free energy. We find, as expected, that the first solution, $z_1(\beta,\lambda)$ is the stable solution with positive compressibility. We have indeed

$$\beta H(\beta,z) = \rho(\beta,z)[1 - \frac{1}{4}\lambda(\beta,z)]^{-1}$$

and we note that it is precisely at $\lambda = \lambda_c = 4$ that H becomes infinite. We observe also that for $\lambda > 4$ where the canonical free energy and pressure are still well-defined $\psi(\beta,z)$ behaves as if we would follow the second solution, $z_2(\beta,\lambda)$.

A completely similar situation is encountered at the thermodynamic level, if, instead of the Debye-Hückel free energy, we would take the Lieb-Narnhofer expression (Eq. 2.2), that is,

$$\chi_{LN}(\beta,\rho) = \rho(1 - \ln \rho\nu) + \frac{9}{10}\rho\Gamma$$

and we recall that

$$\lambda^2 = (\beta e^2 k_D)^2 = (\beta e^2)^2 4\rho \beta e^2 \rho = 3(\beta e^2)^3 a^{-3} = 3\Gamma^3 \quad .$$

In this case we have

$$\rho(\beta,z) = z\nu^{-1} e^{\frac{6}{5}\Gamma(\beta,\rho)}$$

$$\Gamma_c = \frac{5}{2}$$

$$z_c = \frac{3}{4\pi}(\beta e^2)^{-3} (\frac{5}{2})^3 e^{-3}$$

$$\psi(\beta,z) = \rho(\beta,z)[1 - \frac{3}{10}\Gamma(\beta,z)]$$

and

$$\beta H(\beta,z) = \rho(\beta,z)[1 - \frac{2}{5}\Gamma(\beta,z)]^{-1}$$

The same exercise can be made with our upper bounds (Eqs. 2.16 and 2.18) and lastly with the DeWitt interpolation formula. The general result is that, for homogenous states, the fugacity $z(\beta,\rho)$ is a double-valued function of ρ for a fixed β with a maximum z_c at $\Gamma_c \simeq 3$ and that the grand ensemble and canonical ensemble are equivalent, as expected, on the stable portion $P[z,(\Gamma)]$ for $0 \leq \Gamma \leq \Gamma_c$. We will report elsewhere about the subtle question concerning the (in)-equivalence between the different definitions of the pressure and, leaving here these thermodynamic considerations we proceed with the analysis of the Dyson equation in the canonical ensemble.

The linear Debye-Hückel approximation produces screening but it is well-known to yield poor short distance behavior of $h(|x|)$ which becomes <-1! Already the next approximation changes the short distance behavior of $h(|x|)$ radically. From Eqs. (3.32) and (3.26) we have for the first contribution to $\Gamma_{ir}(|x - y|)$

$$(-1) \sum_{n=2}^{\infty} \frac{[-D(|x - y|)]^n}{n!} = e^{-D(|x - y|)} - 1 + D(|x - y|) \tag{3.34}$$

and we shall show below that in this approximation

$$h(|x|) = e^{-D(|x|)} - 1 \geq -1 \quad .$$

The corresponding equation for $D(|x|)$ constitutes the self-consistent version of the so-called "watermelon" approximation. It is also the field theoretic analogue of Debye-Hückel nonlinear theory.

D. The Strong Coupling Limit

We have briefly discussed the first and second approximations to $\Gamma(k,D)$. Numerical and analytical investigations show that, in comparison to other approaches, they are valid for values of the plasma parameter $\Gamma \leq 1$. Since we are interested here in a very strongly coupled plasma, we ask: Can we carry out an infinite resummation of the contributions to $\Gamma(k,D)$ and then relate this to equations in the theory of liquids (these are given in the next lecture)? The answer is Yes. To do so, we note from Eq. (3.23) that we can write

$$S(k) = D_o(k)^{-1} - D_o(k)^{-1}[1 + D_o(k)\ \Gamma(k,D)]^{-1}$$

$$= \Gamma(k,D)[1 + D_o(k)\ \Gamma(k,D)]^{-1} \tag{3.35}$$

$$= \Gamma(k,D) - \Gamma(k,D)\ D(k)\ \Gamma(k,D) \tag{3.35'}$$

and

$$h(k) = \Gamma(k,D) - 1 - \Gamma(k,D)\ D(k)\ \Gamma(k,D)$$

$$= \sigma(k,D) - [1 + \sigma(k,D)]\ D(k)[1 + \sigma(k,D)] \tag{3.36}$$

This way of decomposing $h(k)$ shows the occurrence of a new combination of Γ and D in the form

$$L(k) = [1 + \sigma(k)]\, D(k)\, [1 + \sigma(k)]$$

which says that the two root points x and y can be directly connected with a line $D(|x - y|)$, or indirectly via one or two polarizability corrections $\sigma(|x - z|)$, $\sigma(|t - y|)$, and we call $L(|x - y|)$ this combination. Apart from this, Eq. (3.36) shows that $\sigma(|x - y|)$ plays, in some sense, the same role as $h(|x - y|)$ in the theory of fluids. In the latter, we know that in order to sum up irreducible contributions to $h(|x - y|)$, one must first introduce the direct correlation function through

$$h(|x - y|) = c(|x - y|) + \int \rho dz\, c(|x - z|)\, h(|z - y|) \quad (3.37)$$

Here, $\sigma(|x - y|)$ is already composed of irreducible terms and a decomposition as in Eq. (3.37) can only be made through the resolution of the identity

$$\sigma(|x - y|) = \gamma(|x - y|) + \int \rho dz\, \gamma(|x - z|)\, \sigma(|z - y|)$$

that is

$$\gamma(k) = \sigma(k)[1 + \sigma(k)]^{-1} \quad (3.38)$$

Then, by using $L(k)$ and $\gamma(k)$ just defined, the resummation of all effects occurring in parallel between the two root points x and y can be carried out. This produces the implicit equation

$$\sigma(|x - y|) = e^{-L(|x-y|) + \sigma\cdot\gamma(|x-y|) + \tilde{b}(|x-y|)} + L(|x - y|) - 1 \quad (3.39)$$

where

$$\sigma\cdot\gamma(s) \equiv \rho \int dt\, \sigma(|t|)\, \gamma(|s - t|)$$

and where $\tilde{b}(|x|,L)$ represents the effect of the collection of bridge graphs calculated with the rule indicated before.

Now, in order to make contact with the theory of liquids, we note from Eq. (3.35) that $h(k)$ can also be written as

$$h(k) = s(k) - 1 = \Gamma(k)[1 + D_o(k)\, \Gamma(k)]^{-1} - 1$$

$$= \frac{\Gamma(k) - 1 + D_o(k)\, \Gamma(k)}{1 + D_o(k)\, \Gamma(k)}$$

$$= \frac{-D_o(k) + \sigma(k)[1 + \sigma(k)]^{-1}}{1 + D_o(k) - \sigma(k)[1 + \sigma(k)]^{-1}} \quad (3.40)$$

But this is precisely the O.Z. relation, and therefore, we find the very interesting relation between the correction to c(k) from its Debye-Hückel approximation, $-k_D^2/k^2$, and the polarizability correction $\sigma(k)$. (Note that, with this relation, we make contact with the GKS approach! [Golden, this paper; Kalman, this paper]. Since $D_o(k) = k_D^2/k^2$ we have

$$c(k) = -k_D^2/k^2 + \sigma(k)[1 + \sigma(k)]^{-1} \tag{3.41}$$

$$= -k_D^2/k^2 + \gamma(k) \tag{3.41'}$$

Thus, we arrive at a formally exact equation for the pair correlation function $g(|x - y|)$ expressed in terms of the field-field correlation functions $D(|x - y|)$, namely

$$g(|x - y|) = e^{-\Gamma\cdot D\cdot\Gamma(|x-y|) + \sigma\cdot\gamma(|x-y|) + \tilde{b}(|x-y|,L)} \tag{3.42}$$

where the dots mean convolutions as in Eq. (3.39).

We must compare Eq. (3.42) with the equation in the theory of fluids applied to the OCP model, that is

$$g(|x - y|) = e^{-\beta e^2|x-y|^{-1} + h\cdot c(|x-y|) + b(|x-y|,h)} \tag{3.43}$$

where $b(x,h)$, a functional of $h(|x|)$, represent the effect of the collection of irreducible bridge graphs calculated with one additional rule. This is that no pair of vertices x_i, x_j be directly linked by more than one $h(x_i,x_j)$ bond.

The correspondence between the two approached will be closed if we show that

$$-\beta e^2|x - y|^{-1} + h\cdot c(|x - y|) = -\Gamma\cdot D\cdot\Gamma(|x - y|) + \sigma\cdot\gamma(|x - y|)$$

Indeed, in this case $\tilde{b}[L(D)]$ will become $b(h)$ and conversely.

With

$$h(k) = \frac{\sigma(k) - D_o(k)\ \Gamma(k)}{1 + D_o(k)\ \Gamma(k)}$$

and

$$c(k) = -D_o(k) + \gamma(k)$$

$$= \frac{\sigma(k) - D_o(k)\ \Gamma(k)}{1 + \sigma(k)}$$

we find in deleting for simplicity the arguments, k, of the functions

$$-L(k) + \sigma(k)\,\gamma(k) = -\frac{\Gamma\, D_o\, \Gamma}{1 + D_o\, \Gamma} + \frac{\sigma^2}{1+\sigma}$$

$$= -D_o + \frac{D_o(1 + D_o\,\Gamma)\,\Gamma - \Gamma\, D_o\, \Gamma^2 + \sigma^2(1 + D_o\,\Gamma)}{(1 + D_o\,\Gamma)(1+\sigma)}$$

$$= -D_o + \frac{\sigma^2 + D_o^2\,\Gamma^2 + D_o\,\Gamma(1 - \Gamma^2 + \sigma^2)}{(1 + D_o\,\Gamma)(1+\sigma)}$$

$$= -D_o + \frac{\sigma^2 + D_o^2\,\Gamma^2 - 2D_o\,\sigma\,\Gamma}{(1 + D_o\,\Gamma)(1+\sigma)}$$

$$= -D_o(k) + h(k)\,c(k)$$

as anticipated. The extensive numerical results which exist for $\gamma(k) = c(k) + k_D^2/k^2$ enable a numerical determination of $\sigma(k)$ through Eqs. (3.41) and (3.41'). It is the first time, we believe, that for a given model of matter, so much information concerning the exact Dyson equation will be available.

E. The Weak Coupling Limit

This subject is as important as it is interesting. However, it does not belong to the main theme of this Institute. Therefore, I will restrict myself to making a few comments on some unpublished work done on the OCP and on the symmetric TCP in collaboration with H. Kunz.

The field theoretic formulation discussed in the preceeding sections is also very well-suited to deal with subtle effects occurring in the weak coupling limit where the parameter $\lambda = \beta e^2\, k_D$ is << 1. For the OCP, for example, it permits us to identify and to sum up corrections of the form

$$\lambda^p (\ell n\,\frac{1}{\lambda}\,)^q$$

with $p > 1$ and $q \geq 1$, both integers, to the bare Debye screening length and also to the thermodynamic properties of interest.

The methods of investigation are those used in the theory of critical phenomena, more precisely in the study of the logarithmic corrections to the mean-field theory of tri-critical points

[Stephen and Abrahams, 1975]. The connection made in the preceeding section between the Dyson equation and the classical theory of ionic fluids reveals in particular that the leading logarithmic corrections are contained in the HNC approximation. Of course, the equilibrium properties of plasma are not critical in the weak coupling limit, but they are weakly singular at $\lambda = 0$ as indicated above for the OCP.

The motivation for studying these effects is to push the Debye-Hückel theory to its limit of universality, i.e., to the limit within which the effects of the ionic cores do not come into play. A very useful guideline in this search is provided by the renormalization group idea. In the case of plasmas, it tells us to telescope the space variables $x = bx'$ with $b > 1$, and to rescale the field variable $\psi(x) = b^{-3}b\psi'(x')$ so that the gaussian measure $d\mu(\psi)$ remains invariant in the transformation or, speaking in physical terms, that the electrostatic energy of a confirugation be preserved. Through this transformation, we have that, $z \to z' = b^3 z$,

$$\beta^{1/2}\, e \to (\beta^{1/2}\, e)' = b^{-1/2}\, \beta^{1/2} e \qquad ,$$

$b = b^{5/2}$ and the Wick ordered functional $P[z,\beta^{1/2}e\psi(x)]$ in Eq. (3.16) or (3.18) becomes a function of the scaling parameter b. When this functional is expanded in a power series in $:\psi':$, one finds that the term in $(:\psi':)^n$ is proportional $b^{3-n/2}$. Hence, the limit of universality is associated to a $P_{n=6}(\psi)$ model. Note finally that this transformation leaves invariant the graininess parameter

$$g = \varepsilon_o^{-2}\, \beta e^2\, k_D \qquad .$$

IV. ON THE BBGKY HIERARCHY

A. Introduction

Consider a system of particles in thermal equilibrium interacting pair-wise with a potential $v(|x - y|)$ and subject to an external potential $u(x)$. Assume existence of the thermodynamic limit of the thermodynamic potentials and of the correlation functions. Let $\rho(x)$, $\rho_2(x,y)$, $\rho_3(x,y,z)$ be the one-, two-, and three-particle densities. Introduce the usual pair and triplet correlation functions $g(x,y)$ and $g_3(x,y,z)$ through $\rho_2(x,y) = \rho(x)\,\rho(y)\,g(x,y)$ and $\rho_3(x,y,z) = \rho(x)\,\rho(y)\,\rho(z)\,g(x,y)\,g(x,z)\,g(y,z)\cdot g_3(x,y,z)$. Let ρ be the particle density of the homogenous system, $\beta = (k_BT)^{-1}$, and $F(|x - y|) = -\nabla v(|x - y|)$. Throughout this lecture, it is understood that the gradient operator acts on the first argument of the first term on its right only. With these definitions, the first two equations of the hierarchy are

$$\nabla\rho(x) = -\beta\nabla u(x)\ \rho(x) + \int \rho(x)\ \beta F(|x - z|)\ g(x,z)\ \rho(z)\ dz \tag{4.1}$$

$$\nabla g(x,y) = \beta F(|x - y|)\ g(x,y) + \int \beta F(|x - z|)\ \rho(z)\ g(x,z)\ g(x,y) \cdot [g(z,y)\ g_3(x,y,z) - 1]\ dz \tag{4.2}$$

Furthermore, define one-, two- and three-particle net correlation functions $f(x)$, $h(x,y)$ and $h_3(x,y,z)$ through $\rho(x) = \rho[1 + f(x)]$, $g(x,y) = 1 + h(x,y)$ and $g_3(x,y,z) = 1 + h_3(c,y,z)$.

Approximate schemes such as those presented during this Institute are generally used to investigate solutions of Eqs. (4.1) and (4.2). In the case of homogenous systems for which Eq. (4.1) is identically satisfied, the procedure has been to make some Ansatz for $g_3(x,y,z)$ expressed in terms of the net pair correlation function. This yields a closed nonlinear equation for $h(|x - y|)$.

For inhomogenous systems, the procedure has been to prescribe a homogenous equilibrium pair correlation function and solve for $\rho(x)$. One difficulty of this approach is to appreciate in a systematic way what is left over from the approximate schemes.

Alternative approaches have been developed in which integral equations (generally nonlinear) have been derived by resumming density expansions for the correlation functions. In fact, as early as 1947 Mayer [1947] gave a procedure from which some of the presently known equations can be derived. Moreover, many of these equations can be shown to be derived from a variation principle and this was done initially by Hiroke and DeDominicis [1962]. The method involves expressing the grand partition function of a system with one- and two-body forces as a stationary functional of, and only of, the one- and two-particle correlation functions. The equations contain functional derivatives with respect to $\rho(x)$ and $h(x,y)$ of a functional $S[\{\rho(h)\},\{h(x,y)\}]$ given as an infinite series of terms associated with graphs of ring and bridge type [Hiroke and DeDominicis, 1962]. It is tacitly assumed that $S(\rho,h)$ converges in a certain domain of the thermodynamic variables. In practice of course, $S(\rho,h)$ is truncated somewhere, e.g., neglecting bridge contributions as in the HNC scheme, but there is in principle room for systematic work.

To my knowledge, explicit connections between the various approaches are scarce and, in this lecture, it is my purpose to pursue this problem. The principle idea is simply to use the formally exact implicit equations for $\rho(x)$ and $h(x,y)$ to generate the right-hand members of the BBGKY equations in calculating $\nabla\rho(x)$, $\nabla h(x,y)$, ... This will permit us to systematically construct $g_3(x,y,z)$ in terms of $h(x,y)$ and to identify some of the approximate schemes which have been employed. A further application of this idea is to write down an equation for $\nabla c(x,y)$, called the "grad-c equation", where $c(x,y)$ is the direct correlation function defined by,

$$c(x,y) = h(x,y) - \int h(x,z)\ \rho(z)\ c(z,y)\ dz \tag{4.3}$$

This will give rise to a unified description of the various schemes (Singwi, Tosi, Land and Sjolander [1968], Totsuji and Ichimaru [1974], HNC) presented by some of the participants at this Institute. Lastly, some important questions pertaining to the onset of long-range ordering (l.r.o.) in neutral and Coulomb systems will be formulated in discussing the first BBGKY equation.

The first application, to $\nabla\rho(x)$, is nearly trivial. One applies the gradient operator to $\rho[x,\{h(x,y),\rho(y)\}]$ and then recovers the first BBGKY equation after noting that there is exact cancellation of terms from $\delta S/\delta\rho(x)$ and $\delta S/\delta h(x,y)$. The problem of the linearization of Eq. (4.1) for small inhomogeneities $f(x)$ is more intricate and implies experience gained with the other applications to $\nabla h(x,y)$ and $\nabla c(x,y)$, which are treated below.

Before taking up this subject, however, we need some definitions and fundamental relations. To begin with, consider the inverse of Eq. (4.3). It defines the well-known Ornstein-Zernicke (OZ) relation

$$h(x,y) = c(x,y) + \int c(x,z)\ \rho(z)\ h(z,y)\ dz \tag{4.4}$$

Define also

$$t(x,y) = \int c(x,z)\ \rho(z)\ h(z,y)\ dz = h(x,y) - c(x,y) \tag{4.5}$$

Consider, next, the ring and bridge parts of $S(\rho,h)$: The ring past is defined by the logarithmic series of convolutions

$$(-1)^n\ n^{-1} \int \rho(x_1)\ h(x_1,x_2)\ \rho(x_2)\ h(x_2)\ \ldots\ \rho(x_n)\ h(x_n,x_1)$$

$$\cdot\ dx_1\ dx_2\ \ldots\ dx_n$$

with $n \geq 3$, each integrand being represented by a polygon of n vertices equipped with factors $\rho(x_i)$ and linked by n bonds $h(x_i,x_{i+1})$. As to the bridge part of $S(\rho,h)$, it is given by the sum of contributions associated with irreducible bridge graphs. Recall that an irreducible bridge graph is a graph with a number of vertices ≥ 4 (also equipped with $\rho(x_i)$) and multiply connected by $h(x_i,x_j)$ bonds in such a way that from each vertex emerge at least three bonds; no pair (x_i,x_j) of vertices is linked directly by more than one $h(x_i,x_j)$; no graph can be split into two pieces by cutting two bonds.

In order to obtain an expansion for the right-hand-side of $\ell n[1 + h(x,y)]$, delete in all possible ways a bond $h(x,y)$ together with the vertex factors $\rho(x)$ and $\rho(y)$ to all the graphs of $S(\rho,h)$. For the ring part of $S(\rho,h)$ this operation produces a geometric series, with alternating signs, which when summed gives $t(x,y)$. For the bridge part, no such closed form exists and we simply define by $b(x,y)$ the result of the above operation. The resulting implicit equations for $h(x,y)$ and $c(x,y)$ are

$$h(x,y) = e^{-\beta v(|x - y|) + t(x,y) + b(x,y)} - 1 \tag{4.6}$$

and

$$c(x,y) = e^{-\beta v(|x - y|) + t(x,y) + b(x,y)} - t(x,y) - 1 \tag{4.7}$$

For homogenous systems, and given some approximate $S(\rho,h)$, the OZ relation, Eq. (4.4), and Eq. (4.7), constitute the basic pair of equations of the model considered.

Assuming the applicability of Eqs. (4.4) and (4.6) to the OCP model, DeWitt [this paper] has systematically exploited Hansen's [this paper] numerical data to produce $|x|$ and $|k|$ dependent plots of h and c for Γ values up to 160. Moreover, knowing c and h, thus t, DeWitt was in a position to isolate b from Eq. (4.6). That is, he is able to study the global effect of the bridge terms. It is the first time, I believe, that for a given model of matter treated by computer simulation methods, both qualitative and quantitative information concerning $b(|x|)$ has become available. It is observed that $b(|x|)$ is smaller than $t(|x|)$ by more than one order of magnitude for all $|x|$ and Γ values for which reliable data are available. This supports the evidence gained otherwise that the HNC scheme provides a remarkably good description of the OCP in its homogenous phase.

B. The Second BBGKY Equation

We start from Eq. (4.6) and calculate

$$\nabla h(x,y) = [\beta F(|x - y|) + \nabla t(x,y) + \nabla b(x,y)]\ g(x,y) \tag{4.8}$$

for homogenous systems. With $t(x,y)$ defined by Eq. (4.5), we have

$$\nabla t(x,y) = \int \nabla h(x,z)\ c(z,y)\ \rho dz$$

Next, to calculate $\nabla b(x,y)$ which is a functional of $h(y,z)$ only we introduce the functional derivative

$$\Lambda(x|z,y) \equiv \frac{\delta b(x,y)}{\delta h(x,z)} \tag{4.9}$$

which says that one h-bond has been deleted from the vertex x in all the bridge graphs in all possible ways. Hence,

$$\nabla b(x,y) = \int \nabla h(x,z)\ \Lambda(x|z,y)\ dz \tag{4.10}$$

and Eq. (4.8) becomes

$$\nabla h(x,y) = \beta F(|x - y|)\ g(x,y) + \int \nabla h(x,z)[\rho c(z,y) + \Lambda(x|z,y)]\ g(x,y)\ dz \tag{4.11}$$

This equation is exact and constitutes the starting point of the program outlined above. First let us make contact with the second BBGKY equation and for this purpose we rewrite Eq. (4.11) as

$$\int \nabla h(x,z)[\delta(z - y) - X(x\ z,y)]\ dz = \beta F(|x - y|)\ g(|x - y|) \tag{4.12}$$

where

$$X(x|z,y) = \rho c(z,y) + \Lambda(x|z,y)[1 + h(x,y)] \tag{4.13}$$

The kernel on the left-hand-side of Eq. (4.12) with $X(x|z,y)$ given by Eq. (4.13) is interesting and deserves more attention than we shall give here. Let us nevertheless remark that in the limit $x \to \infty$ with y and z fixed, then since $g(x,y) \to 1$ and $\Lambda(x\ yz) \to 0$, $\delta(z - y) - X(x|z,y) \to \delta(z - y) - \rho c(z,y)$ which is the denominator of h. Excluding singular behavior of $h(x,y)$ and of $\nabla h(x,y)$, using the OZ relation and the identity

$$\int [\delta(x - z) - \rho c(x,z)][\delta(z - y) + \rho h(z,y)]\ dz = \delta(x - y) \tag{4.14}$$

we have also,

$$\delta(z - y) - X(x|z,y) = \int dt[\delta(z - t) - \rho c(z,t)][\delta(t - y) - Y(x|t,y)] \tag{4.15}$$

where*

$$Y(x|z,y) = h(z,y)\ h(x,y) + \Lambda(x|z,y)\ g(x,y) + \int h(z,t)\ \Lambda(x|t,y)\ g(x,y)\ \rho dt \qquad (4.16)$$

Assuming the inverse kernel of Eq. (4.12) to exist, using Eq. (4.15), defining the inverse kernel

$$K(x|z,y) = \delta(z - y) + \int Y(x|zt)\ K(x|t,y)\ dt \qquad (4.17)$$

and using the OZ identity, we find

$$\nabla h(x,y) = \int \rho dz\ \beta F(|x - z|)\ g(x,z) \cdot \int dt\ K(x|z,t)[\delta(t - y) + \rho h(t,y)] \qquad (4.18)$$

This is the second BBGKY equation obtained from the implicit h(x,y) given in Eq. (4.6). Since Eq. (4.18) is valid for any $\beta F(|x - y|)\ g(x,y)$ in the class covered by our assumptions, identification with the right-hand-side of Eq. (4.2) can be made term by term for that class, and we find for the three particle density

$$\rho_3(x,y,z) = \rho^3\Big\{1 + h(x,z) + h(x,y) + h(x,z)\ h(x,y) + [1 + h(x,z)]\Big[h(z,y) + \int \rho dt\ du\ Y(x|z,t)\ K(x|t,u) \cdot [\delta(u - y) + h(u,y)]\Big]\Big\} \qquad (4.19)$$

This equation is useful in identifying approximations and to check their consistency and also to develop new approximations [Raveche and Mauntain, in press]. By playing a detective game we observe that: $K(x|z,y) = 0$ or $h(z,y)$ alone in the square bracket produces STLS; $K(x|t,u) = \delta(t - y)$, the first term in Eq. (4.17), $\Lambda(x|t,y) = 0$ or $h(zy)\ h(xy)$ alone in $Y(x|z,y)$ <u>and</u> neglect of

$$\int h(z,t)\ h(x,t)\ h(y,t)\ \rho dt$$

produces the Kirkwood superposition approximation (KSA); $K(x|t,u) = \delta(t - y)$, $\Lambda(x|z,y) = 0$ in $Y(x|z,y)$ <u>and</u> neglect of $h(x,z)h(z,y)h(x,y)$

*To derive Eq. (4.15) it is convenient to write Eqs. (4.11), (4.13), (4.14) and (4.16) in a matrix form: $X(x|z,y)$ becomes the (z,y) matrix element of a matrix X(x) with a fixed x and g(x,y) becomes a diagonal matrix g(x) with respect to the second variable y, since x is fixed; also h(y,z) and c(y,z) are (y,z) matrix elements of h and c while h(x,y) is a diagonal with respect to y since x is fixed.

produces TI; finally we note that $\Lambda(x|z,y) = 0$ in Eq. (4.16) produces the HNC approximation. We also observe that $\rho_3(x,y,z)$ is symmetric in its three arguments in the TI scheme and in the SA evidently whereas it is not symmetric in (x,y,z) in the STLS and HNC schemes. However, both the SA and TI approximations are inconsistent in that they neglect one contribution which is tri-linear in h. We will see below that the first consistent approximation to Eq. (4.19) is

$$\rho_3(x,y,z) = \rho^3\Big\{1 + h(x,y) + h(y,z) + h(x,z) + h(x,y)\,h(x,z) + h(y,z)\,h(z,x) + h(z,x)\,h(x,y) + h(x,y)\,h(y,z)\,h(z,x) + \int h(x,t)\,h(y,t)\,h(z,t)\,\rho dt + O(h^4)\Big\} \quad (4.20)$$

However, before proposing new schemes, we feel that a deeper understanding of these observations is needed and this is what we wish to achieve next.

C. A Unified Description of Approximation Schemes

We start from Eq. (4.7), the implicit equation for c(x,y) and calculate

$$\nabla c(x,y) = \beta F(|x - y|)\,g(x,y) + \nabla t(x,y)[g(x,y) - 1] + \nabla b(x,y)\,g(x,y) \quad .$$

With

$$\nabla t(x,y) = \int \nabla c(x,z)\,h(z,y)\,\rho dz$$

and

$$\nabla b(x,y) = \int \nabla h(x,z)\,\Lambda(x|z,y)\,dz = \int \nabla c(x,t)[\delta(t - u) + \rho h(t,u)]\,\Lambda(x|u,y)\,\rho dt\,du$$

we obtain, in having identified the function $Y(x|z,y)$ of Eq. (4.16),

$$\nabla c(x,y) = \beta F(|x - y|)\,g(x,y) + \int \nabla c(x,z)\,Y(x|z,y)\,\rho dz \quad (4.21)$$

$$= \int \beta F(|x - z|)\,g(x,z)\,K(x|z,y)\,dz \quad (4.21')$$

This equation is exact and the three schemes considered can be understood as follows: Put $Y(x|z,y) = 0$ in Eq. (4.31) and we obtain STLS; put $\Lambda(x|z,y) = 0$ and approximate $\nabla c(x,z)$ on the right-hand-side of Eq. (4.21) by $\beta F(|x - z|)$ and we obtain TI; put $\Lambda(x|z,y) = 0$ and we obtain HNC. That this is really so for the STLS and TI schemes is best demonstrated by working out the application of Eq. (4.21) to the OCP model in the HNC approximation. For any F(x) occurring in the text define the dimensionless Fourier transformation,

$$F(k) = \rho \int e^{-ikx} F(x)\, dx$$

hence,

$$F(x) = e^{-1} \int \frac{dk}{(2\pi)^3} e^{ikx} F(k)$$

For homogenous states, Eq. (4.2) tells us that only two-body forces come into play. In the present case $\beta v(|x|) = \beta e^2 |x|^{-1}$ and

$$\beta v(k) = 4\pi\, \beta e^2\, e|k|^{-2} = k_D^2 |k|^{-2}$$

Next we Fourier transform Eq. (4.21), multiply scalarly by ik, put $Y(x|z,y) = h(z,y)\, h(x,y)$, and obtain, with $g(x,y) = 1 + h(|x - y|)$

$$-k^2 c(k) = k_D^2 + k_D^2\, e^{-1} \int \frac{dq}{(2\pi)^3} \frac{(k\cdot q)}{|q|^2} h(k - q)$$

$$- e^{-1} \int \frac{dq}{(2\pi)^3} (k\cdot q)\, h(k - q)\, c(q)\, h(q) \qquad (4.22)$$

Next define with $\lambda_D = k_D^{-1}$ a source function to c(x), namely S(x) such that

$$S(k) = -\lambda_D^2 |k|^2\, c(k) \qquad (4.23)$$

and note that $S(k) - 1$ is precisely the function, introduced in the TI scheme. Hence, Eq. (4.22) becomes

$$S(k) = 1 + e^{-1} \int \frac{dq}{(2\pi)^3} \frac{(k\cdot q)}{q^2} h(k - q)[1 + S(q)\, h(q)] \qquad (4.24)$$

and the OZ relation is

$$h(k) = c(k)[1 - c(k)]^{-1} = -S(k)[\lambda_D^2 |k|^2 + S(k)]^{-1} \qquad (4.25)$$

Equations (4.24) and (4.25) define the HNC scheme in a formulation suited to numerical integration in k-space and perhaps to theoretical analysis as well. Now, replace on the right-hand-side of Eq. (4.24) S(k) by some S*(k), i.e., write

$$S(k) = 1 + \rho^{-1} \int \frac{dq}{(2\pi)^3} \frac{(k\cdot q)}{q^2} h(k - q)[1 + S^*(q)\ h(q)] \qquad (4.24')$$

but keep the OZ relation (Eq. 4.25) unchanged. Then, the three schemes turn out to be most simply characterized by

$$\begin{aligned} \text{STLS:}\quad & S^*(q) = 0 \\ \text{TI:}\quad & S^*(q) = 1 \\ \text{HNC:}\quad & S^*(q) = S(q) \end{aligned} \qquad (4.24'')$$

In conclusion consider a more general state of affairs. The approach followed here for investigating equilibrium pair correlation functions of homogenous systems is based on: the functional $S(\rho,h)$ from which $b(x,y)$ and $\Lambda(x|z,y)$ are derived; the OZ relation; the grad-c equation:

$$\nabla c(x,y) = \int \beta F(|x - z|)\ g(x,z)\ K(x|z,y)\ dz \qquad (4.21')$$

with $K(x|z,y)$ defined in Eq. (4.17) as the inverse kernel of $\delta(z - y) - Y(x|z,y)$ where $Y(x|z,y)$ is given in Eq. (4.16). Together with Eq. (4.18), the equation for $\nabla h(x,y)$, and Eq. (4.19), the equation for $\rho_3(x,y,z)$, we have a set of relations permitting consistency checks to given order in powers of h. We ask: Does a given scheme satisfy the OZ relation? Does it produce a $\rho_3(x,y,z)$ symmetric in x,y,z? Is it consistent with an expansion of $\nabla c(x,y)$, $\nabla h(x,y)$ and of $\rho_3(x,y,z)$ in powers of h to a given order? Table II illustrates the situation.

TABLE II

	OZ	Sym.	∇c_{exp}	∇h_{exp}	$\rho_{3\ exp}$
KSA	no	yes	no	no	no
HNC	yes	no	no	no	no
STLS	yes	no	yes	no	no
TI	yes	yes	yes	no	no
Eq. (4.20)	yes	yes	yes	yes	yes

To go beyond the example given in Eq. (4.20) bridge terms must be included. This is readily shown by the graphical analysis of the $Y(x|z,y)$ expansion in Eq. (4.19). Last but not least, it would be interesting to obtain estimates for bounds on $Y(x|z,y)$ for studying the properties of the integral equation, Eq. (4.17), for $K(x|z,y)$.

D. Linearization of the First BBGKY for the OCP

We start from an equivalent version of Eq. (4.1), namely,

$$\nabla \ln \rho(x) = -\beta\nabla u(x) + \int \beta F(|x - y|)\, g(x,y)\, \rho(y)\, dy$$

and consider small deviations from homogeneity. With $\rho(x) = \rho[1 + \varepsilon f(x)]$ and g^o for the homogenous g we have, to first order in ε,

$$\nabla F(x) = \int \rho\beta F(|x - y|)\, g^o(|x - y|)\, f(y)\, dy$$

$$+ \int \rho\beta F(|x - y|)\, \gamma(x,y|z)\, F(z)\, dy\, dz \tag{4.25}$$

where

$$\gamma(x,y|z) = \left.\frac{\Delta g(x,y)}{\varepsilon\Delta F(z)}\right|_{\varepsilon=o} \tag{4.26}$$

is the total functional derivative of $g(x,y)$ with respect to $\rho(x)/\rho$ at $\rho(x) = \rho$. With

$$g(x,y) = e^{-\beta v(|x-y|) + t(x,y) + b(x,y)}$$

we have

$$\Delta g(x,y) = g(x,y)[\Delta t(x,y) + \Delta b(x,y)]$$

To calculate $\Delta t(x,y)$ it is expedient to use the matrix notation; let $\rho(x)$ denote a diagonal matrix and let the dots represent convolutions

$$t = h\cdot\rho\cdot c = h\cdot\rho\cdot h - h\cdot\rho\cdot h\cdot\rho\cdot h + \ldots$$

then, to first order in ε

$$\Delta t = c^o\cdot\Delta\rho\cdot c^o + \Delta h\cdot c^o + c^o\cdot\Delta h - c^o\cdot\Delta h\cdot c^o \tag{4.27}$$

To calculate $\Delta b(x,y)$ we need the partial functional derivatives $[\delta b(x,y)/\delta h(x,z)]_\rho$, $[\delta b(x,y)/\delta h(z,t)]_\rho$ and $[\delta b(x,y)/\delta\rho(z)]_h$. Of these, only the first one has been encountered here before, namely $\Lambda(x|z,y)$ in Eq. (4.9). An example of $[\delta b(x,y)/\delta h(z,t)]_\rho$ can be given by considering the first bridge term; in this case we find $\rho^2 h(x,z)\, h(x,t)\, h(y,z)\, h(y,t)$ and the same example for $[\delta b(x,y)/\delta\rho(z)]_h$ results in $2\int h(x,z)\, h(x,t)\, h(y,z)\, h(z,t)\, h(t,y)\, \rho\, dt$. It is obvious that a complete linearization of the first BBGKY is out of reach. Relying upon the evidence of the accuracy of the HNC scheme for the OCP model, we proceed here with this approximation. Then only

$$\Delta t(x,y)/\Delta\, \ell n\, \rho(z)\Big|_{\varepsilon=o}$$

is needed and $\gamma(\text{HNC}) = \hat{\gamma}$ is seen to satisfy the integral equation

$$\begin{aligned}\hat{\gamma}(x,y|z) = g^o(|x-y|)\Big\{ & c^o(|x-z|)\, c^o(|z-y|) \\ & \cdot \int \hat{\gamma}(x,y'|z)\, c^o(|y'-y|)\, dy' \\ & + \int c^o(|x-x'|)\, \gamma(x',y|z)\, dx' \\ & - \int c^o(|x-x'|)\, \hat{\gamma}(x',y'|z)\, c^o(|y'-y|)\, dx'\, dy'\Big\}\end{aligned} \tag{4.28}$$

Since the homogenous correlation functions are euclidean invariant, we have in general that $\gamma(x_1,x_2|x_3)$ is a function $|r_{13}| = |x_1 - x_3|$, $|r_{23}| = |x_2 - x_3|$ and of $r_{13}\cdot r_{23} = (x_1 - x_3,\ x_2 - x_3)$. Define the double Fourier transform

$$\gamma(k,p) = \rho^2 \int e^{-ikr_{13}-ipr_{23}}\, \gamma(x_1,x_2|x_3)\, dr_{13}\, dr_{23}$$

that is,

$$\gamma(r_{13},r_{23}) = e^{-2} \int \frac{dk}{(2\pi)^3}\frac{dp}{(2\pi)^3}\, e^{+ikr_{13}+ipr_{13}}$$

In particular, with

$$g^o(q) = \frac{1}{\rho(2\pi)^3}\,\delta(q) + h(q)$$

and gathering together the terms in $\delta(q)$ to the left-hand-side, we find that

$$[1 - c^o(k)][1 - c^o(p)\ \hat{\gamma}(k,p) - c^o(k)\ c^o(p)$$

$$= \frac{1}{\rho} \int \frac{dq}{(2\pi)^3}\ h(q)\quad c(k - q)\ c(p + q)$$

$$+ \Big\{1 - [1 - c(k - q)][1 - c(p + q)]\Big\}\ \hat{\gamma}(k - q,\ p + q) \tag{4.29}$$

On the other hand, if we Fourier transform Eq. (4.25), multiply it scalarly by ik we find for the OCP model that the exact linearized homogenous first BBGKY equation is

$$\left\{k^2 + k_D^2 \frac{1}{\rho} \int \frac{dq}{(2\pi)^3} \frac{(k \cdot q)}{|q|^2} [h(k - q) + \gamma(k - q,q)]\right\} F(k)$$

$$\equiv k^2\ \eta(k,\gamma)\ F(k) = 0 \tag{4.30}$$

Clearly, an external disturbance would have added an inhomogenous term to this equation. The function $\eta(k,\gamma)$ just defined is the linear response function of the system to a spontaneous spatial density or charge density inhomogeneity.

It is obviously interesting to compare $\eta(k,\gamma)$ with the function $e(k) = 1 - c(k)$ which is the response function of the system to an external probe. Note that

$$(k) = 1 + \frac{k_D^2}{|k|^2}\ S(k) \tag{4.31}$$

With this purpose in mind, let us try to make contact with the subject of the preceeding section. The first approximation to $\eta(k,\gamma)$ is to neglect γ altogether. This reproduces the scheme based on the linearized first BBGKY equation with the closure $g(x,y) = g^o(x,y)$, an approximation frequently used, From Eqs. (4.31), (4.24") and (4.24'), we find that

$$\eta(k,\gamma{=}o) = e(k)_{\text{STLS}} \quad !$$

We also find this identification to be true for neutral systems including the hard core model: the above closure is equivalent to approximating $\nabla(|x - y|)$ by $\beta F(|x - y|)\ g^o(|x - y|)$, a feature which has been suspected for some time. As a next approximation to $\eta(k,\gamma)$ let us neglect the right-hand-side of Eq. (4.29), then

$$\hat{\gamma}(k,p) \simeq \overline{\gamma}(k,p) = \frac{c^o(k)\ c^o(p)}{[1 - c^o(k)][1 - c^o(p)]} = h(k)\ h(p)$$

or $\bar{\gamma}(k - q, q) = h(k - q)\, h(q)$. Now, From Eqs. (4.31), (4.24') and (4.24") we find surprisingly enough that for the OCP,

$$\eta(k,\bar{\gamma}) = e_{TI}(k) \quad !!$$

What does this mean? It means that in applying the STLS and TI schemes consistently to the study of the linearized first BBGKY equation, the eigenvalue equations $\eta(k)\, F(k) = 0$ and $e(k)\, F(k) = 0$ become identical. From this, we could infer:

(i) the possibility of a second-order phase transition

(ii) that the search for zeros of $e(k)$ and for real $k \neq 0$ is central for predicting l.r.o. in Coulomb systems (and similarly with $1 - c(k) = 0$ for neutral systems). This is compatible with a rather generally accepted belief that we must question.

Concerning (i), it is true that second-order phase transitions are associated with $1 - c(0) = 0$ but $\eta(0,\gamma)$ need not be zero since $F(0)$ is zero down to the critical point. In this case, however, Eq. (4.30) could be used to study density profiles at phase boundaries or interfaces.

Concerning (ii), let us first point out that statistical mechanics tells us that $1 - c(k) = 0$ or $e(k) = 0$ for real k is a sufficient condition to provoke some kind of l.r.o. It is our view that for continuous systems and for l.r.o. with $k \neq 0$ the fundamental response function is $\eta(k,\gamma)$, for which there is no evidence that it coincides with $e(k)$ beyond the quadratic h-dependence of γ. Furthermore, the continuation of the first BBGKY equation in the nonlinear regime is necessary. One reason for this is that, except for possible (and in this case extremely interesting situation), coincidences of $\eta(k,\gamma) = 0 = e(k)$, the emerging small excess free energy is proportional to $f(k)[1 - c(k)\, F(-k)]$ and it will be 0 which precludes small non-zero $F(k)$ to be present in thermodynamic equilibrium. Higher-order free-energy terms will come into play which imply knowledge of $\Delta c(x,y)/\Delta\rho(z)$! It is this approach which, we believe, will shed new light on the theory of inhomogenous states and of first-order transitions.

REFERENCES

Brydges, D. C., preprint.

Choquard, Ph., 1967, Frontiers in Physics, Chapter 3, W. A. Benjamin, New York.

Choquard, Ph., 1972, Helvetica Phys. Acta 45, 913.

Choquard, Ph. and R. R. Sari, 1972, Phys. Lett. 40A.

DeWitt, H. E., this Volume.

Fetter, A. L. and J. D. Walecka, 1971, Quantum Theory of Many-Particle Systems, McGraw-Hill, New York.

Frölich, J. and Y. N. Park, preprint.

Golden, K. I., this Volume.

Hansen, J. P., this Volume.

Kalman, G., this Volume.

Lieb, E., 1976, Rev. Mod. Phys. 48, 4.

Mayer, J., 1947, J. Chem. Phys. 15, 187.

Mermin, N. D., 1968, Phys. Rev. 171.

Raveche, H. J. and R. D. Mauntain, article on "Triplet correlations" to appear in Progress in Liquid State Physics, edited by C. A. Croxton, John Wiley & Sons, New York.

Sari, R. R., D. Merlini and R. Calinon, 1976, J. Phys. A9.

Singwi, K. S., M. P. Tosi, R. H. Land and A. Sjölander, 1968, Phys. Rev. 176, 589.

Stell, G., 1962, J. Math. Phys. 3, 983-1002; see also the article by G. Stell in The Equilibrium Theory of Classical Fluids, edited by H. C. Fritsch and J. L. Lebowitz, W. A. Benjamin, New York, 1964.

Stephen, M. J. and E. Abrahams, 1975, Phys. Rev. B12, 256.

Totsuji, H. and S. Ichimaru, 1974, Progr. Theor. Phys. 52, 42.

STRONGLY CORRELATED PLASMAS AND ASTROPHYSICS

E. Schatzman

Observatoire de Paris
Section D'Astrophysique
92190, Meudon, France

TABLE OF CONTENTS

STRONGLY CORRELATED PLASMAS AND ASTROPHYSICS

E. Schatzman

Observatoire de Paris
Section D'Astrophysique
92190, Meudon, France

I. OBJECTS OF INTEREST

We shall first consider the physical conditions prevailing in various astrophysical objects, in order to show for which of these objects the physics of strongly correlated plasmas is relevant. To that effect, we plot in a plane log T, log N_A the regions corresponding to several astrophysical objects. Due to the fact that the relevant quantity

$$\Gamma = \frac{Z^2 e^2}{a_s k_B T} \tag{1.1}$$

depends on the chemical composition, it is necessary to consider the (log N_A, log T) diagram for various chemical compositions. One more complication is due to the fact that at high densities quantum effects in the solid become important, so much that quantum fusion can take place inside a star. Finally, a quick look at the (log T, log N_A) diagram shows that beta-capture can take place for some elements, with the consequence that a transition in chemical composition can take place.

Let us first consider the (lot T, log N_A) diagram for ^{12}C and for ^{16}O (Figure 1a,b). We plot first the transition line between the solid and fluid phase, according to the paper of Pollock and Hansen [1973].

The classical melting takes place along a line given by the relation with the Lindemann parameter γ:

$$\gamma^2 = \frac{r_s}{6} \left(\frac{3}{\Pi} \right)^{2/3} (kT)_{Ry} \cdot 12.998 \tag{1.2}$$

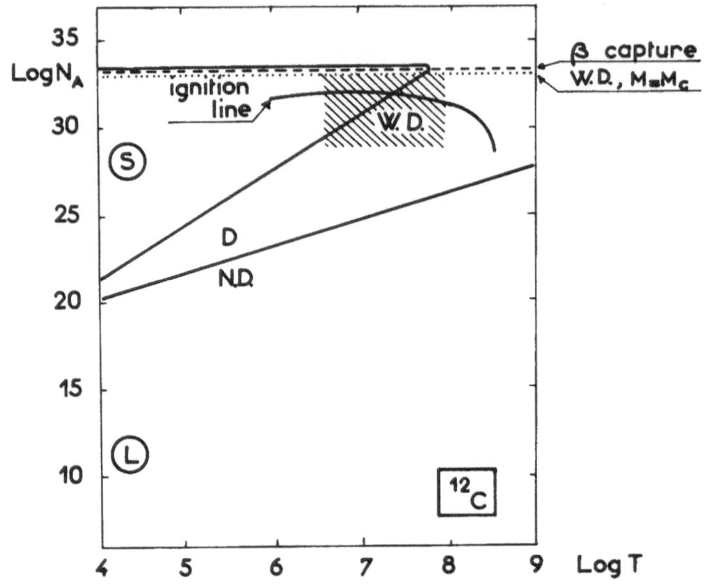

Figure 1(a)

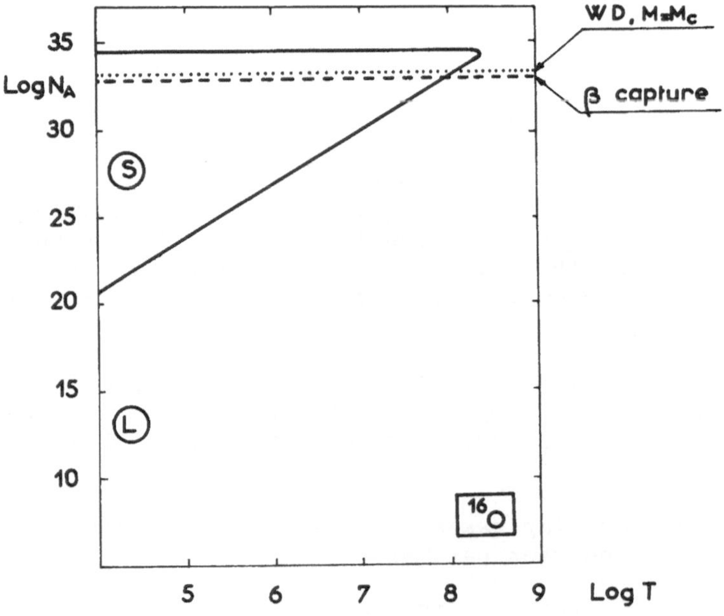

Figure 1(b)

and the quantum fusion takes place along a line

$$\gamma^2 = \frac{1}{2(3r_s)^{1/2}} \left(\frac{3}{\Pi}\right)^{2/3} \cdot 2.8 \tag{1.3}$$

where $r_s = (r/a)$, and

$$a = \frac{h^2}{\mu Z^2 e^2} \tag{1.4}$$

is the ionic Bohr radius and

$$Ry = \frac{M Z^4 e^4}{2h^2} \tag{1.5}$$

is the ionic Rydberg. M is AH, where A is the atomic number of the ions.

Pollock and Hansen give for the Lindemann parameter γ

$$\gamma = 0.17 \tag{1.6}$$

With this value, the quantum fusion takes place for ^{12}C for $\log \rho = 34.191$, corresponding to $\log Ne = 34.191$. As we shall see later, the equilibrium of ^{12}Be with the fully degenerate electron gas takes place for $\log Ne = 34.020$. As the value of γ is not so well known, it is not clear whether quantum fusion takes place for ^{12}C or for ^{12}Be.

For ^{16}O, the quantum fusion takes place along the line $\log \rho = 34.529$, whereas the beta capture takes place for $\log \rho = 32.856$, well below the quantum fusion.

The classical fusion takes place for $\rho = (T^3/Z^6)\ 10^{14.19}$. For ^{12}C, this gives $\rho = 10^{9.52}\ T^3$, and for ^{16}O, this gives $\rho = 10^{8.78}\ T^3$.

We plot on the same diagram, the approximate area for the center of white dwarfs [Schatzman, 1958], and the degeneracy condition, for $(A/Z) = 2$,

$$\frac{\varepsilon_F}{kT} = \frac{2.97 \cdot 10^5}{T} (NZH)^{2/3} >> 1 \tag{1.7}$$

or

$$N >> 10^{14.2}\ T^{3/2} \tag{1.8}$$

As we are especially interested by strongly correlated plasmas, we are likely to have to consider very dense white dwarfs. If we include the general relativity effects [Chandrasekhar and Tooper,

1964], we can give an approximate expression for the mass-radius relation, close to the limit [Canal, Schatzman, 1976].

$$M = M_{lim}\left(1 - \frac{5.04}{y_o^2} - 8.3 \cdot 10^{-4}\, y_o\right) \tag{1.9}$$

$$R = 2.68\, y_o^{-1}\, 10^9 \text{ cm} \tag{1.10}$$

In fact, this does not give the exact value of the radius. Canal and Schatzman [1976] give

$M_c = 1.366\, M_\odot$ $\qquad$ $R_c = 996$ km

$\rho_c = 2.495 \cdot 10^{10}$ g cm^{-3} $\qquad$ $\mu_e = 2$

We include in Figure 2 the (R,M) relation for various chemical compositions in ^{12}C and ^{16}O. The configuration becomes unstable with respect to inverse beta-decay as soon as the concentration of $X(^{16}O)$ (in weight) exceeds about 0.06. However, it can be shown that the unstable configurations are not dynamically unstable, due to the slow rate of electron capture, but are secularly unstable with a time scale which is nevertheless very short, of the order of a second.

On the same (log T, log N_A) diagram, we plot an approximate ignition line [Mittler, 1977] where the pycnonuclear energy generation overtakes the neutrino losses (Figure 1a).

It is interesting to give an estimate of Γ for iron in the conditions of central pressure and temperature of the Sun. We obtain for $\rho = 120$ g cm^{-3}, $T = 13 \cdot 10^6$°K, Γ (iron) $\simeq 41$. This is probably not sufficient to produce grains of solid iron inside the Sun (Figure 3). However, the clustering of iron atoms may have a strong influence on the absorption coefficient of solar matter and then on the internal structure of the Sun.

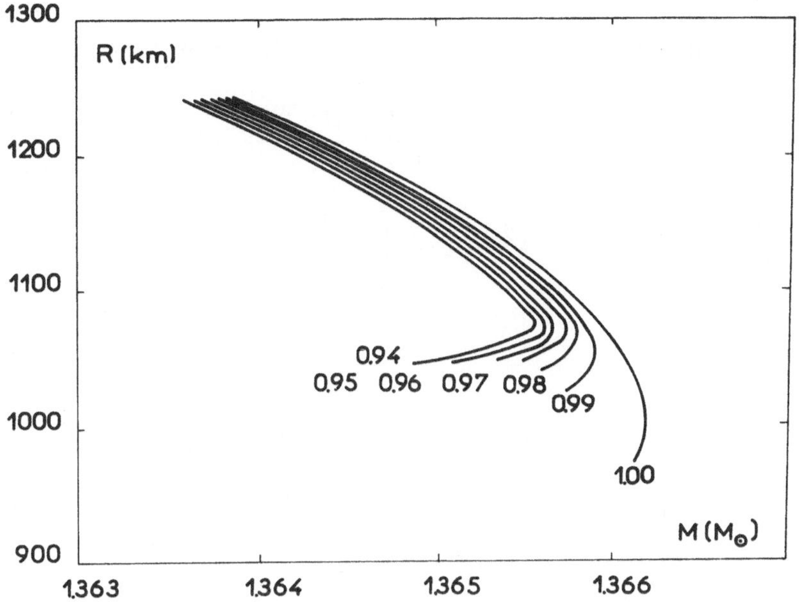

Figure 2. Mass-radius relation for Oxygen-Carbon white dwarfs of various chemical composition. The numbers on the curves give the Carbon concentration in the mixture.

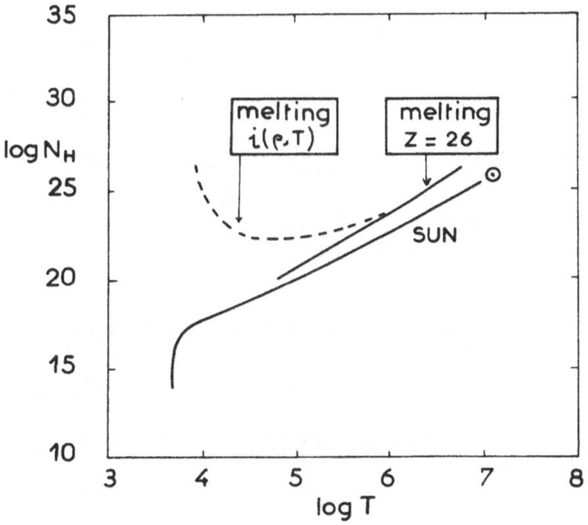

Figure 3

II. NUCLEAR REACTIONS

Close to the critical mass, we expect large values of Γ.

Let us first consider the problem of nuclear reactions at high density. From the paper of Jancovici [1977], we derive that the major correction to the rate of nuclear reaction is given by

$$I = \exp\left|C - \frac{1}{h}\int_0^{\beta h}\left|\frac{1}{4} M \dot{r}^2 + v(r)\right| dt\right|$$

$$= \exp\left|C - \frac{S}{h}\right| \tag{2.1}$$

where $C = 1.0531\ \Gamma + 2.2931\ \Gamma^{1/4} - 0.5551\ h\Gamma - 2.35$ is given in Table I.

TABLE I

Γ	C/r	C
1	0.9962	0.9962
2	1.0648	2.1296
5	1.1047	5.5235
10	1.1084	11.084
20	1.10167	22.0334
50	1.08815	54.4075
100	1.07862	107.862

S is the action integral of a particle of mass (M/2) following a classical trajectory from R back to R in a time βh (R is the nuclear radius) in the potential with a reverse sign.

With $x = (r/a)$, a defined by $(4\Pi/3)\ a^3\ N_A = 1$, we write the potential of the average force

$$v(r) = \Gamma\ k\ T\ \{(1/x) + f(x)\} \tag{2.2}$$

and we know that when x is small, $f(x) = -(C/\Gamma) + (x^2/4)$ and that when $x \gtrsim 0.3$, $f(x) = -B + a_1 x - a_3 x^3$, with $B \simeq 1.25$ and $a_1 = 0.39$, $a_3 = 0.004$. As it is easily found, the case of large Γ with values of x such that $(dv/dr) \simeq 0$, is the difficult and interesting case.

It is then sufficient to consider only the $a_1 x$ term and to forget about the term $(x^2/4)$ for x small. Then a particle is moving along a straight line from 0 to x_o, and we write

$$k^2 = a_1 \, x_o^{\;2} \tag{2.3}$$

Defining two functions J(k) and Q(k),

$$J(k) = \int_o^{\Pi/2} \Delta^{-1} \sin^2\psi \, d\psi \tag{2.4}$$

$$Q(k) = \int_o^{\Pi/2} \Delta^{-1} \left[1 - \frac{1}{2}(1 + k^2) \sin^2\psi + k^2 \sin^4\psi\right] d\psi \tag{2.5}$$

$$\Delta = (1 - k^2 \sin^2\psi)^{1/2} \tag{2.6}$$

we find the relations defining k and the action integral S,

$$\frac{\beta\hbar}{a} (\Gamma \; kT/M)^{1/2} = 2a_1^{\;-3/4} \; k^{3/2} \; J(k) \tag{2.7}$$

$$(S/\hbar) = \Gamma \; a_1^{\;1/2} \frac{1}{k} \frac{Q(k)}{J(k)} \tag{2.8}$$

classically,

$$J(K) = (1/k^2) \; [K(k) - E(k)] \tag{2.9}$$

where K(k) is the complete elliptic integral of the first kind, E(k) is the complete elliptic integral of the second kind.

Similarly, Q(k) can be expressed as a function of K and E:

$$Q(k) = \frac{1}{6k^2} \left| (1 + 5k^2) \; K - (1 + k^2) \; E \right| \tag{2.10}$$

Then

$$\frac{S}{\hbar} = \Gamma \; a_1^{\;1/2} \frac{1}{6k} \frac{(1 + 5k^2) \; K - (1 + k^2) \; E}{K - E}$$

$$= \left(\frac{4Ry}{kT}\right)^{1/3} \left(\frac{K - E}{k^2}\right)^{2/3} \frac{1}{6} \frac{(1 + 5k^2) \; K - (1 + k^2) \; E}{K - E} \tag{2.11}$$

where $Ry = (Z^4 e^4 M/\hbar^2)$ is the ionic Rydberg.

When k goes to zero, we find

$$\frac{S(o)}{\hbar} = \frac{3}{2}\left|\frac{\Pi}{2}\frac{Z^2 e^2}{\hbar}\left(\frac{M}{kT}\right)^{1/2}\right|^{2/3} \quad (2.12)$$

The correction to the increased rate factor is given by

$$\exp\left|-\left(\frac{S}{\hbar}-\frac{S(o)}{\hbar}\right)\right| . \quad (2.13)$$

When k goes to 1, the correction to the increased rate factor is

$$\exp\left|-\left(\frac{4Ry}{kT}\right)^{1/3}\left\{\left(\frac{1}{2}\log\frac{16}{1-k^2}-1\right)^{2/3}-\frac{3}{2}\left(\frac{\Pi}{4}\right)^{2/3}\right\}\right| \quad (2.14)$$

Table II gives Γ, C and the corrected term $C - \delta\left(\frac{S}{\hbar}\right)$ for a few values of the density.

TABLE II

Γ, C, $C - \delta(S/\hbar)$

	$T/10^6$			
log ρ	30	50	62.5	100
10.3			155.4 166. 121.	97.1 104.7 85.5
9.3	150.2 160.4 136.	90.1 97.31 85.7	72.1 78.1 69.	45.1 49.1 45.5
8.3	69.7 75.5 63.8	52.7 57.3	33.5 36.7	20.9 23.

The importance of the exact value of the rate of nuclear reactions at high densities will be emphasized when considering the problem of the evolution of white dwarfs.

The rate of nuclear reactions in the solid phase has been considered by Salpeter and Van Horn. At this point, we must notice that we probably should expect a discontinuity in the rate of nuclear reactions at the transition from the liquid phase to the solid phase. The result of Salpeter and Van Horn includes the effect of the temperature on the rate of reactions, but ignores the influence of the temperature and quantum effects on the lattice.

Mittler [1977] has attempted to fit the increased rate factor in the fluid and the solid phase. It should be noticed that the increased rate factors given by $\exp[C - (\delta S/\hbar)]$ are comparable to those given by Mittler, whereas the factor $\exp C$ are appreciably larger.

Salpeter and Van Horn [1969] give at $T = 0$ for the pycnonuclear reaction rate,

$$P = \frac{\rho}{A} A^2 Z^4 S \begin{pmatrix} 3.90 \\ 4.76 \end{pmatrix} 10^{46} \lambda^{7/4} \exp\left| -\lambda^{1/2} \begin{pmatrix} 2.638 \\ 2.516 \end{pmatrix} \right| \tag{2.15}$$

which corresponds to a time scale

$$t = \begin{pmatrix} 3.90 \\ 4.76 \end{pmatrix}^{-1} 10^{-46} \lambda^{-7/4} \exp\left| \lambda^{1/2} \begin{pmatrix} 2.638 \\ 2.516 \end{pmatrix} \right| \frac{1}{A\ H\ Z^4\ S} \tag{2.16}$$

The time scale of heating is an important quantity, due to the small value of the specific heat. As an order of magnitude, Canal and Schatzman [1976] give

$$t' \simeq 2 \cdot 10^{-4}\ t \tag{2.17}$$

The other parameters have the following meaning; λ is defined by the relation:

$$\lambda = \frac{1}{A} \left\{ \frac{1}{Z^7 \mu_e} \frac{\rho}{1.36 \cdot 10^{11}\ \mathrm{g\ cm^{-3}}} \right\}^{1/3} \tag{2.18}$$

S, the cross section, is given in Mev Barns. Due to the fact that the reactions take place at low energies, it is necessary to extrapolate S to S(o). Fowler and Zimmermann [1975] give

$$S(o) = 8.83 \cdot 10^{16}\ \text{Mev Barns}$$

whereas Michaud [1972] considers an uncertainty by a factor of 10.

With the value of Fowler and Zimmermann, we obtain at $\rho = 2.328 \cdot 10^{10}\ \mathrm{g\ cm^{-3}}$

$$t' = 1.2 \cdot 10^{-4}\ \text{years} \tag{2.19}$$

comparatively, we obtain for the rate in the fluid phase, at the same density, and $T = 10^8$°K, with the screening correction factor,

$$t' = 2 \cdot 10^{-4}\ t$$

$$t = 7.8\ 16\ 10^{-33} \frac{1}{\rho} \frac{Z^2 A^2}{H} \frac{1}{S(o)} \tau^{-2} e^{\tau} e^{-[C - \delta(S/h)]} \tag{2.20}$$

$$\tau^3 = \frac{27\Pi^2}{4} \frac{Ry}{kT} \tag{2.21}$$

with $\tau = 181.34$, $C - \delta(S/h) = 85.5$, we obtain

$$t = 5.40 \cdot 10^6 \text{ sec}$$

$$t' \simeq 3 \cdot 10^5 \text{ years}$$

III. STRUCTURE OF WHITE DWARFS AND EVOLUTION

Close to the critical mass, a small variation in the mass produces an appreciable change in the central density. In fact, a change by a factor 2 in the density (from $2.38 \cdot 10^{10}$ g cm^{-3} to $1.19 \cdot 10^{10}$ g cm^{-3}) corresponds to a decrease of the mass of the order $3 \cdot 10^{-3}$ $M_\odot$.

The time scale of the nuclear reactions, due to the large dependence on the screening factor, can be increased easily by a factor of the order of $10^5 - 10^8$, but the conclusion is that a pure carbon white dwarf cannot exist an appreciable time close to the critical mass, due to the rate of pycnonuclear reactions, unless the increased rate factor were appreciably smaller than presently believed. It should be noticed that a small improvement in the determination of the exponent $C - \delta(S/h)$ can be of the greatest importance, as far as the existence of dense white dwarfs is concerned.

The situation is naturally different for ^{16}O white dwarfs, as the beta capture takes place much before the critical mass, and as the $^{16}O - ^{16}O$ reactions, having a much larger potential barrier, has a much slower rate.

On the other hand, pure helium white dwarfs cannot exist beyond a certain mass, due to increased rate of the 3α reactions, leading to ^{12}C. As an order of magnitude, the density for which the rate of pycnonuclear reactions becomes dangerous for the life of the star is proportional to A^3Z^6, proportional itself to Z^9, then of the order of 10^6 g cm^{-3}. The corresponding value of the mass is $M(He)_{crit} \simeq 0.7$ $M_\odot$. This would have naturally to be determined more precisely.

The situation in fact can be very different if the white dwarf is a mixture of oxygen and carbon. It should be noticed that the melting temperature is proportional to Z^2. To a melting temperature of $50 \cdot 10^6$°K for ^{12}C corresponds a melting temperature of $88 \cdot 10^6$°K for ^{16}O. If cooling takes place very slowly it is likely that ^{16}O would become solid first, forming grains of solid oxygen in liquid carbon [Kovetz and Shaviv, 1976].

The growth of the grains is essentially ruled by a diffusion process, the diffusion of the oxygen issues in the liquid carbon. However, in the solidification regime, oxygen atoms hit the forming grain and evaporate from the forming grain. Thermodynamics give an indication of the size of the grains.

According to Landau and Lifshitz [1958], the probability of appearance of a nucleus is w, proportional to

$$w \sim \exp \left\{ - \frac{16\Pi \, \gamma^3 \, v'^2 T_o}{3q^2 \, k(\delta T)^2} \right\} \tag{3.1}$$

where v' is the molecular volume of the solid phase, T_o the melting temperature, γ the coefficient of surface tension, q the latent heat $q \simeq ckT$, where c is of the order of a few units and δT the temperature difference from equilibrium.

The coefficient of surface tension γ can be estimated from Hirschfelder [1964]. Introducing an efficiency parameter ξ, we find

$$\gamma \simeq \frac{3}{32} \left(\frac{\delta n}{n}\right)^2 \xi \, Z^2 \, e^2 \tag{3.2}$$

where $(\delta n/n)$ is the relative change of volume in the phase transition.

Nuclei appear when $(\delta T/T)$ exceeds a value given, in order of magnitude by

$$\left(\frac{\delta T}{T}\right)^2 = \left(\frac{3}{32}\right)^3 \left(\frac{\delta n}{n}\right)^6 \frac{\xi^3}{c^2} \Gamma^3 \tag{3.3}$$

For $\Gamma = 10^2$, $c \simeq 3$, $\xi \simeq 1$, $(\delta n/n)$ $\quad 3 \cdot 10^{-4}$, we find

$$(\delta T/T) \simeq 3 \cdot 10^{-9} \tag{3.4}$$

We can conclude that seeds of grains appear very early during cooling. The critical radius is then

$$R_{cr} = \frac{2\gamma v'}{q}\frac{1}{\Theta} , \qquad (3.5)$$

or

$$\frac{r_{crit}}{a} = 2\left(\frac{32}{3}\right)^{1/2} \frac{1}{\xi^{1/2}} \frac{1}{\Gamma^{1/2}} \left(\frac{n}{\delta n}\right) \qquad (3.6)$$

with the same values,

$$\frac{r_{crit}}{a} = 2 \cdot 10^3 , \quad r_{crit} \simeq 1.25 \cdot 10^{-8} \qquad (3.7)$$

$$n_{crit} \simeq 10^{10} \text{ atoms}$$

The grains can fall according to the Stokes law. The direction of the motion of the oxygen grains depend on the magnitude of the change of volume due to the phase transition. The weight per baryon is slightly less for oxygen than for carbon,

$$\frac{\rho(^{16}O) - \rho(^{12}C)}{\rho} \simeq -3.2\ 10^{-4} . \qquad (3.8)$$

It is clear that if

$$\frac{\rho(^{16}O \text{ solid}) - \rho(^{16}O \text{ liquid})}{\rho} > 3.2 \cdot 10^{-4} , \qquad (3.9)$$

the grains of solid oxygen will drop inside. If, on the contrary

$$\frac{\rho(\text{solid}) - \rho(\text{liquid})}{\rho} \lesssim 3.2 \cdot 10^{-4} , \qquad (3.10)$$

the grains of solid oxygen will move up where they will melt again. It turns out that the relative change of volume at the phase transition is of the order of a few 10^{-4}, but is not accurately known. The time scale of the motion of the grains depends on their size. For the critical readius, the time scale is of the order of 10^3 years. In fact, as soon as there are seeds, they will grow.

As a conclusion of this section, we do not know yet whether, during solidification, oxygen and carbon will remain fully mixed, will separate in grains of oxygen and carbon, will experience sedimentation of the oxygen grains, these grains falling in or floating.

IV. FORMATION OF NEUTRON STARS

By capture of mass in a binary system, a white dwarf can reach the domain of instability, and then, by inverse beta process experience a slow collapse (driven first by the time scale of electron capture) followed by neutronization, can become a neutron star.

TABLE III

Threshold Density for Electron Capture

(μ_e = 2)

Reaction	Density (10^{10} g cm^{-3})
$^{12}C(e^-,\nu)^{12}BE$	3.49
$^{12}Be(e^-,\nu)^{12}B$	2.34
$^{16}O(e^-,\nu)^{16}N$	1.92
$^{20}Ne(e^-,\nu)^{20}F$	0.62
$^{24}Mg(e^-,\nu)^{24}Ne$	0.4 (allowed)
$^{24}Na(e^-,\nu)^{24}Ne$	0.06
$^{23}Na(e^-,\nu)^{25}Na$	0.17
$^{25}Mg(e^-,\nu)^{25}Na$	0.116
$^{28}Si(e^-,\nu)^{28}Al$	0.2

However, close to the domain of instability, the rate of pycnonuclear reactions in the center of the star may become greater than the rate of increase of the central density due to the change of mass of the star by accretion.

The exact situation depends on different factors: (1) the rate of accretion. For capture with spherical symmetry, there is a well known limit to the luminosity, the Eddington limit, $L = (4\pi GcM/\sigma)$, which corresponds to a maximum rate of accretion, $\dot{m} = (LR/GM)$, where R is the radius of the accreting star. However, this limit can be greatly exceeded in the case of non-spherical symmetry. That accretion is a disc seems to be the rule, and the limit goes up like ($4\Pi/\Omega$), where Ω is the solid angle under which the disc is seen from the accreting star. Anyhow, a rate of accretion appreciably larger than 10^{-4} $M_\odot$ year^{-1} has to be reached, at least for a short interval of time, if the star is going to bypass the pycnonuclear run away; (2) the structure and chemical composition. If the star is homogeneous and the concentration of oxygen is larger than 0.06 in weight, collapse begins for a central density $\rho = 1.92 \cdot 10^{10}$. The rate of ($^{12}C$ - ^{12}C) reaction depends also if the center of the star is fluid or solid. In a fluid, the rate of reaction is certainly proportional to the product of the number density of ^{12}C. In a solid, where only the nearest neighbor can react, this has to be multiplied by the probability $|N(C)/[N(C) + N(0)]|$ that the nearest neighbor is a carbon nucleus.

In a homogeneous mixture, the rate of pycnonuclear reactions can be appreciably decreased in a star with a low concentration of carbon.

If grains of oxygen and carbon are thoroughly mixed, the situation is about as bad as in a pure carbon star, the rate of pycnonuclear reactions being the same than in pure carbon, only the rate of heating being decreased in the $|N(C)/[N(C) + N(0)]|$ ratio or about.

If oxygen has settled down, collapse can overtake the pycnonuclear regime; if oxygen has settled up, we are back to the problem of the carbon star.

Altogether, depending on the structure and chemical composition of the star, accretion on a white dwarf can lead either to a supernova outburst or to the formation of a binary star with a neutron star as one of the members of the couple. The supernova outburst can well produce a runaway neutron star, and this can well explain the existence of a sub-class of high velocity pulsars [Katz, 1975; Canal and Schatzman, 1976].

V. MAGNETIC FIELDS

A simple model of magnetic white dwarfs can be obtained in the following way. Let us assume that the magnetic pressure is everywhere proportional to the electron pressure, $P = P_e + P_B$, forgetting about anisotropies.

A standard notation is

$$P = \frac{P_e}{1 - \beta} = \frac{P_B}{\beta} \tag{5.1}$$

For the case of relativistic degeneracy, we can write

$$P_e = K\rho^{4/3}$$

$$P = K' \rho^{4/3} \quad , \quad K' = K(1 - \beta)^{-1}$$

Ignoring general relativity, it is easy to find the new Chandrasekhar limit,

$$M'_c = M_c(1 - \beta)^{-3/2} \tag{5.2}$$

In order to include the general relativity effects, we can use an elementary argument due to Levy Leblond [1969]. Let us call U the specific internal energy of the gas. The gravitational mass is

$$M_G = M_B + \frac{U}{c^2} M_B$$

where M_B is the baryonic mass. We then obtain the total energy of the order of

$$W = -G(M_B + \frac{U}{c^2} M_B) \frac{1}{R} + U M_B$$

with

$$M_B = R^3 \rho_B$$

where ρ_B is the baryonic mass. Using the usual parameter $x = (p_F/m_e c)$, U is given by

$$U = K' \rho^{4/3} + L\rho + M\rho^{2/3}$$

or

$$U = A(x^4 - \frac{4}{3} x^3 + x^2) \quad \text{with } \rho = Bx^3$$

Calculating the maximum of W, then the maximum of M_B, one finds

$$M_B \sim (3K'/G)^{3/2}$$

$$\rho_B \simeq \frac{3}{4} \frac{M_c^2}{K'^2} = \frac{3}{4} \frac{B^{2/3} c^2}{K'}$$

and then, the similarity relations

$$M_{B\,max} = M_{B_o max} (1 - \beta)^{-3/2}$$

$$\rho_{B\ max} = \rho_{B_o max} (1 - \beta)$$

$$R_{B\ max} = R_{B_o max} (1 - \beta)^{-5/6} \qquad (5.3)$$

As a check, it is possible to use the results of Hartwick and Ostriker [1968]. From their paper, it is possible to deduce a relation

$$(R/R_o) = (1 - \beta)^{-0.89}$$

The exponent has to be compared to the exponent in the relation (5.3), equal to -0.833. The difference is probably meaningless.

Let us consider the value of β for a 10% increase of the critical mass

$$(1 - \beta)^{-3/2} = 1.1$$

The corresponding central magnetic field is $B \simeq 9.7\ B_{crit}$ with $B_{crit} = (m^2\ c^3/e\ h) = 4.414 \cdot 10^{13}$ gauss.

The Fermi energy of the electrons is smaller and, with $\rho_B = 1.86 \cdot 10^{10}$ g cm^{-3} the density is smaller than the density for electron capture on oxygen. At the same density, the rate of the pycnonuclear reactions (^{12}C - ^{12}C) has been decreased by a factor 5 to 10.

Magnetic white dwarfs could then very well be the parents of neutron stars in binary systems. Non-explosive collapse after accretion would be followed by the formation of a neutron star having about the mass of the parent white dwarfs, about 1.6 $M_\odot$. This corresponds to the mass actually suggested for the neutron stars in binary X-ray sources.

REFERENCES

Canal, R., E. Schatzman, 1976, Astron. Astrophys. 46, 229.

Chandrasekhar, S., R. F. Tooper, 1964, Ap. J. 139, 1396.

Fowler, W. A., G. R. Caughlan, B. A. Zimmermann, 1975, Ann. Rev. Astron. Astrophys. 13, 69.

Hartwick, F. D. A., J. P. Ostriker, 1968, Ap. J. 153, 806.

Hirschfelder, J. O., C. F. Curtiss, and R. B. Bird, 1964, Molecular of Gases and Liquids, John Wiley & Sons, New York.

Jancovici, B., 1977, preprint.

Katz, J. I., 1975, Nature 253, 698.

Landau, L. D., and E. M. Lifshitz, 1958, Statistical Physics, Pergamon Press, New York.

Levy-Leblond, J. M., 1969, Physique et Astrophysique.

Michaud, G., 1972, Ap. J. 175, 751.

Mittler, H. E., 1977, Ap. J. 212, 513.

Pollock, E. L., and J. P. Hansen, 1973, Phys. Rev. A8, 3110.

Salpeter, E. E. and H. M. Van Horn, 1969, Ap. J. 155, 183.

Schatzman, E., 1958, White Dwarfs, North Holland Hand. d. Phys. LXI Springer.

Shaviv, G. and A. Kovetz, 1976, Astron. Astrophys. 51, 383.

THE STATUS OF LASER FUSION*

R. A. Grandey

KMS Fusion, Inc.
Ann Arbor, Michigan

*Work supported (in part) by the U. S. Energy Research and Development Administration under Contract No. ES-77-C-02-4149.

TABLE OF CONTENTS

THE STATUS OF LASER FUSION

R. A. Grandey

KMS Fusion, Inc.
Ann Arbor, Michigan

I. INTRODUCTION

Laser-induced fusion is a branch of the inertial confinement approach to thermonuclear fusion. The objective is to use a high-intensity laser beam to implode a pellet, containing fuel, to the ignition point. The possible application of lasers to thermonuclear fusion was discussed in the mid-1960's by Basov [1964], Dawson [1964] and by Daiber, Hertzberg and Wittliff [1966]. The use of lasers to compress overdense thermonuclear fuel by ablating the surface of a pellet was first reviewed in the early 1970's [Nuckolls, Wood, Thiessen, and Zimmerman, 1972; Brueckner and Jorna, 1974].

Most concepts are based on the reactions of deuterium, tritium and their products. The principal reactions are

$$D + T \rightarrow n + \alpha + 17.6 \text{ MeV}$$

$$D + D \rightarrow T + p + 4.0 \text{ Mev}$$

$$D + D \rightarrow He^3 + n + 3.3 \text{ MeV}$$

$$D + He^3 \rightarrow \alpha + p + 18.3 \text{ Mev}$$

The two D-D reactions occur with approximately equal probability. The reaction rate $\overline{\sigma V}$ cm^3/sec for the D-T reaction is plotted in Figure 1 averaged over a Mexwellian distribution of ions at a

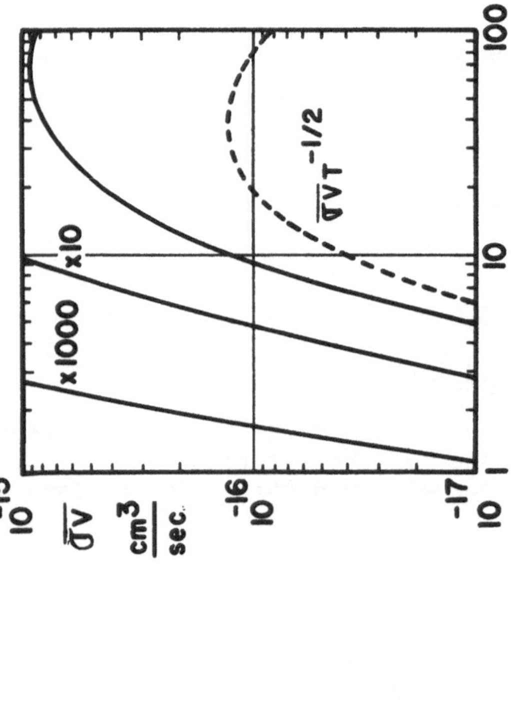

Figure 1. DT reaction rate vs. ion temperature.

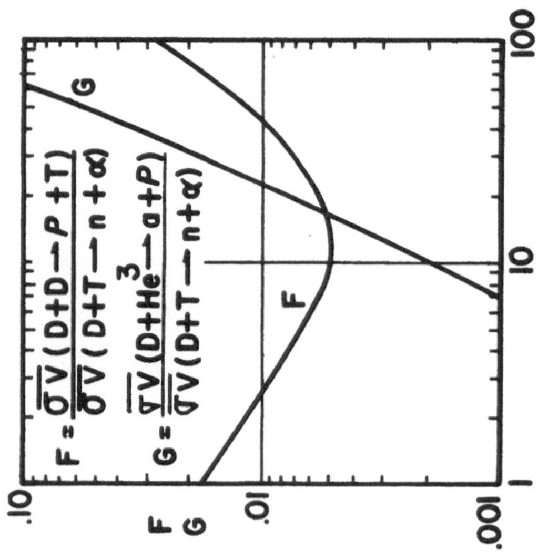

Figure 2. Reaction rate ratios vs. ion temperature.

temperature T_i. Figure 2 is a plot of the ratios $\overline{\sigma V}$[D(D,T)p]/$\overline{\sigma V}$[D(t,α)n] and $\overline{\sigma V}$[D(He^3,α)p]/$\overline{\sigma V}$[D(T,α)n] also averaged over Maxwellian ion distributions.

It is apparent from Figure 2 that the D-T reaction will dominate if the fuel contains more than a few atomic percent of tritium. If a fuel can be brought to conditions wherein the D-D reactions contribute strongly to the energy release, the product T and He^3 will react promptly with the D, yielding an overall energy release per unit burned mass the same as that for the DT reaction, i.e., 3.5 MeV/nucleon.

A fundamental difference between inertial and magnetic confinement systems is that inertial confinement allows the fuel to be compressed enough for reaction products to deposit an appreciable fraction of their energy in the fuel, providing additional heating and an increase in the reaction rate. The hydrodynamic confinement time τ for a fuel reacting at a temperature T_i is inversely proportional to the sound speed, i.e., to $T_i^{-1/2}$. The product $\overline{\delta V}$ τ for D-T has a broad maximum (Figure 1) in the range of 20 to 80 keV. Ignition occurs in an inertial confinement system if the fuel deposits nuclear energy rapidly enough to overcome conductive losses and raise the temperature to about 20 keV before hydrodynamic disassembly has started. The ignition condition depends on the initial temperature T_o, the density N_i cm^{-3} and the time τ_c during which the compression is near its peak value. Computed values for DT are given in Table 1.

Table 1

$N_i\tau_c$ for Ignition vs. Initial Temperature

T_o(keV)	$N_i\tau_c$ (10^{14} sec/cm^3)
2	39.1
3	16.0
4	8.75
5	5.57
6	3.91
7	2.92
8	2.26
9	1.80
10	1.46

Since two ions are involved in each DT reaction, an average of 8.8 f_B MeV per ion will be released in burning a fraction f_B of DT fuel. The initial energy in the fuel per ion is $3T_o + E_D + E_p$, where E_D is the electron degeneracy energy and E_p the remaining potential energy per ion. The energy gain in a configuration in which an initial fraction of the laser energy f_I is coupled uniformly into the fuel is thus

$$G = 8800\ f_I f_B/(3T_o + E_D + E_p)$$

where T_o, E_D and E_p are all measured in keV. An upper limit to f_I and f_B, from numerical simulations at KMSF, are $f_I \approx 0.3$, $f_B \approx 0.5$. IF $T_o \gtrsim 2$ keV then $3T_o \gtrsim E_D + E_p$ and $G \lesssim 440/T_o$, so that even with ignition the gain of a configuration in which the fuel is uniformly heated does not much exceed 100. Higher gain configurations arise from igniting the center of a highly compressed, degenerate fuel in which $3T_o \ll E_D$ and allowing a burn front to propagate outwards. The degeneracy energy per electron is $E_D \approx .0078\rho^{2/3}$ keV, where ρ is the D-T density. (Most simulation models do not include E_p (ion-ion correlation energy, etc.). As emphasized in this volume [DeWitt; Hansen; Rogers] E_p is a significant quantity and strong coupling effects make the fuel equation of state softer than expected on the basis of degeneracy effects alone. Simulations, which have not yet been correlated with data, suggest that $G \gtrsim 10^3$ can be obtained by igniting a spherically divergent detonation in a degenerate D-T fuel, if the fuel can be compressed sufficiently adiabatically to a density of several hundred g/cm^3.

The remainder of this paper outlines the phenomena and processes involved in using lasers to compress thermonuclear fuel; presents the basic model, including transfer coefficients, in the KMSF hydrodynamic code used to make quantitative analyses of the laser-fusion process; discusses the basic diagnostic techniques, recent data and their significance; and concludes with a discussion of recent theoretical modeling and areas in which the experimental data are not yet understood.

II. LASER-INDUCED FUSION PHENOMENOLOGY

As a preface to consideration of how to efficiently transfer incident laser radiation into internal energy of a compressed, partially degenerate fuel, the phenomena involved in the process will be sketched. When a pellet containing thermonuclear fuel is illuminated with a laser, surface breakdown and ionization begins as the laser power density exceeds approximately 10^{10} W/cm^2. Once a plasma has formed, further absorption occurs in the corona, that is, in the vicinity of and outside the critical density surface. The laser-plasma interaction processes are inverse bremsstrahlung and various collective interactions between the laser field and the plasma. Part of the absorbed energy is transferred, primarily by electron conduction, into the overdense plasma inside the critical surface. The penetration of energy is accompanied by a hydrodynamic rarefaction wave following the thermal front. Consequently much of the energy deposited by the laser is transferred into outward directed kinetic energy of the plasma. Some energy is lost from the thermal front by radiation and some is coupled, primarily hydrodynamically and secondarily by radiative and electron transport, into the remnant pellet causing it to implode and to be preheated.

The simplest pellet configuration is a homogeneous, solid sphere of thermonuclear fuel, e.g., solid D-T, lithium, hydride, a deuterated-tritiated polymer, etc. If a solid sphere is to be compressed adiabatically the driving pressure must increase rapidly with time so that acoustic signals from the driving surface arrive at the center of the pellet simultaneously. The driving pressure must increase with time as $(t - t_c)^{-n}$ where t_c is the collapse time and $n \approx 5$. Several analyses of such an implosion have appeared in the literature [Brueckner and Jorna, 1974; Kidder, 1976; Yabe and Niu, 1977]. The impossibility of obtaining such strongly shaped pulses with present lasers and the uncertainty of the practicality of this approach in the foreseeable future promptly led to consideration of non-uniform pellet configurations in which most of the mass to be imploded lies in a spherical shell. Such a configuration increases the volume on which work is to be done, thereby reducing the required driving pressure; and energy delivered into the implosion preferentially flows toward the inner free surface of the shell as a natural consequence of spherical convergence. Almost all of the experimental work has been done on variations of spherical shell pellets and the remaining discussion will be restricted to such systems.

A. Radiative Transitions

The several radiative electron-ion collisional processes will be seen to strongly influence laser-fusion pellet design. Before attempting a more than qualitative discussion of the implosion process, some radiative functions will be presented.

The most significant processes are free-free (bremsstrahlung) and free-bound (recombination radiation and radiative ionization) transitions. The coefficient k_ν^{ff} for the free-free absorption of radiation of frequency ν by a plasma with the electrons in a Maxwellian distribution at a temperature T_e, in cgs units, is

$$k_\nu^{ff} \approx 3.4 \times 10^8 g_\nu \, Z^2 N_e N_i T_e^{-1/2} \, \nu^{-3} \text{ cm}^{-1} \qquad (2.1)$$

where

$N_i(\text{cm}^{-3})$ = ion number density

$N_e(cm^{-3}) = Z\ N_i$ = electron number density

Z = average ion charge

g_ν = Gaunt factor

When the effects of stimulated emission, in the limit $h\nu << kT_e \equiv \theta_e$, and the dielectric properties of the plasma are taken into account the coefficient K_{ff} for free-free absorption of laser light of wavelength λ derives from Eq. (2.1). The result is

$$K_{ff} = 2 \times 10^3\ g_\nu\ Z\ \eta^2\ T_e^{-3/2}\ \lambda^{-2}(1 - \eta)^{-1/2}\ cm^{-1} \tag{2.2}$$

where

$\eta = N_e/N_c$

$N_c = \pi m_e c^2/e^2\lambda^2$ = critical electron density

c = light velocity in vacuum

m_e = electron charge

h = Planck's constant

k = Boltzmann's constant

$Z = \overline{Z^2}/\overline{Z}$ averaged over all ions.

In the interior of the pellet the emission coefficients are of concern. The total bremsstrahlung emission, with the assumption of local thermodynamic equilibrium (LTE), is [Cox, 1965]

$$J_{ff} \approx 1.2 \times 10^{-19}\ N_e\ N_i\ \theta_e^{1/2}\ Z^2\ \overline{g}_{ff}\ ergs/cm^3\ sec \tag{2.3}$$

where $\overline{g}_{ff}$ is the average Gaunt factor. The total recombination radiation emission, also derived under the assumption of LTE, is [Cox, 1965]

$$J_{fb} \approx 2.65 \times 10^{-30}\ N_e\ N_i\ \theta_e^{-1/2} \sum_j Z_j^4\ j^{-5}\ \Delta_j\ \overline{g}_j\ ergs/cm^3 - sec \tag{2.4}$$

where Z_j and Δ_j are the (screened) charge and number of vacancies for the jth principal atomic level.

B. Coronal and Thermal Front Processes

Some of the effects in the corona and the thermal conduction front can be estimated to obtain a semi-quantitative energy partition among the various processes. In the corona electron energy transport is so rapid that the electrons are approximately isothermal. Electron-ion collisional energy exchange is so slow that $\theta_i << \theta_e$. In this limit the hydrodynamic equations of motion and continuity are

$$m_i N_i \left(\frac{\partial}{\partial t} + U \frac{\partial}{\partial r} \right) U = -\theta_e Z \frac{\partial N_i}{\theta r}$$

$$\frac{\partial N_i}{\partial t} + \frac{1}{r^2} \frac{\partial (r^2 N_i U)}{\partial r} = 0$$

where r is the radial coordinate and U the local flow velocity. An approximate solution to these equations, strictly valid in the limit

$$\int_0^t c_T \, dt' << r \quad ,$$

but sufficiently accurate for the present estimate is

$$U = c_T \left[1 + (r - r_s) / \int_0^t c_T \, dt' \right] \tag{2.5}$$

$$N_i = N_s (r_s/r)^2 \exp \left[-(r - r_s) / \int_0^t c_T \, dt' \right] \tag{2.6}$$

$$= N_s (r_s/r)^2 \exp \left[(c_T - U)/c_T \right]$$

$$c_T = (\theta_e \, Z/m_i)^{1/2} \quad .$$

c_T is the isothermal sound speed. r_s and N_s are the radial coordinate and ion density at the sonic point. The total energy per unit area in the corona is easily seen to be

$$E_c = N_c \, \theta_e \int_0^t c_T \, dt'$$

Under conditions such that

$$\dot{\theta}_e \int_0^t c_T \, dt' << \theta_e \, c_T$$

which quickly develop for a constant rate of laser energy deposition, the rate of increase of coronal energy per unit area is

$$\dot{E}_c = 4N_c \, \theta_e \, c_T \approx 4\pi m_e c^2 (2e^4 m_p)^{-1/2} \lambda^{-2} \theta_e^{3/2}$$

$$\equiv \beta\theta_e^{3/2} \lambda^{-2} \text{ erg/cm}^2 \text{ sec} \quad .$$

The mass of the ion has been approximated by $m_i \approx 2Z\, m_p$ where m_p is the proton mass.

The total rate of energy change in the plasma is $\dot{E}_c/f$ where f is the fraction of energy going into the corona. Both experimentally and theoretically, $f \gtrsim 0.5$ and is a slowly varying function of the experimental conditions. Thus the energy balance during a constant rate of absorption of laser energy ω_L is

$$\omega_L = 10^{-7} \beta \lambda^{-2} \theta_e^{3/2}/f \text{ W/cm}^2 \quad , \quad f \approx 0.5 \quad .$$

The absorbed energy ω_L is found by integrating the flux equation $d\omega/dr = -K\omega$ from the outside of the plasma where $N_e = 0$ up to the critical point and back out. If the laser absorption process is dominated by inverse bremsstrahlung, the flux equation can be approximately integrated using Eq. (2.2) for K_{ff} and Eq. (2.6) for the dependence of N_e on radius. The fractional absorptivity is

$$\overline{KL} = 1 - \exp\left(-2 \int_{\eta=0}^{\eta=1} K_{ff} \, dr\right)$$

where

$$2 \int_{\eta=0}^{\eta=1} K_{ff} \, dr \approx \frac{8}{3} \cdot 2 \times 10^3 \, g_\nu \, Z \, T_e^{-3/2} \lambda^{-2} L$$

$$L^{-1} \approx 2/r_c + 1/\int_0^t c_T \, dt' \quad .$$

r_c is the radius of the critical surface. In the limit of small absorptivity ($\overline{KL} \lesssim 0.5$)

$$\overline{KL} \approx 5 \times 10^3 \, g_\nu \, Z \, T_e^{-3/2} \, \lambda^{-2} \, L \equiv \alpha \, Z \, \theta_e^{-3/2} \, \lambda^{-2} \, L \quad .$$

In this limit the energy balance relation for the absorption of a temporally flat laser pulse of intensity ω_I W/cm^2 is

$$\omega_I \, \alpha \, Z \, \theta_e^{-3/2} \, \lambda^{-2} \, L = 10^{-7} \, \beta \, \theta_e^{3/2} \, \lambda^{-2} \, f^{-1}$$

or

$$\omega_L \, \lambda^2 = 10^{-7} \, \beta \, \theta_e^{3/2}/f = (10^{-7} \, \omega_I \, \alpha \, \beta \, ZL/f)^{1/2} \quad . \tag{2.7}$$

Equation (2.6) shows that the isothermal sound speed and hence θ_e can be obtained from a measurement of $N_i(U)$. $N_i(U)$ and f can be measured using experimental techniques which will be discussed later. An experimental departure from $\omega_L \lambda^2 = 10^{-7} \beta \theta_e^{3/2}/f$ means that the energy balance analysis leading to Eq. (2.6), and hence the value inferred for θ_e, is invalid. A departure from $\omega_L \lambda^2 = (10^{-7} \omega_I \alpha \beta ZL/f)^{1/2}$ while $\omega_L \lambda^2 = 10^{-7} \beta \theta_e^{3/2} f$ is satisfied indicates that inverse bremsstrahlung does not dominate the laser absorption process.

The behavior of the flow inside the critical surface is more complicated, involving a simultaneous solution of the hydrodynamic and electron thermal conduction equations. The most complete analysis which can be used to obtain semi-quantitative results for the fractional energy coupling into the implosion appears to be that of Morse [Gitomer, Morse and Newberger, 1977]. The remainder of this section, however, addresses a more specialized portion of the thermal front problem -- the importance of radiative processes and their consequence on material selection for laser-induced fusion pellets.

The density and electron temperature distribution in the coronal/thermal front region are sketched in Figure 3. The ablation surface $r = R_a$ is the surface of peak pressure (maximum ρT). The surface $r = R_r$ is the surface of peak bremsstrahlung radiation (maximum $\rho T_e^{1/2}$). Inside the critical surface, ion-electron thermal equilibrium is approximately attained and the ablation pressure p_a at $r = R_a$ is given by

$$p_a = (1 + Z) \, N_i \, kT_e \Big|_{r=R_a} \quad .$$

The rate at which hydrodynamic work is being done on the imploding remnant shell, moving inward with velocity U at radius $r = R$ is

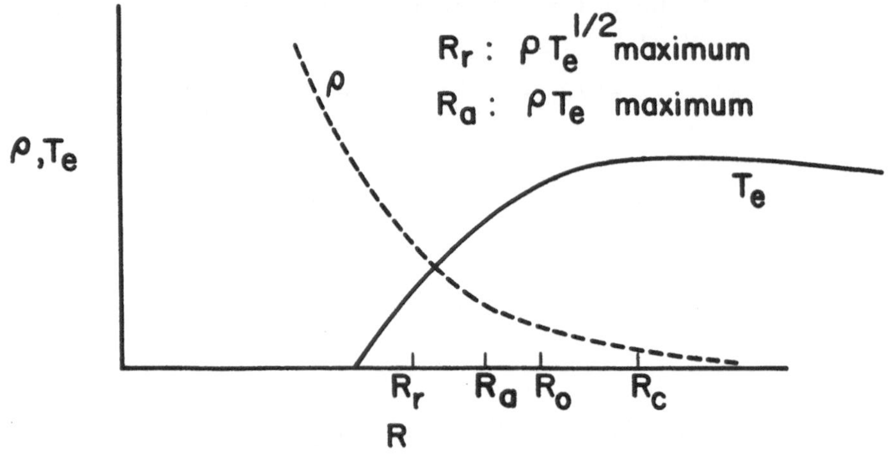

Figure 3. Mass Density and Electron Temperature in Ablation Region.

$$\dot{E}_{hydro} \approx 3p_a U/R = 3(1 + Z)(N_i\ kT_e)\Big|_{r=R_a} U/R \text{ ergs/cm}^3 \text{ sec}$$

Although the maximum bremsstrahlung radiation rate occurs at the surface of maximum $\rho T_e^{1/2}$, recombination radiation has an integrated $\rho T_e^{-1/2}$ dependence. For moderate Z ions and under conditions of laser fusion where temperatures are of order 1 keV, recombination radiation is dominated by transitions into the K-shell and hence the radiation rate will peak at the surface where K-shell ionization has just been completed, i.e., from the surface on which $T_e \approx T_k$, the temperature for 50% occupation of the last level. The total radiation loss rate is thus approximately

$$\dot{E}_r = 1.2 \times 10^{-19}\ N_e\ N_i\ (\theta_e^{1/2}\ Z^2\ \bar{g}_{ff})\Big|_{r=R_a}$$

$$+\ 2.65 \times 10^{-30}\ \Sigma(N_e\ N_Z\ \theta_e^{-1/2}\ Z^4 g)\Big|_{r=R_Z}$$

where the sum is over the various ions, each of which has associated its own surface of K-shell ionization. If θ_e is expressed in keV, the ratio of the rate of radiative energy loss to hydrodynamic work can be written

$$\dot{E}_r/E_{hydro} = \frac{10^{-15}(N_e N_i \theta_e^{1/2} Z^2)_{r=R_r} + 2.7x10^{-17}\Sigma(N_e N_i \theta_e^{-1/2} Z^4)_{r=R_Z}}{(1 + Z)(N_i \theta_e)_{r=R_a} \quad U/R} \quad . \tag{2.8}$$

In typical implosions, N_e in the thermal front is a few times $10^{22}cm^{-3}$ and the implosion velocity U a few times 10^7 cm/sec (to provide the energy required to drive the fuel to a fraction of a kilovolt); thus, very roughly

$$\dot{E}_r/\dot{E}_{hydro} \approx RZ\ \theta_e^{-1/2}(1 + .027\ Z^2/\theta_e)$$

where the atomic numbers and temperatures are appropriate averages for compounds. It is apparent that for R in the range from .01 cm (current experiments) to 0.1 cm (reactor size pellets) radiative losses become noticeable during the implosion process for Be, appreciable for C, for example, and very important when materials of higher atomic number are present in the ablative region of the pellet.

C. Reaction Region Phenomena

Consider an implosion whose result has been to bring a mass M_F gm of (DT) fuel to a density ρ_F, a radius R_F, related by

$$M_F = \frac{4}{3}\pi\ \rho_F\ R_F^3 \quad ,$$

and a temperature T_o. The compressed, hot fuel will cool by electron thermal conduction across the surface $r = R_F$ and, after a time $\tau_h = R_F/C$, where C is the sound speed, will rapidly expand and cool further. Consider the fuel to be enclosed by a tamper of atomic number Z.

The electron thermal conductivity K of a plasma of atomic number Z and electron temperature T_e is given to within 20% of the classical value by $K \approx 7.5 \times 10^{-6}\ T_e^{5/2}/(4 + Z)$ in cgs units, ignoring degeneracy effects. (A more accurate expression for K will be discussed in the following section.) In a time τ a thermal front at a temperature T_o will have moved a distance X given by

$$X^2 \approx \sigma_o T_o^{5/2}\ \tau$$

where

$$\sigma_o = \frac{2}{7}\ T_o^{-5/2}\ K/\rho C_v = 10^{10}/N_e(4 + Z) \quad .$$

The energy which has been removed from the fuel across the surface $r = R_F$ into the tamper is $4\pi R_F^2 \rho C_v T_o X$. A time τ_c can be defined as the time required for a fraction f of the energy in the fuel to be conducted into the tamper. If the temperature is measured in units of keV,

$$\tau_c \approx 4 \times 10^{-29} N_e(4 + Z) f^2 R_F^2 T_o^{-5/2} \text{ sec} \quad .$$

The hydrodynamic confinement time τ_h, with T_o in keV is

$$\tau_h \approx 2.5 \times 10^{-8} R_F T_o^{-1/2} \text{ sec} \quad .$$

The ratio $\tau_c/\tau_h \approx 400f^2(4 + Z) T_o^{-2} \rho_F R_F$. For present experiments

$$\rho_F R_F \approx 2 \times 10^{-3} \text{ gm/cm}^2 \quad \text{and} \quad T_o \approx 3 \text{ keV} \quad ;$$

accordingly, $\tau_c/\tau_h \approx .1 f^2(4 + Z)$. It can be seen from Figure 1 that a 10% change in T_o causes a factor of two change in the DT reaction rate for $T_o \approx 3$ keV. Thus f cannot be much larger than 0.1 in present experiments; the reaction time is dominated by thermal conductive cooling and hence proportional to 4 + Z. This implies that the tamper should be of as high atomic number as possible.

The previous considerations led to the conclusion that the ablative region of the pellet should be of as low atomic number as possible. A generic composite spherical pellet thus has the basic features sketched in Figure 4. A low Z ablator, thick enough to keep the thermal front from penetrating during the implosion process, overlays a high Z tamper and the fuel. The most general fuel configuration is a (cryogenic) shell filled with a vapor mixture of D-T-He^3. Some He^3 will always be present as a tritium decay product and an additional amount may be added for diagnostic or other design considerations.

Two additional advantages of using a high density, high Z tamper between the ablator and fuel are that the overall compression of the tamper is reduced, thereby reducing the compressive energy stored in the tamper and hence unavailable to the fuel and that fuel preheat by high energy electron transport from the corona is reduced.

A modification of this general composite pellet has been discussed for reactor-size configurations [Afanes'ev et al, 1975]. Two high Z tampers are proposed, one very thin and inserted into the fuel to allow that small fraction of the fuel which triggers a detonation to be thermally insulated from the fuel proper. This allows the main fuel mass to be compressed more adiabatically than would be the case if it were in direct thermal contact with the inner portion of the fuel which is most strongly heated by the final

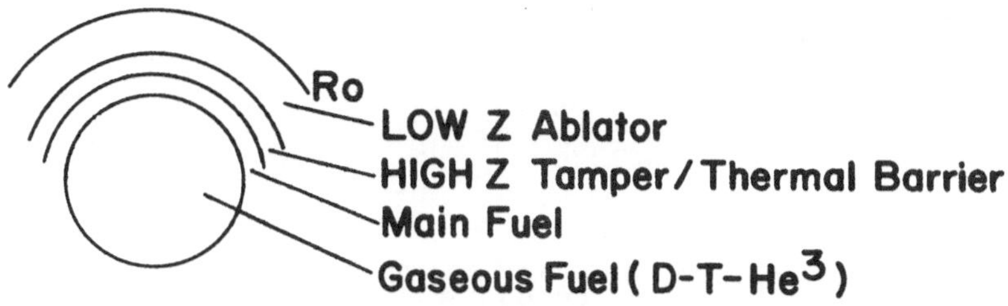

Figure 4. Generic composite pellet.

spherical convergence.

III. CLASSICAL HYDRODYNAMIC MODEL AND TRANSFER COEFFICIENTS

Several "proof-of-principle" experimental programs throughout the world have been established to obtain data on the three phases of the laser-fusion process: reaction regime, implosion physics and laser/plasma interaction. The numerical hydrodynamic codes which are used to design reactor-sized pellet configurations are currently being (interactively) tested against the complete set of experimental data from the various laboratories. Most codes presently in use, including those at KMSF, are in the Lagrange formulation; that is, the motion of mass elements is followed. Generally, the codes describe the medium as a one-fluid, two-temperature plasma which implies that all relevant collisonal mean free paths are short compared to the natural scale lengths in the pellets and collision times are short compared to the several natural scale times including plasma/electromagnetic-field interactions. Both TRHYD (a one-dimensional formulation) and HYRAD (a two-dimensional formulation treat hydrodynamic flow, electron-ion collisional energy exchange, electron and ion thermal conduction and plasma/laser interaction through inverse bremsstrahlung. In addition, TRHYD includes radiative transfer via electron free-free, free-bound and bound-bound transitions, and deposition from thermonuclear reaction products. In this section the basic classical hydrodynamic model and transfer coefficients are given in sufficient detail to provide a background for a discussion of the modifications which have been made to both the basic model and coefficients as a consequence of the experimental data which will be presented in the next section.

The Lagrange hydrodynamic equations are:

Transformation Equation

$$\frac{dr}{dt} = U$$

Equation of Motion

$$\frac{dU}{dt} = - \frac{1}{\rho} \frac{\partial}{\partial r} [P - \frac{4\mu}{3} r^{-2} \frac{\partial}{\partial r} (r^2 U)] - \frac{4U}{r} \frac{\partial \mu}{\partial r} \tag{3.1}$$

Continuity Equation

$$\rho \frac{dr^3}{dm} = 3$$

where U is the plasma flow velocity, ρ the density and r the spatial coordinate. $P = P_e + P_i$ is the sum of the ion and electron partial pressures, and μ is the first coefficient of viscosity. The hydrodynamic energy transfer mechanisms are PdV work and viscous dissipation, W_{visc}.

In the basic two-temperature model, internal energy equations are required for both ions and electrons. The ion energy equation is

$$\frac{dE_i}{dt} = -P_i \frac{dV}{dt} + W_{visc} + \frac{C_V^i}{\tau} (T_e - T_i) + \frac{1}{\rho r^2} \frac{\partial}{\partial r} \left(K_i r^2 \frac{\partial T_i}{\partial r}\right) + S_i \tag{3.2}$$

where E_i is the ion internal energy per unit mass, $V = 1/\rho$ is the specific volume and C_V^i is the ion specific heat. The first two terms are hydrodynamic work and viscous heating, the third is electron-ion collisional exchange. T_e and T_i are electron and ion temperatures and τ the collisional energy exchange time. The fourth term gives the ion thermal conduction, K_i being the ionic conductivity, and the last term gives the rate of nuclear energy deposition in the ions. The electron equation is similarly

$$\frac{dE_e}{dt} = -P_e \frac{dV}{dt} - \frac{C_V^i}{\tau} (T_e - T_i) + \frac{1}{\rho r^2} \frac{\partial}{\partial r} \left(K_e r^2 \frac{\partial T_e}{\partial r}\right)$$

$$+ \frac{1}{\rho} \int_0^{\infty} (c\, \sigma_\nu\, E^\nu - J^\nu)\, d\nu + S_e + \phi K_L \tag{3.3}$$

The electron energy equation differs from the ion equation in that electron viscosity has been neglected while radiative absorption and emission terms have been added. E^ν is the plasma generated spectral radiation density, σ_ν the absorption coefficient and J^ν the emission function. The laser flux is ϕ and K_L the laser absorption coefficient.

The viscous heating term in Eq. (3.2) must be consistent with the viscous contribution to the momentum equation. For spherical symmetry,

$$W_{visc} = \frac{4}{3} \mu \left[r \frac{\partial}{\partial r} (U/r)\right]^2 \quad .$$

The above equations are the basic set solved in the hydro simulation codes. (In the electron transport formulation described in the following chapter, the electron energy equation does not explicitly appear. The term $(1/\rho)(\partial P_e/\partial r)$ appearing in the momentum equation and the ion-electron collisional exchange term in the ion energy equation are replaced by respective integrals over the electron distribution function.)

A. The Transfer Coefficients

It is anticipated that laser-induced fusion will involve near adiabatic compression of thermonuclear fuel to densities which can exceed 10^3 gm/cm^3. Not only must the equation of state used for the electrons extend to the Fermi limit, but the transfer coefficients, i.e., the electron conductivity, radiative cross-sections, electron-ion collisional exchange coefficient, and the contributions of electrons to charged reactants energy loss must be obtained for a Fermi distribution. Radiative cross-sections have not yet been modified. Discussions of the modifications made in the other coefficients have been published [Brysk, Campbell and Hammerling, 1975] and will only be outlined here.

Electron Conductivity. The non-degenerate electron thermal conductivity is based on a Chapman-Enskog solution of the Balescu-Lenard equation by Lampe [1968]. The result can be put into a prescription for the $\varepsilon\,\delta_T$ correction to the classical Lorentz conductivity [Spitzer, 1962]

$$\varepsilon\,\delta_T = (15\pi/256)(45X + 433X^2)/(9 + 151X + 217X^2)$$

$$X = 2^{-5/2}\, Z\, \ell n\, \Lambda_{ei}/\ell n\, \Lambda_e$$

where Z is the atomic number, and $\ell n\, \Lambda_{ei}$ and $\ell n\, \Lambda_e$ are the electron-ion and electron-electron Coulomb logarithms.

Hubbard [1966] derived the conductivity of a highly degenerate Lorentz gas to be

$$K_e = \frac{2^{1/2}\pi}{3} \frac{k(kT_e)\, E_F^{3/2}}{m_e^{1/2} e^4 Z} G_\Gamma\, (\kappa_F)$$

with E_F the zero temperature Fermi energy,

$$E_F = (\hbar^2/2m_e)(3\pi^2 N_e)^{2/3} \equiv \frac{3}{2}\, kT_F \quad .$$

G_Γ is a function of the ion coupling parameter $\Gamma = Z^2e^2/RkT_i$ where R is the interionic distance. For materials with atomic weight $A \approx 2Z_0$ (where Z_0 is the atomic number), Γ, which is the ratio of the interionic Coulomb interaction energy to the ion thermal energy, is approximately $.015\ Z^2(\rho/Z_0)^{1/3}/T_i$, where T_i is measured in keV. For hydrogen, Γ can exceed unity for $\rho > 10^3$ gm/cm^3 and $T_i < 1$ keV, while for SiO_2 pellets, $\Gamma > 1$ for densities and temperatures of order 1. Laser-fusion implosions typically tend to proceed within regions where $\Gamma > 1$ so that further ion-coupling corrections will be required.

The classical and degenerate conductivities intersect near the point where the electron temperature and Fermi temperature T_F are comparable. TRHYD uses a root mean square average. K_e is plotted in Figure 5 for hydrogen.

Electron-Ion Equilibration and Ion Conduction. The expression for the ion-electron equilibration time, τ, for a Maxwellian distribution of ions and a Fermi distribution of electrons was derived by Brysk [1974]:

$$\tau = 3\pi\, m_i\, \hbar^3(1 + A)/8m_e^2\, Z^2e^4\, \ell n\, \Lambda_{ei}$$

where A is the implicit Fermi normalization factor.

The ion conductivity is that of Hochstim and Massel [1969]

$$K_i = 3.28\, \left(\frac{2}{\pi}\right)^{3/2} k(kT_i)^{5/2}/m_i^{1/2}\, c^4\, Z^4\, \ell n\, \Lambda \quad .$$

Both the expressions for K_i and τ should be used with caution when $\Gamma > 1$.

Collision Logarithms. All Coulomb logarithms are expressed in the form

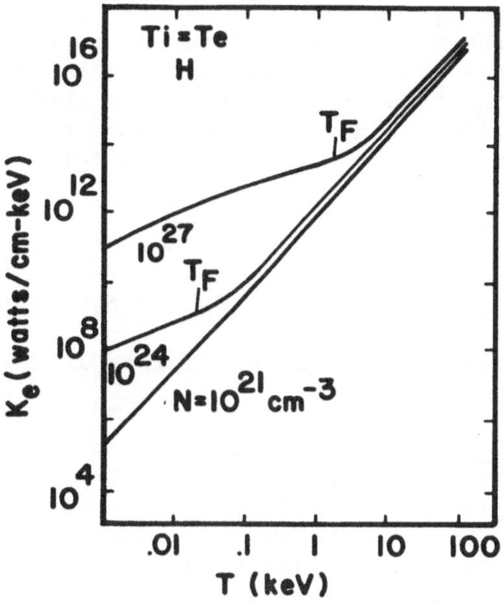

Figure 5. Electron thermal conductivity for H.

$$\ell n \; \Lambda = \frac{1}{2} \; \ell n(1 + b^2_{max}/b^2_{min})$$

where b_{max} and b_{min} are cutoffs on the Coulomb scattering impact parameter.

b_{max} is set by screening. It is normally taken to be the maximum of the Debye length and the interionic distance, R. For an electron distribution, the Debye length D_e is given by

$$D_e^{-2} = f_D \; 4\pi \; N_e e^2/kT_e$$

while for an electron-ion plasma it is D_{ei},

$$D_{ei}^{-2} = D_e^{-2} + 4\pi \; N_e \; Ze^2/kT_i$$

f_D is a degeneracy correction factor. Salpeter [1954] derived it as being given by the logarithmic derivative of the Fermi integral. To within 5% it can be approximated by

$$f_D \approx T_e/(T_e^2 + T_F^2)^{1/2} \quad .$$

The lower limit b_{min} is taken to be the minimum of the interionic distance and the maximum of the impact parameter b_0 for a 90° collision and one-half the deBroglie wavelength. For electrons, it can be argued that b_0 is never an appropriate choice [Brysk, Campbell and Hammerline, 1975]. The root mean square of the classical deBroglie wavelength and its Fermi counterpart is used:

$$b^e_{min} = \min\,[R,\ \hbar^2/12\ m_e k(T_e^2 + T_F)^{1/2}] \quad .$$

Figure 6 displays $\ell n\ \Lambda_{ei}$ and $\ell n\ \Lambda_e$ for hydrogen.

The $\ell n\ \Lambda_i$ for ion conduction is obtained in a similar manner, ignoring degeneracy corrections.

Electron Equation of State. The (glass and polymer) shells used in present experiments are in a state of partial ionization over much of the implosion history. The electron equation of state must therefore be keyed to this regime. The following model is used to predict the average ionization state and ionization energy of a partially ionized plasma. The objective is to express the significant features of the ionization processes in terms of a few simple atomic constants.

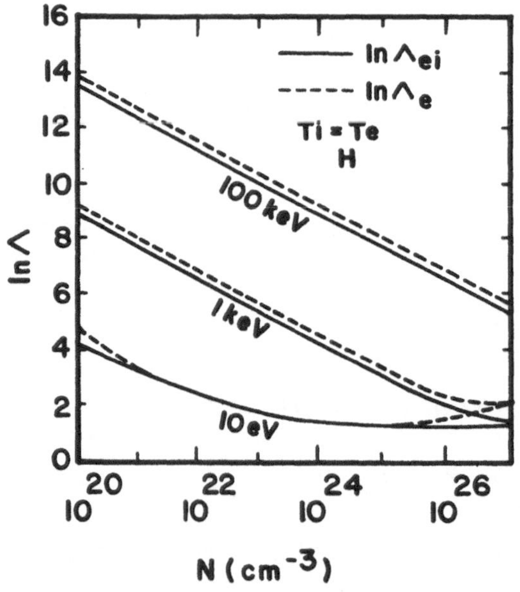

Figure 6. Coulomb logarithms for H.

The ionization model is expressed in the form of a set of Saha equations [Dewan, 1961]

$$\frac{N(j+1)\,N_e}{N(j)} = f(\theta_e)\,\exp\,\{-[E(j) - \Delta E(j)]/\theta_e\}$$

where

$N(j)$ is the ion number density in the jth ionization state.

$E(j)$ is the ionization potential from state j to state j + 1.

$\Delta E(j)$ is the change in the ionization potential $E(j)$ from density effects, the "pressure shift".

$f(\theta_e)$ is a complicated expression that contains parameters of the electron gas and weight functions for the ionic states j and j + 1.

The substance of the model comes from assumptions about the atomic processes, from the term $\Delta E(j)$ that modifies the ionization potentials and from the contents of $f(\theta_e)$ that represent the partition functions of the interacting states.

The first assumption is the standard one of thermodynamic equilibrium. The set of Saha equations can then be solved. This yields a set of ionic populations for a given density and temperature. Thermodynamic averages are formed from this set.

The second assumption is that the system is an ionic model rather than an "average atom" model [Cox, 1965]. The information about ionic states is represented through the experimental values of the ionization potentials $E(j)$. Energy levels are not assigned to the individual electrons of a particular atomic state as in an "average atom" model. Even the ionic model is complex inasmuch as many different atomic configurations form a single ionic state. Choosing the configurations is the major problem addressed by most of the detailed work on atomic structure and opacity [Stewart and Pyatt, 1961; Cohen, Parks and Petschek, 1971].

At this point, a drastic simplification is introduced. It is assumed that only the ground state exists at each ionization level. This means that the weight functions for the ionic states in the function $f(\theta_e)$ are simple constants rather than complicated partition functions over many atomic configurations. The model becomes computationally a simple one, although the ability to calculate any effects that depend on individual atomic level structure, such as radiative transitions is lost.

The third major assumption concerns the distribution of the free electron gas. For dilute gases, with particle densities of 10^{16} - 10^{18}, Stewart and Pyatt [1961] assumed that the electrons are Maxwellian. Then the equations have the Saha form, the pressure shifts in the ionization potential become small, and the behavior is close to that of an ideal gas. The function $f(\theta_e)$ contains a factor of $\theta_e^{3/2}$ which comes from the usual assumption of collisional ionization and recombination. The present interest is in moderately higher electron densities, where some effects of electron degeneracy enter. The work of Cohen, Parks and Petschek [1971] extended the Saha equations to the case where the electrons have a Fermi distribution at finite temperature. This work has been generalized, leading to a different form of the gas parameters in $f(\theta_e)$, but retaining the Maxwellian result at low density.

The last major assumption in the model concerns the change of the ionization potential with temperature and density. The standard value of the ionization potential is for an isolated atom. In a gas, however, the effects of neighboring atoms decrease the effective potential that a given electron experiences. This leads to a lower binding energy and, thus, a lower ionization potential. The analysis of Stewart and Pyatt [1965] has been followed. They derive an expression for ionization-potential lowering from a Thomas-Fermi model. It serves as an extrapolation method from a low-density to a moderate-density range, somewhat like a generalization of Debye screening. The approach is not completely consistent from a thermodynamic viewpoint [Parks et al, 1967].

This ionization model is used over a temperature range from about 1 eV to almost 100 keV and over a number density range from 10^{16} to 10^{25}. The model is best in the temperature range of a few hundred eV to a few keV and a density range from the lower limit to about normal solid densities. In these ranges the ionization energies are comparable to the thermal electron energies and the specific heat is quite different from that of an ideal gas.

B. Radiation Transport

Radiative transfer is handled in TRHYD by a moment expansion of the angularly integrated Boltzmann transport equation. Let $U(\vec{r},t,\nu,\omega)\,d\nu\,d\omega$ be the energy density per unit volume of photons in the frequency interval ν to $\nu + d\nu$ traveling in the solid angle defined by ω and $d\omega$, measured at position $\vec{r}$ and time t. In TRHYD, slab or spherical symmetry is assumed. $U_{\nu,\omega}$ is a function of only one coordinate, r, and the angle θ between the direction of symmetry and the direction of travel of the photon: $U_{\nu,\omega} = U_{\nu,\mu}(r,t,\nu,\mu \equiv \cos\theta)$. In a medium of unit refractive index, neglecting scattering and ionic transitions, and assuming that the plasma is in local thermodynamic equilibrium, the Boltzmann equation for photons can be written as

$$\frac{dU_{\nu,\omega}}{dt} \equiv \frac{\partial U_{\nu,\omega}}{\partial t} + c(\omega \cdot \nabla)\, U_{\nu,\omega} = c(B_\nu - U_{\nu,\omega})/\lambda_\nu^a \qquad (3.4)$$

where

$$\lambda_\nu^a = [\rho K_\nu^a(1 - e^{-h\nu/kT_e})]^{-1}$$

$$B_\nu = 2h\nu^3[c^3(e^{h\nu/kT_e} - 1)]^{-1}$$

and

K_ν^a is the absorption coefficient.

In slab geometry (spherical geometry does not change the nature of the solution), the integration of Eq. (3.4) over angle gives

$$\frac{\partial E_\nu}{\partial t} + \frac{\partial F_\nu}{\partial r} = c(4\pi\, B_\nu - E_\nu)/\lambda_\nu \quad .$$

$E_\nu \equiv 2\pi \int_{-1}^{1} U_{\nu,\mu}\, d\mu$ is the spectral radiation density

and

$\vec{F}_\nu = F_r \equiv 2\pi\, c \int_{-1}^{1} \mu U_{\nu,\mu}\, d\mu$ is the spectral flux.

By forming $\partial F_\nu/\partial t$, the first moment equation is obtained:

$$\frac{\partial F_\nu}{\partial t} = -cF_\nu/\lambda_\nu = 2\pi\, c^2 \int_{-1}^{1} \mu^2 \frac{\partial U_{\nu,\mu}}{\partial r}\, d\mu \equiv -cF_\nu/\lambda_\nu - c\,\frac{\partial}{\partial r}\,(f_\nu E_\nu). \qquad (3.5)$$

The integral defining the Eddington factor f,

$$f_\nu E_\nu \equiv 2\pi \int_{-1}^{1} U_{\nu,\mu}\mu^2 \, d\mu = P_\nu$$

is the component of the radiative stress tensor in the direction of symmetry.

In TRHYD, the moment expansion is truncated with Eq. (3.5) supplemented by calculating a local approximation to the Eddington factor f_ν from a simple physical-geometric model representative of the local configuration. The approach is called the variable Eddington method and was developed by Freeman and Spillman [1968]. The electron transport treatment discussed in the following chapter is also based on this approach.

The free-free and free-bound absorption coefficients are taken from the work of Cox [1965] while bound-bound coefficients are calculated following the analysis of Griem [1964]. Details of the entire radiation treatment can be found in KMSF internal documents [Campbell, Kubis and Mitrovich, 1976; Campbell, 1973].

C. Deficiencies in the Classical Model and Coefficients

The discussion of transfer coefficients has noted that strong coupling effects must be more adequately modeled when

$$\Gamma \simeq .015\ Z^{5/3}\rho^{1/3}/\theta$$

becomes of order one (θ in keV). A transition density σ_ρ can be defined from $\Gamma(\rho_p) = 1$:

$$\rho_p \simeq (\theta/.015)^3\ Z^{-5} \approx 3 \times 10^5\ \theta^3\ Z^{-5}$$

It has also been noted that the electron conductivity and Coulomb logarithms depart abruptly from non-degenerate values when $\gamma \equiv E_D/\theta \simeq .0078\rho^{2/3}/\theta$ approaches unity. A degeneracy transition density ρ_D can be defined from $\gamma(\rho_D) = 1$:

$$\rho_D \approx (\theta/.0078)^{3/2} \simeq 1.5 \times 10^3\ \theta^{3/2} \quad .$$

It is apparent that, whereas degeneracy effects normally occur in hydrogen, before ion-coupling effects, ion-coupling effects occur earlier when $Z \gtrsim 2$ and, indeed, occur at normal densities for $Z \gtrsim 10$, that is for the glass pellets used in current proof-of-principle experiments and for the even higher Z materials which will be used for the tamper/thermal barrier in forthcoming experiments. The ion-ion correlation energy is about a 10% effect on the total equation of state for SiO_2 under present experimental conditions, but can easily dominate the behavior of higher Z materials. Fortunately, the recent work by DeWitt [this volume], Hansen [this volume] and particularly Rogers [this volume] is beginning to provide calculational results necessary for the hydrodynamic codes. Their work must be extended to the set of materials of interest to laser-fusion.

Ion coupling effects are but one of a series of anomalous plasma phenomena. A "classical" plasma may be defined as one (I) in which the local energy density is dominated by thermal energy, (II) in which two- or at most three-body processes are the dominant interactions and (III) which has existed in a large enough volume for a long enough time for "almost all" particles to have made many collisions.

(I) Charged particles interact through their Coulomb fields. As a result there is always some degree of arbitrariness as to whether a phenomenon is characterized as a plasma potential energy effect or an electromagnetic field effect. Bearing this in mind, the following energy states exist in a plasma:

1. Thermal energy characterized by θ_i and θ_e (including degeneracy energy).
2. Flow energy characterized by $\frac{1}{2} m_i U^2$ where m_i is an average ion mass.
3. Energy density of an external electromagnetic field, E_r. For an incident laser power density ϕ_I, $E_r = \phi_I/c$.
4. Energy density of the self-induced electric field characterized in terms of the (radial) electric field ε.
5. Potential energy from strong coupling effects.

In a laser-induced fusion plasma, the last four states can all be comparable to or larger than the thermal energies for significant portions of the plasma and for significant periods of time:

(a) $\frac{1}{2} m_i U^2/(\theta_i,\theta_e)$ exceeds unity when the flow is supersonic, which happens in the corona and much of the ablation region as well as in the spherically convergent implosion front. The hydrodynamic equations generally handle this flow properly; however, the coefficient of viscosity, in particular, deviates grossly from the classical value for highly supersonic flow in which $\nabla \cdot \vec{U}$ is large.

(b) Two generic effects arise when $E_r/\theta_e N_e \gtrsim 1$. The energy of oscillation of an electron in the free-space electric field of the incident radiation is then not small relative to θ_e, leading to the possibility of one or more collective mechanisms between the laser field, electrons and ions becoming significant. Secondly, the laser field directly affects the fluid flow through the ponderomotive force.

(c) In a laser-heated plasma most of the absorbed energy appears in flow energy of the coronal plasma. Most of the hydrodynamic work is done not by electron-ion collisions but via the self-induced electric field arising from the collective rearrangement of the plasma to maintain charge neutrality.

(d) The potential importance of strong ion-coupling effects has already been discussed.

(II) When (radiative) electron-ion and electron-electron collisions dominate the behavior of the Boltzmann transport equations, quasi-thermodynamic equilibrium distributions, f_e, f_i result -- Fermi distributions for the electrons and Maxwellian distributions for the ions. However, when $\dot{f}_{Field}$ is not small compared to $\dot{f}_{collisions}$ for all significant regions of the phase space, the resultant distributions can become very non-thermodynamic and all relevant transfer coefficients must be recomputed over the actual distribution functions. It will be seen later that the electron thermal conductivity, for example, can be one to two orders of magnitude reduced in the corona of a laser-heated plasma.

(III) Most of the above phenomena can be modeled as corrections to the various transfer coefficients. When the mean free path, λ, for any significant process is comparable to or larger than the pertinent scale-length, L, spatial transport phenomena which cannot be treated by the basic hydrodynamic/diffusion model of Eqs. (3.1), (3.2) and (3.3) can enter. Although finite-rate effects in addition to those discussed above are not too important, K-shell collisional ionization rates are not high enough under present experimental conditions to keep the thermal front in equilibrium and rate corrections have a noticeable effect on the pellet behavior.

IV. EXPERIMENTAL PROGRAMS

An experimental program to study the potential of laser-fusion through ablation-driven compression of thermonuclear fuel was initiated at KMSF in 1972 and experiments began in 1973. Materials technology at that time did not exist to fabricate the generic spherical composite pellets discussed above and the early experimental program was keyed to the use of existing materials, particularly to D and DT gas-filled glass shells. Computer calculations suggested that, with the 0.2 TW laser power then available, these configurations would produce neutron yields of the same order of magnitude as more complex configurations. Materials programs initiated at that time have since provided the capacity to produce polymer shells and cryogenic composites with the requisite quality control.

The first "implosion" neutrons from the laser-induced fusion process were obtained in 1974 using deuterium-filled glass shells [Charatis et al, 1974]. The yield was 10^4. Since that time, the neutron yield at KMSF from DT-filled glass shells has been increased to 10^8 with a Nd (λ = 1.06 μm) laser power of 0.4 TW. Neutron yields of 10^9 have been obtained at the Lawrence Livermore Laboratory [Storm, 1976], with a Nd laser power of 2 TW and a yield of 3 x 10^6 neutrons from deuterium-filled glass shells has been reported at 0.04 TW by the Lebedev Institute [Basov, 1977]. Additional experimental programs using multi-terawatt laser systems are in progress throughout the world with the specific objective of studying the laser-induced fusion process while many other programs are studying various facets of the laser/plasma interaction process.

A. Experimental Arrangement

Although multi-beam systems are under construction [Basov, 1977], all spherical implosion experiments to date have used pellets illuminated with two beams of light. Most laboratories use two lenses with f-numbers of about 1, overfilling the target slightly. The arrangement is sketched in Figure 7. KMSF uses a two-lens/two (elliptic)-mirror illumination as shown in Figure 8 to illuminate the target pellet from over 70% of the total solid angle. Concomitant with the more spherically uniform illumination provided by the mirror system, however, is a restricted Direct View Diagnostic Region (DVDR) available for the several diagnostics which must have an unobstructed view of the target.

B. Diagnostic Techniques

Diagnostic techniques have been developed to measure incident and absorbed energy, x-radiation, implosion density and temperature, and to obtain data on coronal phenomena. As in any new field being studied by imaginative experimentalists, diagnostics have proliferated in both quantity and quality and the following discussion is by no means exhaustive.

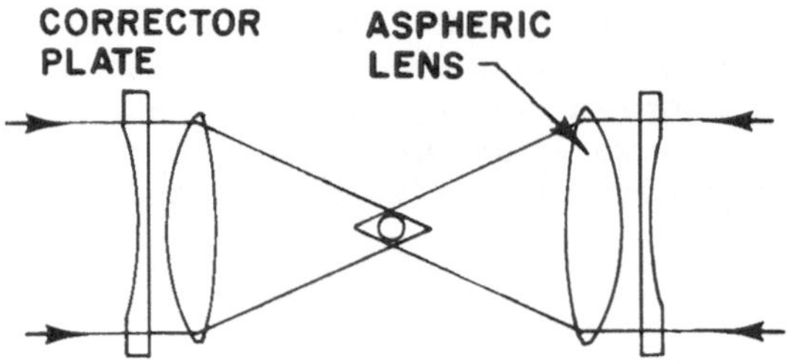

Figure 7. Typical two-lens illumination system.

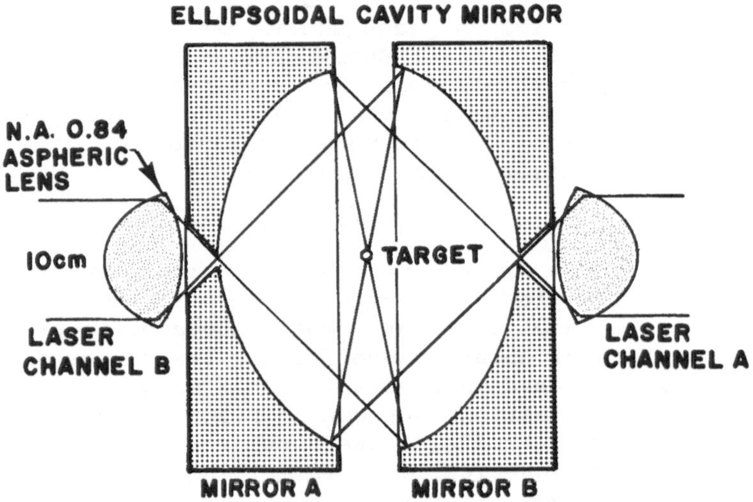

Figure 8. KMSF two-mirror illumination system.

In a typical experiment in which a DT gas-filled glass shell is illuminated with 1.06 μm light, 15% to 40% of the incident light is absorbed by the pellet; of the absorbed energy, 10% to 20% is reradiated as x-rays and 10% is coupled into the imploding remnant shell and fuel. Approximately one-half of the x-radiation occurs during the run-in phase and one-half during the final spherically convergent compression of the tamper and fuel. The remaining 60% to 85% of the incident light is refracted and reflected and provides a high energy content background. Diagnostic methods must be considered in this context. The diagnostics and data to be discussed below are from KMSF unless otherwise noted.

Total Energy. A completely experimental determination of the fractional energy absorption, sufficiently accurate to resolve questions of the relevant absorption processes, is one of the most difficult experimental problems in laser-fusion. To begin, it is experimentally impractical to measure the angular intensity distribution of the incident radiation over the entire beam. This problem is even more difficult with the lens/mirror system used at KMSF. The incident laser irradiation is characterized by measuring the (time-integrated) intensity distribution in several planes normal to the axis of symmetry [Thomas, 1976]. This data is correlated with additional data on the illumination pattern from the laser and ray-tracing calculations to infer an effective fractional laser energy on target as a function of laser power and target size. The uncertainty in fractional energy on target is about 20%.

In most experiments, three separate determinations of absorbed energy are made. The first determination is by optical energy balance -- the refracted and reflected energy is subtracted from the incident energy to determine an upper limit to the absorbed energy [Charatis et al, 1974]. Total absorbed energy is also measured with differential calorimeters -- the laser energy and total energy coming from the target are separately measured, the difference being the absorbed energy [Charatis et al, 1974]. A box calorimeter technique is also commonly used [Manes, Ahlstrom, Haas and Holzrichter, to be published] but is not practical with a lens/mirror illumination system. In some experiments, absorbed energy measurements are made by enclosing the target in a plastic bubble, transparent to the laser light, and measuring infrared radiation from the bubble after it has been heated by absorbing the secondary radiation and plasma from the target [Mayer, Siebert and Simpson, 1976]. This technique precludes other diagnostics except neutron measurements, however, and is only used periodically. The third routine measurement of absorbed energy is obtained from integrating the x-radiation and plasma energy spectral measurements which will be discussed below. The overall uncertainty in absorbed energy is also about 20% for a total uncertainty in fractional energy absorption of 30%.

X-radiation. X-radiation is detected by CaF_2 (Dy) thermoluminescent dosimeters (TLD) [Charatis et al, 1974], silicon PIN x-ray diodes and photographic film using sets of thin foil attenuators to provide spectral information. At KMSF, TLD detectors are used to observe the spectrum below about 10 keV and silicon diodes for the harder portion of the spectrum.

As a result of recent film calibrations by Stanford Research Institute [Armistead, 1974] and the Naval Research Laboratory [Brown, Criss and Birks, 1976; Dozier, Brown, Birks, Lyons and Benjamin, 1976], quantitative spatially resolved spectral data can be recorded photographically. The time-integrated spatial distribution of radiation can be photographed with a pinhole camera. The simplest model suggests that the time available for radiation at a given radius is inversely proportional to the velocity of the implosion at that radius. The x-ray image, for a constant emission rate, would thus be most intense near the initial radius of the pellet where the implosion velocity is small. A narrow well-defined ring near the original pellet radius implies little motion of the conduction front during the laser pulse.

If sufficient energy has been coupled into the implosion, further radiation occurs during the collapse phase. This radiation, principally from the outer portion of the compressed imploded shell, is visible in a pinhole picture as a second inner (compression) ring. If the implosion is sufficiently intense, the inner ring will fill within the resolution of the camera, forming what is called an "implosion spike".

Typically, each pinhole camera takes four pictures, with varying thicknesses of Be or Al filters, to provide some data on the spectral distribution radius. Quantitative results are obtained by scanning the pictures with a micro-densitometer and using the calibrated film response to convert to radiation intensity. Figure 9 is a pinhole picture and densitometer trace of a double-ring structure. The pellet was an unfilled glass shell (containing a background filling of $\sim$0.5 atm of water vapor) with an initial diameter of 198 μm and a wall thickness of 1.8 μm illuminated with 76 J of 1.06 μm light delivered in 270 nsec. The inner void is a measure of the diameter of the compressed water vapor. The associated volumetric compression of 7100 is the highest resolved compression that has been measured.

Time-resolved x-ray spectral data has been obtained by Livermore using an x-ray streak camera with an optical image intensifier [Atwood, Coleman, Larsen and Storm, 1976]. A temporal resolution of 15-psec is obtained. By using the streak camera in conjunction with a pinhole, spatially and temporally resolved data have also been obtained [Atwood et al, 1976]. The x-ray streak camera will become a popular diagnostic tool as it provides direct data on the collapse time, one of the basic parameters of the implosion.

Plasma Flow Velocity Distribution. The velocity distribution of the final expanding plasma is measured primarily by ion charge collectors [Charatis et al, 1974]. Data on the ion current vs. time is reduced to provide a mass vs. flow velocity distribution. The resultant data provides rather detailed information on the supersonic region of the flow. Information as to the energy transferred into the implosion is quite indirect both because electron thermal conduction and radiation transfer energy out of the imploding shell during final spherical convergence and because the relatively cold imploding shell, which is only 10% to 20% of the original mass, substantially recombines before it has expanded sufficiently to freeze its charge state.

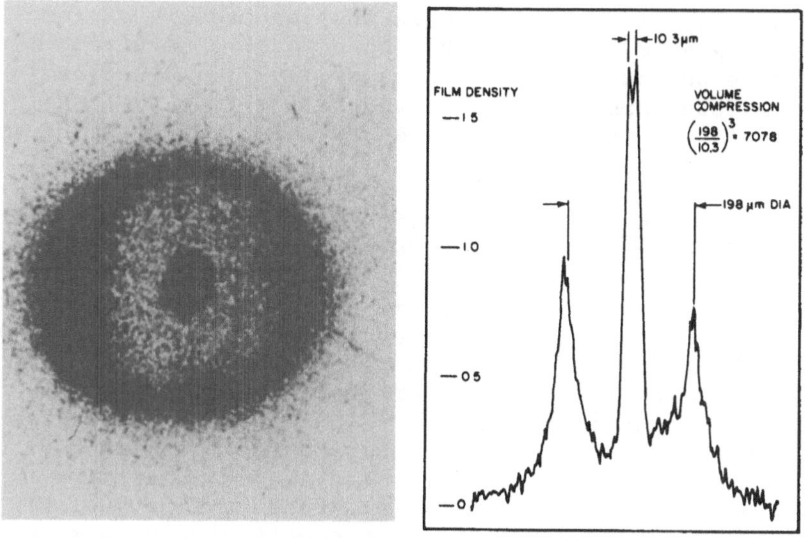

Figure 9. Double ring x-ray pinhole picture.

An electrostatic mass spectrometer is used to measure ion species and energy in the plasma [Charatis et al, 1974]. Several points on the ion flow energy per unit charge (E/Z) distribution of each charge per nucleon (Z/A) species can be determined. This data provides a second measurement of the mass vs. flow velocity at a few points and also provides a value for $\overline{(Z/A)}$ as a function of flow velocity to calibrate the charge collector data.

If the electron temperature of the (isothermal) corona, θ_e, is measured in keV, the isothermal sound speed is

$$c_T \approx 7 \times 10^7 \ \theta_e^{1/2} \text{cm/sec} \quad .$$

It is seen from Eqs. (2.5) and (2.6) that flow velocities up to a few times c_T carry significant energy, i.e., that the plasma mass distribution as a function of flow velocity must be measured out to a value of

$$U \approx (2 \text{ to } 3) \times 10^8 \ \theta_e^{1/2} \text{cm/sec} \quad .$$

(The flow energy per nucleon is given by $E/A \approx 5.2 \times 10^{-16} \ U^2$ keV/nucleon, so that energies of order $50\theta_e$ keV/nucleon are involved.) At these energies, the secondary emission correction to the raw charge collector data becomes questionable [Charatis et al, 1974] and the ions arrive at the charge collector before the photoelectric signal from the reflected laser pulse has decayed. A second measurement of the plasma flow distribution beyond a flow velocity of about 2 x 10^8 cm/sec is thus necessary. A (non-focussing) magnetic spectrograph with a passive detection madium is used for this purpose [Slater and Mayer, 1977]. Cellulose nitrate-coated foil detectors are used to record the tracks of the plasma ions. Heavy ions with velocity greater than 2 x 10^8 cm/sec can be seen.

Reaction Products. The principal reaction products used for diagnostic purposes are the 14.1 MeV neutron and 3.5 MeV α-particle from the D(T,α)n reaction and the 3 MeV proton from the D(D,T)p reaction. In the near future, it is anticipated that He^3 will be added to the fuel to make the 15 MeV proton from the D(He^3,α)p reaction available. The early experiments used time-of-flight measurements employing scintillator/photo-multiplier (SPM) detectors to identify the reaction products. Alpha particles were first observed at the Lawrence Livermore Laboratories [Slivinsky et al, 1976] and have since been observed at the Los Alamos Scientific Laboratory [McCall, Tai Ho Tan and Williams, 1975] and at KMSF [Goforth, Mayer, Brysk and Cover, 1976]. Proton measurements have been reported by the two latter groups.

The center-of-mass energies of the reaction products are thermally broadened in the laboratory system. The spectrum is thus a function of the reaction temperature T_i. In addition, the charged reaction products lose energy continuously as they traverse the cold, dense tamper, the loss being a function of the tamper mass per unit area $(\rho R)_t$ and electron temperature T_t. Since the relative importance of these effects is reaction product dependent, an examination of the energy spectra of the several products can allow values for T_i, $(\rho R)_t$ and T_t to be inferred. It is also apparent from Figure 2 that the ratio of the number of α-particles to protons can be used to infer the fuel temperature for $T_i \lesssim 7$ keV and that the D-He^3 to D-T reaction ratio can be used for $T_i \gtrsim 7$ keV.

For the yields ($\lesssim 10^9$) presently available, the long flight paths required to obtain the requisite energy resolution are inconsistent with statistical accuracy. A weakly focussing magnetic spectrograph is used to provide an order of magnitude improvement in resolution over the SPM detectors [Slater and Mayer, 1977]. Cellulose nitrate foils are used as detectors for alpha particles and nuclear emulsion plates will be used for the 3 MeV protons.

It has been noted that the peak compression of the gas within the imploding tamper can be determined by measuring the diameter of the inner surface of the inner ring in the x-ray pinhole picture. The fuel compression at peak reaction time has been directly determined by two independent measurements performed at the Lawrence Livermore Laboratories. The alpha particles from the fuel have been photographed with a pinhole camera, using cellulose nitrate film [Slivinsky et al, 1977] and an alpha particle image of the burning fuel has been constructed using a coded imaging technique, zone-plate-coded imaging [Ceglio and Coleman, 1977] (ZPCI).

Additional Coronal Diagnostics. The possibility of gross electron density profile modification in the vicinity of the critical surface as a consequence of the laser ponderomotive force was noted among the several potentially important deficiencies in the classical model. Since both the classical and anomalous laser/plasma interaction processes can be very sensitive to such modifications, it is important to obtain data from which the coronal electron density distribution can be inferred. Optical techniques have been developed using both the scattered primary laser radiation and secondary probe radiation. The Lawrence Livermore Laboratories have reported measurements of the critical density radius using a 2660 Å probe beam [Atwood, 1976] and measurements of the density scale length near the critical surface using holographic interferometry at 2660 Å [Atwood, Sweeney, Auerbach and Lee, 1978]. Measurements of the temporal evolution of the critical and quarter critical radius have also been reported by the University of Rochester [Jackel, Perry and Lubin, 1976] and by KMSF [Leonard and Cover, 1977] by observing the 5320 Å and 7070 Å light resulting from stimulated scattering at the critical and quarter critical surfaces.

V. EXPERIMENTAL RESULTS AND THEORETICAL IMPLICATIONS

This section will discuss data which have been reported by KMSF, the Lawrence Livermore Laboratory (LLL), the Los Alamos Scientific Laboratory (LASL) and the P. N. Lebedev Physical Institute (LPI) on the response to laser illumination of spherical composite targets containing thermonuclear fuel. Data from typical implosion experiments at KMSF will be presented along with data on the coronal electron temperature vs. absorbed flux density (see Eqs. (2.6) and (2.7)) from several laboratories. This provides a data base from which the basic classical model described earlier can be criticized. Results from further experiments performed to obtain data relative to specific nonclassical processes will be discussed along with parametric modifications of the basic hydrodynamic simulation code to model various anomalous phenomena.

The following data on the response of a 54-μm-diameter glass shell with a wall thickness of 0.8 μm filled with 0.002 g/cm^3 of a 60/40 deuterium-tritium mixture to 21 J of 1.06-μm radiation delivered in a 70-psec pulse are typical of a large number of experiments. Total absorbed energy, as measured by the sum of the plasma energy (charge collector) and the x-ray energy (TLD) was $3 \pm .8$ J. The experiment produced $(2.2 \pm 0.9) \times 10^6$ neutrons as measured by the silver activation counter.

Figure 10 is a digitized charge-collector trace showing the pronounced "fast-ion" peak (i.e., ions traveling at velocities greater than 2 x 10^8 cm/sec.) typical of experiments in which the absorbed laser flux exceeds 10^{14} W/cm^2. The solid curve of Figure 11 is a plot of the ion velocity spectrum deduced from the charge-collector trace. Points on the spectrum beyond a velocity of 2 x 10^8, obtained from the magnetic spectrograph, are also shown. The dashed curve presents the results of a numerical simulation which will be discussed later. The coronal electron temperature deduced from the slope beyond a velocity of 2 x 10^8 is 10 $\pm$ 2 keV.

The above experiment was one of the first for which ion spectral data were obtained from the magnetic spectrograph. Neutron yield was too low to obtain an accurate α-spectrum. Figure 12 is an α-particle spectrum typical of experiments with a neutron yield exceeding 10^7. The fuel-ion temperature deduced from the width of the α-spectrum is 2.0 $\pm$ 0.3 keV.

Figure 13 is a plot of coronal electron temperature, as deduced from either the high energy x-radiation spectral distribution or the slope of the fast component of the ion spectrum, vs. the product of absorbed laser flux and the square of the laser wavelength (θ_e vs $P_L\lambda^2$), as suggested by Eq. (2.7). The data, at 1.06 μm and at 10.6 μm, are from several laboratories.

The above data set, from a representative experiment at KMSF and from several laboratories on coronal temperature vs. absorbed flux, will be used to discuss coronal phenomenology. As a preface, some observations will be made regarding electron collisional processes. The two dominant electron collisional processes are electron-ion (90°) scattering and electron-electron energy exchange.

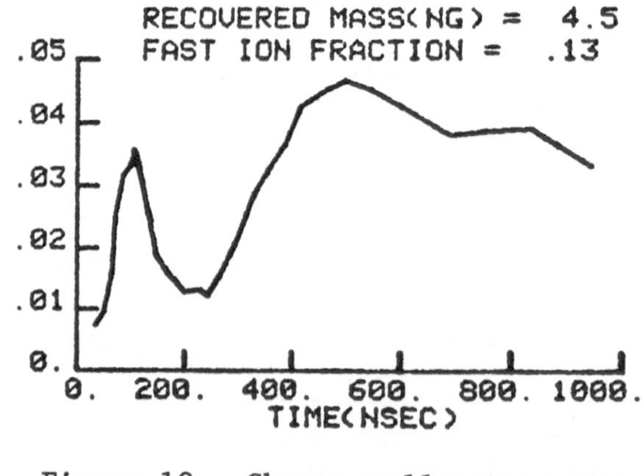

Figure 10. Charge collector trace.

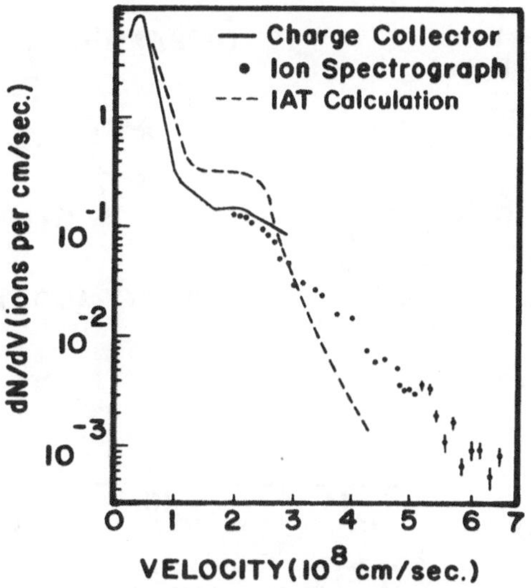

Figure 11. Ion velocity spectrum.

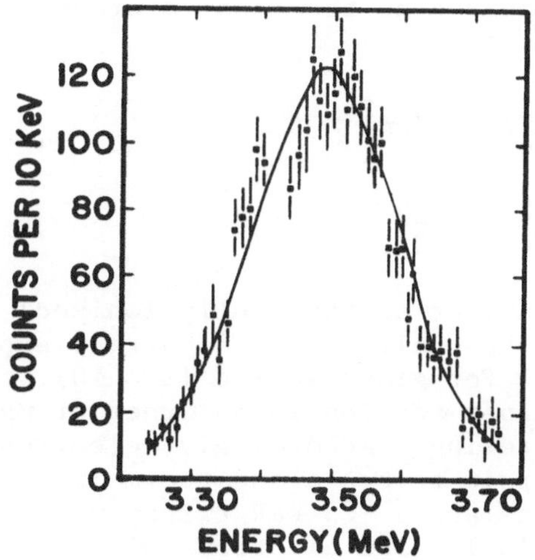

Figure 12. Alpha particle spectrum.

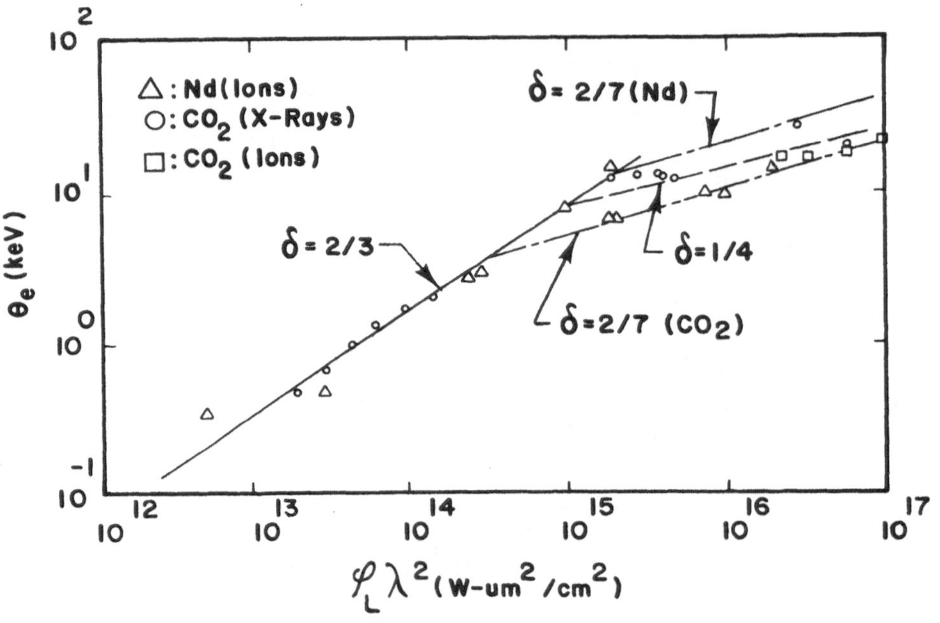

Figure 13. Coronal electron temperature vs. $\phi_L\lambda^2$.

The respective mean free masses, $\rho\lambda$ gm/cm^2, for an electron with energy E_e keV large compared to the local average thermal energy, for the two processes are

$$\rho\lambda_{ei} \approx 3 \times 10^{-6} E_e^2/Z \text{ gm/cm}^2$$

$$\rho\lambda_{ee} \approx 3 \times 10^{-6} E_e^2 \text{ gm/cm}^2$$

where Z is the atomic number of the (fully ionized) plasma. The mass per unit area of the shells used in current experiments is of order 2×10^{-4} gm/cm^2. For glass shells ($Z \approx 10$), the electron-electron energy loss cross-section is an order of magnitude less than that for 90° scattering. Accordingly, electrons with energy $\lesssim$ 6 keV diffuse in coordinate space and thermalize within typical scale lengths, electrons above $\sim$25 keV transport freely, losing energy by collisions slowly, and electrons with energies 6 keV $\lesssim E_e \lesssim$ 25 keV more or less diffuse in space while slowly losing energy. As a consequence, electron transport phenomena might be expected to manifest themselves (as preheat of the imploding tamper and fuel, for example) when the coronal electron temperature exceeds about 2 keV and to become increasingly important as the coronal temperature approaches 10 keV.

A. X-ray Spectral Data

Below the K-shell ionization energy of silicon, the x-radiation from SiO_2 is dominated by line emission. It has been noted that electron transport effects can be expected for electron energies above 6 keV and also that the x-radiation during the implosion process is predominantly recombination radiation from that portion of the thermal front where K-shell ionization is nearly complete. This temperature can be measured by examining the x-ray spectrum in the region 2 keV $< h\nu <$ 6-8 keV, i.e., above the line radiation but below the region where transport effects ensue. The so-called "cold component" x-ray temperature is typically 0.5 keV to 0.7 keV in agreement with the K-shell ionization temperature of silicon at a density of order 0.1 g/cm^3, typical of conditions near the thermal conduction front.

The imploding shell becomes increasingly transparent to electrons with energy greater than 6 keV, with the electron energy distribution above 20 to 25 keV being an increasingly accurate reflection of the distribution in the corona. It might therefore be expected that the high energy portion of the coronal electron distribution can be inferred by examining the (bremsstrahlung and recombination) radiation coming from the transport of this distribution through the high density imploding shell. In particular, if the coronal electron distribution is Maxwellian, the temperature should be deducible by examining the x-ray spectrum in the region beyond 20 to 25 keV. In a spherically convergent experiment, however, such an analysis is complicated by radiation from the imploded glass tamper resulting from the transport of the high energy tail of the DT fuel electron distribution. This radiation can dominate radiation which originated during the implosion out to photon energies of about 50 keV. Figure 14 is a plot of the temperature deduced from a fit to the tail of the x-ray spectrum vs. the coronal temperature deduced from the ion spectra for a large number of experiments at KMSF. Hand estimates suggest that the above phenomenon is responsible for the discrepancy. The hydrodynamic simulation codes do not exhibit such effects since the thermal electron distribution from the imploded fuel is not transported through the glass in any existing code. Without a much more detailed treatment of electron transport than previously available, extreme care must be exercised in the analysis of x-ray spectral data.

B. Ion Spectral Data

The absorbed energy as calculated by the basic classical model described earlier is 3.0 J. The calculated coronal electron temperature is 2.5 keV as compared to the experimental value of 10 keV deduced from the slope of the fast-ion spectrum beyond a velocity of 2×10^8 cm/sec. It is obvious that the basic classical model does not describe the behavior of the corona at this incident laser flux level (3×10^{15} W/cm^2). Apparently the basic model grossly overestimates electron energy transport out of the corona into the ablation region.

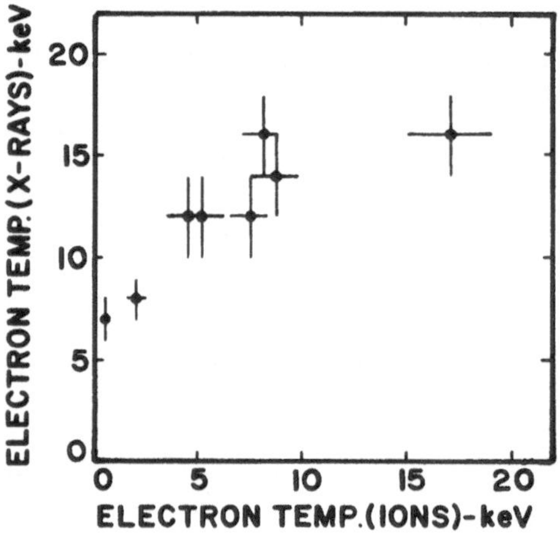

Figure 14. Coronal electron temperature as deduced from x-ray and ion spectra.

Additionally, the fact that classical inverse bremsstrahlung absorption of laser light varies as $T_e^{-3/2}$ (Eq. (2.2)) coupled with the fact that the basic calculation gave approximately the correct absorption when T_e = 2.5 keV implies that, at the observed coronal temperature of 10 keV, inverse bremsstrahlung within the basic model is inadequate to explain the observed absorption.

Two problems must be addressed:

(1) What additional absorption mechanisms must be considered and/or what additional coronal mechanisms might enhance inverse bremsstrahlung absorption?

(2) What mechanisms might reduce the electron energy flux inward from the corona?

C. Absorption Processes

As noted, when the laser energy density is not small compared to the thermal energy density, collective processes may become significant. These processes involve the parametric excitation of plasma instabilities and are often called anomalous phenomena since they occur only when the laser energy density exceeds a threshold value. (Brueckner and Jorna [1974] review the several categories and their paper contains an extensive bibliography.) The anomalous processes which may enhance laser absorption are oscillating two-stream and ion-acoustic (also called parametric decay) instabilities. In addition, stimulated Brillouin and Raman scattering may reduce laser absorption.

The inverse-bremsstrahlung absorption coefficient given in Eq. (2.2) is derived by treating laser propagation in the classical geometric optics limit. Two corrections must be made for finite values of k_oL, where $k_o = 2\pi/\lambda$ is the free-space wave number and L the electron density scale-length in the vicinity of the critical surface. Consider laser light incident on a plasma at an angle θ to the local density gradient. The electric field of the laser light can be resolved into a P component E_P in the plane of incidence and an S component E_S normal to this plane (and parallel to the critical surface). Zhekulin [1934] showed that these two components propagate independently. Brysk [1976] has shown that the S component can be described in the geometric optics limit quite adequately until $k_oL < 1$; contributions from the evanescent wave inside the critical surface cancel effects outside the critical surface. Denisov [1956] showed that the P component satisfies an equation which has a resonance at the critical surface. The strength of the resonance is inversely proportional to the electron collision frequency with the result that the collisional absorption rate is approximately independent of the collision frequency. Friedberg et al [1972] gave the name resonant absorption to this phenomenon and analyzed it in context with laser fusion. The strength of the resonance disappears for both normal and large-angle incidence, reaching a maximum at an angle given by $(k_oL)^{1/3} \sin\theta \approx 0.7$. The absorption fraction at the optimum angle is in the range 0.3 to 0.6 so that resonant absorption can be an important process in laser fusion.

One of the most definitive experiments in laser fusion was performed at LLL to measure resonant absorption [Manes, Rupert, Auerbach, Lee and Swain, 1977]. Planar parylene (C_8H_8) disk targets were irradiated with ~100 GW of 1.06-μm light from the Janus laser facility, using an aspheric f/10 lens to focus the light onto the target. The incident light was polarized so that only S or P components were present. For P-polarized light, the maximum absorption of 0.4 to 0.45 occurs at an angle of incidence $\theta \approx 25°$. This implies a scale-length of $L \approx 1$ to 2 μm in agreement with results obtained by holographic interferometry [Atwood, Sweeney, Auerbach and Lee, 1978]. For S-polarized light, the absorption varies from 0.3 at normal incidence, to 0.25 at $\theta \approx 25°$ and to 0.15 at $\theta = 60°$.

The enhancement of the S component of the field at the critical surface for normal incidence and of the P component for a finite angle of incidence implies that the ratio of the laser field to thermal energy densities $E_r/\theta_e N_e$ can easily exceed one for an incident flux exceeding 10^{15} W/cm^2. Several calculations of profile modification by ponderomotive force have been performed [Lee, Forslund, Kindel and Lindman, 1977; Mulser and VanKessel, 1977; Virmont, Pellat and Mora, 1977]. All yield scale-lengths in the vicinity of the critical surface of order 1 μm, in agreement with the above experimental values.

The observed absorption for S-polarized light is in itself still larger than expected from inverse bremsstrahlung. Estimates of the absorption expected from parametric decay instability using the observed scale-lengths (of order 1 μm) are at most 0.1 to 0.2 [Kruer, Haas, Mead, Phillion and Rupert, 1977]. The additional absorption of 0.1 to 0.2 may result from a combination of resonant absorption off ripples in the critical density surface [Atwood, Sweeney, Auerbach and Lee, 1978; Kruer, Haas, Mead, Phillion and Rupert, 1977] and an oscillating two-stream instability. Normally, the two-stream instability is ineffective as an anomalous absorption mechanism, in part because it occurs inside the critical surface where the laser field is small if $k_oL >> 1$. Short scale-length ($k_oL \lesssim 5$) however, implies that the evanescent wave penetrating the critical surface may have an energy density larger than outside the critical surface [Brysk, 1976] and accordingly enhance two-stream absorption. Resonant and parametric-decay absorption can result in feeding a large fraction of the absorbed energy into high energy ($\gtrsim$ 100 keV) electrons. The two-stream instability tends to form a more thermal distribution. In view of the programmatic importance of high-energy-electron preheat, further experiments carefully coordinated with theoretical analyses, such as the LLL resonant-absorption experiments, are required to define adequately the absorption mechanisms. In particular, submicron resolution of the electron density distribution in the vicinity of the critical surface is required along with spatially and temporally resolved data on the high energy (> 10 keV) x-ray spectrum.

It has been suggested that the reduced electron energy transport and enhanced laser absorption may both be a result of enhanced electron-ion collisions resulting from an ion-acoustic instability [Malone, McCrory and Morse, 1975; Manheimer, 1977]. The instability develops turbulence and ion-clumping resulting in strongly enhanced scattering. KMSF has implemented such a model into TRHYD and indeed the basic features of the ion-spectrum appear [Campbell, Johnson, Mayer, Powers and Slater, 1977]. The dashed curve of Figure 11 is a calculated ion-spectrum for an absorbed energy of 3 J. The spectra beyond a velocity of 3 x 10^8 cm/sec do not agree. Inclusion of a transporting fast-electron component, arising from resonant absorption, for example, can match the spectral data as well as improving the agreement with the x-ray pinhole picture data. This inclusion has not been done consistently with the ion-acoustic turbulence model, however. In particular, it is not obvious that the electron preheating required to give agreement with the x-ray pinhole data is consistent with the very high electron-ion collisional rates rquired to give the observed coronal temperatures.

The LASL has developed a model based on resonant absorption and profile modification caused by ponderomotive force which also fits the observed ion-spectral data [Bezzerides, Dubois, Forslund, Kindel, Lee and Lindman, 1976]. The model, which is based on a two-temperature electron distribution in the corona, leads to a hot coronal electron temperature proportional to the 1/4 power of the absorbed laser flux. The normalized curve is shown as the dashed line of Figure 13.

The average coronal electron temperature is the temperature obtained from the slope of the fast-ion spectrum. The hot component temperature is measured by the tail of the x-ray spectrum. The general agreement between these two numbers (Figure 14) is difficult to reconcile with a two-temperature coronal model. Some further considerations of electron transport processes, which suggest that the observed phenomena may be explained in terms of simple models of (resonant and instability) absorption and electron transport through a non-turbulent plasma, will be presented below.

It is noteworthy that the departure in θ_e vs. $\phi_L\lambda^2$, as shown in Figure 13, from the 2/3 power given by Eq. (2.7) occurs at a temperature around 8 keV. This is the temperature at which over half of the electron energy is in that portion of the distribution with a mean free path larger than the gross coronal electron density scale-length (and much larger than the local fine scale-length at the critical surface arising from profile modification). For temperatures such that $\lambda_{ei} > L$, the isothermal region does not abruptly terminate at the critical surface, but extends inside to a density where $\lambda \approx L$. Since the mean free path is proportional to θ_e^2/ρ, the isothermal region extends in to a density ρ_i which is proportional to θ_e^2/L. This implies that the energy content in the isothermal corona, which is proportional to $\rho_i\theta_e{}^{3/2}$ varies as $\theta_e{}^{7/2}/L$. Quantitatively, for Z = 10 and including electron-electron collisions and the electric field in a manner consistent with the $\varepsilon\delta$ corrections to thermal conduction, the two broken line fits shown in Figure 13 are obtained for 1.06 μm and 10.6 μm irradiation. It is apparent that this simple $\delta = 2/7$ transport model; where $\theta_e \propto \phi_L{}^\delta$ fits the θ_e vs $\phi_L\lambda^2$ data adequately, as does the LASL $\delta = 1/4$ model.

Before proceeding further with such arguments, some additional experiments to obtain further data on transport processes will be presented. Fast ions have been observed on the unilluminated side of planar targets by a number of groups. Two sets of experiments were performed at KMSF to determine whether these ions originated at the unilluminated surface or transported from the corona. In the first experiment, on glass hemispheres, the ion magnetic spectrograph was used to detect fast ions from the inner surface using one-sided illumination of approximately two-thirds of the hemisphere. Figure 15 presents the spectrum from the inside along with a spectrum from a fully-illuminated, spherical companion target.

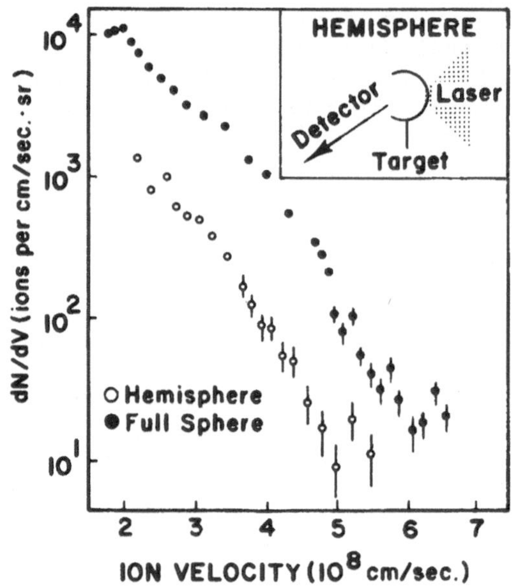

Figure 15. Front and rear surface ion spectra.

In a second series of experiments, the magnetic spectrograph was replaced by a Thomson parabola spectrograph so that the charge-to-mass ratio of the fast ions could be determined. An 0.1 μm layer of B_4C was evaporated onto the outside surface of three glass hemisphere targets. The Thomson parabola detected oxygen, carbon and hydrogen but no boron. In a companion experiment, the coronal fast ions consisted only of boron and carbon.

The mean free masses noted earlier apply to both electrons and hydrogen ions, i.e., to any $Z = 1$ particle. Another set of experiments was performed to measure hydrogen transport through the glass shell. The experiments used tritium filled shells coated with CH_2 and CD_2. The tritium was contaminated with 0.45% of deuterium as measured by gas chromatography. In consecutive experiments, CD_2 and CH_2 coated targets were irradiated. Four out of five data sets show an enhanced neutron yield for the CD_2 coated target. About 2% of the deuterium available in the CD_2 would be sufficient to explain the data if it were compressed and heated along with the tritium. This fraction is consistent with diffusion of deuterons in the high energy tail of the 1 to 2 keV (ion temperature) plasma in the corona and thermal front. The possibility of direct mixing as a consequence of gross shell breakup cannot yet be ruled out as an explanation, but the data does support the position that hydrogen ion transport may be a significant material and energy transfer mechanism and should ultimately be included in the hydro codes along with electron transport.

Pinhole camera pictures have been taken of the fast ions at KMSF using cellulose acetate film. If the fast ions are radially directed, the pinhole camera should see only an image of that portion of the target in line with the pinhole and of about the same area. Instead, the picture (Figure 16) looks very similar to an x-ray pinhole picture, implying that the fast ions have an angular distribution with a width of order π radians. The distribution extends over less than π radians since no limb brightening is observed. Several processes which might artificially broaden the ion image have been considered, but the possibility that the fast ions acquire tangential velocities comparable to their radial velocities cannot yet be eliminated.

The concluding discussion of coronal and transport phenomena will present some results of calculations of the quasi-equilibrium coronal electron distribution in the Fokker-Planck approximation and some calculations of a more realistic least upper bound to electron energy transport than provided by the usual flux-limit.

The data certainly suggests that, at laser fluxes beyond 10^{14} W/cm^2 for 1.06 μm light and beyond 10^{12} W/cm^2 for 10.6 μm light, resonant and instability processes determine the laser/plasma interaction. The inner-surface fast ion data, the deuterium diffusion data, and the agreement of the ion and x-ray deduced coronal temperatures suggest, at least to the author, that energy transport from the corona into the imploding shell may be dominated by classical collisional processes and the associated radial electric field induced to maintain charge neutrality. The transport must be calculated by examining the Boltzmann equation outside the usual diffusion limit, however, particularly since the absorption processes may well feed a large fraction of the absorbed energy initially into electrons with energies of order 100 keV.

As noted, whenever the laser field is depositing energy into the plasma and/or work is being done on the ions faster than the electrons can collisionally exchange energy, the quasi-equilibrium electron distribution can become very non-Maxwellian. The electron distribution has been calculated in the Fokker-Planck approximation (see the following chapter) for a situation typical of experiments described earlier. For pellets with a diameter of 50 to 80 μm, the calculation deposits approximately 3×10^{15} W/cm^2 in electrons with an energy 10 times the plasma "temperature", to simulate resonant or parametric decay deposition, and deposits approximately 1×10^{15} W/cm^2 by inverse bremsstrahlung. This is balanced by allowing 3×10^{15} W/cm^2 to be lost by "PdV" work and 1×10^{15} W/cm^2 to be lost by frictional energy exchange to the body of the pellet. In experiments such as described earlier, the problem describes a corona with a "temperature" of 10 keV with energy being deposited

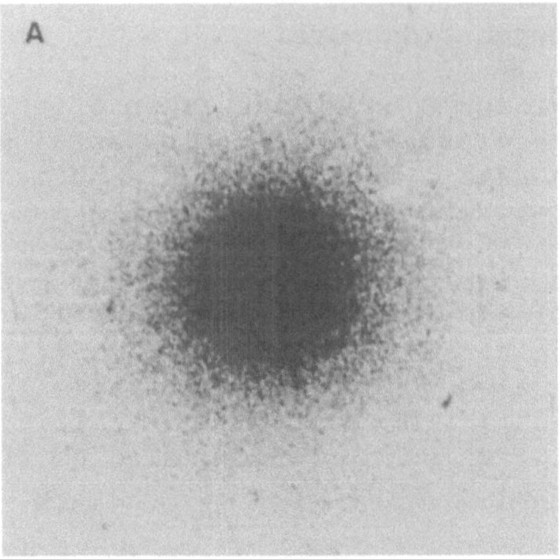

Figure 16. Fast ion pinhole picture.

in 100 keV electrons.

The resultant quasi-equilibrium number and energy density in energy space are plotted in Figure 17 along with the corresponding Maxwellian distribution. The Fokker-Planck (F-P) energy distribution is depressed by "PdV" and frictional losses between 2 and 5.5 times the "temperature", with the result that the transporting part of the distribution (beyond $W \equiv E/\theta = 3$) has approximately the same energy content for both distributions. The integrated rate of frictional energy loss (preheating) is proportional to the integral of the velocity times the energy distribution times the frictional cross-section. Since the cross-section is proportional to W^{-2}, the frictional loss is proportional to $\int F dW$, the integral, in this case, extending from $W \approx 3$ to ∞. The amount of preheating in the F-P distribution is about a factor of two less than for a Maxwellian distribution.

As the electron mean-free path becomes longer than the electron temperature scale-length, the diffusion approximation for electron thermal conduction becomes increasingly invalid. In typical experiments this occurs for electrons with energies greater than 6 keV as noted, so that for the 10 keV coronas experimentally observed, nearly the entire electron distribution function lies outside the diffusion regime. The range of applicability of a diffusion approximation is commonly extended somewhat by limiting the flux to be less than the (vacuum) free-streaming limit. The upper limit to the flux can be written in the form

$$F_{max} = f N_e V_e \theta_e \quad ; \quad \frac{1}{2} m_e V_e^2 \equiv \theta_e \quad .$$

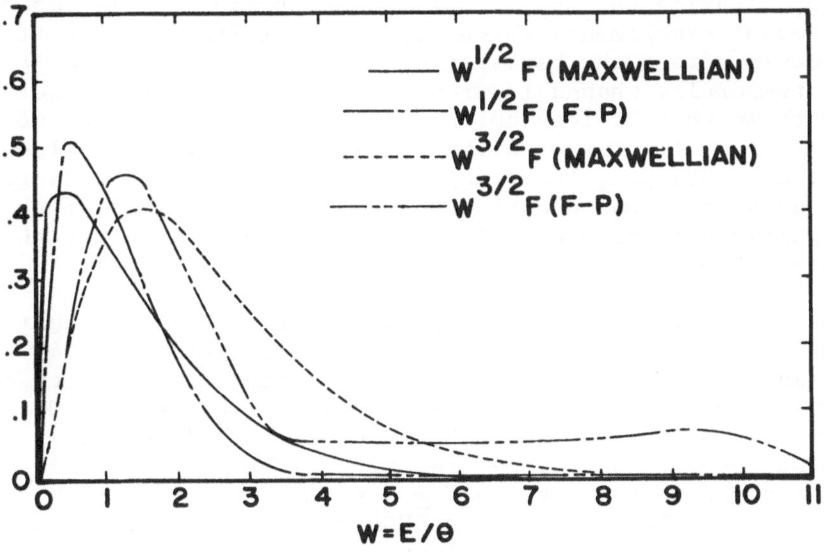

Figure 17. Fokker-Planck calculated coronal electron distribution.

For an isotropic Maxwellian electron distribution, $f = \pi^{-1/2}$.

The free-streaming limit assumes that charge neutrality is preserved by returning a current of cold electrons. For conditions in the vicinity of the critical surface, it is probably more appropriate to at least return electrons with the same average energy. Under this condition, $f = \pi^{-1/2}/4 \approx .14$. Using this value of f in the calculation of the experiment described earlier, the corona temperature is increased from 2.5 keV to 4 keV and a fast ion double-hump distribution is formed.

Although this is probably a more realistic upper limit to f than the vacuum free-streaming limit, it still greatly overestimates the flux since the return current for electrons with energies beyond 25 keV is an (attenuated) distribution from the opposite side of the pellet. For such electrons, $f \approx \delta_o/\lambda_o$ where δ_o and λ_o are the total pellet thickness and the electron mean-free path, respectively. Upon integrating over the electron distribution functions, values of f ranging from 10^{-3} to 10^{-2} can be obtained, varying with the quasi-equilibrium distribution function and assumptions as to the radial profile of the electrons in the 6 to 25 keV range. The uncertainty of such estimates indicates that a much more thorough treatment of electron transport is required in the hydrodynamic computer codes. The next chapter describes such a treatment.

Before leaving the subject of corona physics, it is important to note recent experiments which suggest that Brillouin scattering may seriously affect the behavior of large pellets exposed to strongly temporally shaped laser pulses. NRL performed a set of experiments in which a long corona was formed with a prepulse and the target then illuminated with a short high-flux main pulse [Ripin, Young, Stamper, Armstrong, Decoste, McLean and Bodner, 1977]. Backscatter was greatly enhanced and absorption of the main pulse significantly reduced. Further experiments to study this effect are currently in progress.

Finally, an (unpublished) experiment at the Lebedev Institute should be noted [Basov, 1977]. A glass pellet, diameter 140 μm, wall thickness 2.2 μm was filled with deuterium at a density of 5.4 mg/cm^3. The pellet was illuminated with 0.04 TW of 1.06 μm light for 2.5 nsec. The observed D-D neutron yield was $(4.5 \pm 1.5) \times 10^6$. The absorbed energy was reported to be 22.5 ± 2.5 J. The fuel density was measured to be 7 ± 1 g/cm^3 from x-ray pinhole camera pictures. The Soviet code LUCH was used to calculate the experiment and predicted a fuel density of 9.7 g/cm^3, a fuel ion temperature of .67 keV and a neutron yield of 4.4×10^6. LUCH is said to include a "real" equation-of-state.

Simulations of this experiment at KMSF give a fuel density of 8 g/cm^3, but an ion temperature of only 0.5 keV with a resultant neutron yield of 3×10^4. Interestingly, the code predicts an imploded glass peak density of about 150 g/cm^3 with a temperature of 50 to 100 eV. This places the glass into the strongly-coupled regime where the KMSF equation-of-state is grossly inaccurate. If the ion bonding energy were available to the implosion, the fuel ion temperature could easily be elevated to the two-thirds of a keV required to match the observed neutron yield. It is not known whether LUCH includes strong-coupling effects.

A significant aspect of the experiment is obtaining a volume compression of nearly 2000 from such a low laser flux. The corona temperature calculated by TRHYD is only 0.7 keV. It is very important to find that hydrodynamic stability of the imploding shell was maintained at such a low acceleration.

VI. SUMMARY AND CONCLUDING REMARKS

The general mood of the laser-induced fusion community is one of cautious optimism. The recently reported experiments of Lebedev on the implosion of 140 μm diameter D-filled glass pellets with only 40 GW of 1.06 μm light suggest that implosion stability is not as severe a problem as had been feared early in the program. The forthcoming experiments using the LLL Shiva laser will, of course, be crucial to the future of laser fusion.

Although the numerical hydrodynamic codes used to design and analyze experiments to date can be made to agree with the data with one or more of the noted parametric modifications, the models must more properly treat several phenomena to improve their reliability when applied to extrapolations from existing data:

1. The transport in both coordinate and energy space of the entire electron distribution should be included, as should transport of hydrogen ions with energies exceeding several keV;

2. A self-consistent treatment of ponderomotive force and a numerically efficient treatment of the resultant profile modification in the vicinity of the critical surface is required;

3. The electron transport equations must (parametrically) include the effects of all relevant absorption processes;

4. The effects of strong-coupling phenomena on the equation-of-state and the several transfer coefficients must be included before more adiabatic implosions, such as reported by the Lebedev group and as required for reactor-sized configurations, can be accurately analyzed.

REFERENCES

Afanes'ev, Y. V. et al, 1975, ZhETF Pis. Red. 21, 150 [Sov. Phys. JETP Lett. 21, 68).

Armistead, R. A., 1974, Rev. Sci. Inst. 45, 996.

Atwood, D. T., L. W. Coleman, J. T. Larsen and E. K. Storm, 1976, Phys. Rev. Lett. 37, 499.

Atwood, D. T. et al, 1976, "The Correlation of X-ray Temporal Signatures, Neutron Yields, and Laser Performance", Proc. of the 10th Euc. Conf. of Laser Intn. with Matter, Paliseau, France.

Atwood, D. T., August 1976, "Ultraviolet Probing of Laser Produced Plasmas with Picosecond Pulses", Proc. of the 12th Int'l Cong. on High Speed Photography, Toronto, Canada.

Atwood, D. T., D. W. Sweeney, J. M. Auerbach and P. H. Y. Lee, 1978, Phys. Rev. Lett. 40, 184.

Basov, N. G. and O. N. Krokhin, 1964, Zh. Eksp. Toer. Fiz. 46, 171 [Sov. Phys. JETP 19, 123].

Basov, N. G., November 1977, unpublished, "Prospects and Problems of Laser Thermonuclear Fusion for Future Energetics", Int'l Scientific Forum for an Acceptable Nuclear Energy for the World, Fort Lauderdale, Florida.

Bezzerides, B., D. F. Dubois, D. W. Forslund, J. M. Kindel, K. Lee and E. L. Lindman, October 1976, IAEA Symposium on Plasma Physics and Controlled Nuclear Fusion Research, Berchtesgaden, Germany.

Brown, D. B., J. W. Criss and L. W. Birks, 1976, J. Appl. Phys. 47, 3722.

Brueckner, K. A. and S. Jorna, 1974, Rev. Mod. Phys. 46, 325.

Brysk, H., 1974, Plasma Phys. 16, 927.

Brysk, H., 1976, Radio Science 11, 861.

Brysk, H., P. M. Campbell and P. Hammerling, 1975, Plasma Phys. 17, 473.

Campbell, P. M., 1973, unpublished, "Energy and Momentum Transfer in Dense Thermonuclear Plasmas" - KMSF U-105.

Campbell, P., R. Johnson, F. Mayer, L. Powers and D. Slater, 1977, Phys. Rev. Lett. 39, 274.

Campbell, P. M., J. J. Kubis and D. Mitrovich, 1976, unpublished, "Radiative Processes in a Laser-Fusion Plasma" - KMSF U-457.

Ceglio, N. M. and L. W. Coleman, 1977, Phys. Rev. Lett. 39, 20.

Charatis, G. et al, (Proc. Int. Conf., Tokyo, 1974), IAEA Vienna, 1975, Vol. II, 317, 1974, "Plasma Physics and Controlled Nuclear Fusion Research".

Cohen, H. D., D. E. Parks and A. G. Petschek, 1971, unpublished, "A Study of Equations of State/Opacity of Ionized Gases", Systems, Science and Software Reports No. 3SR-465 and 3SR-859-1.

Cox, A. N., 1965, Chapter 3, Stellar Structure, edited by L. Aller and D. McLaughlin, University of Chicago Press.

Daiber, J. W., A. Hertzberg and C. E. Wittliff, 1966, Phys. Fluids 9, 617.

Dawson, J. M., 1964, Phys. Fluids 7, 981.

Denisov, N. G., 1957, Zh. Eksp. Teor. Fiz. 31, 609 [Sov. Phys. JETP 4, 544.

Dewan, E. M., 1961, Phys. Fluids 4, 759.

DeWitt, H. E., this Volume.

Dozier, C. M., D. B. Brown, L. S. Birks, P. B. Lyons and R. F. Benjamin, 1976, J. Appl. Phys. 47, 3732.

Freeman, B. E. et al, 1968, unpublished, "The VERA Code: A One-Dimensional Radiative Hydrodynamic Program" - DASA Report No. 2135.

Freidberg, J. P., R. W. Mitchell, R. L. Morse and L. I. Rudsinski, 1972, Phys. Rev. Lett. 28, 795.

Gitomer, S. J., R. L. Morse and B. S. Newberger, 1977, Phys. Fluids 20, 234.

Goforth, R. R., F. J. Mayer, H. Brysk and R. A. Cover, 1976, J. Appl. Phys. 47, 4850.

Griem, H. R., 1964, Chapter 8, Plasma Spectroscopy, McGraw-Hill, New York.

Hansen, J. P., this Volume.

Hochstim, A. R. and G. A. Massel, 1969, "Transport Coefficients in Ionized Gases" in Kinetic Processes in Gases and Plasmas, edited by A. R. Hochstim, Academic Press, New York.

Hubbard, W., 1966, Astrophys. J. 146, 858.

Jackel, S., B. Perry and M. Lubin, 1976, Phys. Rev. Lett. 37, 95.

Kidder, R. E., 1976, Nuclear Fusion 16, 3.

Kruer, W. L., R. A. Haas, M. W. Mead, D. W. Phillion and V. C. Rupert, 1977, Plasma Physics, edited by H. Wilhelmsson, Plenum Press, New York.

Lampe, M., 1968, Phys. Rev. 170, 306; 174, 276.

Lee, K., D. W. Forslund, J. M. Kindel and E. L. Lindman, 1977, Phys. Fluids 20, 51.

Leonard T. A. and R. A. Cover, 1977, unpublished, Bull. Am. Phys. Soc. 22, 1077.

Malone, R., R. McCrory and R. Morse, 1975, Phys. Rev. Lett. 34, 721.

Mayer, F. J., L. D. Siebert and J. Simpson, 1976, Proc. IEEE Conf. on Plasma Science, Austin.

Manes, K. R., V. C. Rupert, J. M. Auerbach, P. H. Y. Lee and J. E. Swain, 1977, Phys. Rev. Lett. 39, 381.

Manes, K. R., H. G. Ahlstrom, R. A. Haas and J. F. Holzrichter, to be published, J. Opt. Soc. Am.

Manheimer, W., 1977, Phys. Fluids 20, 265.

McCall, G. H., Tai Ho Tan and H. A. Williams, 1975, Bull. Am. Phys. Soc. 20, 1318.

Mulser, P. and C. VanKessel, 1977, Phys. Rev. Lett. 38, 902.

Nuckolls, J., L. Woods, A. Thiessen and G. Zimmermann, 1972, Nature 239, 139.

Parks, D. E. et al, 1967 unpublished, "Optical Constants of Uranium Plasma", NASA Lewis Research Center, Report NASA-CR-72348.

Ripin, B. H., F. C. Young, J. A. Stamper, C. M. Armstrong, R. Decoste, E. A. McLean and S. E. Bodner, 1977, Phys. Rev. Lett. 39, 611.

Rogers, F. J., this Volume.

Salpeter, E. E., 1954, Aust. J. Phys. 7, 373.

Slater, D. C. and F. J. Mayer, 1977, "Laser Interaction and Related Plasma Phenomena" (4th Int'l Workshop Conf. 1976), edited by H. J. Schwarz and H. Hora, Vol. 4B, 603, Plenum Press, New York.

Slivinsky, V. W. et al, 1976, Phys. Rev. Lett. 16, 1803.

Slivinsky, V. W. et al, 1977, Appl. Phys. Lett. 30, 555.

Spitzer, L., 1962, Physics of Fully Ionized Gases, 2nd Ed., Interscience, New York.

Stewart, J. S. and K. D. Pyatt, 1961, unpublished, "Theoretical Studies of Optical Properties", Air Force Special Weapons Center Report AFSWC-TR-61-71, Vol. 1.

Stewart, J. C. and K. D. Pyatt, 1965, Appl. J. 144, 1203.

Storm, E. K., 1976, Bull. Am. Phys. Soc. 21, 1173.

Thomas, C. E., May 1976, unpublished, Conf. on Laser and Electro-optical Systems, San Diego, California.

Virmont, J., P. Pellat and P. Mora, 1977, submitted to Phys. Fluids.

Yabe, T. and K. Niu, 1977, Nuclear Fusion 17, 269.

Zhekulin, L. A., 1934, Zh. Eksp. Teor. Fiz. 4, 76 [Sov. Phys. JETP 4, 76.

ELECTRON TRANSPORT IN A SPHERICALLY SYMMETRIC PLASMA

R. A. Grandey

KMS Fusion, Inc.
Ann Arbor, Michigan

TABLE OF CONTENTS

ELECTRON TRANSPORT IN A SPHERICALLY SYMMETRIC PLASMA

R. A. Grandey

KMS Fusion, Inc.

Ann Arbor, Michigan 48106

I. INTRODUCTION

The previous paper has outlined the phenomena involved in laser-induced fusion and has indicated that the numerical models used for hydrodynamic simulation must incorporate a rather complete treatment of electron-transport, not restricted to the diffusion approximation, before final resolution of coronal physics mechanisms can be made.

This chapter discusses such a treatment. The following points should be kept in mind while considering the formulation:

(1) In laser-induced fusion plasmas, charge neutrality is preserved throughout the sensible portion of the plasma, at least for laser wavelengths less than a few μm. Thus normally, the radial electric field can be obtained by requiring no net current flow across a radial surface.

(2) For the same "temperature", ion thermal velocities are a factor of the square root of the ion to electron mass ratio ($\sim 60Z^{1/2}$) smaller than electron thermal velocities. As a consequence, transport effects resulting from the thermal portion of the distribution can usually be neglected. The current resulting from the flow velocity may, however, be important.

(3) The ion-ion equilibration rate is approximately $Z^3/(60Z^{1/2})$ times that of the electron-electron equilibration rate. Thus for $Z \gtrsim 5$, ions equilibrate more rapidly than electrons. For this reason, the ion distribution can be assumed to be Maxwellian, transporting with the local flow velocity in most instances with the ion

density, flow velocity, and internal energy found from the usual macroscopic transfer equations of continuity, momentum and energy [Spitzer, 1962].

(4) The numerical hydro codes perform a temporal integration of the ion momentum equation and electron and ion internal energy equations. A multivelocity-group numerical integration of the electron Boltzmann equation must be performed and appropriate integrals made over velocity space for incorporation into the ion-momentum and internal energy equations.

(5) Consistent with the (tacit) assumptions of most hydrodynamic simulation codes, the various ion species are characterized as a Maxwellian distribution transporting at the local flow velocity, U. Inter-ion-species friction is assumed to be sufficient to allow a common flow velocity to be used for all ions, so that insofar as the ion continuity, momentum and energy equations are concerned, the ions can be characterized by an average mass m_i, charge Z and ion density N_i such that $ZN_i = N_e$, where N_e is the integrated electron density. Electron-ion collision integrals will be obtained by averaging over the several ion species. The numerical hydro code into which the treatment of electron transport is to be incorporated is in the Lagrange frame, i.e. moving with the flow velocity U. In this frame

$$\left(\frac{d}{dt} \equiv \frac{\partial}{\partial t} + U \frac{\partial}{\partial r} \right)$$

the ion continuity and momentum equations become

$$\frac{d \ln \rho}{dt} + \frac{1}{r^2} \frac{\partial(r^2 U)}{\partial r} = 0 \tag{1.1}$$

$$\rho \frac{dU}{dt} = - \frac{\partial P_i}{\partial r} + ZeN_i \varepsilon + \frac{\partial P_{ei}}{\partial t} \tag{1.2}$$

where $\rho = m_i N_i$ is the ion mass density, r is the spherical coordinate, ε the (radial) electric field and $(\partial P_{ei}/\partial t)$ the rate of collisional momentum transfer from electrons to ions across the surface $\vec{r} = r$. P_i, the ion stress, is taken to be $N_i kT_i$.

II. THE BOLTZMANN EQUATIONS FOR A SPHERICALLY SYMMETRIC PLASMA

The following paragraphs address the form of the phase space distribution functions $f_k(\vec{r},\vec{V},t)$ of plasma species k in the presence of other species j [Hochstim, 1969]. In spherical symmetry, f_k is a function of the position vector $\vec{r}$ only through the spherical coordinate, r, and a function of the velocity vector, $\vec{V}$, only through its magnitude V, and the direction cosine $\mu = \cos\theta$, where θ is the angle between $\vec{V}$ and the radial vector:

$$f_k(\vec{r},\vec{V},t) = f_k(r,V,\mu,t), \text{ with the usual normalization}$$

$$\int f_k d\vec{V}_k = N_k = 2\pi \int_{-1}^{1}\int_{0}^{\infty} f_k(r,V,\mu,t)\, V^2 dV\, d\mu$$

where N_k is the particle density per unit volume of the kth species.

The Boltzmann equation for species k is used in the form

$$\frac{Df_k}{Dt} = \sum_j J_{kj} + \sum_{j,j'} J_{kjj'} + ----$$

$$\frac{Df_k}{Dt} \equiv \frac{\partial f_k}{\partial t} + \vec{V}_k \cdot \nabla_r f_k + \vec{a} \cdot \nabla_V f_k$$

$$J_{kj} \equiv \int \left[f'_k(V'_k)\, f'_j(V'_j) - f_k(V_k)\, f_j(V_j) \right] \vec{V}_{kj} \frac{d\sigma_{kj}}{d\Omega}\, d\Omega\, d\vec{V}_j \quad .$$

In the binary collision integral J_{kj}, V'_k and V'_j are the final velocities of two particles that have collided with initial velocities V_k and V_j, i.e., with relative velocity $\vec{V}_{kj} = \vec{V}_k - \vec{V}_j$ and with a differential cross section $(d\sigma_{kj}/d\Omega)$. The additional terms represent three-body and higher order collisional processes.

In spherical geometry, the quantity

$$\vec{V}_k \cdot \nabla_r f_k = V\left(\mu \frac{\partial}{\partial r} + \frac{1-\mu^2}{r}\frac{\partial}{\partial \mu}\right) f_k \quad .$$

The contribution to the acceleration, $\vec{a}$, due to external forces, arising from a radially directed electric field, ε, (which might be internally generated) is given by

$$a_\varepsilon \cdot \nabla_V f_k = \frac{q_k \varepsilon}{m_k}\left(\mu \frac{\partial f_k}{\partial V} + \frac{(1-\mu^2)}{V}\frac{\partial f_k}{\partial \mu}\right)$$

where q_k and m_k are the charge and mass of the k specie. In the absence of additional external forces the Boltzmann equation becomes

$$\frac{\partial f_k}{\partial t} + V\mu \frac{\partial f_k}{\partial r} + V\frac{(1-\mu^2)}{r} \cdot \frac{\partial f_k}{\partial \mu} + a_\varepsilon \cdot \nabla_V f_k = \sum_j J_{kj} + --- \tag{2.1}$$

III. THE VARIABLE EDDINGTON APPROXIMATION

The Variable Eddington Approximation was developed for use in problems of radiative transport [Freeman, et al, 1968]. It is a moment expansion of the complete distribution function, f_k, truncated after the first moment. First define an angularly integrated distribution function by

$$F_k(r,V,t) \equiv 2\pi \int_{-1}^{1} f_k(r,V,\mu,t)\, d\mu$$

and a flux density distribution function by

$$G_k(r,V,t) \equiv 2\pi V \int_{-1}^{1} \mu\, f_k(r,V,\mu,t)\, d\mu \quad .$$

Equation (2.1) integrated over angle gives the zero-moment equation

$$\frac{\partial F_k}{\partial t} + \frac{1}{r^2}\frac{\partial(r^2 G_k)}{\partial r} + \frac{q_k \varepsilon}{m_k V^2}\frac{\partial(G_k V)}{\partial V} = \sum_j \int_0^{2\pi}\int_{-1}^{1} J_{kj}\, d\mu d\psi + \text{---} \equiv J_0 \quad . \tag{3.1}$$

The first-moment equation for G_k is obtained by multiplying Eq. (2.1) by μV and integrating over angle:

$$\frac{\partial G_k}{\partial t} + V^2 \frac{\partial (f_E F_k)}{\partial r} + \frac{V^2}{r}(3f_E - 1)\, F_k + \frac{\vec{q}_k \varepsilon}{m_k}\left[V \frac{\partial(f_E F_k)}{\partial V} + (3f_E - 1)F_K\right]$$

$$= V \sum_j \int_0^{2\pi}\int_{-1}^{1} \mu J_{kj}\, d\mu d\psi + \text{---} \equiv \overline{V\mu J} \quad . \tag{3.2}$$

The key point of the Variable Eddington approximation is the introduction of physical and geometric arguments to approximate the Eddington factor

$$f_E \equiv \int_{-1}^{1} \mu^2 f_k\, d\mu \Bigg/ \int_{-1}^{1} f_k\, d\mu \quad .$$

f_E varies between the limits of 1/3 for diffusive flow and 1 for streaming flow.

For species k electrons, the collision integral J_0 is dominated by electron-electron energy exchanging collisions and $\overline{V\mu J}$ by electron-ion momentum exchanging collisions, for which the dominant term can be written in the form $-\nu_{ei}\, G_e$ where ν_{ei} is the electron-ion momentum exchange collision frequency [Hochstim, 1969]. The right hand side of Eq. (3.2) for electrons becomes

$$\overline{V\mu J} = -\nu_{ei}\, G_e + \overline{V\mu J_r}$$

where $\overline{V\mu J_r}$ represents a sum over all other collisional processes. Introducing this form for $\overline{V\mu J}$ into Eq. (3.2) and expressing both Eqs. (3.1) and (3.2) in terms of (d/dt), using the ion continuity equation (1.1) to rearrange terms, the zero- and first-moment equations for electrons in the Lagrange frame become:

$$\rho \frac{d(F_e/\rho)}{dt} = R \equiv -\frac{1}{r^2}\frac{\partial}{\partial r}\left[r^2(G_e - UF_e)\right] + \frac{e\varepsilon}{m_e V_e^2}\frac{\partial}{\partial V_e}(V_e G_e) + J_0 \tag{3.3}$$

$$\rho \frac{d(G_e/\rho)}{dt} + \frac{1}{2} V_e^2 \frac{\partial (1 - f_E)F_e}{\partial r}$$

$$+ \frac{1}{r^2} \frac{\partial\left[\frac{1}{2}(3f_E - 1)\, V_e^2 F_e - UG_e\right] r^2}{\partial r}$$

$$- \frac{e\varepsilon}{2m_e}\left[V_e \frac{\partial(1 - f_E)F_e}{\partial V_e} + \frac{1}{V_e}\frac{\partial(3f_E - 1)\, V_e^2 F_e}{\partial V_e}\right]$$

$$= -\nu_{ei} G_e + \overline{V\mu J_r} \quad . \tag{3.4}$$

Equation (3.3) can be partially integrated from time t_0 to t:

$$F_e(t) = [\rho(t)/\rho(t_0)]\left[F_e(t_0) + \rho(t_0)\int_{t_0}^{t} R(t')/\rho(t')\, dt'\right] \tag{3.5}$$

This partial integration conserves electrons exactly.

The first-moment equation can be further partially integrated by using the integrating factor

$$\exp\left(\int_{t_0}^{t} \nu_{ei} dt'\right) \equiv \exp\,(\tau - \tau_0) \quad .$$

$$G_e(r_L, t) = \frac{\rho(t)}{\rho(t_0)}\left[G_e(t_0)e^{-\Delta\tau} - \rho(t_0)V_e \int_{\tau_0}^{\tau} I_L/\rho(t')e^{\tau' - \tau} d\tau'\right] \tag{3.6}$$

where

$$I_L = \lambda_{ei}\left(\frac{1}{2}\frac{\partial(1 - f_E)F_e}{\partial r} + \frac{1}{r^2}\frac{\partial}{\partial r} r^2\left[\frac{1}{2}(3f_E - 1)F_e - \frac{UG_e}{V_e^2}\right] - \frac{1}{V_e}\overline{\mu J_r} - \frac{e\varepsilon}{2m_e V_e}\left[\frac{\partial(1 - f_E)F_e}{\partial V_e} + \frac{1}{V_e^2}\frac{\partial(3f_E - 1)V_e^2 F_e}{\partial V_e}\right]\right)$$

and $\lambda_{ei} = V_e \nu_{ei}^{-1}$ is the electron-ion momentum-exchange mean free path. r_L denotes the radial coordinate of Lagrange element L. The integral is evaluated by assuming that I_L/ρ is slowly varying in the interval from t_0 to t:

$$\rho(t_0)G_e(r_L,t) = \rho(t)\left[G_e(t_0)e^{-\Delta\tau} - V_e\overline{I}_L(1 - e^{-\Delta\tau})\right] \qquad (3.7)$$

The electron flux Φ_e across the surface $\vec{r} = r$ is given by

$$\Phi_e = \int_0^\infty G_e V_e^2 \, dV_e$$

The ion flux is N_i U. The net current across the surface is thus $e\Phi_e$ - Ze N_i U. The assumption is now made that the local scale lengths are sufficiently larger than the Debye length that local charge neutrality can be assumed ($Z N_i = N_e$). This means that the radial electric field, ε, can be obtained by requiring no net current across the surface $\vec{r} = r$:

$$\int_0^\infty G_e V_e^2 \, dV_e = N_e U \qquad (3.8)$$

Introducing the solution for $G_{e\_}(t)$ from Eq. (3.6) and ignoring terms of order U^2,

$$\frac{e\overline{\varepsilon}}{m_e}\int_0^\infty \lambda_{ei}\left(1 - e^{\tau_0-\tau}\right)\left(V_e^2\frac{\partial(f_E F_e)}{\partial V_e} + (3f_E - 1)V_e F_e\right)dV_e$$

$$= N_e(t_0)U + \int_0^\infty V_e^3 \lambda_{ei}\left(1 - e^{\tau_0-\tau}\right)\left(\frac{\partial(f_E F_e)}{\partial r} + \frac{1}{r}(3f_E - 1)F_e - \frac{1}{V_e}\overline{\mu J_r}\right)dV_e - \int_0^\infty G_e(t_0)\, e^{\tau_0-\tau}\, V_e^2 \, dV_e \qquad (3.9)$$

The rate of collisional momentum transfer from electrons to ions across the surface $\vec{r} = r$ is given by

$$\frac{\partial p_{ei}}{\partial t} = \int_0^\infty \nu_{ei}\, m_e G_e V_e^2 \, dV_e - \int_0^\infty V_e^3 \, m_e \, \overline{\mu J_r} \, dV_e \quad .$$

The integrated ion-momentum Equation (1.2) is accordingly

$$\rho \frac{dU}{dt} = - \nabla P_i - e\varepsilon \int_0^\infty e^{\tau_0 - \tau} \left(V_e^3 \frac{\partial (f_E F_e)}{\partial V_e} + V_e^2 \, (3f_E - 1) \, F_e \right) dV_e$$

$$+ m_e \int_0^\infty \nu_{ei} \, G_e(t_0) \, e^{\tau_0 - \tau} \, V_e^2 \, dV_e - m_e \int_0^\infty V_e^3 \, \overline{\mu J_r} \, e^{\tau_0 - \tau} \, dV_e$$

$$- m_e \int_0^\infty V_e^4 \left(1 - e^{\tau_0 - \tau}\right) \left(\frac{\partial (f_E F_e)}{\partial r} + \frac{1}{r} \, (3f_E - 1) \, F_e \right) dV_e \qquad (3.10)$$

In both the diffusion limit $\tau - \tau_0 >> 1$ for all V_e and the collisionless limit $\tau - \tau_0 << 1$, the ion-momentum equation approaches

$$\rho \frac{dU}{dt} = - \frac{\partial P_i}{\partial r} - m_e \int_0^\infty V_e^4 \left(\frac{\partial (f_E F_e)}{\partial r} + \frac{1}{r} \, (3f_E - 1) \, F_e \right) dV_e$$

with the momentum being exchanged (and hence "PdV" work being done) via the induced field, ε, in the collisionless limit and via electron-ion collisions in the diffusion limit. Although the net rate of "PdV" energy exchange from electron internal energy to ion flow energy is the same in both limits, the differential rates are different and hence the quasi-equilibrium electron distribution can be grossly different in the two limits.

The objective now is to (numerically) simultaneously solve Eqs. (3.5), (3.7) and (3.10) using the velocity-integrated Eq. (3.9) to eliminate the induced electric field and using a supplemental dynamic prescription for the Eddington factor f_E. Before discussing finite-difference approximations to these equations, however, a discussion of the dominant contributions to the zero-moment collision integral, J_0, will be given.

IV. THE COLLISION INTEGRALS J_{kj} IN THE FOKKER-PLANCK APPROXIMATION

The following paragraphs address the form of the contributions to J_{kj} from Coulomb collisions in the local center-of-mass frame, i.e., from the isotropic portion of the distributions. In terms of the energy variable

$$E = \frac{1}{2} m V^2 \quad ,$$

the collision integral for the time rate of change of particles of species k as the result of collisions with particles of species j is

$$J_{kj} = \int \left[f_k(E_k - \Delta E)\, f_j(E_j + \Delta E) - f_k(E_k)\, f_j(E_j) \right] d\vec{V}_j\, \vec{V}_{kj}$$

$$\cdot \frac{d\sigma_{kj}}{d\Omega}\, d\Omega \quad . \tag{4.1}$$

where $f_k(E_k)$ is still the density function per unit volume in velocity space $\vec{V}_k$. For Coulomb collisions, the energy transfers can be assumed to be small in most instances, giving rise to the Fokker-Planck (F-P) approximation:

$$J_{kj}^{F-P} \simeq \int \Bigg[\Delta E \left(f_k \frac{\partial f_j}{\partial E_j} - f_j \frac{\partial f_k}{\partial E_k} \right) + \frac{\Delta E^2}{2} \cdot \left(f_k \frac{\partial^2 f_j}{\partial E_j^2} + f_j \frac{\partial^2 f_k}{\partial E_k^2} - 2 \frac{\partial f_k}{\partial E_k} \frac{\partial f_j}{\partial E_j} \right) \Bigg] \cdot \vec{V}_{kj} \frac{d\sigma_{kj}}{d\Omega}\, d\Omega\, d\vec{V}_j \tag{4.2}$$

The integration over $d\Omega$ has been discussed [Rosenbluth, MacDonald, and Judd, 1957; Holt and Haskell, 1965]. By using the relation

$$f_j(m_j E_k/m_k) = -\frac{m_k}{m_j} \frac{\partial}{\partial E_k} \int_{m_j E_k/m_k}^{\infty} f_j\, dE_j \quad ,$$

the result of the angular integration of Eq. (4.2) can be cast as an exact differential, demonstrating conservation of particles and suggesting finite-difference schemes;

$$J_{kj}^{F-P} = \nu_{kj}^{F-P}\, E_k^{-1/2} \frac{\partial H_{kj}}{\partial E_k} \tag{4.3}$$

where

$$\nu_{kj}^{F-P} \equiv (4\pi)^2 Z_k^2 Z_j^2 e^4 \ln \Lambda_{kj} (m_k/m_j)^{3/2}/(m_k/m_j)$$

and

$$H_{kj} \equiv f_k \int_0^{m_j E_k/m_k} E_j^{1/2} \left(f_j + \frac{\partial \ln f_k}{\partial E_k} \int_{E_j}^{\infty} f_j'\, dE_j' \right) dE_j$$

In the above m_k, m_j, Z_k and Z_j are the masses and charges of the k and j species and $\ell n\ \Lambda_{kj}$ is the Coulomb logarithm for collisions of species k with j. The first term in the integral for the flux H_{kj} is often called the frictional flux and the second term the diffusive flux [Holt and Haskell, 1965].

The rate of change of particles due to collisions in the Fokker-Planck approximation is proportional to

$$\int_0^\infty E_k^{1/2} \frac{df_k^{F-P}}{dt} dE_k \quad \text{,i.e., to} \quad \int_0^\infty \frac{\partial H_{kj}}{\partial E_k} dE_k$$

which vanishes as long as $f_k(\infty) = 0$ and $f_k(0)$ is finite. The rate of change of energy is proportional to

$$\int_0^\infty E_k \frac{\partial H_{kj}}{\partial E_k} dE_k = - \int_0^\infty H_{kj}\ dE_k \quad .$$

It can be shown by integration by parts that this integral vanishes, i.e., that energy is conserved in the F-P approximation, for collisions within a specie.

When specie k is an electron and specie j is an ion, $m_j >> m_k$. For most of the electron distribution, the various ion integrals can then be approximated by

$$\int_0^{m_j E_k / m_k} E_j^{1/2} f_j\ dE_j \simeq \left(\frac{m_j}{2\pi}\right)^{3/2} N_j \frac{\pi^{1/2}}{2} \quad .$$

For a Maxwellian distribution of ion species j = i,

$$H_{ei} \simeq \left(\frac{m_i}{2\pi}\right)^{3/2} N_i \frac{\pi^{1/2}}{2} \left(f_e + \theta_i \frac{\partial f_e}{\partial E_e}\right)$$

where θ_i is the ion temperature.

The electron-ion collisional energy-exchange rate is proportional to

$$\int_0^\infty E_e \frac{\partial H_{ei}}{\partial E_e} dE_e \simeq \left(\frac{m_i}{2\pi}\right)^{3/2} N_i \frac{\pi^{1/2}}{2} \left\{- \int_0^\infty f_e\ dE_e + \theta_i\ f_e(0)\right\} \quad .$$

For a Maxwellian distribution of electrons, one has the usual result that the electron-ion collisional energy-exchange rate is proportional to

$$\theta_e^{-3/2} (\theta_i - \theta_e)$$

and that energy flows from the "hotter" to "cooler" species. For a non-Maxwellian electron distribution (i.e. not in thermodynamic equilibrium) it is apparent that this result may no longer be true. For example, if the cold part of the electron distribution is depressed by inverse-bremsstrahlung absorption of laser energy, additional heat may flow into the ions, the electrons acting as an energy transfer medium between the source and the ions.

The following two chapters present finite difference approximations to the above equations, first for the collision integrals in the Fokker-Planck approximation and then for the remaining "Vlasov" terms incorporating radial gradients and electric fields.

V. FINITE DIFFERENCE METHODS FOR THE FOKKER-PLANCK APPROXIMATION

Finite difference methods have been applied to the Fokker-Planck approximation for several years [Killene and Marx, 1970; Whitney, 1970]. Numerical approximations to the equation

$$\frac{df_k^{F-P}}{dt} = \nu_{kj}^{F-P} E_k^{-1/2} \frac{\partial H_{kj}}{\partial E_k} \tag{5.1}$$

should satisfy three requirements:

(1) Conservation of particles

(2) Conservation of energy

(3) $\dfrac{df_k^{F-P}}{dt} = 0$ for $f_k = \left(\dfrac{m_k}{2\pi\theta_k}\right)^{3/2} N_k e^{-E_k/\theta_k}$,

$$f_j = \left(\frac{m_j}{2\pi\theta_j}\right)^{3/2} N_j e^{-E_j/\theta_j} , \quad \theta_j = \theta_k .$$

Consider a finite difference scheme in which f is defined at mesh-points i and the flux H_{kj} at mid-points i $\pm$ (1/2). It is obvious that the generic differencing

$$E_k^{-1/2} \left.\frac{\partial H_{kj}}{\partial E_k}\right|_i = \frac{3}{2} \frac{H_{i+(1/2)} - H_{i-(1/2)}}{E_{i+(1/2)}^{3/2} - E_{i-(1/2)}^{3/2}} ,$$

will conserve the quantity

$$\sum_i f_i \left(E^{3/2}_{i+(1/2)} - E^{3/2}_{i-(1/2)}\right)$$

as long as H = 0 at both boundaries. (The subscripts k and j have been suppressed.) The deviation of this quantity from the value

$$6\pi^{1/2} \, (m_k/8\pi)^{3/2} \, N_k$$

will be a measure of the fundamental accuracy of the numerical integration.

The flux H is centered at $i + \frac{1}{2}$:

$$H_{i+(1/2)} = f_k\Big|_{i+(1/2)} \int_0^{m_j E_{i+(1/2)}/m_k}$$

$$\cdot \; E_j^{1/2} \left(f_j + \frac{\partial \, \ell n \, f_k}{\partial \, E_k} \Bigg|_{i+(1/2)} \int_{E_j}^{\infty} f_j' \, dE_j' \right) dE_j$$

If the derivative is differenced as

$$(\ell n \, f_k|_{i+1} - \ell n \, f_k|_i)/(E_{i+1} - E_i)$$

it will equal $-\theta_k^{-1}$ for f_k Maxwellian. In order to have $H_{i+(1/2)} = 0$ for $\theta_j = \theta_k$, the

$$\int_{E_j}^{\infty} f_j' \, dE_j'$$

must then equal $f_j(E_j)\theta_j$ when f_j is Maxwellian. This suggests the approximation that $\ell n \, f_j$ is linear in the interval a - b in regions where f_j is a decreasing function of E_j:

$$\int_{E_a}^{E_b} f_j' \, dE_j' \simeq \frac{(E_b - E_a)[f_j(E_b) - f_j(E_a)]}{\ell n \, f_j(E_b) - \ell n \, f_j(E_a)}$$

Consistent with the approximation that ℓn f is linear in the interval between mesh points

$$f_k|_{i+(1/2)} = \left(f_k|_i \times f_k|_{i+1}\right)^{1/2}$$

and the two integrals are approximated by the generic form

$$\int_{E_1}^{E_2} E^{1/2} g\,dE \simeq X\left[E_1^{1/2} g_1 - E_2^{1/2} g_2 + \frac{\pi^{1/2}}{2}\, g_1 \exp\,(E_1/X)\; X^{1/2}\right.$$

$$\left.\left\{\mathrm{erf}\,\left((E_2/X)^{1/2}\right) - \mathrm{erf}\,\left((E_1/X)^{1/2}\right)\right\}\right]$$

where $X = (E_2 - E_1)/(\ell n\ g_1 - \ell n\ g_2)$.

Although satisfactory results have been obtained using an explicit, conditionally stable, backward time-differenced scheme for problems in which the average energy of the distribution does not change rapidly, the time interval can be substantially increased by using an implicit, forward time-differenced scheme. Denoting the two integrals in the flux H_{kj} by R and S,

$$H_{kj}\Big|_{i+(1/2)}^{n+1} \simeq f^n\Big|_{i+(1/2)} \left[\left(R^n_{i+(1/2)} + \frac{\partial\ \ell n\ f}{\partial E}\Big|^n_{i+(1/2)} S^n_{i+(1/2)}\right)\right.$$

$$\cdot\left(1 + \frac{1}{2}\,\frac{\Delta f_i}{f_i^n} + \frac{1}{2}\,\frac{\Delta f_{i+1}}{f^n_{i+1}}\right)$$

$$\left. + \frac{S^n_{i+(1/2)}}{E_{i+1} - E_i}\left(\frac{\Delta f_{i+1}}{n_{i+1}} - \frac{\Delta f_i}{f_i^n}\right)\right] \qquad (5.2)$$

where $\Delta f = f^{n+1} - f^n$. The subscripts j,k have been suppressed on the right hand side.

The key elements of the above scheme are factoring into the form of Eq. (6.3), with the concomitant straightforward differencing of $\partial\ \ell n\ f_k/\partial E_k$, and the assumption that $\ell n\ f_j$ is linear between mesh points.

VI. INTEGRATION OF THE "VLASOV TERMS"

A brief discussion of the integrated electron continuity and energy equations is given here as a preface to a discussion of finite difference approximations to the transport terms. The electron continuity equation is obtained by multiplying Eq. (3.3) by V_e^2 and integrating over dV_e. Using Eq. (3.8) and the boundary conditions $G(\infty) = G(0) = 0$, it simply becomes

$$\rho \frac{d(N_e/\rho)}{dt} = \int_0^\infty V_e^2 J_o \, dV_e \quad . \tag{6.1}$$

The integral of J_o contains contributions from ionization and recombination processes. The total electron kinetic energy per unit volume is

$$T_e = \frac{1}{2} m_e \int V_e F_e \, dV_e = \int E_e V_e^2 F_e \, dV_e \quad .$$

The time rate of change of kinetic energy is found by multiplying Eq. (3.3) by $E_e V_e^2 dV_e$ and integrating over velocity:

$$\rho \frac{d(T_e/\rho)}{dt} = -\frac{1}{r^2} \frac{\partial}{\partial r}\left[r^2\left(\int_0^\infty E_e V_e^2 G_e \, dV_e - UT_e\right)\right] + e\varepsilon N_e U + \int_0^\infty E_e V_e^2 J_o \, dV_e \tag{6.2}$$

where again Eq. (3.8) has been used along with a partial integration of the electric field term.

The first term on the right hand side of Eq. (6.2) includes electron thermal conduction. Since both it and the corresponding term in the microscopic equation are the divergence of a flux, Eq. (3.3) can be consistently differenced to conserve both electrons and energy. The second term on the right hand side of Eq. (6.2) is the rate at which the electric field is transferring energy from the electrons into flow energy of the ions. The differencing of this term must be consistent with the differencing of the term ZeN_i in the ion-momentum equation. The final collisional term in Eq. (6.2) contains terms involving electron-ion thermal energy exchange, electron-ion collisional PdV work and electron-(ion-radiation) interactions. The collisional terms have energy and momentum conservation built into their kinematics and finite difference approximations, as discussed in the previous chapter, will likewise be conservative or at least insofar as total number and energy can be defined over a finite mesh.

Since the difference scheme for the collisional terms is keyed to the assumption that $\partial \ln F_e/\partial E_e$ is slowly varying, it is convenient to recast Eqs. (3.5) and (3.7) in energy space. The term $\lambda_{ei} r^{-2} \, \partial(r^2 U G_e/V_e^2)/\partial r$ appearing in the expression for I_L will be dropped, being of second order in the flow velocity U.

$$I_L = \frac{1}{2} \lambda_{ei} \left[\frac{\partial(1 - f_E)}{\partial r} F_e + \frac{1}{r^2} \frac{\partial}{\partial r} \left(r^2 (3f_E - 1) F_e - 2\, \overline{\mu J_r / V_e} \right) \right.$$

$$\left. - 2e\varepsilon F_e \left(f_E \frac{\partial \ln F_e}{\partial E_e} + \frac{\partial f_E}{\partial E_e} + \frac{1}{2} (3f_E - 1)/E_e \right) \right] \tag{6.3}$$

$$\rho(t_o)\, G_e(r_L, t) = \rho(t) \left[G_e(t_o)\, e^{-\Delta\tau} - (2Ee/m_e)^{1/2}\, \overline{I}_L (1 - e^{-\Delta\tau}) \right] \tag{6.4}$$

$$R(t') = - \frac{1}{r^2} \frac{\partial}{\partial r} \left[r^2\, G_e(r_L, t') - UF_e(r_L, t') \right]$$

$$+ \frac{e\varepsilon}{E_e^{1/2}} \frac{\partial[E_e^{1/2}\, G_e(r_L, t')]}{\partial E_e} + J_o \tag{6.5}$$

The

$$\int_{t_o}^{t} R(t')/\rho(t') dt'$$

appearing in Eq. (3.5) for $F_e(t)$ is evaluated by again assuming that I_L/ρ is slowly varying over the interval from t_o to t but with the time variation in $e^{-\Delta\tau'}$ taken into account. Two integrals arise:

$$\int_{t_o}^{t} e^{-\Delta\tau'}\, dt' = \lambda_{ei}/V_e (1 - e^{-\Delta\tau})$$

$$V_e\, \lambda_{ei} \int_{t_o}^{t} (1 - e^{-\Delta\tau'})\, dt' = \lambda_{ei}^2 (\Delta\tau - 1 + e^{-\Delta\tau})$$

The exponentials are approximated, after the integration, by

$$e^{-\Delta\tau} \simeq (2 - \Delta\tau)/(2 + \Delta\tau) \quad \text{if} \quad \Delta\tau < 2$$

$$\simeq \theta \quad \text{if} \quad \Delta\tau \geq 2 \quad .$$

The final step in numerically solving for F_e is to write

$$R \equiv R_V + J_o, \; \rho \frac{d(F_e/\rho)}{dt} \equiv \left. \frac{d(F_e/\rho)}{dt} \right|_V + \rho \left. \frac{d(F_e/\rho)}{dt} \right|_{coll}$$

and to assume that the solution can be approximated by independently solving the equations

$$\rho \left. \frac{d(F_e/\rho)}{dt} \right|_V = R_V \quad \text{and} \quad \rho \left. \frac{d(F_e/\rho)}{dt} \right|_{coll} = J_o \quad .$$

This assumption is analogous to the alternating gradient technique in two spatial dimensions. Here, the coordinate and energy spaces are assumed to be separable. The electron distribution function F_e appearing in the gradient Vlasov term is forward time-differenced to give the usual implicit, unconditionally stable conduction term.

VII. NUMERICAL RESULTS

At the time of publication, the above scheme had been programmed to include electron-ion momentum exchange with U = 0 and electron-electron energy exchange in the Fokker-Planck approximation. Work was still in progress to include electron-ion collisional PdV work, bremsstrahlung absorption and emission and an internally consistent integration of the electric field Vlasov term and the ion-momentum equation.

Figures 1, 2 and 3 give some results of a calculation of electron transport through a 0.7 μm thick shell of SiO_2 filled with DT gas at a density of .002 gm/cm^3. The SiO_2, with a density of 2.25 gm/cm^3, contains 15 equally spaced radial zones and the DT gas contains 5 equally spaced radial zones. The SiO_2 and DT gas were both originally at a temperature of .07 keV. Energy groups were defined at 0, .03, .1, .3, 1., 2., 3., 4.5, 6., 8., 10., 15., 25., 40., 70. and 100 keV. A 2.0 keV Maxwellian distribution of electrons was maintained in the outer zone of SiO_2 (J=20). The Eddington factor was taken to be 1/3 for all electrons. The diameter of the shell is 55 μm.

Figure 1 is a plot of average electron energy versus zone at times $2 \cdot 10^{-13}$ sec and $4 \cdot 10^{-13}$ sec. At these times, the transporting electrons (for which the SiO_2 is basically transparent) have reached the outer zone of DT gas (J=5) and are heating it, but have

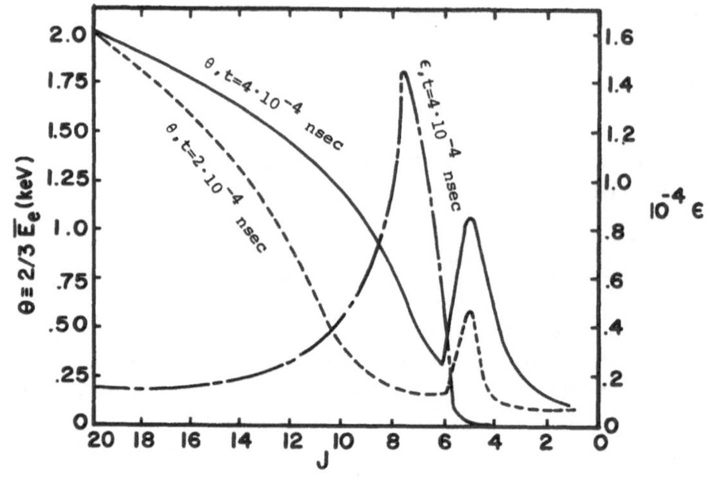

Figure 1. Spatial Distribution of Average Energy and Electric Field

not yet reached the next DT zone. The electric field is shown as a function of zone number at $4 \cdot 10^{-13}$ sec. It is seen that the field peaks rather strongly at the head of the diffusive region of the electron flow. Figures 2 and 3 are plots of the electron distributions in SiO_2, where the flow is comparatively diffusive, and in the relatively transparent DT gas, at points which have comparable average energies. $E_e^{3/2}\ F_e$ is also plotted to show the kinetic energy distribution.

The above results, for an idealized situation, dramatically illustrate the effect that non-diffusive, yet collisionally initiated, electron transport can have, particularly upon low density, low atomic number material inside a more diffusive material. It is hoped that these preliminary results will prompt further interest in such calculations.

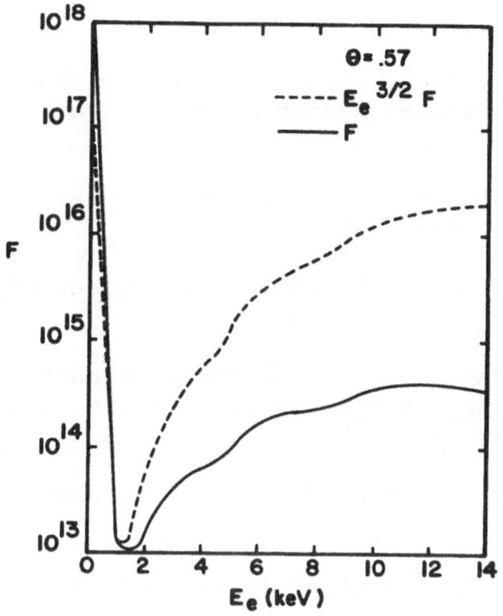

Figure 2. Electron Distribution in DT (J=5), $t = 2 \cdot 10^{-4}$ nsec.

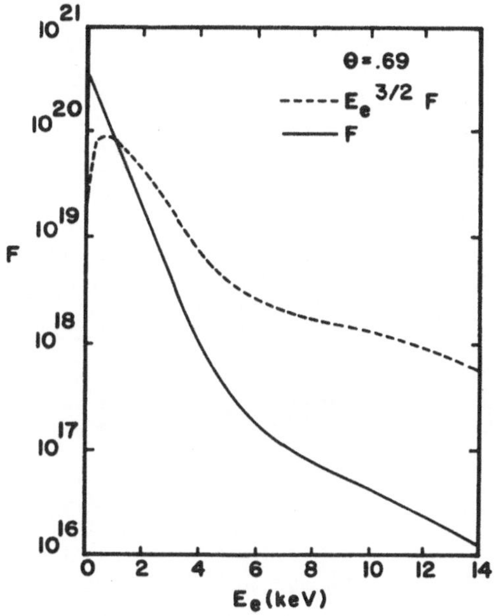

Figure 3. Electron Distribution in SiO_2 (J=11) at $t = 2 \cdot 10^4$ nsec.

REFERENCES

Freeman, B. E. et al, 1968, unpublished, "The VERA Code: A One-Dimensional Radiative Hydrodynamic Program", DASA Report No. 2135.

Hochstim, A., 1969, Ed. Kinetic Processes in Gases and Plasmas, Academic Press, New York.

Holt, E. H. and R. E. Haskell, 1965, Foundations of Plasma Dynamics, McMillan.

Killeen, J. and K. Marx, 1970, "The Fokker-Planck Equation", in Methods in Computational Physics, Vol. 9, Plasma Physics, Academic Press, New York.

Rosenbluth, M. N., W. M. MacDonald and D. L. Judd, 1957, Phys. Rev. 107, 1.

Spitzer, L., 1962, Physics of Fully Ionized Gases, 2nd edition, Interscience, New York.

Whitney, J. C., 1970, J. of Comp. Physics 6, 483.

COMPUTER EXPERIMENTS IN ONE-DIMENSIONAL PLASMAS

Marc R. Feix

CRPE/CNRS
Université d'Orléans
France

TABLE OF CONTENTS

COMPUTER EXPERIMENTS IN ONE-DIMENSIONAL PLASMAS

Marc R. Feix

CRPE/CNRS, Universite D'Orleans, France

(Based on work by R. Navet, E. Bonomi and M. R. Feix)

I. INTRODUCTION

Computer simulation of an electrostatic plasma is now a well developed field. It was started in about 1960 by different authors [Dawson, 1962; Eldridge and Feix, 1962; Feix, 1969]. From the very beginning the question of dimension was found to be a central one. both from a theoretical and a practical point of view, and is closely related to the number of particles which can be treated on the computer (a number much smaller, of course, than in reality). For practical reasons the first simulations were one-dimensional [Dawson, 1962; Eldridge and Feix, 1962]. Two-dimensional plasmas are currently studied and the real-three dimensional plasma is beginning to be investigated. A three-dimensional plasma is composed of point particles, positively and negatively charged; a two-dimensional plasma is a set of parallel rods moving in the plane perpendicular to their direction; finally a one-dimensional plasma is a set of parallel planes moving in a direction perpendicular to themselves, with given charge and mass per unit area.

Moreover, computer simulation, through the dynamics of particles, can be performed with two different philosophies. In the first one, we want to study what happens in a hot plasma where the coupling parameter is small. In fact, we are often interested in the limit where this parameter is so small that it can be considered as vanishing. In that case a "computer particle" is no longer an elementary structure which cannot be destroyed, but a practical way to discretize information in phase space. In that case, the particle method (often called the lagrangian method) must be compared with other (eulerian) methods where we consider the distribution function $f(x,v,t)$ as field described, for example, by Fourier or

Hermite components or sampled values of f in phase space. Indeed we are often interested in decreasing the noise due to the grain structure of our particles and we usually smooth the field created at small distances (concept of "cloud" particles [Dawson, 1970]) as an alternative to the process of increasing the number of particles.

The point of view of the statistical physicist is just the opposite: he wants to study the individual effects, i.e. these effects connected to the finite values of the elementary charge and mass.

Now we can distinguish two kinds of plasmas: the weakly correlated plasmas where this grain effect is small. Their simulation is, in a certain sense, more difficult since we must have long distances (to allow the development of long wavelength fluctuations) and a great number of particles in the Debye cube (square, length) for the three- (two-, one-) dimensional case.

Strongly correlated plasmas, on the other hand, are easier to simulate numerically since we have only a small number (eventually less than one) of particles per Debye length (for a one-dimensional plasma). The total number of particles can consequently be kept reasonable although other difficulties will show up.

II. THE GRAININESS FACTOR

Let us define the graininess factor which is the only parameter in a classical plasma. The computer people quite often use the quantity

$$g = 1/n\ D^d \tag{2.1}$$

where D is the Debye Length

$$D = \sqrt{\varepsilon_o / n e^2\ \beta}\quad ,$$

with n the particle density, e the charge and β the inverse of $k_B T$; d is the dimension. We rewrite Eq. (2.1) as:

$$g = e^d\ \beta^{d/2}\ n^{d/2-1}\ \varepsilon_o^{\ -d/2} \tag{2.2}$$

Equation (2.2) shows that the graininess parameter g:

- increases for d = 1, 2, 3 when T decreases;
- increases when n increases for three-dimensional plasmas;
- is independent of the density for two-dimensional plasmas;

decreases when n increases for one-dimensional plasmas.

The term "graininess parameter" can be given an operational content in computer experiments by noticing that if we cut a particle in two, each part carrying half of the charge and half of the mass but keeping the same velocity (with, as a consequence, the temperature divided by two), then the Debye length, the thermal velocity and also the plasma frequency are invariant while the graininess is divided by two. Consequently, the inverse of the graininess factor is proportional to the number of particles we should take in the simulation.

Now people in statistical physics are used to the ratio of the potential energy to the kinetic energy and in three-dimensions define:

$$\Gamma = (4\pi\ \varepsilon_o)^{-1}\ e^2\ n^{1/3}/k_B T$$

We have

$$g = (4\pi)^{3/2}\ \Gamma^{3/2}$$

If we define a coupling Γ in one dimension by

$$\Gamma = e^2\ n^{-1}/\varepsilon_o\ k_B T$$

we have

$$g = \Gamma^{1/2} \qquad (2.3)$$

III. WHY ONE DIMENSION?

As we have just mentioned, the behavior of a strongly correlated plasma is governed both by the collective and individual effects (i.e. effects where the exact motion of the particles must be correctly treated, including the short distance effects). It is well known that the N body problem is usually quite unstable in three and two dimensions. The crucial point is that the one-dimensional model can be exactly treated (except for unavoidable round-off errors) because of the very simple relation between positions and fields. Figure 1 shows the electric fields for a two-component plasma (TCP) and for a one-component plasma (OCP). In these cases, as long as a particle does not cross one of its neighbors, the fields experienced by this particle are respectively

$$E = E_o$$

$$E = E_o + \frac{n\sigma}{\varepsilon_o}\ x \qquad (3.1)$$

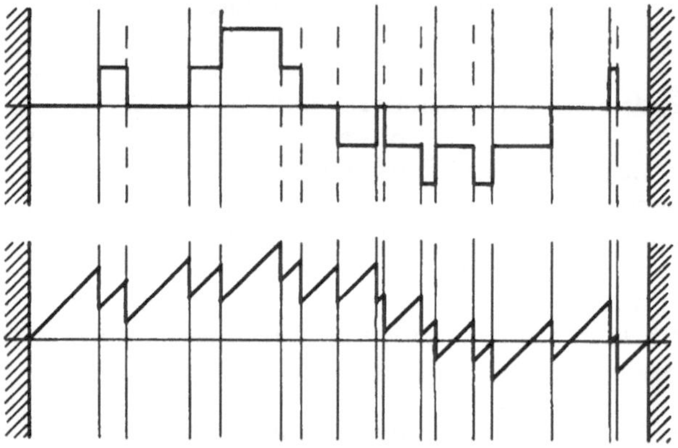

Figure 1. Electric field profile for two-component (upper part) and one-component (lower part) plasma. In both cases the plasma is strictly neutral.

where $n\sigma$ is the charge density of the background. The trajectories are consequently a succession of arcs of parabolas or parts of sinusoids in the TCP and OCP cases, respectively.

When two particles cross, we have no reason to add another field and will consider that they pass through each other if they are of different types. If they are of the same type, the addition of a hard repulsive core for an infinitessimal distance is equivalent to an exchange of the names of the particles with no physicl consequence. (In the appendix, E. Bonomi adopts this point of view in an OCP).

In these one-dimensional problems, the exact treatment of the system is consequently based on the precise bookkeeping of the sequence of particles and has been described previously [Eldridge and Feix, 1962].

This possibility of solving exactly the equations of motion of the system in the 2N-dimensional Liouville phase space Γ is nicely illustrated when we try on our program the most difficult test, i.e. the test of microreversibility (since, as we previously stated, the trajectory in the Γ space is highly unstable). Nevertheless, if the code is precise enough, the reversal at $t = T$ of all velocities should bring the system back to its initial state at time $t = 2T$. Figure 2 shows that indeed all curves are precisely reproduced after an inversion which takes place at $T = 10\pi\ \omega_p^{-1}$.

These interesting numerical properties of the model are matched

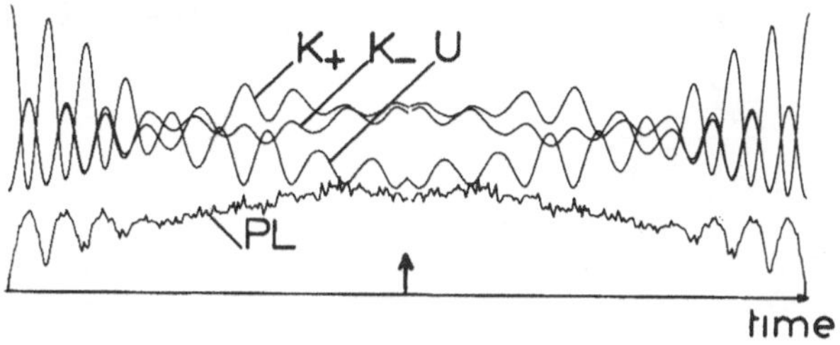

Figure 2. Test of the microreversibility of the code. At time indicated by the arrow the velocities are reversed K_+ kinetic energy (ions), K_- kinetic energy (electrons), U potential energy and PL product pressure X length. (Arbitrary scales, origins shifted).

by equally remarkable properties of the theoretical description. Lenard [1961; 1962] and Prager [1961] for the TCP, and Baxter [1964; 1963] and Kunz [1974] for the OCP have computed the thermodynamic functions of these systems taking into account the possibility of a description through the increasing sequence of the positions. We will come back to these points later.

Now we want to return to the crucial question of what can be learned from one-dimensional simulations of real (three-dimensional) systems. Again we must make a distinction between weakly and strongly correlated plasmas.

In a weakly correlated plasma, a central hypothesis is the random phase approximation which states that if we consider the collective variable ρ_k with (for an OCP)

$$\rho_{\vec{k}} = \sum_{i=1}^{N} \exp i\vec{k} \cdot \vec{x}_i$$

we can take all the ρ_k as completely independent variables describing the complex relations between the positions of the particles, while strictly speaking, only 3N $\rho_{\vec{k}}$ are independent. Now both in three and one dimensions the statistical properties of $\rho_{\vec{k}}$ are given by the dielectric constant formalism, exactly identical in the two cases (at least for isotropic three-dimensional plasmas). Now if the properties in $\vec{k}$ space are the same, properties involving summation in $\vec{k}$ may be different since in the integration process the $\vec{k}$ are weighted differently.

The best example is the computation of the three-dimensional and one-dimensional OCP correlation function between particles. From the same relations for the average fluctuating values of $<\rho_{\vec{k}}\, \rho_{\vec{k}}^{*}>$:

$$<\rho_{\vec{k}}\, \rho_{\vec{k}}^{*}> = N \frac{k^2 D^2}{1 + k^2 D^2}$$

valid for three and one dimension, we deduce by integration for three-dimensional OCP:

$$n_{12}/n^2 = 1 - g/(r/D) \exp - r/D \tag{3.2}$$

for one-dimensional OCP:

$$n_{12}/n^2 = 1 - g/2 \exp - |x|/D \tag{3.3}$$

In Eqs. (3.2) and (3.3) r and x are the distance between particles 1 and 2.

We notice that in the three-dimensional case the correlation is not correctly computed at short distance (where, due to the $1/r^2$ character of the Coulomb field, the coupling cannot be weak, while the one-dimensional plasma does not introduce any difficulty). Nevertheless, a check on Eq. (3.3) will also be a confirmation of Eq. (3.2) since the central complex hypothesis on the $\vec{k}$ independence and the general philosophy of the computation is the same to obtain the dielectric constant through the linearized Vlasov equation).

For a strongly coupled plasma, things become more difficult since we do not have a precise scheme. Moreover, the beautiful independence of the collective variables $\rho_{\vec{k}}$ is no longer valid since we now have an interaction between the different $\rho_{\vec{k}}$, implying a nonlinear formalism.

Nevertheless, some very general "ansatz" in statistical physics remain unchanged in three and one dimension. An example is the central ergodic hypothesis which states that the time average on one system must reproduce the ensemble average through which the thermodynamic quantities are obtained. Moreover, the different hypotheses to cut the hierarchy quite often do not depend on the dimensionality of the problem. If after such a hypothesis the problem can be rigorously solved both for three and one dimensions, the computer will eventually prove or disprove the hypothesis. Unfortunately, to solve the complex nonlinear equations it will be necessary to use new mathematical simplifications which could well be connected to the dimensionality of the problem.

Finally, some phenomena may be similar, like the crystallization, the melting, the pair correlation behavior.

IV. THE TWO-COMPONENT PLASMA

A. Thermodynamics

The general theory has been given by Lenard. We have two interesting limits:

In the weakly coupled case ($g \to 0$), i.e. the Vlasov limit, the gas becomes a perfect gas with a ratio of the potential energy to the kinetic energy going to zero and a pressure given by $n\,k_BT$.

In the opposite strongly coupled case ($g \to \infty$) we see the formation of neutral pairs (ion-electron). Figure 3 shows the picture of a part of the system for $nD = .429$. Most of the electrons are clearly paired with one ion except on a small zone where we have a clustering of three ions and three electrons. Going to more quantitative results, it can be shown that in the $g \to \infty$ limit:

$$U = \frac{\varepsilon_o}{2} \int_o^L E^2\, dx$$

goes to zero as the kinetic energy $K = 1/2\ N\,k_BT$. Consequently, using the virial theorem

$$PL = 2K - U \tag{4.1}$$

we see that PL goes to zero as $1/2\ n\,k_BT$, i.e. as for a perfect gas of N/2 atoms.

Figure 4 gives a computational check of the Lenard result. The agreement is quite good. We must outline that here we did not measure the pressure directly but used Eq. (4.1).

In the TCP model there is no ambiguity in the definition of U and consequently no difficulty in defining the pressure by means of the virial theorem.

B. Fluctuations

The following problem was suggested by J. Leibowitz. What are the charge fluctuations in a box of length Δ immersed in a much longer plasma? It provides an interesting insight on the test particle picture concept. The theory goes as follows: Figure 5 shows the box of length Δ and a particle situated at $x = x_o$ inside the box. This particle is dressed by its cloud, as given by Eq. (3.3), and the charge density is:

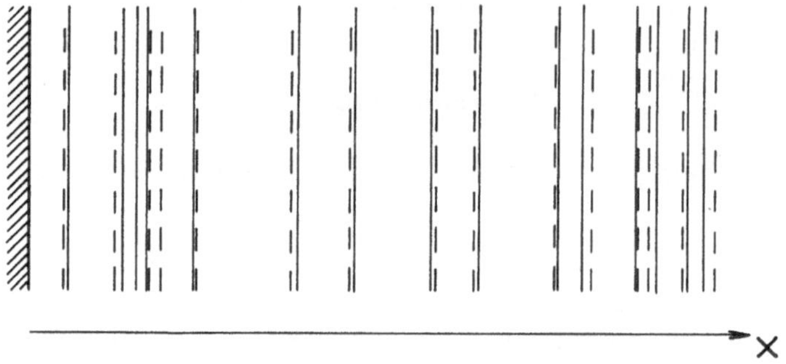

Figure 3. Formation of pairs (electron-ion) in a TCP (nD = .43).

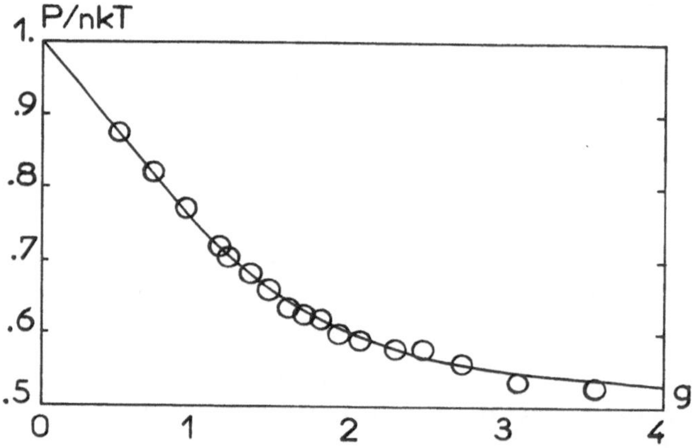

Figure 4. TCP: $P/N\ k_BT$ function of $g = (nD)^{-1}$. (Theoretical curve from Lenard).

$$n(x)\sigma = \sigma\delta(x - x_o) - \frac{\sigma}{2D} \exp - \left|\frac{x - x_o}{D}\right| \tag{4.2}$$

The charge included in the box is

$$q(x_o) = \int_o^\Delta \sigma n(x)\, dx = \frac{\sigma}{2}\left[\exp - \frac{x_o}{D} + \exp - \frac{\Delta - x_o}{D}\right] \tag{4.3}$$

The average charge of a test particle located inside the box, including the cloud, is:

$$\overline{q} = \frac{1}{\Delta}\int_o^\Delta q(x_o)\, dx = \sigma\,\frac{D}{\Delta}\,(1 - \exp - \Delta/D)$$

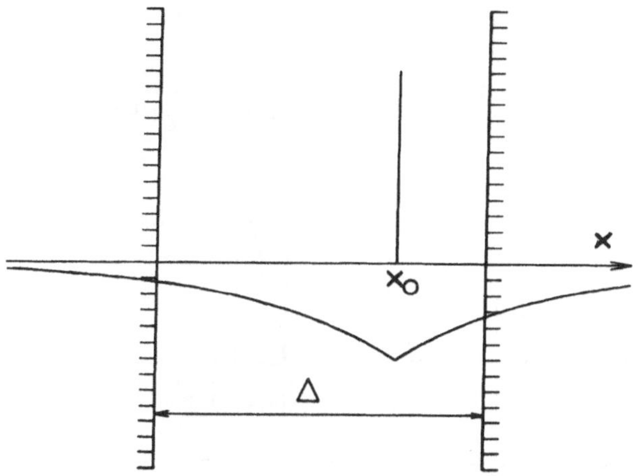

Figure 5. Test particle, located at $x = x_o$, and its cloud in a box of length Δ.

Introducing the hypothesis of independence of dressed test particles (nΔ in the box of length Δ) we obtain for the average values of $(N_e - N_i)^2$

$$\langle(N_e - N_i)^2\rangle = nD(1 - \exp - \Delta/D) \tag{4.4}$$

In Eq. (4.4) n is the total density (ion + electrons) per unit length. Equation (4.4) exhibits two interesting limits:

For small Δ (more precisely $\Delta \ll D$), $\langle(N_e - N_i)^2\rangle \to n\Delta$ and the fluctuations vary as the square root of the total number of particles in the box, a result identical to the case of noninteracting particles. For a box smaller than the Debye length the interacting character of the particles does not appear.

For large Δ ($\Delta \gg D$), $\langle(N_e - N_i)^2\rangle$ goes to a finite limit. It is clear that the method can be generalized to three dimensions by considering test particles with their clouds located randomly inside a cube. The result will be that $\langle(N_e - N_i)^2\rangle$ will vary for large volumes as the area, i.e. as $V^{2/3}$ (V volume).

Computer experiments have been performed on this problem and are given in Figure 6. For nD = 2.7 and .79 (although in this last case we are already dealing with a strongly coupled plasma) the theoretical prediction of Eq. (4.4) agrees very well with the computer simulation; for smaller nD the computer simulation exhibits a stronger screening, indicating that the hypothesis of independence

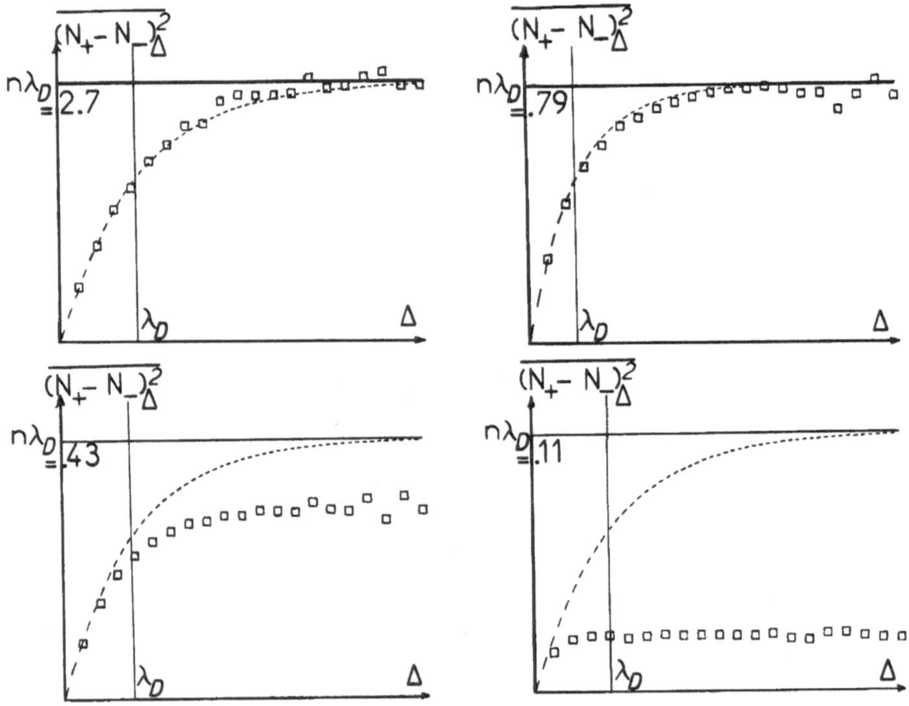

Figure 6. TCP: Fluctuations of the charge density as a function of the length of the box for different values of nD.

of the dressed test particles is no longer valid (i.e. that it is insufficient to surround each particle by its cloud. Moreover, the linear calculation of the cloud as it is given in Eq. (4.2) is certainly not possible any more). The property of $<(N_e - N_i)^2>$ going to a much smaller limit can be explained by the fact that now we are dealing with a system of neutral atoms (as shown explicitly in Figure 3). Deviation from perfect neutrality in the box will occur only when one of the pair will have its electron inside the box and its ion outside (or vice versa), an event of probability proportional to the size of the neutral atoms. Dimensional arguments show that this size will be of the order of Dg^{-1} and the number of atoms sitting on the edges of the box will be of order

$$n\Delta \ \frac{Dg^{-1}}{\Delta} = nD\ g^{-1} \quad .$$

We can check that indeed for large g and large box $<(N_e - N_i)^2>/nD$ varies as g^{-1}.

Unfortunately this kind of result obtained for large g cannot be transposed to the three-dimensional case since the treatment of the pair formation must involve quantum mechanics. (In purely classical physics we have an unavoidable divergence for electron-ion correlation). But slight deviation from the Debye Hückel limit as given by second order treatment of the BBGKY hierarchy can be tested on the one-dimension scheme and transposed to three dimensions.

V. THE ONE-COMPONENT PLASMA

A. The Minimum Energy Configuration

The one-component plasma is a favorite model for theoreticians, the main reason being that it has simpler mathematics. The model is a motionless continuum background of one sign in which evolve N particles. Some justifications can be found in the fact that some very dense plasmas have their electrons completely degenerate (and consequently not able to move). Moreover, they are diluted on a distance characterized by the DeBroglie wavelength, much larger than the interparticle distance. Ions, due to their large mass, can still be considered as classical.

Unfortunately the OCP (sometimes called Jellium) must be handled with care when we use the thermodynamic formalism, especially when we derive with respect to the volume, since the continuous background can exchange momentum and energy with the system of N particles. On the other hand, the background is not only submitted to the field created by the particles and itself (but also to constraints which keep it uniform). Its motion in an expansion (increase of length of the system) is very artificial. Consequently, we will refrain from applying theorems such as the virial theorem which imply a global balance of momentum between the entire system and the walls.

Now the OCP behavior is dominated by the existence of a minimum energy configuration for a box of length L which imposes precise positions for the N particles. This property is irrespective of the length of the box.

We consider such a box of length L with N particles and a background of density N/L in order to assume a strict neutrality of the system. To avoid difficulties connected to the definition of the potential energy we write for this last quantity U

$$U = \frac{1}{2\varepsilon_o} \int_o^L E^2 \, dx \tag{5.1}$$

where E is the electric field. After some algebra this last quantity is transformed into:

$$U = \frac{1}{2} m \Omega_p^2 \sum_{i=1}^{N} (X_i - a_i)^2 + \frac{\sigma^2 L}{24\varepsilon_o} \qquad (5.2)$$

with

$$a_i = (2i - 1) \frac{L}{2N}$$

$$\Omega_p^2 = \frac{N\sigma^2}{\varepsilon_o\ Lm} \qquad (5.3)$$

Equation (5.2) shows that the energy is minimum if the particles are located at the center of the N segments dividing the box into N equal parts. Moreover, Eq. (5.2) looks as if the problem could be reduced to N uncoupled harmonic oscillators. We must remember that the coupling comes from the fact that the X_i must be classified in ascending order. Here it can be useful to adopt the point of view of particles including also a hard repulsive core and exchanging their velocities at each collision (or crossing).

B. Kinetic Versus Potential Energy

Figure 8 shows the behavior of the total energy $E/N\ k_BT$ with the total energy E given by

$$E = U + K - \frac{\sigma^2 L}{24\varepsilon_o} \quad .$$

The theoretical curve is given by Baxter. Again we see the two limits:

$T \to \infty$, the kinetic energy (much larger than the potential energy) is $N\ k_BT/2$ and the curve starts for $g = 0$ from the value 1/2.

$T \to 0$, particles are more and more located close to the values a_i and the crossings become less and less frequent (only a few particles belonging to the tail of the Maxwellian distribution can escape their center of force). The coupling due to the "exchange" scheme goes to zero and basically we have a set of N harmonic oscillators with equipartition of the kinetic and potential energy redefined as

$$\frac{1}{2} m \Omega_p^2 \Sigma(x_i - a_i)^2 \quad .$$

As can be seen, Baxter's theoretical values are fully supported by the computer simulation.

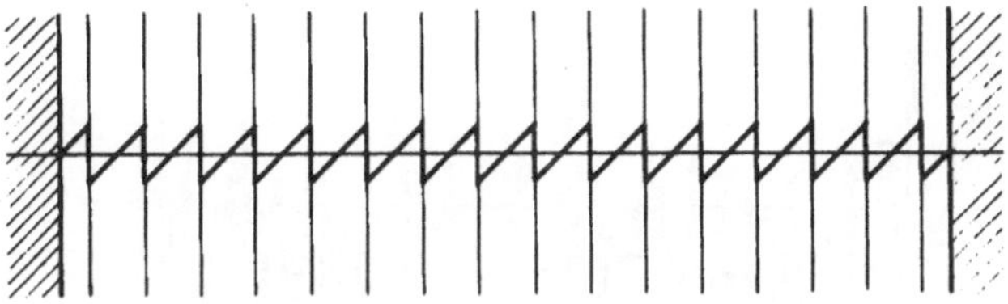

Figure 7. OCP: Electric field in the minimum energy configuration for a 16-particle system.

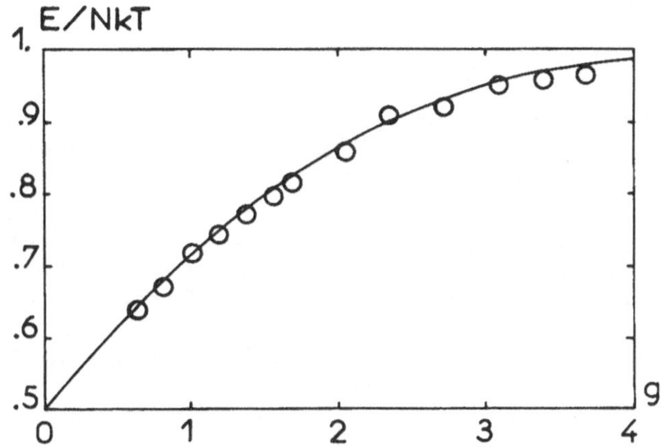

Figure 8. OCP: Total energy as a function of $g = (nD)^{-1}$. (Theoretical curve from Baxter).

C. One-Body and Two-Body Correlations

As we have already seen, the system is not invariant for an arbitrary translation and particles tend to stay around the points a_i. The distribution function n(x) obtained by superposing a great number of pictures of the system at different times should consequently exhibit a periodic behavior with maxima at points $x = a_i$. This is the meaning of the result obtained by Kunz although it is difficult to extract from his paper the amplitude of the periodic behavior compared to the average density N/L.

Figure 9 shows what happens for a very small system (taking eight particles) and dividing the total length in 160 smaller boxes of equal length. The average is taken 10.000 times (at each ω_p^{-1}) which gives an average value of 500 samples in each box. This

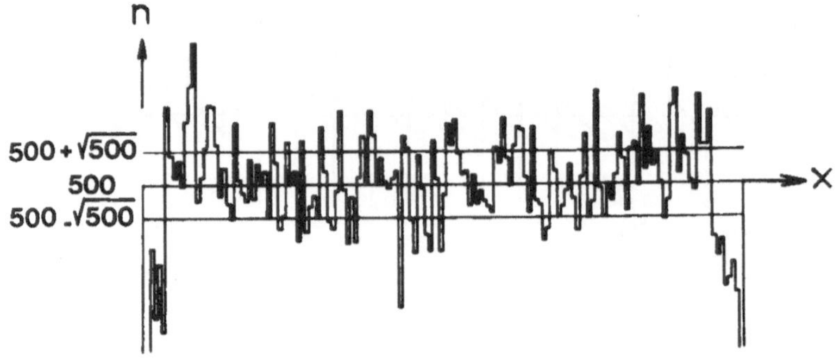

Figure 9. OCP: Density of charges in the box (eight particle system).

corresponds to the central line of Figure 9. The two other lines corresponding to the value $500 \pm \sqrt{500}$. The corresponding g is g = 1.47. Two remarks are in order:

No periodic behavior appears on n(x). Although for $g \to \infty$ we will obtain a trivial sum of δ functions (since the particles must be located at $x = a_i$), nothing appears for g = 1.47.

On the extremities we definitely have a lesser number of particles.

These results are of course very preliminary. We must work with greater g but in that case the interesting coupling (i.e. the crossing) is due to the very few energetic particles of the tail of the maxwellian distribution. To improve statistics we must deal with a greater number of particles (to have enough crossing and consequently mixing in the Liouville phase space).

This absence of periodic behavior is somewhat surprising since Kunz predicts a crystal behavior at any temperature. Of course, in Figure 9 we may have a periodic behavior which is completely masked by thermal fluctuations and of very small amplitude.

Since we have only one parameter g in this problem we were expecting that the crystal behavior would have appeared around g = 1. In fact we must go to higher values of g. This is due to the expression of the residual energy $\sigma^2 L/24\varepsilon_o$. Let us suppose that the total energy is just equal to twice this value and that we have an equipartition of particle energies between the kinetic and the $m/2\Omega_p^2 \, \Sigma(x_i - a_i)^2$ potential energy, consequently.

$$N \frac{k_B T}{2} = \frac{\sigma^2 L}{24\varepsilon_o} \tag{5.4}$$

indicating a $g^2 = 24$ or $g \simeq 5$. For this value of g the energy available is just equal to the irreducible part $U_o = \sigma^2 L/24\varepsilon_o$. We could expect that around this value g = 5 crystallization effects will begin to be important.

Two-body correlation functions show, of course, the same lack of periodic behavior for nD small but not very small. In these experiments we have measured the average number of particles present at a distance x_{12} of a given particle. Figure 10 shows the result for nD = .9 where the experimental curve fits pretty well the Debye-Hückel result (although g is greater than 1: as in the TCP we see that the linearized Debye-Hückel theory is quite good not only for g << 1 but up to g = 1 at least, a somewhat surprising result). Figure 11 shows the result for nD = .43. Again no periodic behavior is obtained. In Figures 12 and 13 (respectively for nD = .37 and nD = .25) the periodic component (with maximum at $x_{12} = n^{-1}$, $2n^{-1}$, $3n^{-1}$) begins to appear (in Figure 12) and becomes very important (in Figure 13). Unfortunately these curves are somewhat sensitive to the initial conditions and to the time of averaging. It seems that we are not reaching the thermodynamic equilibrium and that the system is trapped in a region of the Liouville phase space from which it cannot escape. We must work with a much higher number of particles but qualitatively Figures 12 and 13 indicate that indeed around g = 3 - 4 the crystal character begins to be important.

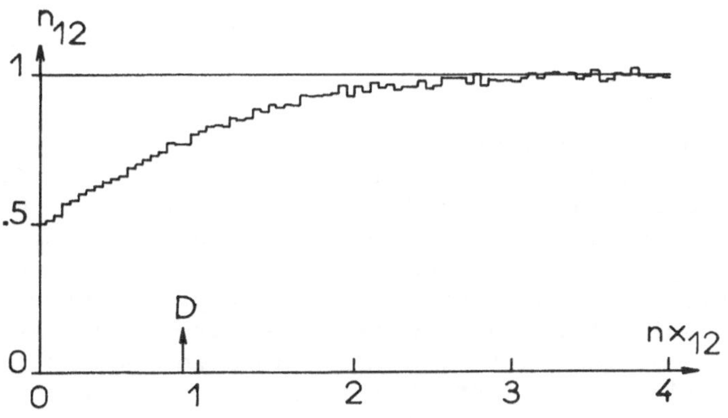

Figure 10. OCP: nD = .9, 40 particles time average on $10^4\ \omega_p^{-1}$.

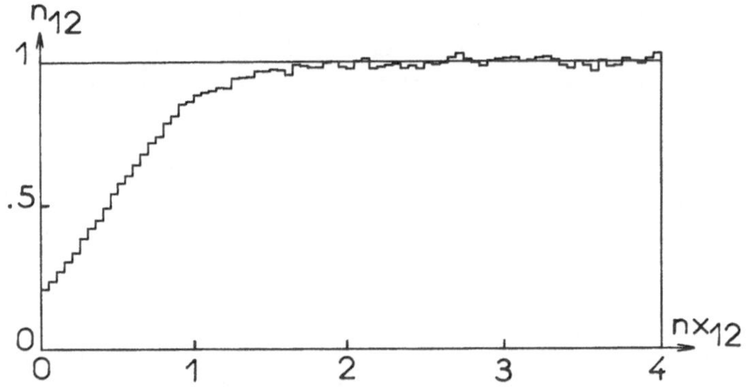

Figure 11. OCP: nD = .43, 40 particles time average on 10^3 ω_p^{-1}, ensemble average on 32 systems.

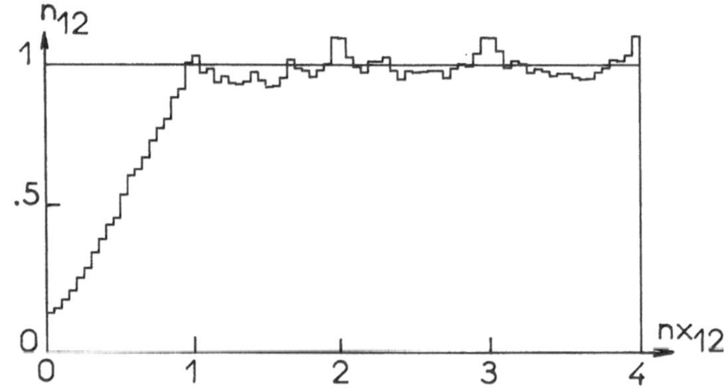

Figure 12. OCP: nD = .37, 40 particles time average on 10^3 ω_p^{-1}, ensemble average on 32 systems.

This trapping of the system in a region of the Γ space from which it cannot escape is shown in Figure 14. We simulate a crystallization process with boundaries at temperature T = 0. Any particle falling on the boundary is reintroduced with a velocity equal to zero. Energy is consequently taken away but we see that for a 16-particle system we arrive at a situation where the amplitudes of the oscillations of the particles are such that no crossing and no striking of the wall take place anymore. Although the system still has kinetic energy left, the crystallization process has been stopped. Of course, a much higher number of particles would allow further exchanges and very likely a trapping of the system in a region of smaller energy.

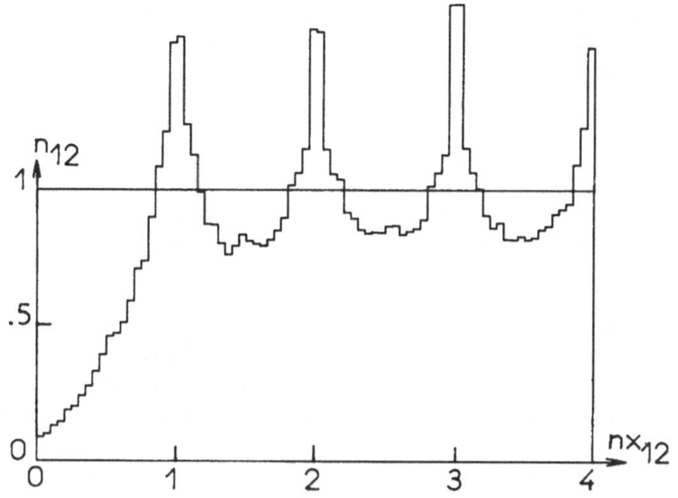

Figure 13. OCP: nD = .25, 40 particles time average on $10^3\ \omega_p^{-1}$, ensemble average on 32 systems.

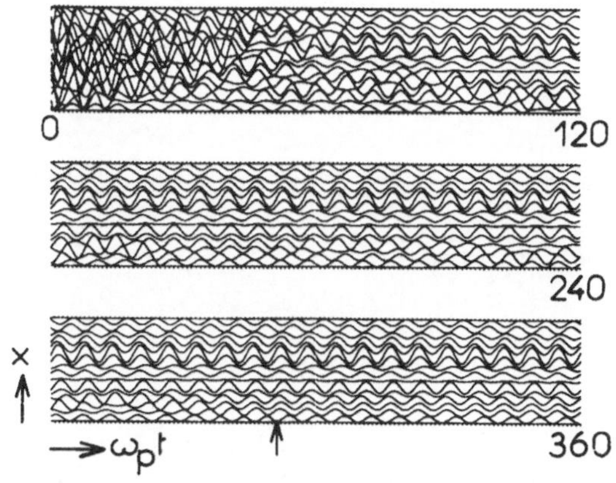

Figure 14. Crystallization for an OCP (16 particles) with walls at T = 0. The arrow indicates the definitive trapping of the system. The time varies from 0 to 360 ω_p^{-1}.

VI. CONCLUSION

What did we get from one-dimensional computer experiments? First of all, a complete and precise check of the theory of Lenard and Baxter respectively for TCP and OCP on the computation of thermodynamic quantities (potential energy, equation of state, at least for a TCP; For an OCP the pressure must be computed avoiding the volume derivative).

The second interesting point is both for TCP and OCP the validity of the Debye-Hückel theory and related concepts (such as test particle, random phase approximation) based on a linearized treatment of the Vlasov equation. They give good quantitative results up to $g = (nD)^{-1} \simeq 1$; second-order effects in g may consequently be very difficult to measure.

The third point concerns the appearance of a periodic structure for a cold OCP. Studies of such systems through particle dynamics methods must be undertaken with a rather large number of particles in order to prevent the system from stopping its ergodic motion in the Liouville phase space, remaining trapped in a region of this space. We feel that this remark may also be valid in three dimensions and invite more careful study of the computer model both in Monte Carlo and particles dynamics methods.

Other problems can be studied with this model. For example the partition of particle energies and the formation of sheaths for non-neutral plasmas.

REFERENCES

Baxter, R. J., 1964, Phys. of Fluids 7, 38.

Baxter, R. J., 1963, Proc. Camb. Phil. Soc. 59, 779.

Dawson, J., 1962, Phys. of Fluids 5, 445.

Dawson, J., 1970, Methods of Comput. Phys. 9, 1.

Eldridge, O. C. and M. Feix, 1962, Phys. of Fluids 5, 1076.

Feix, M., 1969, "Mathematical models of plasma" in Nonlinear Effects in Plasma, edited by G. Kalman and M. Feix, p. 151, Gordon and Breach, New York.

Kunz, H., 1974, Ann. Phys. 85, 803.

Lenard, A., 1961, J. Math. Phys. 2, 682.

Prager, S., 1961, "The one-dimensional plasma" in Advances in Chemical Physics, Vol. IV. p. 201, Wiley Interscience, New York.

APPENDIX

E. Bonomi
Ecole Polytechnique Federale
Lausanne, Switzerland

I. INTRODUCTION

In this appendix we study more precisely the dynamical aspects of the one-dimensional jellium. We are interested in knowing what happens if we consider the system in its ground state (T = 0), when we excite only one particle. We want to investigate how this perturbation will propagate in the system. As we have seen in the paper we can compare this system to a set of linear oscillators of the same frequency ω_p. Although the form of the hamiltonian looks like decoupled oscillators, a coupling both at the dynamical and thermodynamic levels exists through the need of constantly keeping in order the sequence q_i of increasing values.

In the first section, we shall describe the dynamics of these oscillators.

In the second section, we shall give a rule, confirmed by computer experiments, which allows us to know the number of plasma oscillations we can excite when perturbing one particle, the system being in its ground state. This number is roughly proportional to the initial energy of the particle.

II. TIME EVOLUTION

As we have seen in the main paper, the hamiltonian of a particle of rank i attracted by the ith potential well is:

$$h(p_i;\ q_i) = \frac{p_i^2}{2m} + \frac{m\,\omega_p^2}{2}(q_i - a_i)^2$$

$h(q_i;\ p_i)$ being a constant as long as this particle is not crossed by any other.

Let us define for times between two shocks:

$$P_i(t) = \frac{\rho}{m\,\omega_p}\,p_i(t)$$

$$Q_i(t) = \rho\,q_i(t)$$

$$t\varepsilon[t_1;\ t_2 + \Delta t] \tag{A.1}$$

Note that in this annex we adopt the concept of hard elastic collisions (supplementing the Coulomb interaction) between particles to keep the same order of labelling: $0 \leq q_1 \leq q_2 \leq \ldots \leq q_N \leq +L$. Of course, for one species problem "this point of view is strictly equivalent to the concept of crossing" and the adopted point of view is a matter of taste. From A.1 $h(p_i, q_i)$ becomes:

$$h\left\{P_i[p_i(t)];\ Q_i[q_i(t)]\right\} = \frac{m\omega^2}{2\rho^2} \cdot [P_i^2 + (Q_i - \bar{a}_i)^2]$$

$$\bar{a}_i = a_i\rho \qquad \text{and} \qquad |\bar{a}_i - \bar{a}_{i+1}| = 1 \tag{A.2}$$

From the canonical equations we have:

$$\dot{P}_i(t) = -\omega_p \cdot [Q_i(t) - \bar{a}_i]$$

$$\dot{Q}_i(t) = \omega_p \cdot P_i(t)$$

Therefore we can write, keeping in mind conditions (A.1)

$$\begin{pmatrix} P_i(t) \\ Q_i(t) - \bar{a}_i \end{pmatrix} = \phi_t \begin{pmatrix} P_i(t - t_1) \\ Q_i(t = t_1) - \bar{a}_i \end{pmatrix}$$

$$\phi_t = \begin{pmatrix} \cos\omega_p t & -\sin\omega_p t \\ \sin\omega_p t & \cos\omega_p t \end{pmatrix}$$

Consequently, using the transformation (A.1), the trajectory of a particle in its own phase space, $\forall t\ [t_1;\ t_1 + \Delta t]$, is an arc of a circle of radius R_i, centered in a_i:

$$R_i^2 = \frac{h[P_i(t);\ Q_i(t)]}{\omega_p^2 \cdot m} \cdot 2\rho^2 \tag{A.3}$$

It should be noted that R_i^2 is proportional to the energy of the particle in the ith potential well. To describe the dynamics of shocks we show all the orbits $[P_i(t);\ Q_i(t) - \bar{a}_i]$ in the same space EcR^2.

We are interested in the time-evolution equations of the neighboring particle on the right associated with the $(i+1)^{th}$ potential well. If a shock for particle i occurs first with particle i-1 (on the left) it will be described by the dynamical evolution of this last particle. In $t\varepsilon[t_1; t_1 + \Delta t]$ this evolution is given by:

$$\begin{pmatrix} P_i(t) \\ Q_i(t) - \bar{a}_i \end{pmatrix} = \phi_t \begin{pmatrix} P_i(t_1) \\ Q_i(t_1) - \bar{a}_i \end{pmatrix} \quad ;$$

$$\begin{pmatrix} P_{i+1}(t) \\ Q_{i+1}(t) - \bar{a}_{i+1} \end{pmatrix} = \phi_t \begin{pmatrix} P_{i+1}(t_1) \\ Q_{i+1}(t_1) - \bar{a}_{i+1} \end{pmatrix}$$

After a shock at time $t_2 = t_1 + \Delta t$ we shall have for the following interval of time without shocks $[t_2; t_1 + \Delta t]$:

$$\begin{pmatrix} P_i(t) \\ Q_i(t) - \bar{a}_i \end{pmatrix} = \phi_t \begin{pmatrix} P_{i+1}(t_2) \\ Q_i(t_2) - \bar{a}_i \end{pmatrix} \quad ;$$

$$\begin{pmatrix} P_{i+1}(t) \\ Q_{i+1}(t) - \bar{a}_i \end{pmatrix} = \phi_t \begin{pmatrix} P_i(t_2) \\ Q_{i+1}(t_2) - \bar{a}_{i+1} \end{pmatrix} \qquad (A.4)$$

where

$$Q_i(t_2) = Q_{i+1}(t_2) \quad .$$

See Figure A.1. Notice we have simply interchanged the velocities at time t_2 of particles i and i+1.

It is clear that if for every t there are no shocks, no energy is exchanged and the trajectory of the particle will be a circle centered in $\bar{a}_i$ of radius R_i.

III. GROUND STATE AND ENERGY PROPAGATION

Let us consider the system in its ground state (T = 0):

$$(P_i = 0; Q_i = \bar{a}_i) \; \forall_i = 1, 2, \ldots, N.$$

Exciting the kth particle we want to know how the energy of the perturbation will propagate and distribute among the different particles of the system. Let us assume that the system is long enough not to take into account collisions with the boundaries of the box.

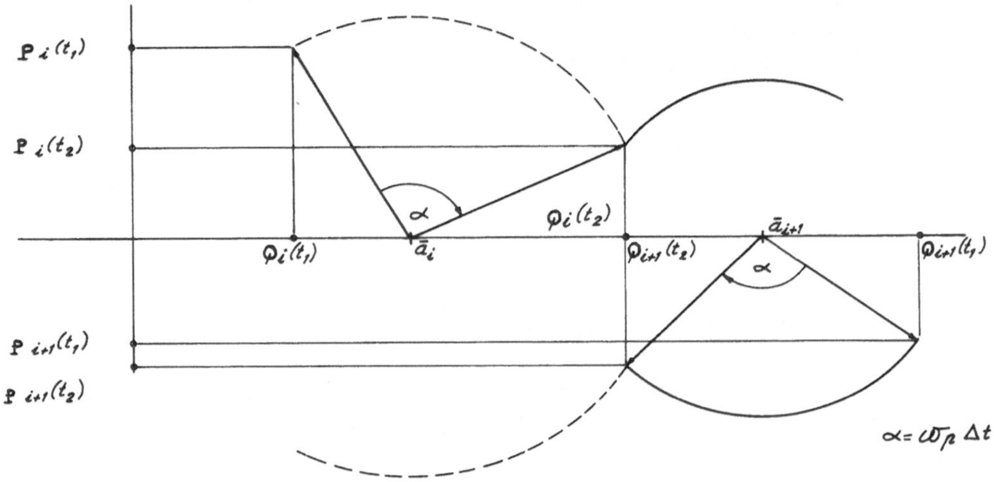

Figure A.1. Orbit in the E-space of the ith particle (dashed line) before and after a shock with the (i+1)-particle (full line) at $t = t_2$.

According to Eq. (A.2):

(a) If $R_k \leq 1$ at $t = 0$, the kth particle will vibrate around its equilibrium position $\bar{a}_k$ at the plasma frequency ω_p without perturbing the system. We shall call such an oscillation a "plasmon" although it involves only one particle.

(b) If $R_k > 1$ at $t = 0$, according to the mechanism previously described in Eq. (A.4), the kth particle will perturb the system, communicating a part of its initial energy to particle (k+1).

Our purpose is to give a rule for this last case.

Let $Im = [\sqrt{n+1}; \sqrt{n+2}]$ $n \in N$ ($I_n \subset R_+$), and let $R_{n,j} \varepsilon I_n$ the radius of the orbit of the particle attracted by the jth potential well. We note that every $R_{n,k} \varepsilon I_n$ may be written:

$$R_{n,k} = \sqrt{n+x^2} \quad \text{with} \quad x \varepsilon\ I_o = [1; \sqrt{2}] \tag{A.5}$$

Moreover, we can write:

$$\sqrt{R^2_{n,k} - 1} = \sqrt{n - 1 + x^2}\ \varepsilon\ I_{n-1} \quad ,$$

but in the case of our initial configuration, according to Eq. (A.4) we have:

$$\sqrt{R_{n,k}^2 - 1} = R_{n-1,k+1} \quad .$$

Note that without loss of generality we can always suppose the initial perturbation is in the positive direction (i.e., $P_k(0) > 0$). As before, we can write:

$$\sqrt{R_{n-1,k+1}^2 - 1} = \sqrt{n - 2 + x^2} \in I_{n-2}$$

and

$$\sqrt{R_{n-1,k+1}^2 - 1} = R_{n-2,k+2}$$

Iterating this up to n = 0, we see that: $x = R_{o,k+n}$ and Eq. (A.5) becomes:

$$R_{n,k} = \sqrt{n + R_{o,k+n}^2} \qquad R_{n,k} \in I_n \tag{A.6}$$

$R_{o,k+n}$ is the radius of the orbit imposed by the front of the perturbation on the particle fixed at the equilibrium position $\bar{a}_{k+n}$. Equation (A.6) suggests the following simple rule.

If at t = 0 the kth particle has a radius $R_{n,k}$, we shall excite exactly n plasmons, after which the $(k+n)^{th}$ particle will start an odd movement throughout the rest of the system with a radius equal to $R_{o,k+n}$.

In terms of energy repartition, if we give to the kth particle an energy equal to

$$E = \omega_p \sqrt{m/2} \cdot 1/\rho \cdot R_{n,k} \quad ,$$

one part of it will generate n plasmons of equal energy

$$E = \omega_p \sqrt{m/2} \cdot 1/\rho$$

and will be confined in a region of finite length, while the rest, proportional to $R_{o,n+k}$, will propagate towards the boundary of the box. See Figure A.2.

We have also detected an interesting particular case: namely, if we give to the kth particle an energy corresponding to a radius of $R_k = \sqrt{2}$, we obtain a propagation of finite length involving only

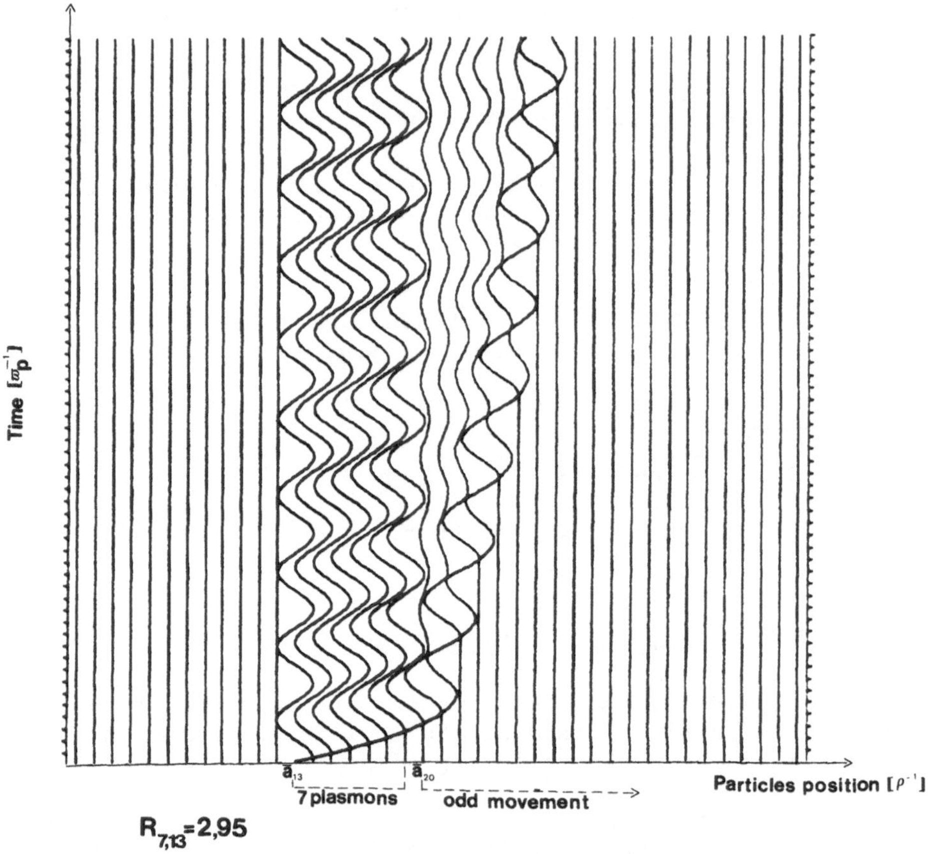

Figure A.2. Particle trajectories showing that seven plasmons of equal energy $E = \omega_p \sqrt{m/2}$ have been confined in a region of finite length while the rest of the initial energy is propagating towards the boundary of the box leaving the system lightly perturbed behind it.

two particles and traveling throughout the whole system, leaving it unperturbed with respect to the ground state. See Figures A.3a,b. As before, if at t = 0, $R_k = \sqrt{n}$, we shall generate (n-2) plasmons and one excitation of the type $R = \sqrt{2}$. See Figure A.4.

These results have been checked by the computer codes described in the paper. Figures A.2, A.3b and A.4 are the computer simulation results for each of these cases.

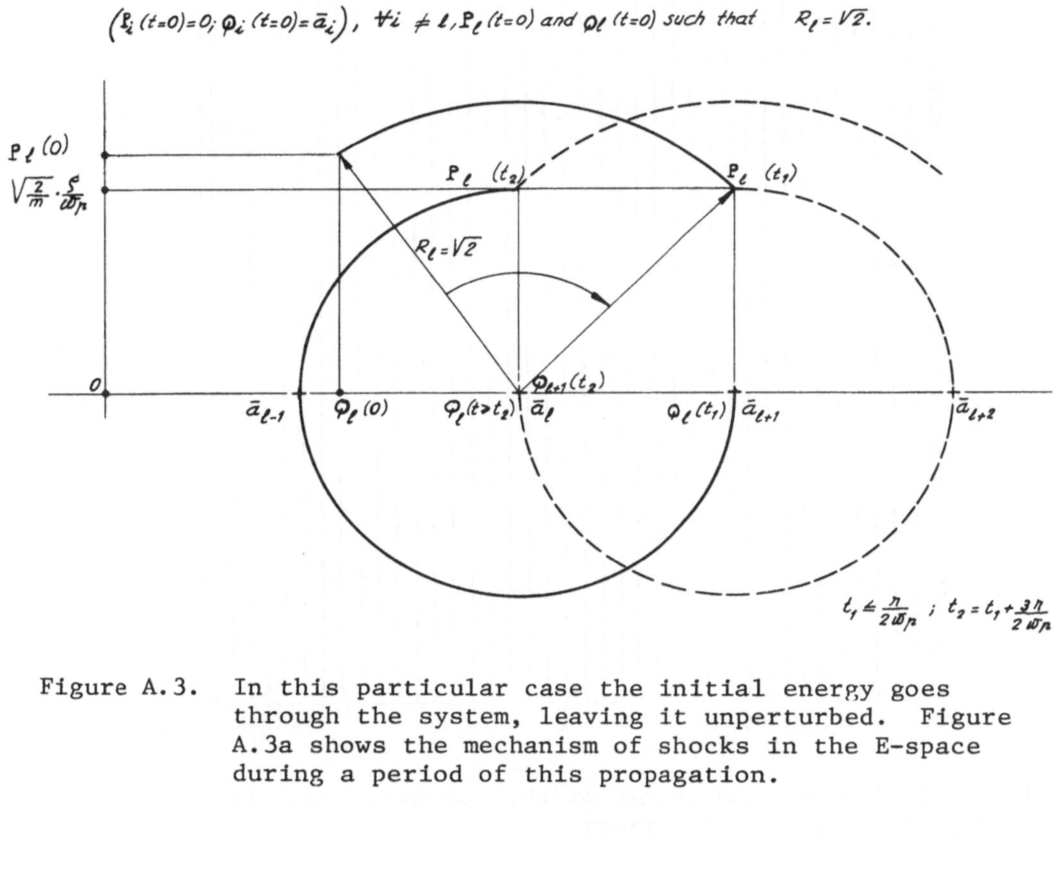

Figure A.3. In this particular case the initial energy goes through the system, leaving it unperturbed. Figure A.3a shows the mechanism of shocks in the E-space during a period of this propagation.

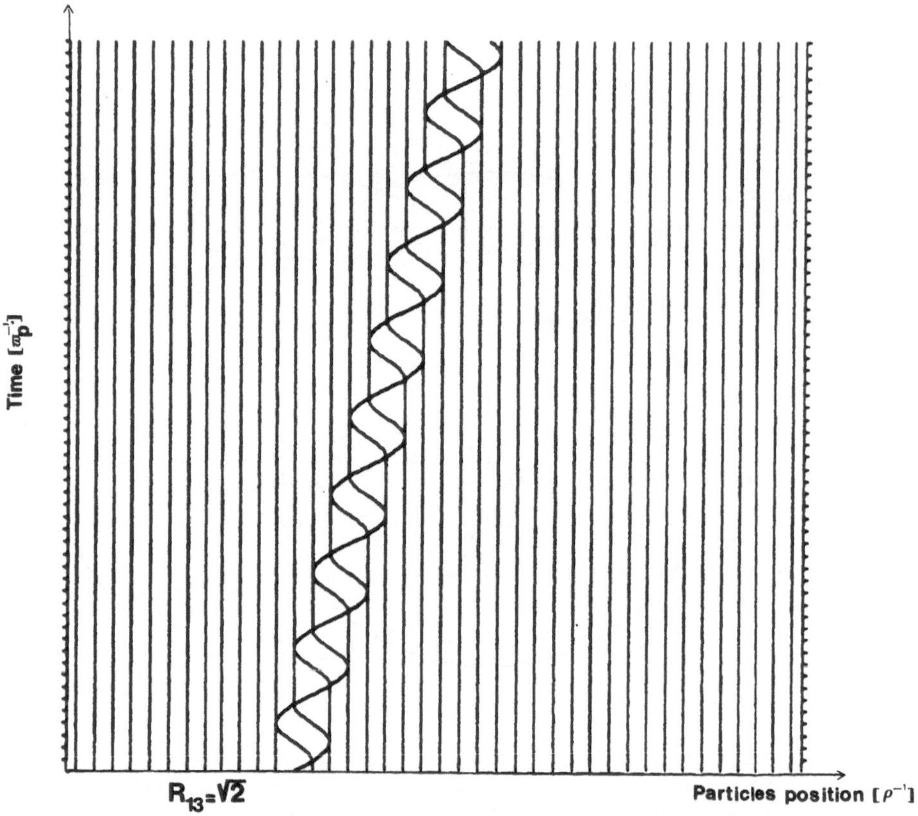

Figure A.3b gives the result of the computer simulation for this same propagation characteristic.

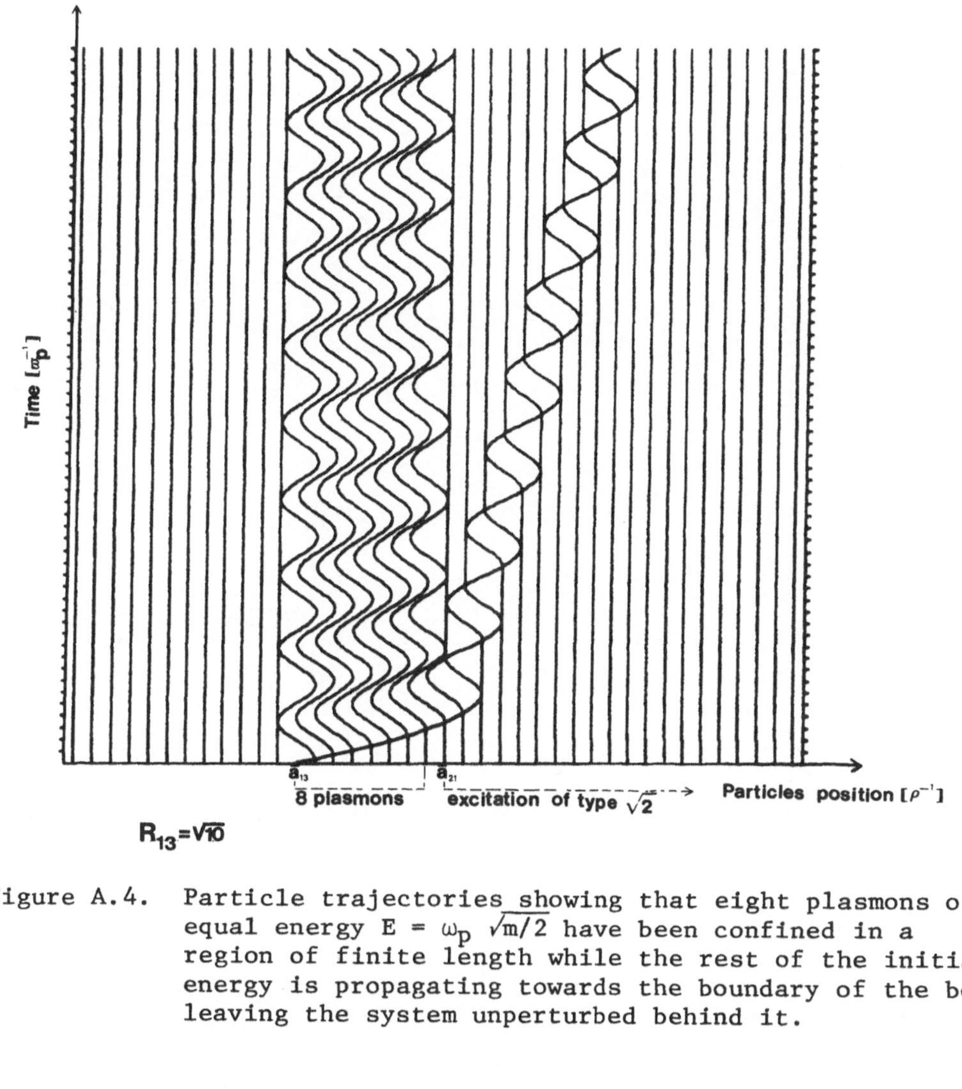

Figure A.4. Particle trajectories showing that eight plasmons of equal energy $E = \omega_p \sqrt{m/2}$ have been confined in a region of finite length while the rest of the initial energy is propagating towards the boundary of the box leaving the system unperturbed behind it.

ACKNOWLEDGEMENTS

It is a pleasure to thank Professor Ph. Choquard and my friends M. Robert and D. Merlini for stimulating discussions. I cannot forget Annica R. for her useful help.

SEMINARS

CONFIGURATION SPACE FORMULATION AND IMPLICATIONS FOR APPROXIMATION SCHEMES FOR STRONGLY COUPLED PLASMAS

Pradip Bakshi

STRONG LASER PLASMA COUPLING

Som Krishan

ASYMPTOTIC BEHAVIOR OF THE CORRELATION FUNCTION OF DENSE PLASMAS

Michel Lavaud

SOME EQUILIBRIUM PROPERTIES OF THE 2D STRONGLY COUPLED PLASMA

D. Merlini

GROUP TRANSFORMATION FOR PHASE SPACE FLUIDS

J. R. Burgan, J. Gutierrez, A. Munier, E. Fijalkow, M. R. Feix

EQUATION OF STATE OF REACTING STRONGLY COUPLED PLASMAS

F. J. Rogers

CONFIGURATION SPACE FORMULATION AND IMPLICATIONS FOR APPROXIMATION SCHEMES FOR STRONGLY COUPLED PLASMAS*

Pradip Bakshi

Department of Physics
Boston College
Chestnut Hill, MA 02167

I. INTRODUCTION

Some of the recent approximation schemes [Singwi, Tosi, Land and Sjölander, 1968 (STLS); Totsuji and Ichimaru, 1973 (TI), Golden, Kalman and Silevitch, 1974 (GKS)] for strongly coupled plasmas have been formulated in terms of integral equations in k-space for the pair correlation function, or the structure function, or other related quantities. The integrand, one observes, often has (almost) a convolution structure, which strongly suggests that a transition to the configuration space may simplify the problem and offer some specific advantages. We have so transformed [Bakshi, Kalman, Silevitch and Golden, 1976] the STLS and TI equations (Section II), which appear now as Poisson equations for the direct correlation function, with distinctive *source* terms involving the pair correlation function. This, in turn, suggests that the direct correlation function (properly normalized) must be an *effective potential* due to an effective charge density of the source. Indeed, the source terms, in both cases, obey the same normalization condition which essentially reflects the requirement of perfect screening. Various other physical requirements and their implications for the correlation functions and the source function are also noted in Section III.

This formulation allows a unified comparison of various approximation schemes, including the earlier, originally fluids-oriented approaches [Percus and Yevick, 1968 (PY), the hypernetted chain (HNC)], in terms of their source structure (Section IV). It also suggests *source phenomenology*, where one directly introduces model source terms with adjustable parameters, subject only to the

*Part of this work has been supported by AFOSR Grant 76-2960.

rigorous physical requirements the source (and the resulting correlation functions) must satisfy. The essential problem is to solve simultaneously the source equation and the Ornstein-Zernike relation connecting the direct and the pair correlation functions. Explicit solutions can be obtained for simple sources.

Another advantage of the r-space formulation is that it allows an analytical determination of the small-r behavior of the correlation functions. The method is outlined in Section V, and various schemes are found to have anomalous small-r behaviors. This also raises some questions about the mathematical consistency of the corresponding equations.

Finally, some conjectures, and conclusions are given in Section VI.

II. CONFIGURATION SPACE FORMULATION

The k-space integral equation for the pair correlation function G(k) in the STLS scheme is

$$\frac{G(k)}{1+G(k)} = -\frac{\kappa^2}{k^2}\left\{1+\frac{1}{(2\pi)^3 n}\int d\vec{q}\,\frac{\vec{k}\cdot\vec{q}}{q^2}\,G(|\vec{k}-\vec{q}|)\right\}, \tag{2.1}$$

$$\kappa^2 = 4\pi\, n\, \beta e^2 \quad , \qquad \beta = (k_B T)^{-1} \quad .$$

Let

$$ng(r) = G(r) = \frac{1}{(2\pi)^3}\int d\vec{q}\; e^{i\vec{q}\cdot\vec{r}}\, G(q) \quad , \tag{2.2}$$

and noting g(o) = -1, (since 1 + g(r), being proportional to the probability of finding a pair of particles with separation r, must vanish for r = o), we obtain the normalization condition

$$-1 = \frac{1}{(2\pi)^3 n}\int d\vec{q}\; G(q) \quad . \tag{2.3}$$

Combining Eqs. (2.1) and (2.3), and defining the direct correlation function

$$C(k) = \frac{G(k)}{1+G(k)} \quad , \tag{2.4}$$

we find the convolution structure

$$-k^2\ C(k) = \frac{\kappa^2}{(2\pi)^3 n} \int \frac{\vec{q} \cdot (\vec{k} - \vec{q})}{q^2}\ G(|\vec{k} - \vec{q}|)\ d\vec{q} \quad , \tag{2.5}$$

and the equivalent configuration space equation,

$$\nabla^2 C(r) = -b\,\vec{\nabla} G(r) \cdot \vec{\nabla} \frac{1}{r} \quad , \tag{2.6}$$

$$b = \frac{\kappa^2}{4\pi n} = \beta e^2 \quad .$$

This equation is in the form of a Poisson equation for the 'potential' C, with a 'source' on the right-hand-side. The configuration space analog of Eq. (2.4) is the Ornstein-Zernike relation

$$C(r) = G(r) - (C * G)(r) \quad , \tag{2.7}$$

where * denotes convolution. Now Eqs. (2.6) and (2.7) form a coupled set of equations for determining C(r) and G(r).

For the TI scheme, the configuration space version is found to be

$$\nabla^2 C(r) = -b\ \vec{\nabla} G \cdot \vec{\nabla} \frac{1}{r} - b(\vec{\nabla} \frac{1}{r} * G) \cdot \vec{\nabla} G + 4\pi\ b\ G^2 \quad . \tag{2.8}$$

III. EFFECTIVE POTENTIAL AND SOURCE

If we define $C(r) = -n\ \beta\ e\ \psi(r)$, Eq. (2.6) reduces to a proper Poisson equation for a potential $\psi(r)$,

$$\nabla^2\ \psi(r) = e\ \vec{\nabla}\ g(r) \cdot \vec{\nabla} \frac{1}{r} \equiv -4\pi\ e\ S(r) \tag{3.1}$$

with the source function S(r) and the normalization

$$\int S(r)\ d\vec{r} = \int_0^\infty g'(r)\ dr = -g(o) = 1 \quad . \tag{3.2}$$

Then the STLS source function is

$$S(r) = -\frac{1}{4\pi}\ \vec{\nabla}\ g(r) \cdot \vec{\nabla} \frac{1}{r} \quad . \tag{3.3}$$

Similarly, the TI source function is

$$S(r) = -\frac{1}{4\pi}\left\{\vec{\nabla} g \cdot \vec{\nabla}\frac{1}{r} + n(\vec{\nabla}\frac{1}{r} * g) \cdot \vec{\nabla} g - 4\pi n g^2\right\} \quad (3.4)$$

and it has the same normalization (3.2), as the last two terms in Eq. (3.4) form a perfect differential and cancel out upon integration.

The source function S(r) formally represents the spatial distribution of a unit source charge. In the Debye approximation, the direct correlation function is proportional to the coulomb potential, and the source is a delta function $S(r) = \delta(\vec{r})$.

Our $\psi(r)$, defined here though it was from a different point of view, is just the effective potential in the mean field description [Nelkin and Ranganathan, 1967]. The meaning of the normalization condition (3.2) becomes clear if we recognize that it represents $S(k = o) = 1$, which implies

$$\lim_{k\to o} [k^2 C(k)] = -\kappa^2 \quad .$$

By virtue of the fluctuation-dissipation theorem, this condition is the same as the condition for perfect screening, that the dielectric function $\varepsilon(k) \to \infty$ as $k \to o$. This follows from $1 + G = (\alpha/\varepsilon)(1/\alpha_o)$, $\alpha = \varepsilon - 1$, $\alpha_o = \kappa^2/k^2$, and Eq. (2.4). This condition is also equivalent to the Stillinger-Lovett condition which specifies the second moment of G(r), $G''(k \to o) = (2/\kappa^2)$, and it is also implied by the compressibility sum rule.

Thus we conclude that Eq. (3.2) is a general physical requirement for S(r). Another requirement, this one in terms of the correlation function, is the condition $1 + g(r \to o) = o$, as mentioned in the previous section. More generally, $1 + g(r)$ should be non-negative for all r since it is proportional to the probability of finding two particles with separation r. A stronger requirement for small r is $1 + g(r) \sim Ae^{-b/r}$ as $r \to o$, since at very small separations the two-particle configuration is governed primarily by their coulomb interaction. Other physical requirements such as exact sum rules and any other exact results for the system should be added to this list.

IV. SOURCE PHENOMENOLOGY

Various approximation schemes for strongly coupled plasmas can now be compared uniformly in terms of the source formulation,

$$\nabla^2 C(r) = \kappa^2 S(r) \quad . \quad (4.1)$$

Each scheme is characterized by its distinctive source function,

S(r). This also suggests a *new way* of looking at the problem of static equilibrium properties of strongly coupled plasmas: The source function implicitly embodies *all* the specific information of the approximation scheme; then one can construct *new approximation schemes* simply by an ad hoc prescription of the source term!

In order that a source describes the physical system, it should satisfy all the physical requirements of the type described in the previous section. If there are adjustable parameters in the source model, these are determined by the various physical constraints, on the source as well as the correlation functions. Philosophically, this approach is akin to the axiomatic type approach, familiar from some other areas of physics. Whether it proves to be advantageous in practice will depend, of course, on its ability to provide new insights, or new results about the physical system.

Besides the physical constraints, one would be guided by *simplicity* in the choice of S. The simplest choice, which obeys the normalization requirement

$$\int S(\vec{r})\, d\vec{r} = 1 \quad , \tag{4.2}$$

is

$$S(r) = \delta(\vec{r}) \quad . \tag{4.3}$$

The solution for the direct correlation function is immediately found from Eq. (4.1),

$$C(r) = -\frac{\kappa^2}{4\pi r} = -\frac{nb}{r} \quad , \tag{4.4}$$

and we discover that the simplest source corresponds to the Debye-Hückle solution! In k-space, taking the Fourier transform and using Eq. (2.4), we find

$$C(k) = -\frac{\kappa^2}{k^2} \quad , \qquad G(k) = -\frac{\kappa^2}{\kappa^2 + k^2} \quad , \tag{4.5}$$

and with an inverse Fourier transform,

$$G(r) = -\frac{\kappa^2}{4\pi r}\, e^{-\kappa r} \quad , \qquad g(r) = -\frac{b}{r}\, e^{-\kappa r} \quad . \tag{4.6}$$

This model strongly violates the requirement $g(r \to o) = -1$. It should be noted that because the source was prescribed to be an explicit function of r, we were able to solve Eq. (4.1) directly to obtain C(r). Whenever we make a direct model for the source function as an explicit function of r, we are in essence specifying C(r) directly, and Eq. (2.4) is then merely used in an auxilliary fashion to determine G(r) from the known C(r).

At the next level of complexity, if the source is prescribed to be a functional of C(r), then again Eq. (4.1) can be solved directly without involing Eq. (2.4), to determine C(r). An example of this type is the model

$$S(r) = -\frac{1}{4\pi n} \vec{\nabla} C(r) \cdot \vec{\nabla} \frac{1}{r} \quad , \tag{4.7}$$

which, along with Eq. (4.1) provides the equation for C,

$$\nabla^2 C(r) = -b \vec{\nabla} C \cdot \vec{\nabla} \frac{1}{r} \quad . \tag{4.8}$$

The model source (4.7) can be viewed to be a linearized version of the STLS source (3.3), where G is linearized with respect to C. However, it will violate the source normalization condition $\int S\, d\vec{r} = 1$, since $\int S\, d\vec{r} = -C(r = o)/n$ from Eq. (4.7), and by virtue of Eq. (2.7) and Eq. (2.3), $C(r = o) = -n - \int C(k)\, G(k)\, d\vec{k} \equiv -n\mu$, $\mu \neq 1$. A model which is structurally similar to Eq. (4.7) and which satisfies the source normalization condition is given by

$$S(r) = -\frac{1}{4\pi\, n\mu} \vec{\nabla} C \cdot \vec{\nabla} \frac{1}{r} \quad , \tag{4.9a}$$

$$\mu = -\frac{1}{n} C(r = o) \quad . \tag{4.9b}$$

The solution of

$$\nabla^2 C = \kappa^2 S = -\frac{b}{\mu} \vec{\nabla} C \cdot \vec{\nabla} \frac{1}{r} \quad , \tag{4.10}$$

satisfying the boundary condition (4.9b) and the physical requirement $C(r \to \infty) \to o$ is

$$c(r) \equiv n^{-1} C(r) = -\mu(1 - e^{-(b/\mu r)}) \quad . \tag{4.11}$$

This model correctly predicts $c(r) \to -b/r$ for large r. The constant μ is determined by the second physical requirement, $g(r = o) = -1$, by first taking the Fourier transform of Eq. (4.11) which expresses C(k) in closed form in terms of Bessel functions K_2 of complex arguments, and then using Eq. (2.4), and a numerical Fourier transform to obtain g(r). The details of this study, for various coupling strengths, will be reported elsewhere [Bakshi, Kalman and Silevitch, to be published].

Finally, the source may involve G and its derivatives, or be a functional of G. Examples in this category are the two approximation schemes, STLS and TI, considered in Section III, as can be seen from Eqs. (3.3) and (3.4). The source equation which relates C and G through a specific model, and the Ornstein-Zernike identity (2.7), together form the coupled set of equations for the unknowns G and C. While explicit solutions are not feasible for these models, it is possible to infer analytically their short-range behavior as shown in Section V.

The approximation schemes such as the PY and HNC can also be cast in terms of the source formulation, and the source terms for various schemes can then be compared uniformly. Schematically, the relation between c(r) and g(r) is an algebraic one:

$$c = (1 + g)(1 - e^{b/r}) \quad , \quad \text{PY},$$
$$c = g - \ln(1 + g) - \frac{b}{r} \quad , \quad \text{HNC}. \qquad (4.12)$$

The source functions are easily obtained by forming the Laplacian of Eq. (4.12) and their structure can then be compared with those of other schemes [Bakshi, Kalman and Silevitch, to be published].

The normalization condition (4.2) can be cast in the form

$$\kappa^2 = \int \nabla^2 C \, d\vec{r} = 4\pi[r^2 C'(r)]\Big|_o^\infty \qquad (4.13)$$

and tested directly for any algebraic scheme such as Eq. (4.12). From Eq. (2.7), $C'(r \to o) = o$, and Eq. (4.13) reduces to a constraint on the long-range behavior of c(r),

$$r^2 c'(r) \to \frac{\kappa^2}{4\pi n} = b \quad , \quad c(r) \to -\frac{b}{r} \quad , \quad (r \to \infty). \qquad (4.14)$$

This imparts yet another meaning to the source normalization condition. Then, for explicit models like PY and HNC, checking the source normalization condition is equivalent to checking directly whether $c(r) \to$ coulomb interaction as $r \to \infty$. This is obviously satisfied in both these models (4.14).

V. SHORT-RANGE BEHAVIOR

The r-space formulation enables one to determine analytically the short-range behavior of g(r) for the STLS, TI and similar approximation schemes. We illustrate this for the STLS model. Taking the Laplacian of Eq. (2.7) and using Eq. (2.6), we find

$$\nabla^2 G = -b(\vec{\nabla} G \cdot \vec{\nabla} \frac{1}{r}) - b(\vec{\nabla} G \cdot \vec{\nabla} \frac{1}{r}) * G \tag{5.1}$$

or

$$\frac{1}{r^2} \frac{d}{dr} r^2 G' = \frac{b}{r^2} G' + b \int_o^\infty dr' \int d\Omega' \, G(|\vec{r} - \vec{r}'|) \, G'(r') \, .$$

Multiplying by r^2 and one integration provides

$$G(r) - G(o) = \frac{r^2}{b} G'(r) - \int_o^r r^2 \, dr \int_o^\infty dr' \int d\Omega' \, G(|\vec{r} - \vec{r}'|)$$

$$\cdot \, G'(r') \quad . \tag{5.2}$$

As $r \to o$, the $\vec{r}'$ integrations in the last term simplify to

$$\iint dr' \, G(r') \, G'(r') \, d\Omega' = 2\pi \, G^2(r')\Big|_o^\infty = -2\pi \, G^2(o)$$

yielding the simplified differential equation for $r \to o$,

$$G(r) - G(o) = \frac{r^2}{b} G' + \frac{2\pi}{3} G^2(o) \, r^3 + \ldots \tag{5.3}$$

The homogeneous part provides the singular solution proportional to $\exp(-b/r)$, which vanishes faster than any power of r, and can be completely ignored as $r \to o$. With $G(r) = ng(r)$ and $g(o) = -1$, we then find

$$g(r) = -1 + \frac{2\pi n}{3} r^3 + \frac{r^2}{b} g'(r) + \ldots$$

$$= -1 + \frac{1}{2} \left(\frac{r}{a} \right)^3 + \frac{3}{2\Gamma} \left(\frac{r}{a} \right)^4 + \ldots \tag{5.4}$$

where a is the ion sphere radius defined by $(4\pi \, n \, a^3/3) = 1$, and Γ is the strong coupling parameter $\Gamma = (b/a) = (\beta e^2/a)$.

Since we expect $1 + g(r) \sim e^{-b/r}$ as $r \to o$ on physical grounds, the appearance of the powers of r in Eq. (5.4) is quite disturbing. It shows that the STLS approximation does not describe the short-range behavior of g(r) correctly. The computational studies of this scheme apparently have not probed this region with sufficient accuracy, and the power law behavior was missed.

The relative strength of the cubic and the fourth power terms determines the domain where the pure cubic behavior can be seen,

$$(r/a) << \Gamma \quad , \qquad r << b \quad .$$

This is also the domain where the homogeneous term with the essential singularity becomes vanishingly small.

Such an anomalous behavior for small r turns out to be a common feature for other approximation schemes as well. For the TI scheme we find, with a little more algebra, the short-range result

$$\begin{aligned} g(r) &= -1 + \frac{2\pi n}{3} r^3 \left\{ -1 - n \int g^3 \, d\vec{r} \right\} + \frac{r^2}{b} g'(r) + \ldots \\ &= -1 + \frac{2\pi n^2}{3} r^3 \left\{ \int (g - g^3) \, d\vec{r} \right\} + O(r^4) \quad . \end{aligned} \tag{5.5}$$

The second version has the advantage of showing that the anomalous behavior is weakened by the compensating contributions from g and g^3. For small coupling γ, ($\gamma = \kappa^3/4\pi n = b\kappa$), g is negative for all r and $g^2 < 1$, which makes the coefficient of the r^3 term become negative. This is an additional, and more serious problem, since it implies a negative probability, $1 + g(r) < o$, for sufficiently small r. For large γ, g can become positive for some range of r and it would be a matter for detailed calculation to see whether the net sign still remains negative.

Another approximation scheme which also displays the r^3 behavior for g(r) for small r is the Frieman-Book [1963] model,

$$\nabla^2 G = -b \vec{\nabla} G \cdot \vec{\nabla} \frac{1}{r} + \kappa^2 G \quad , \tag{5.6}$$

which leads to

$$g(r) = -1 + \frac{4\pi n}{3} r^3 + O(r^4) \quad . \tag{5.7}$$

Apparently, the r^3 behavior is the rule, rather than the exception, for approximation schemes based on hierarchial truncation. The coefficients of the r^3 terms differ, according to the truncation procedure, and it may be possible to develop a truncation scheme that might make the coefficient of the r^3 term vanish. However, then the higher powers of r will emerge as the anomalous terms. This phenomenon suggests that it may be extremely difficult to avoid the power-law anomalies for small r in the context of hierarchical truncation schemes. On the other hand, the PY and HNC models, as can be seen from Eq. (4.12), have built in the correct small r behavior.

For a two-dimensional plasma, one can write the analogous k-space and r-space equations for these schemes, and study the short-range behavior for g(r). Again one finds anomalous behavior, now beginning with r^2. The power index is the same as the dimensionality of space, since the inhomogeneous term produces different powers in different dimensions. For the STLS scheme, we find, in two dimensions, with $\gamma = \beta e^2$,

$$g(r) = \begin{cases} -1 - \dfrac{\kappa^2 r^2}{4(2-\gamma)} + Cr^\gamma + \ldots, & \gamma \neq 2 \\[2ex] -1 + \dfrac{\kappa^2 r^2}{4}\left(\dfrac{1}{2} - \ln r\right) + C_2\, r^2 + \ldots, & \gamma = 2 \end{cases} \tag{5.8}$$

The r^γ term represents the homogeneous solution and it dominates over the inhomogeneous solution r^2 for $\gamma < 2$. Anomalous behavior sets in for $\gamma \geq 2$, where the inhomogeneous term provides the dominant term. Again, 1 + g(r) remains positive throughout, for the STLS model in two dimensions as well. The TI scheme in two dimensions also displays the anomalous r^2 behavior.

From these results we can conclude that the short-range anomaly is quite widespread, afflicting many models, in two- as well as three-dimensional plasmas. This behavior also raises an important question about the mathematical consistency of these schemes. In three dimensions, one can see that $\nabla^4\, g(r) \sim r^{-1}$ will be singular as $r \to 0$. This implies $\int d\vec{k}\, k^4\, G(k) \to \infty$, or $G(k) \sim k^{-7+\varepsilon}$ as $k \to \infty$, $\varepsilon \geq 0$. On the other hand, for the STLS scheme, the integral operator I in the k-space equation

$$\frac{G}{1+G} = I\, G \tag{5.9}$$

seems to possess a power raising property for large k, which would make it impossible to have a power law behavior for G(k) for large k, and thus contradicts the previous inference based on r-space considerations. This suggests that the nonlinear equation (5.9) may not possess a consistent solution in the mathematical sense. This question should be carefully examined for this, and other similar schemes.

VI. CONJECTURES AND CONCLUSIONS

As we have seen in the preceding sections, the r-space formulation provides a useful framework for the classification and study of various approximation schemes. It also suggests direct construction of new *source* models.

From the lectures at this Institute [DeWitt, this Volume] we learn that a well-defined liquid state sets in for $\Gamma \gtrsim 1$ and persists until the phase transition around $\Gamma \approx 155$. A remarkable feature of this regime is the relative invariance of the shape of the scaled function $c(x)/\Gamma$, $x = r/a$. This, in turn, implies a corresponding source function with a similar relative invariance. An important feature is its finite range in r. All this suggests that the effective expansion parameter is $1/\Gamma$ and that the regime from $\Gamma = 20$ to 150 would show only minor changes in some of the (properly scaled) functions. On the other hand, the height of the first maximum of g(r), around $r \approx 1.7a$ keeps increasing with Γ. These features suggest that there are two scales involved in the problem of solving the coupled equations for G and C, and that the small $1/\Gamma$ associated with the highest derivative induces a singular perturbation over a narrow band of r. A formal multi-scale analysis of these coupled equations may prove useful. These ideas will be further developed elsewhere.

It also appears that the phenomenological source size for the liquid state saturates at $r_o \approx 1.7a$. An interesting question is how the source varies for small Γ. For the Debye source, the size is 0, and $\Gamma = o$. For the liquid state $r_o/a \sim 1$ is reached for $\Gamma \gtrsim 1$. This suggests that the actual source size is some increasing function of Γ, possibly a linear one. This can be examined by constructing a simple model source of range r_o, with a step function character, and normalized to unit integral. The second physical constraint $g(r = o) = -1$ will then relate r_o to Γ, and one can determine* how the source width increases with Γ.

The most practical use of the r-space formulation so far has been in determining the short-range behavior of g(r). This can be developed for many schemes and in different dimensions. Corresponding behavior of c(r) and S(r) can also be obtained quite similarly. The most important question that has been raised is regarding the mathematical consistency of some of the schemes in three dimensions, and it should be examined rigorously.

---

*This problem has been subsequently solved, and was presented in Bakshi, Silevitch and Kalman [1977]. We find $(r_o/a) \sim (3\Gamma/2)$ for small Γ.

ACKNOWLEDGEMENT

A more detailed version of many of the topics presented here will appear in a paper under preparation with G. Kalman and M. Silevitch, to whom I am much indebted for their extensive collaboration on this program. I also wish to thank K. Golden for many useful conversations at the initial stages of this program.

REFERENCES

Bakshi, P., G. Kalman and M. B. Silevitch, to be published.

Bakshi, P., G. Kalman, M. B. Silevitch and K. I. Golden, 1976, Bull. Amer. Phys. Soc. 21, 1086.

Bakshi, P., M. B. Silevitch and G. Kalman, 1977, Bull. Amer. Phys. Soc. 22, 1208.

DeWitt, H. E., this Volume.

Frieman, E. and D. Book, 1963, Phys. Fluids 6, 1700.

Golden, K. I., G. Kalman and M. B. Silevitch, 1974, Phys. Rev. Lett. 33, 1544.

Nelkin, M. and S. Ranganathan, 1967, Phys. Rev. 164, 222.

Percus, J. and G. Yevick, 1958, Phys. Rev. 110, 1.

Singwi, K., M. P. Tosi, R. H. Land and A. Sjölander, 1968, Phys. Rev. 176, 589.

Totsuji, H. and S. Ichimaru, 1973, Progr. Theor. Phys. 50, 753.

STRONG LASER PLASMA COUPLING

Som Krishan

Department of Physics, Indian Institute of Science

Bangalore-560012, India

I. INTRODUCTION

Many reviews [Schwarz and Hora, eds., 1971; Mulser, Siegel and Witkowski, 1973; Boyer, 1973; Nuckolls, Emmet and Wood, 1973; Brueckner and Jorna, 1974; Giovantelli and Godwin, unpublished] and papers [Freidberg, Mitchell, Morse and Rudsinski, 1972; Godwin, 1972; Mueller, 1973; Klein and Manheimer, 1974; Thomson, Faehl and Kruer, 1973; Seely and Harris, 1973; Seely, 1974; Silin, 1965; Bethe, 1972; Kaw and Dawson, 1970; Kruer and Dawson, 1972] on collisionless and collision dominated mechanisms have been written to investigate the laser energy absorbed by the plasma. Along with the theoretical studies, a number of experiments [Floux, 1971] have also been done. Some experimental studies [Billman, Roweley, Stalcop and Presly, 1974] have emphasized the heating by inverse bremsstrahlung [Stallcop, 1974], while other such studies [Fabre and Stenz, 1974] have emphasized the laser heating by nonlinear instabilities [Jorna, 1974; Kaw and Dawson, 1969; V. Krishan, S. Krishan, Sinha and Ganguli, 1976].

In this paper we shall discuss some aspects of laser plasma interaction when the laser is strongly coupled to the plasma. Here by strong coupling we mean $(W_L)/(nT) > 1$, where W_L is the laser energy density, n the particle density and T the temperature in eV. If the coupling is weak, i.e., $(W_L)/(nT) < 1$, the laser can be regarded as a perturbing force on the plasma system. As a consequence the usual theories on the parametric instability will remain reliable. In particular, the renormalization [Bezzerides and Dubois, 1976] of three-wave (laser, Langmuir wave and ion acoustic wave) coupling can be accurately calculated by truncating the higher order effects. When the laser is strong, i.e., $(W_L)/(nT) > 1$, the estimates for the saturation level of the parametric instability and the consequent heating of the particles cannot be calculated by regarding the laser as a perturbation because then the convergence of the series for the coupling parameter for the three wave interaction is in doubt.

The treatment given here is heuristic for want of space. However, it is based on a detailed derivation. The plasma densities considered are in the neighborhood of 10^{19} particles/cm^3 and the interacting laser is a CO_2 laser. To begin with, the laser will be taken in the dipole approximation and then extended to the non-dipole case.

II. LASERS IN THE DIPOLE APPROXIMATION AND DISPERSION RELATION

Consider a particle of charge e and mass m moving in a circularly polarized laser whose vector potential is given by

$$\vec{A}(t) = A_o(\hat{e}_x \cos \omega t + \hat{e}_y \sin \omega t) \quad . \tag{2.1}$$

The solution of the Schrödinger equation in the presence of the above field is given by

$$\psi_{\vec{p}}(\vec{x},t) = \frac{1}{\sqrt{V}} \sum_{n=-\infty}^{\infty} J_n[\beta(p_\perp)] \cdot \exp[-\frac{i}{\hbar} E_n(\vec{p})\, t + \frac{i}{\hbar} \vec{p} \cdot \vec{x} - in\delta(p_\perp) + i\beta(p_\perp) \sin \delta(p_\perp)] \quad , \tag{2.2}$$

where $\vec{p}$ is the momentum and

$$E_n(\vec{p}) = \frac{p^2}{2m} + \frac{e^2A_o^2}{2mc^2} - \hbar n\omega \; , \qquad \delta(p_\perp) = \tan^{-1} p_y/p_x \quad ,$$

$$\beta(p_\perp) = \frac{eA_o p_\perp}{mc\hbar\omega} \quad , \qquad p_\perp = (p_x^2 + p_y^2]^{1/2} \quad ,$$

V = volume, $J_n[\beta(p_\perp)]$ is the Bessel function of first kind. Let us concentrate on the function

$$\phi_{\vec{p},n}(\vec{x},t) = \frac{1}{\sqrt{V}} \exp[-\frac{i}{\hbar} E_n(\vec{p}) + \frac{i}{\hbar} \vec{p} \cdot \vec{x} - in\delta(p_\perp)$$

$$+ i\beta(p_\perp) \sin\delta(p_\perp)] \, J_n[\beta(p_\perp)] \quad , \tag{2.3}$$

occurring in Eq. (2.2). This function is orthogonal with respect to $\vec{p}$, but not with respect to n. If we take the time average of the scalar product

$$<\phi_{\vec{p},n'}|\phi_{\vec{p},n}>$$

for $t > 1/\omega$, we find that it vanishes unless $n = n'$. Thus in the time averaged sense $\phi_{\vec{p},n}$'s are orthogonal with respect to n also. The $\phi_{\vec{p},n}$ can be regarded as an orthogonal basis with energy $E_n(\vec{p})$. The $J_n[\beta(p_\perp)]$ can then be regarded as the probability amplitude of being in the state $|(p,n)>$. The net probability $f(\vec{p})$ for the system to be in the state $|\vec{p}>$ is then given by

$$f(\vec{p}) \equiv \sum_n f_n(\vec{p}) = \frac{1}{N} \sum_{n=-\infty}^{\infty} J_n^2[\beta(p_\perp)] \exp\left[-\frac{E_n(\vec{p})}{T}\right] \quad , \tag{2.4}$$

where N = total number of particles. Now going through the usual steps and using Eqs. (2.1)-(2.4) one obtains the dielectric function given by

$$\varepsilon(\Omega,\vec{q}) = 1 + \sum_s m_s \frac{\omega_{ps}^2}{n_o} \int \sum_{n,\ell} J_n^2 \, J_{n-\ell}^2 \left[\vec{q} \cdot \frac{\partial f_{n,s}}{\partial \vec{p}} (\vec{p}) - \frac{\ell\omega}{T_s} f_{\vec{n},s}(\vec{p})\right] \frac{1}{\left[\Omega - \ell\omega - \frac{\vec{q}\cdot\vec{p}}{m}\right]} \, d^3p \quad , \tag{2.5}$$

where m_s, ω_{ps}, T_s are the species mass, plasma frequency and temperature respectively. The argument of the Bessel functions is $\beta(p_\perp)$, and n_o is the particle density. The largest maxima of the Bessel function occurs when its argument $\beta(p_\perp) \simeq |n|$. Equation (2.5) can then be simplified to obtain -- for the cold plasma --

$$\varepsilon(\Omega,\vec{q}) = 1 - \sum_{a=\pm 1} \sum_{\ell,s} \omega_{ps}^2(f) \left\{ \frac{1}{[\Omega - \ell\omega - u_s \vec{q} \cdot (\hat{e}_x + a\,\hat{e}_y)]^2} \right.$$

$$\left. + \frac{1}{[\Omega - \ell\omega + u_s\vec{q}\cdot(\hat{e}_x + a\,\hat{e}_y)]^2} \right.$$

$$- \frac{\ell\omega m}{T_s q^2} \left[\frac{1}{\Omega - \ell\omega - u_s \vec{q} \cdot (\hat{e}_x + a\,\hat{e}_y)} \right.$$

$$\left. + \frac{1}{\Omega - \ell\omega - u_s \vec{q} \cdot (\hat{e}_x + a\,\hat{e}_y)} \right] \Bigg\} \tag{2.6}$$

where

$$\omega_{ps}(f) = \omega_{ps} J_{N_s - \ell} \left(\frac{e^2 A_o^2}{m_s \hbar \omega c^2} \right) ,$$

$$N_s \simeq \frac{e^2 A_o^2}{m_s \hbar \omega c^2} ,$$

$$u_s = \frac{e A_o}{\sqrt{2}\, m c}$$

In the non-dipole approximation where the laser field $\vec{A}$ is given by

$$\vec{A} = A_o [\hat{e}_x \cos (\vec{K} \cdot \vec{x} - \omega t) + \hat{e}_y \sin (\vec{K} \cdot \vec{x} - \omega t)] \tag{2.7}$$

Therefore in this case when the particles interact the momentum transfer is not $\vec{q}$, but $(\vec{q} - n\vec{K})$ and Eq. (2.5) is then modified to

$$\varepsilon(\Omega,\vec{q}) = 1 + \int \sum_s \sum_{n,\ell} m_s \frac{\omega_{ps}^2}{n_o} \frac{J_n^2 J_{n-\ell}^2}{q^2} \frac{\left[(\vec{q} - \ell\vec{K}) \frac{\partial f_{n,s}(\vec{p})}{\partial \vec{p}} - \frac{\ell\omega}{T_s} \right] d^3p}{\Omega - \ell\omega - \frac{\vec{q} \cdot \vec{p}}{m}} \tag{2.8}$$

III. ELECTROSTATIC OSCILLATIONS

A. Linear Theory

Setting the right hand side of Eq. (2.8) equal to zero and

integrating over $\vec{p}$ and employing the same procedure as was done to arrive at Eq. (2.6), one obtains the dispersion relation

$$1 - \sum_{a=\pm 1} \sum_{\ell,s} \omega_{ps}(f)^2 \frac{(\vec{q} - \ell\vec{K})^2}{q^2} \left\{ \left[\frac{1}{[\Omega - \ell\omega - u_s \vec{q} \cdot (\hat{e}_x + a\,\hat{e}_y)]^2} + \frac{1}{[\Omega - \ell\omega - u_s \vec{q} \cdot (\hat{e}_x + a\,\hat{e}_y)]^2} \right] - \frac{\ell\omega m}{q^2 T_s} \left[\frac{1}{\Omega - u\,\vec{q} \cdot (\hat{e}_x + a\,\hat{e}_y) - \ell\omega} + \frac{1}{\Omega - \ell\omega - u\,q \cdot (\hat{e}_x + a\,\hat{e}_y)} \right] \right\} = 0 \quad . \tag{3.1}$$

We are interested in the unstable low frequency ($\Omega << \omega$) oscillations, therefore only $\ell = 0$ term should contribute most. Further, we shall assume that $\Omega >> q\,u_i$. Now the maximum growth rate is obtained when

$$q\,u_e - \omega_{pe}(f) = \omega_{pi}(f) \tag{3.2}$$

where for simplicity $\vec{q}$ has been taken along the x-axis. Using Eq. (3.2), Eq. (3.1) therefore yields

$$\Omega = \omega_{pi}(f) + i\,\gamma_o \quad , \tag{3.3}$$

where

$$\gamma_o = \left(\frac{Z}{32}\,\frac{m}{M}\right)^{1/4} \omega_{pe}(f)\, J_{N_i}\left(\frac{e^2 A_o^2}{M c^2 \hbar\,\omega}\right) \tag{3.4}$$

and Z is the charged state of the ion.

B. Nonlinear Theory

Nonlinearities are introduced because of the effect of the growing wave back on the particles which causes perturbation in the particle orbits. Taking into account the lowest order perturbations in the particle orbits the nonlinear dispersion becomes

$$1 - \frac{\omega_{pe}(f)^2}{2} \left\{ \left[\Omega - q\, u_e - \frac{5}{2}\, \omega_{Be}^4\, (\Omega - q\, u_e)^{-3} \right]^{-2} + \left[\Omega + q\, u_e - \frac{5}{2}\, \omega_{Be}^4\, (\Omega + q\, u_e)^{-3} \right]^{-2} \right\} - \frac{\omega_{pi}(f)^2}{\Omega^2} = 0 \quad , \tag{3.5}$$

where

$$\omega_{Be} = J_{N_e}^{1/2} \left(\left(\frac{e^2\, A_o^2}{m\, c^2\, \hbar\, \omega} \right) \left(\frac{e\, q\, \omega\, A_o}{m\, c} \right) \right)^{1/2} = \text{effective bounce frequency} \tag{3.6}$$

and the perturbation in the ions has been neglected as it turns out to be much smaller than the electrons. The solution of Eq. (3.6) is given by

$$\Omega = \omega_{pi}(f) + 0.25\, \frac{\omega_{Be}^4}{\gamma_o^{\,3}} + i\, \gamma_o \left(1 - \frac{\omega_{Be}^4}{\gamma_o^{\,4}} \right) \quad . \tag{3.7}$$

We notice that the wave saturates when the imaginary part of Ω vanishes, i.e.,

$$\omega_{Be}^4 = \gamma_o^{\,4} \tag{3.8}$$

We now calculate the energy transferred by the wave to the resonant particles. This process is simply the linear wave particle interaction (Landau Damping) corrected for the perturbation in the particle orbits. Near the saturation we find that the rate equations for ion thermal energy ξ_i and electron thermal energy ξ_e are given by

$$\frac{\partial \xi_i}{\partial t} = \frac{2}{\sqrt{\pi}}\, J_{N_e}^2 \left(\frac{e^2\, A_o^2}{m\, c^2\, \hbar\, \omega} \right) \left[\frac{1}{32} \left(\frac{m}{ZM} \right)^{1/2} \right] \omega_{pi}(f)^2 \left[1 + 0.5\, J_{N_e} \left(\frac{e^2\, A_o^2}{m\, c^2\, \hbar\, \omega} \right) \left(\frac{1}{32}\, \frac{M}{Zm} \right)^{1/4} \right]^{-3/2} \alpha_i^{\,5/2}\, \exp(-\alpha_i) \tag{3.9}$$

$$\frac{\partial \xi_e}{\partial t} = 2\sqrt{\frac{2}{\pi}}\, J_{N_e}^2 \left(\frac{e^2\, A_o^2}{m\, c^2\, \hbar\, \omega}\right)\left[\frac{1}{32}\left(\frac{Zm}{M}\right)^2\right]$$

$$\left(\frac{T_e\, m\, \omega}{e^2}\right)\omega_{pe}^2 \left[1 + J_{N_e}\left(\frac{e^2\, A_o^2}{m\, c^2\, \hbar\, \omega}\right)\left(\frac{1}{32}\,\frac{M}{Zm}\right)^{1/2}\right]$$

$$\alpha_e^{5/2}\, \exp[-\alpha_e] \qquad (3.10)$$

where

$$\alpha_i = \left[1 + 0.5\, J_{N_e}\left(\frac{e^2\, A_o^2}{m\, c^2\, \hbar\, \omega}\right)\left(\frac{1}{32}\,\frac{M}{Zm}\right)^{1/4}\right]$$

$$\cdot \frac{Ze^2\, A_o^2}{2m\, T_i\, c^2\, J_{N_e}^2 \left(\frac{e^2\, A_o^2}{m\, c^2\, \hbar\, \omega}\right)}$$

$$\alpha_e = \frac{e^2\, A_o^2}{4m\, T_e\, c^2}$$

The peak rate of energy absorption for electrons and ions is given by $\alpha_i = \alpha_e = 2.5$. At plasma temperature of $\simeq 10$ eV, (as in the experiment of Fabre and Stenz [1974] with carbon plasma) the maximum absorption rate by electrons occurs at a laser power of $\simeq 10^{12}$ watts/cm^2 whereas for electrons it occurs at $\simeq 7 \times 10^{10}$ watts/cm^2. The peak fractional energy absorbed per second by electrons and ions are

$$\frac{1}{W_L}\,\frac{\partial \xi_i}{\partial t}\,, \qquad \frac{1}{W_L}\,\frac{\partial \xi_e}{\partial t}\,.$$

These quantities attain their maximum values when $\alpha_i = \alpha_e = 1.5$. Calculating the peak fractional energies absorbed by electrons and ions in 4 nanoseconds one finds the dominant contribution to the absorption coefficient comes from ions and it occurs at a laser power of 3.5×10^{10} watts/cm^2 at which, near the critical density, the absorption coefficient = 0.5 which is close to the experimental value of ·0.6. The laser power at 3.5×10^{10} watts/cm^2 when converted to its vacuum value becomes $\simeq 4 \times 10^{10}$ watts/cm^2 which is also very close to the experimental value.

During the interaction of the laser with the plasma, one assumed a constant electron density as was done in the experiment. According to them the electron density due to ionization could have at most gone up by a factor of 1.33 which was, however, compensated by lateral expansion due to heating. Thus it was concluded that the particle density to a good approximation remains constant as was assumed in the present calculations.

REFERENCES

Bethe, H. A., 1972, unpublished, Los Alamos Scientific Laboratory Informal Report No. LA 5031-MS.

Bezzerides, B. and D. F. Dubois, 1976, Phys. Rev. Lett. 36, 729.

Billman, K. W., P. D. Roweley, J. R. Stallcop and L. Presly, 1974, Phys. Fluids 17, 759.

Boyer, K., 1973, Astronaut. Aeronaut. 11, 28.

Brueckner, K. A. and S. Jorna, 1974, Rev. Mod. Phys. 46, 325.

Fabre, E. and C. Stenz, 1974, Phys. Rev. Lett. 32, 823.

Floux, F., 1971, Nucl. Fusion 11, 635; and references quoted therein.

Freidberg, S. P., R. W. Mitchell, R. L. Morse and Rudsinski, 1972, Phys. Rev. Lett. 28, 795.

Giovantelli, D. V. and R. P. Godwin, unpublished, Los Alamos Scientific Laboratory Report No. LA-UR-74-1735.

Godwin, R. P., 1972, Phys. Rev. Lett. 28, 851

Jorna, S., 1974, Phys. Fluids 17, 765.

Kaw, P. and J. Dawson, 1969, Phys. Fluids 12, 2586; 1970, Phys. Fluids 13, 472.

Klein, H. H. and W. M. Manheimer, 1974, Phys. Rev. Lett. 33, 953.

Krishan, V., S. Krishan, K. P. Sinha and A. Ganguli, 1976, Appl. Phys. Lett. 29, 90.

Kruer, W. L. and J. Dawson, 1972, Phys. Fluids 15, 446.

Mueller, M. M., 1973, Phys. Rev. Lett. 30, 582.

Mulser, P. R., Siegel and S. Witkowski, 1973, Phys. Rep. (Amsterdam) 60, 187.

Nuckolls, J. J., Emmet and L. Wood, 1973, Phys. Today, 8, 46.

Schwarz, H. J. and Hora, Eds., 1971, Laser International and Related Plasma Phenomena, Vols. 1-3, Plenum Press, New York.

Seely, J. F. and E. G. Harris, 1973, Phys. Rev. A7, 1064.

Seely, J. F., 1974, Phys. Rev. A10, 1863.

Silin, V. P., 1965, Sov. Phys. JETP 20, 1510.

Stallcop, J. R., 1974, Phys. Fluids 17, 751.

Thomson, J. J., R. J. Faehl and W. L. Kruer, 1973, Phys. Rev. Lett. 31, 918.

ASYMPTOTIC BEHAVIOR OF THE CORRELATION FUNCTION OF DENSE PLASMAS

Michel Lavaud

Centre de Recherches sur la Physique des Hautes Températures
45045 Orleans-Cédex, France

I. INTRODUCTION

The study of the thermodynamic properties of charged point particles in the usual space is an extremely difficult one(1). The two main difficulties are due (i) to the infinite attraction between pairs of oppositely charged particles, and (ii) to the infinite range of the coulombian interaction between pairs of particles of any sign.

A. The One-Component Plasma

The first difficulty can be dealt with by delocalizing the particles to prevent the collapse in pairs of oppositely charged particles. This can be made in various ways, as illustrated in Figure 1. One way is to make use of quantum mechanics(1) [Morita, 1959; Ebeling, 1976]. Another way is to study two-component plasmas in one [Lenard, 1961; Edwards and Lenard, 1962] or two [Hauge and Hemmer, 1971; Deutsch and Lavaud, 1974] dimensions. A third possibility is to replace the real system of point charges by charged hard spheres or by a one-component plasma. It is this latter system which will be studied in this article. It consists of charged classical point particles of a given sign, in a uniform neutralizing background.

B. The Debye Length

The second difficulty can be dealt with by taking into account

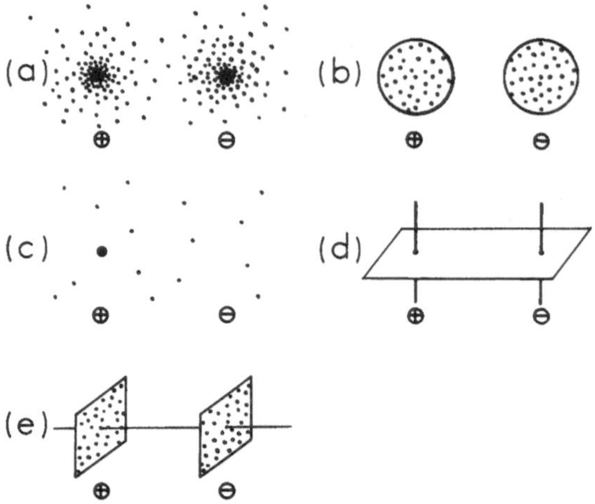

Figure 1. Various ways of delocalizing the particles. (a) In quantum mechanics, both types of particles are delocalized in the three directions and in the whole space. (b) In the case of charged hard spheres, the charges are delocalized in the three directions of space, but only in a finite domain (a point charge at the center of each sphere is equivalent to a uniform density of charge inside the sphere, because the latter is impenetrable). (c) For the one-component plasma, one type of charge is delocalized in the three directions, and in all the space. The charges of the other type are point particles. (d) The two-dimensional two-component plasma. The two types of charges are delocalized in only one direction of space. (e) In the case of the one-dimensional two-component plasma, the charges are delocalized in two directions.

the electroneutrality of the system. This results in the possibility of replacing the Coulomb potential by an effective interaction with a finite range λ_D. More precisely, it has been shown [Friedman, 1963] that the correlation function $h(r) = g(r)-1$ decays exponentially like $\exp(-r/\lambda_D)$. λ_D is called the Debye length. It has a fundamental importance in the study of ionized systems, because it represents the distance at which a given charge begins to be screened by charges of opposite sign.

C. Corrections to the Debye Length for Large Values of the Plasma Parameter

For dilute plasmas (small values of the plasma parameter

$$\varepsilon = [4\pi\ \rho(e^2/4\pi\ \varepsilon_o\ kT)^3]^{1/2} \quad) \quad ,$$

λ_D is equal to $(\varepsilon_o\ kT/\rho e^2)^{1/2}$, where ρ is the density of the system and T its temperature.

For the systems we are interested in, i.e. dense plasmas, it has been suggested [Hirt, 1965; Cohen and Murphy, 1969; DelRio and DeWitt, 1969; Mitchell and Ninham, 1968] that the Debye length must be corrected for large values of the plasma parameter. More precisely, it has been shown [Cohen and Murphy, 1969; DelRio and DeWitt, 1969; Mitchell and Ninham, 1968] that the correlation function h(r) decays exponentially like exp $(-r/\Lambda_D)$, where Λ_D is equal to:

$$\Lambda_D = \lambda_D[1 + \varepsilon(\ell n\ 3)/8 + \ldots]^{-1} \tag{1.1}$$

under the following two assumptions:

(i) The long distance behavior of h(r) is given by the sum of the 2-graphs which are dominant at large distances.

(ii) The 2-graphs dominant at large distances are made of Debye-Hückel lines alternating with Abe-Meeron lines (see Figure 2). These lines represent the two functions:

$$b(r) = -\varepsilon e^{-r}/r \tag{1.2}$$

$$B(r) = e^{b(r)} - 1 - b(r) \tag{1.3}$$

The aim of this article is to give a formal* proof of assumption (ii).

Figure 2. The 2-graphs which are dominant at large distances. Debye-Hückel and Abe-Meeron lines are represented respectively by straight and wiggly lines.

*By formal proof, we mean that we do not worry about 2-graphs which are infinite: we assume that all the divergencies can be cancelled by suitable resummations (see Sections IV and V).

D. The Asymptotic Behavior of 2-Graphs with Debye-Hückel Lines

The main step to prove the previous conjecture consists in finding the behavior at large distances of any 2-graph (or Mayer graph with two root-points) with Debye-Hückel functions (or lines) b(r). In their important work, DelRio and DeWitt [1969] have remarked that the particular 2-graphs made of κ chains in parallel (without points in common) decay exponentially like $e^{-\kappa r}$, that is, one has

$$\lim_{r\to\infty} \frac{1}{r} \ln |\Gamma(r)| = -\kappa \quad .$$

Here, we solve the problem in the general case, by proving that any 2-graph decays exponentially like $e^{-\lambda r}$, where stands for the maximum number of chains linking the root-points, which have no line in common[(2)] [Lavaud, to be published; Lavaud, submitted for publication]. This problem has also been investigated by Deutsch et al for general 2-graphs [Deutsch, Furutani, 1975; Deutsch, Furutani and Gombert, 1976]. But their proof is incorrect[(4)] and their method, even if it could be corrected, could only give upper bounds which decay exponentially like $e^{-\kappa r}$, where κ is the maximum number of chains without points in common[(5)] Moreover, nothing ensures that their bounds are finite.

This paper is organized as follows. In Section II, we recall some definitions and notations. In Section III, we find the exponential decay of any 2-graph with multiple Debye-Huckel lines. In Section IV, we generalize the preceding results to 2-graphs with Debye-Huckel and Abe-Meeron lines and prove assumption (ii). Finally we point out that, to prove assumption (i), an improvement over the Abe-Meeron Theory of ionized systems is necessary.

II. DEFINITIONS AND NOTATION

A. Definition of a 2-Graph

A 2-graph* is a multiple integral of the following type

$$\Gamma(r_{12}) = \int_{\Lambda_\infty^k} \prod_{L\in\mathcal{L}\Gamma} f_L(r_{ij})\, d\vec{r}_3 \ldots d\vec{r}_{k+2} \tag{2.1}$$

*This is usually called a graph with 2 root-points, or a 2-rooted graph. But, in our work, we must make a clear distinction between a graph and the multiple integral it represents. Therefore, we have chosen to keep the name graph with 2 root-points for the graph theoretical concept, and to call its associated integral a 2-graph. Similarly, a graph is denoted by Γ and its associated integral by $\Gamma(r_{12})$.

where the symbols have the following meanings: Γ is a graph with two root-points, k field-points and ℓ lines L joining the points i and j. The set of lines of Γ is denoted by $\mathcal{L}\Gamma$. In Eq. (2.1), the product runs over all lines of $\mathcal{L}\Gamma$, and the integration runs over the k field-points varying in the infinite domain Λ_∞.

$$r_{ij} = |\vec{r}_i - \vec{r}_j|$$

denotes the distance between particles i and j. This will be also denoted by r_L when we will find this convention more convenient.

B. Definition and Notation of the f_L

The corrections to the Debye length for large values of the plasma parameter have been established in the framework of the Abe-Meeron theory of the one-component plasma [Hirt, 1965; Cohen and Murphy, 1969; DelRio and DeWitt, 1969; Mitchell and Ninham, 1968]. Therefore, we also make use of this theory, in this article.

This implies that, in Eq. (2.1), some of the f_L are equal to the Debye-Hückel function (or line) b(r) and the others are equal to the Abe-Meeron function B(r). The graph Γ can therefore have two types of lines. A factor $b(r_{ij})$ in Eq. (2.1) will be represented in Γ by a straight line joining the points i and j, and a factor $B(r_{ij})$ will be represented by a wiggly line. For example, we will have:

$$\int b(r_{13})\, B(r_{32})\, d\vec{r}_3 \equiv \text{Figure Eq. (2.2) (page 582)} \qquad (2.2)$$

Moreover, in the Abe-Meeron theory, there can be no multiple lines joining a pair of points. This means that one can have no factor b^2, B^2 nor, more generally $b^n B^m$ in Eq. (2.1). For example the 2-graph:

$$\int b^2(r_{13})\, B(r_{32})\, d\vec{r}_3 \equiv \text{Figure Eq. (2.3) (page 582)} \qquad (2.3)$$

does not occur in the development of the 2-body distribution function.

Nevertheless, it will not be necessary to impose this restriction before Section IV.B. Therefore, Γ will be allowed to have several lines between a given pair of points i and j, unless explicitly stated otherwise.

C. The Abe-Meeron Development of the Potential of Mean Force

We are interested in the asymptotic behavior of the correlations in a plasma. For such a problem, it is indifferent to study the correlation function h(r), or the potential of mean force w(r), because one has, at large distances:

$$h(r) = e^{-\beta w(r)} - 1 \sim -\beta w(r) \tag{2.4}$$

We will therefore restrict ourselves to w(r).

The development of the potential of mean force reads, in the Abe-Meeron theory [Cohen and Murphy, 1969]:

$$w(r) = b(r) + \sum_{k=1}^{\infty} \beta_k(r)(4\pi\ \varepsilon)^{-k} \tag{2.5}$$

with:

$k!\beta_k(r)$ = sum over all distinct labeled simple irreducible 2-graphs with k field-points, which have single lines joining pairs of points and no chain made only of Debye-Hückel lines.

We have made use here of the following definitions:

Definition 2.1 - A chain is a sequence of lines of the form (i_1,i_2), (i_2,i_3), ... (i_{k-1},i_k) where all points are distinct.

Definition 2.2 - A 2-rooted graph is irreducible if each field-point belongs to a chain of field-points linking the root-points.

Definition 2.3 - An irreducible 2-rooted graph is simple if each pair of field-points is linked by a chain of field-points.

Finally, a simple irreducible 2-graph is a 2-graph which is represented by a simple irreducible 2-rooted graph.

These definitions are illustrated in Figure 3.

The first few terms of the development (2.5) are:

$w(r_{12})$ = Figure Eq. (2.6) (page 582) (2.6)

In the last factor of (2.6), the 2-graphs which are obtained by substituting some Debye-Hückel lines to Abe-Meeron ones have been omitted.

III. EXPONENTIAL DECAY OF 2-GRAPHS WITH MULTIPLE DEBYE-HÜCKEL LINES

As we already said, the aim of this article is to prove that, in the development (2.5) of w(r), the 2-graphs which are dominant at large distances are chains made of Debye-Hückel lines alternating

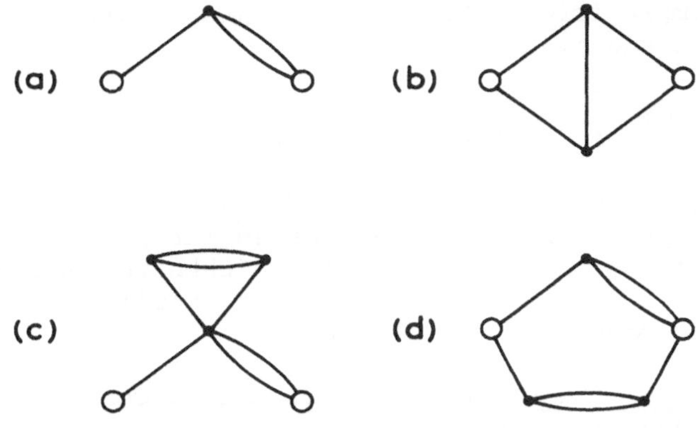

Figure 3. The 2-graphs (a) and (b) are simple irreducible. (c) is reducible because it has an articulation point, and (d) is irreducible, but not simple because the root-points form an articulation set.

with Abe-Meeron lines. To prove this result, we will see in the next section that it is sufficient to find the exponential decay of all the 2-graphs which occur in the development of w(r).

We know that these 2-graphs can have Debye-Hückel and Abe-Meeron lines, and that they have single lines joining pairs of points. But, because of the asymptotic equivalence

$$B(r) \sim \frac{\varepsilon^2}{2} (e^{-r}/r)^2 \quad ,$$

we see that a related and simpler problem is to find the exponential decay of 2-graphs with only Debye-Hückel lines, but with multiple lines joining pairs of points. It is this problem that will be investigated in this section.

We know also that the 2-graphs which occur in the development of w(r) have no chains made of Debye-Hückel lines. But this restriction will not be imposed in this section, because each 2-graph will be considered individually. For the same reason, the factor $-\varepsilon$ in the definition (1.2) of a Debye-Huckel line will be omitted.

In short, in all this section, we will study the exponential decay of 2-graphs of the following type:

$$\Gamma(r_{12}) = \int_{\Lambda_\infty^k} \prod_{L \in \mathcal{L}\Gamma} \frac{e^{-r_L}}{r_L} \, dr_3 \ldots dr_{k+2} \tag{3.1}$$

where Γ is a simple irreducible 2-rooted graph with k_{ij} lines joining points i and j.

A. Reformulation of $\Gamma(r_{12})$

To find the exponential decay of $\Gamma(r_{12})$, it is particularly convenient to express it in the form of an integral of Laplace's type [Erdelyi, 1956]. Indeed, $\Gamma(r_{12})$ can then be evaluated by means of the Laplace method. Let us first recall the philosophy and the simplest results of the latter.

Description of the Laplace Method. An integral of Laplace's type reads, in one dimension:

$$\Gamma(r) = \int_a^b e^{-rh(t)} g(t)\, dt \tag{3.2}$$

Let us suppose that h(t) admits an absolute minimum at a point t_o. The crux of the Laplace method is that, for large values or r, only the neighborhood of t_o contributes to the integral (3.2). This enables one to compute easily $\Gamma(r)$. One finds [Erdelyi, 1956]:

$$\Gamma(r) \sim e^{-rh(t_o)} \sqrt{2\pi/r\,|h''(t_o)|}\; g(t_o) \quad , \quad r \to \infty \tag{3.3}$$

under the following conditions:

(i) h(t) is twice differentiable in (a,b).

(ii) $h'(t_o) = 0$, $h''(t_o) \neq 0$, $a < t_o < b$.

(iii) $h(t) > h(t_o)$, $\forall(t) \neq t_o$ (t_o is an absolute minimum).

(iv) g(t) is continuous in (a,b).

We will say that the principal (or exponential) decay of $\Gamma(r)$ is in $e^{-rh(t_o)}$, and that its complementary (or power) decay is in $r^{-1/2}$.

For multi-dimensional integrals, one has a formula quite analogous to Eq. (3.3), which gives both the principal and the complementary decay, if h(t) has an absolute minimum at a point t_o [Hsu, 1948; 1951] (t is now a multi-dimensional variable), and if the Hessian [Hsu, 1948; 1951] of h(t) is non-null at t_o. Note that, for a different set of conditions (i) to (iv), one can have formulae which differ from Eq. (3.3) by a power of r [Erdelyi, 1956], but the important point for us is that the factor $e^{-rh(t_o)}$, that is the principal decay of $\Gamma(r)$, remains unchanged.

Application of the Laplace Method to 2-Graphs with Debye-Hückel Lines. To reformulate $\Gamma(r_{12})$ in the form of an integral of Laplace's type, let us regroup the exponentials together, and set:

$$\vec{r}_i = \vec{r}_{12}\,\vec{R}_i \tag{3.4}$$

we obtain:

$$\Gamma(r_{12}) = r_{12}^{3k-\ell} \int_{\Lambda_\infty^k} e^{-r_{12} \sum_{L\varepsilon\mathcal{L}\Gamma} |\vec{R}_L|} \prod_{L\varepsilon\mathcal{L}\Gamma} \frac{1}{|\vec{R}_L|}\, d\vec{R}_3 \ldots d\vec{R}_{k+2} \tag{3.5}$$

where $\vec{R}_L$ is defined by:

$$\vec{R}_L = \vec{R}_i - \vec{R}_j \tag{3.6}$$

i and j denoting the endpoints of line L.

$\Gamma(r_{12})$ is effectively of the form (3.2) above. Therefore, the considerations of the preceding paragraph suggest that $\Gamma(r_{12})$ ought to decay exponentially like exp $(-\mu r)$, where μ is the minimum of the quantity:

$$h_\Gamma(\vec{R}_3, \ldots, \vec{R}_{k+2}) = \sum_{L\varepsilon\mathcal{L}\Gamma} |\vec{R}_L| \tag{3.7}$$

This is what we are going to prove in the next paragraphs.

Unluckily, it will be seen below that the minimum of h_Γ is not, in general, reached at a point, but at a set of points, so that it is not possible to get the exponential decay of $\Gamma(r_{12})$ with the usual formulae. Therefore, we will have to study Eq. (3.5) directly. We will find the principal decay of $\Gamma(r_{12})$ by exhibiting upper and lower bounds which decay exponentially like exp $(-\mu r)$.

To this end, a first important step is to compute the value of the minimum of h_Γ.

B. Computation of the Minimum μ of h_Γ

Let us first note that $h_\Gamma(\vec{R}_3, \ldots, \vec{R}_{k+2})$ is a finite sum of non-negative continuous functions, and so its minimum μ does exist, is non-negative, and is reached on a certain set of points

$$\Delta \subset \Lambda_\infty^k \quad .$$

Having seen that μ exists, we want now to prove:

Theorem 3.1 - The minimum μ of $h_\Gamma(\vec{R}_3, \ldots, \vec{R}_{k+2})$ is equal to the local line-connectivity $\lambda(1,2)$ of Γ.

$\lambda(1,2)$ is defined in the following way:

Definition 3.2 - The local line-connectivity $\lambda(1,2)$ is equal to the maximum number of line-disjoint chains (chains without lines in common) linking the root-points 1 and 2.

This is illustrated in Figure 4.

By virtue of the max-flow min-cut theorem [Ford and Fulkerson, 1962; Berge, 1973], the maximal number of line-disjoint chains between 1 and 2 is equal to the minimal number of lines which separate* the root-points.

For the simplest 2-graphs, $\lambda(1,2)$ can then be computed very easily. For, suppose we have found a set of m line-disjoint chains; if we can find also a line-cutset with m lines, the max-flow min-cut theorem ensures us that $m = \lambda(1,2)$. For example, Figures 4a and b show that $\lambda(1,2) = 2$, and Figures 4c and d give $\lambda(1,2) = 3$. In the general case, there are known algorithms to compute $\lambda(1,2)$ [Ford and Fulkerson, 1962; Berge, 1973].

Let us now prove Theorem 3.1.

Lower Bound for μ. We have restricted our study to simple irreducible 2-graphs, and therefore there is at least one chain going from point 1 to point 2. For a given chain C, it is possible to write:

$$\vec{R}_{12} = \sum_{L\in \mathcal{L}C} \vec{R}_L \tag{3.8}$$

where $\vec{R}_L$ was defined in Eq. (3.6) to be:

$$\vec{R}_L = \vec{R}_i - \vec{R}_j$$

But, from Eq. (3.4), we have $R_{12} = 1$. So, by applying the triangular inequality to Eq. (3.8), we obtain, for any chain C linking the root-points:

---

*A set of lines is said to separate the points 1 and 2 if their deletion gives two connected components, one of them containing the point 1 and the other point 2 [Harary, 1969]. A line-cutset is a set of lines which separate the root-points. This latter definition is illustrated in Figure 4b and d.

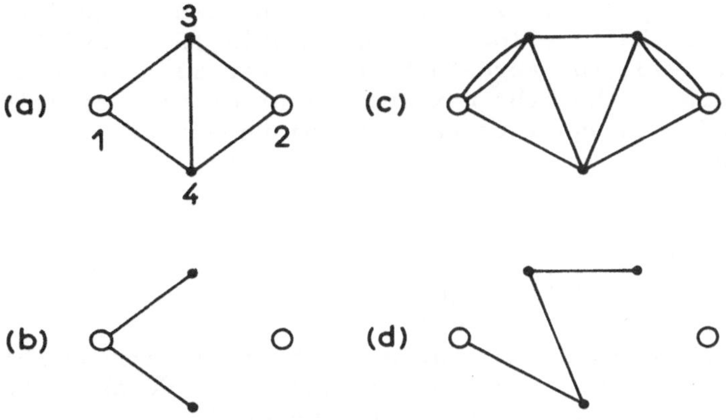

Figure 4. Illustration of the local line-connectivity $\lambda(1,2)$. In the 2-graph (a), one has two line-disjoint chains, one made of the lines (1,3) and (3.2), and the other made of the lines (1,4) and (4,2). (b) represents a line-cutset of (a). It has two lines and so, by the max-flow min-cut theorem, the local line-connectivity of (a) is equal to 2. In the 2-graph (c), one can find three line-disjoint chains, and the line-cutset (d) has three lines. Therefore, for (c), we have $\lambda(1,2) = 3$.

$$1 \leq \sum_{L \varepsilon \mathcal{L}C} |\vec{R}_L| \tag{3.9}$$

Let us now choose a maximal set of line-disjoint chains C_k. We have $LC_m \cap LC_n = \phi$ for any pair of integers m and n, $m \neq n$. This enables us to write:

$$\sum_{m=1}^{\lambda(1,2)} \sum_{L \varepsilon \mathcal{L}C_m} |\vec{R}_L| = \sum_{L \varepsilon \underset{m}{U} \mathcal{L}C_m} |\vec{R}_L| \tag{3.10}$$

But Eq. (3.9) shows that the left-hand-side of Eq. (3.10) is bounded below by $\lambda(1,2)$ and, because

$$\underset{m}{U} \mathcal{L}C_m \subset \mathcal{L}\Gamma \quad ,$$

the right-hand-side of Eq. (3.10) is bounded above by $h_\Gamma(R_3,..,R_{k+2})$. This gives the desired inequality:

$$\lambda(1,2) \leq \mu \tag{3.11}$$

<u>Upper Bound for μ</u>. To prove the converse inequality $\lambda(1,2) \geq \mu$, we need to use the max-flow min-cut theorem [Ford and Fulkerson, 1962; Berge, 1973] (or Ford-Fulkerson theorem). As we already said, it tells us that the maximum number of line-disjoint chains between 1 and 2 is equal to the minimum number of lines which separate 1 and 2.

Let then

$$C = \left\{L_1, L_2, \ldots, L_{\lambda(1,2)}\right\}$$

be a family of such lines. For a given simple irreducible 2-graph, this family is not void, because $\lambda(1,2) \geq 1$. We have then $\Gamma - C = \Gamma_1 \cup \Gamma_2$, with Γ_1 and Γ_2 disconnected. Moreover, one end of line L_i belongs to Γ_1, while the other belongs to Γ_2. The line-cut-set will be displayed by symbolizing Γ in the following way:

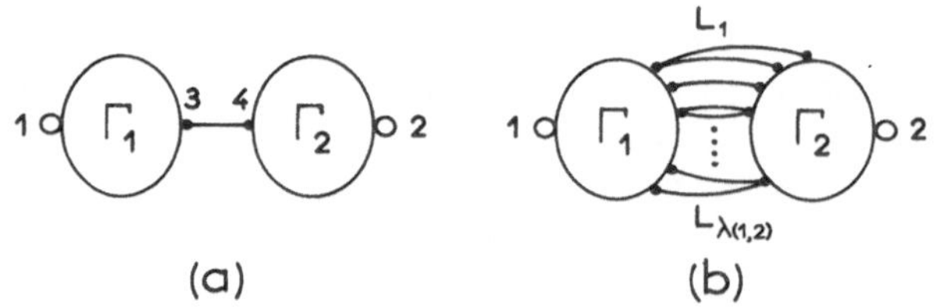

Figure 5. General structure of 2-graphs with a given local line-connectivity. (a) $\lambda(1,2) = 1$. In this case, we have an isthmus, the line (3,4). (b) $\lambda(1,2) = n$.

Let us now call $i_1, j_1, \ldots$ the points of Γ_1, and $i_2, j_2 \ldots$ the points of Γ_2. We can realize the abstract graph Γ in the usual space R^3, by regrouping all the points of Γ_1 in a sphere S_1 centered at point 1 and with radius η, and all the points of Γ_2 in a sphere S_2 centered at point 2, with the same radius η. This is illustrated in Figure 6.

As the subgraphs Γ_1, Γ_2 and C have no line in common, and as their union give back Γ, we can split h_Γ in three parts:

$$h_\Gamma = \sum_{L\in \mathcal{L}\Gamma_1} \left|\vec{R}_{i_1} - \vec{R}_{j_1}\right| + \sum_{L\in \mathcal{L}\Gamma_2} \left|\vec{R}_{i_2} - \vec{R}_{j_2}\right| + \sum_{L\in \mathcal{L}C} \left|\vec{R}_{i_1} - \vec{R}_{j_2}\right| \tag{3.12}$$

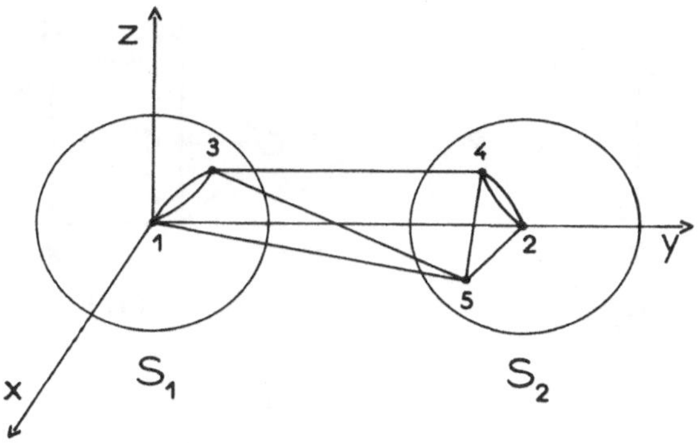

Figure 6. A realization, in the real space R^3, of the 2-graph (c)' of Figure 4.

We easily see that the two first sums can be made negligible provided η is sufficiently small, and that the last one is approximately equal to $\lambda(1,2)|\vec{R}_{12}| = \lambda(1,2)$. More precisely, we have:

$$|h_\Gamma - \lambda(1,2)| \leq 2\ \ell\eta \tag{3.13}$$

We have thus exhibited a set of points $\vec{R}_3, \ldots, \vec{R}_{k+2}$, such that h_Γ is arbitrarily close to $\lambda(1,2)$. This proves that $\mu \leq \lambda(1,2)$, and completes the proof of Theorem 3.1.

We can see now that the minimum of h_Γ is not, in general, reached at a point, but at a set of points. It is sufficient to consider the simplest irreducible 2-graph, that is a chain with two lines. We have then:

$$h_\Gamma(\vec{R}_3) = |\vec{R}_{13}| + |\vec{R}_{32}| \tag{3.14}$$

and the minimum $|\vec{R}_{12}|$ of h_Γ is reached when $\vec{R}_3$ belongs to the segment of line which joins the root-points (in R^3). For the 2-graph of Figure 7a, the minimum is reached when $\vec{R}_3$ and $\vec{R}_4$ come to a same point on the preceding segment. Note nevertheless than μ can be reached sometimes at only one point, as in Figure 3a or in Figure 7b. In the latter case, h_Γ reaches its minimum at the point $(\vec{R}_1, \vec{R}_1, \vec{R}_1, \vec{R}_2, \vec{R}_2, \vec{R}_2)$.

C. Exponential Decay of 2-Graphs with Debye-Hückel Lines

We can now prove the following result:

Theorem 3.2. A given irreducible 2-graph $\Gamma(r_{12})$ with ℓ lines e^{-r}/r,

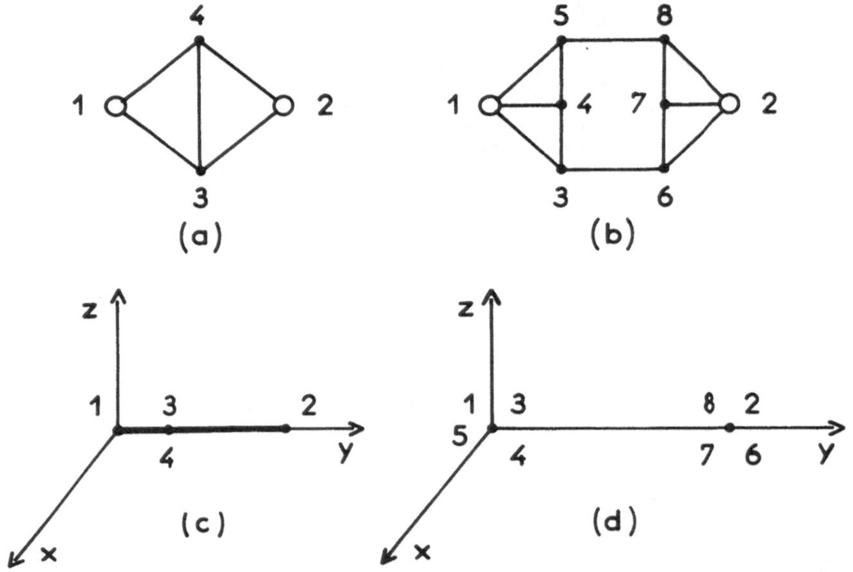

Figure 7. The minimum of h_Γ is reached on a set of points for the 2-graph (a), and at a single point for (b). The realization, in the real space, of the 2-graphs (a) and (b) when h_Γ is minimum, are given respectively in (c) and (d).

k field-points and a local line-connectivity equal to λ, has the following upper and lower bounds:

$$\Gamma(r_{12}) < C_M \Gamma(r_0)\, r_{12}^{3k-\ell}\, e^{-\lambda r_{12}} \quad , \quad \forall r_{12} > r_0 \tag{3.15}$$

$$C_m \quad r_{12}^{-\lambda}\, e^{-\lambda r_{12}} < \Gamma(r_{12}) \quad , \quad \forall r_{12} > r_1 \tag{3.16}$$

with:

$$C_M = r_0^{\ell-3k}\; e^{\lambda r_0} \tag{3.17}$$

$$C_m = C_1(1,1)\; C_2(1,1)\; e^{-2\lambda}\; (1 + 2r_1^{-1})^{-\lambda} \tag{3.18}$$

$C_1(1,1)$ and $C_2(1,1)$ are positive constants defined by Eq. (3.22) below, and r_0, r_1 are arbitrary real positive numbers. We have obtained only upper and lower bounds, but not the exact asymptotic behavior, because the minimum of h_Γ is not reached on a unique point, but on a set of points, as we already noticed.

Let us now turn to the proof of Theorem 3.2.

Upper Bound for $\Gamma(r)$. To obtain an upper bound for $\Gamma(r)$ which decays exponentially like exp $(-\lambda r)$, one will exploit the inequality $\lambda(1,2) \leq h_\Gamma$. But if one tries to introduce it directly into Eq. (3.5), one runs into difficulties which come from the infinite volume of integration, because the function $1/r$ is not integrable in it. To get around this difficulty, we let a convergence factor $e^{-r_o R}$ with each non-integrable factor $1/R$, where r_o is an arbitrary positive number. In other terms, we rewrite Eq. (3.5) in the form:

$$\Gamma(r_{12}) = r_{12}^{3k-\ell} \int_{\Lambda_\infty^k} e^{-(r_{12}-r_0)h_\Gamma} \prod_{L\varepsilon \mathcal{L}\Gamma} \frac{e^{-r_0|\vec{R}_L|}}{|\vec{R}_L|} d\vec{R}_3 \ldots d\vec{R}_{k+2} \tag{3.19}$$

Then, provided $r_{12} \geq r_o$, the theorem of the means [Hardy, Littlewood and Polya, 1937] gives:

$$\Gamma(r_{12}) \leq r_{12}^{3k-\ell} e^{-\lambda r_{12}} r_0^{\ell-3k} e^{\lambda r_0} \Gamma(r_0) \quad , \quad \forall r_{12} > r_o \tag{3.20}$$

This proves the inequality (3.15) with

$$C_M = r_o^{\ell-3k} e^{\lambda r_0}$$

It is essential to remark that our upper bound (3.20) is finite as soon as $\Gamma(r)$ is not infinite everywhere. We see also from Eq. (3.20) that, if $\Gamma(r_0)$ is finite, then $\Gamma(r)$ is finite too, for any $r \geq r_0$.

Lower Bound for $\Gamma(r)$. To obtain a lower bound for $\Gamma(r)$ which decreases exponentially like exp $(-\lambda r)$, one will take advantage of inequality (3.13). One is then led to restrict the domain of integration to the domain of validity of this inequality. This gives actually a lower bound on $\Gamma(r)$ because its integrand is positive. We have:

$$\Gamma(r_{12}) \geq r_{12}^{3k-\ell} \frac{e^{-\lambda r_{12}(1+2\eta)}}{(1+2\eta)^\lambda} C_1(\eta,r_{12}) \, C_2(\eta,r_{12}) \tag{3.21}$$

where $C_1(\eta,r_{12})$ stands for the contribution of the points $\vec{R}_{i_1}$, $\vec{R}_{j_1}$, ... with domain restricted to the sphere $S_1(\eta)$, and similarly for $C_2(\eta, r_{12})$:

$$C_1(\eta, r_{12}) = \int_{[S_1(h)]^{k_1}} \prod_{L \in \mathcal{L}\Gamma_1} \frac{e^{-r_{12}|\vec{R}_L|}}{|\vec{R}_L|} d\vec{R}_{i_1} \ldots d\vec{R}_{j_1} \tag{3.22}$$

For the lower bound Eq. (3.21) to have the required behavior at infinity, we can choose $\eta = r_{12}^{-1}$. Then, by homogeneity considerations, we obtain:

$$C_1(r_{12}^{-1}, r_{12}) = r_{12}^{\ell_1 - 3k_1} C_1(1,1) \tag{3.23}$$

where ℓ_1 and k_1 stand respectively for the number of lines and field-points of Γ_1. By putting this relation into Eq. (3.21) and making use of the relations $k = k_1 + k_2$ and $\ell = \ell_1 + \ell_2 + \lambda$, we obtain the following lower bound, valid for any r_1:

$$\Gamma(r_{12}) \geq C_m\, r_{12}^{-\lambda}\, e^{-\lambda r_{12}} \tag{3.24}$$

with C_m defined by Eq. (3.18). This completes the proof of Theorem 3.2.

As a consequence of Theorem 3.2, we see that an irreducible 2-graph $\Gamma(r_{12})$ with Debye-Hückel lines e^{-r}/r decays exponentially like exp $(-\lambda r_{12})$, where λ is the local line-connectivity of Γ, provided $\Gamma(r_{12})$ is not infinite everywhere. Conversely, a 2-graph which decays exponentially like exp $(-nr_{12})$, with n an integer, is made of two disjoint connected 2-graphs Γ_1 and Γ_2 linked by n lines, and which contain respectively the root-points 1 and 2 (see Figure 5b). Let us note that λ is an integer, contrary to what has been stated by Deutsch et al[3].

IV. ASYMPTOTIC BEHAVIOR OF THE POTENTIAL OF MEAN FORCE

In the preceding section, we have found the exponential decay of any 2-graph with Debye-Hückel lines. In this section, we find first the exponential decay of any 2-graph with Debye-Hückel and Abe-Meeron lines. From this result, we deduce that the 2-graphs of the development of w(r) which are dominant at large distances must have an isthmus. The 2-graphs dominant at large distances are then obtained by finding the complementary decay of 2-graphs with isthmuses. This enables us to prove assumption (ii).

A. Exponential Decay of 2-Graphs with Debye-Hückel and Abe-Meeron Lines

The generalization of the results of the preceding section to 2-graphs which have Debye-Hückel and Abe-Meeron lines is a simple consequence of Theorem 3.2 and of the equivalence

$$B(r) \sim \frac{\varepsilon^2}{2} (e^{-r}/r)^2 \quad ,$$

at large distances.

Theorem 4.1. A given simple irreducible 2-graph $\Gamma(r_{12})$ with Debye-Hückel and Abe-Meeron lines decays exponentially like $\exp(-\lambda r_{12})$, where λ is the local line-connectivity of the 2-rooted graph γ obtained by replacing each Abe-Meeron line by two Debye-Hückel lines in parallel. One has the inequalities:

$$|\Gamma(r_{12})| \le C'_M \, |\gamma(r_o)| \, r_{12}^{3k-\ell} \, e^{-\lambda r_{12}} \quad , \quad \forall r_{12} > r_o \tag{4.1}$$

$$C'_m (\varepsilon/r_{12})^{\lambda} \, e^{-\lambda r_{12}} < |\Gamma(r_{12})| \quad , \quad \forall r_{12} > r_1 \tag{4.2}$$

Here, ℓ denotes the total number of lines of γ, and C'_M and C'_m are constants which are analogous to the constants C_M and C_m defined in Eqs. (3.17) and (3.18). For example, one has:

$$C_M = 2^{-\ell_2} \, r_o^{\ell-3k} \, e^{\lambda r_o} \tag{4.3}$$

where ℓ_2 denotes the number of Abe-Meeron lines of Γ.

We will not write down the expression of C'_m, which is a little more complicated. The main point about these constants is that they are positive, and depend only on the topological structure of Γ, but not on r_{12}.

Remark 1. One has $|\Gamma(r_{12})| = (-1)^{\ell_1} \Gamma(r_{12})$, where ℓ_1 is the number of Debye-Hückel lines in Γ, because $B(r)$ is postiive everywhere.

Remark 2. $|\gamma(r_o)|$ is of order ε^{ℓ}, according to the definition (1.2).

Proof. The upper bound (4.1) is obtained by bounding $|\Gamma(r_{12})|$ by means of $|\gamma(r_{12})|$, owing to the inequality:

$$B(r) \le \frac{\varepsilon^2}{2} (e^{-r}/r)^2 \qquad \forall r > 0 \tag{4.4}$$

and then applying theorem 3.2 to $\gamma(r_{12})$.

The lower bound (4.2) is obtained by a simple adaptation of the reasoning that was used in the paragraph entitled Lower Bound for $\Gamma(r)$ in Chapter III, Section C. This is possible, because the Abe-Meeron line is positive everywhere, and thus the integrand of $\Gamma(r_{12})$ has a constant sign for any values of the variables. There is only one extra complication, due to the fact that the partition of the points of Γ, which is used to obtain the lower bound (see Lower Bound for $\Gamma(r)$), is induced by a line-cutset C of γ and not of Γ. This is illustrated in Figure 8. The Abe-Meeron lines which have an endpoint in the neighborhood $S_1(\eta)$ of root-point 1, and the other endpoint in the neighborhood $S_2(\eta)$ of root-point 2, are bounded below by means of the inequality:

$$(1 - u)\,\frac{\varepsilon^2}{2}\,(e^{-r}/r)^2 < B(r) \quad , \quad \forall r > r_1 \tag{4.5}$$

This inequality is valid for a given positive number u as small as we want, and r_1 sufficiently large.

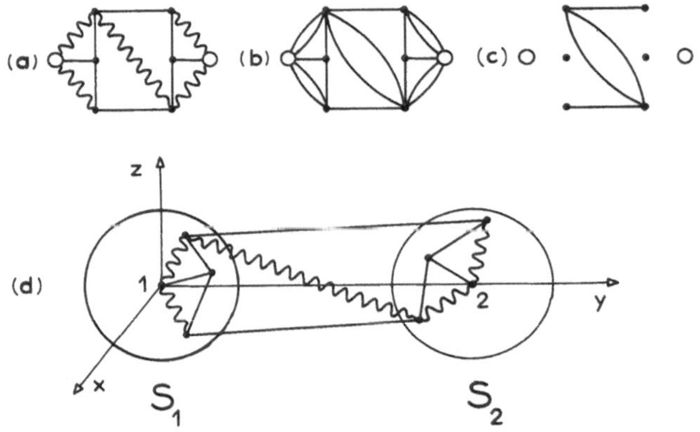

Figure 8. The exponential decay of the 2-graph (a), which has Debye-Hückel and Abe-Meeron lines, is obtained by finding the local line-connectivity λ of the 2-rooted graph (b). The latter is obtained from (a) by replacing each Abe-Meeron line by two lines in parallel. (c) is a minimal line-cutset of (b). (c) shows that $\lambda = 4$, and thus the 2-graph (a) decays exponentially like exp (-4r). (Note that the local line-connectivity of (a) is equal to 3). The partition of the points of (a), which is used to obtain a lower bound decaying exponentially like exp (-4r), is indicated in (d). This partition is induced by (c), the minimal line-cutset of (b).

B. Asymptotic Behavior and Topological Structure of the 2-Graphs Dominant at Large Distances

We first find the exponential decay of the 2-graphs of w(r) which are dominant at large distances. Then, we find also their complementary decay by combining Theorem 3.2 to an argument of DelRio and DeWitt [1969].

Corollary 4.2. The simple irreducible 2-prototypes* which are dominant at large distances decay exponentially like exp (-r) and have at least one isthmus.

Proof. Simple irreducible 2-graphs satisfy the condition $\lambda \geq 1$. Moreover, the class of simple irreducible 2-prototypes decaying exponentially like exp (-r) is not void, because it contains the class of chains made of Debye-Hückel lines alternating with Abe-Meeron lines (see Figure 1). Finally, a 2-graph which decays exponentially like exp (-r) satisfies $\lambda = 1$, and thus has an isthmus, by the max-flow min-cut theorem [Ford and Fulkerson, 1962].

Theorem 4.3. A simple irreducible 2-prototype which has exactly i isthmuses between the root-points decays asymptotically like r^{i-2} exp (-r).

Proof. An irreducible 2-graph which has exactly i isthmuses has the structure indicated in Figure 9.

The Fourier transform of $\Gamma(r)$ is thus:

$$\Gamma(k) = \frac{i}{(1 + k^2)^i} \prod_{j=1}^{i+1} \Gamma_j(k) \tag{4.6}$$

Figure 9. Structure of an irreducible 2-graph which has exactly i isthmuses. The 2-rooted graphs γ_j associated to Γ_j have a local line-connectivity $\lambda \geq 2$, except possibly γ_1 or γ_{i+1} (or both of them), which can be reduced to a single point if the lines (3,4) or (2i+1, 2i+2) (or both of them) are linked directly to a root-point.

*Here, by 2-prototype, we will mean a 2-graph with Abe-Meeron and Debye-Hückel lines, which has no chain of Debye-Hückel lines.

The important point to note here is that, because the local line-connectivity of all the γ_j associated to the Γ_j is larger than or equal to 2, the Γ_j decay exponentially at least as fast as exp (-2r). Therefore, the Fourier transforms $\Gamma_j(k)$ are analytic for Im k < 2 and the pole k = i is isolated. By making exactly the same analysis of the residue of this pole, in the general case, that was made by DelRio and DeWitt [1969] in the particular case where

$$\Gamma_j(k) = \frac{1}{k} \operatorname{arctg} \frac{k}{2} \quad ,$$

one finds that $\Gamma(r)$ decays like $r^{i-2} \exp(-r)$.

C. Formal Proof of Assumption (ii)

To prove assumption (ii), there remains only [Hirt, 1965; Cohen and Murphy, 1969; DelRio and DeWitt, 1969; Mitchell and Ninham, 1968] to order the development (2.5) in powers of ε, and then to extract out of each coefficient of this new development, the 2-graphs which are dominant at large distances.

Lemma 4.4. One has

$$\lim_{\varepsilon\to 0} \Gamma(r_{12}) \Big/ \gamma(r_{12}) = 1 \quad ,$$

provided $\gamma(r_{12})$ is finite. (The quantities ℓ and $\gamma(r_{12})$ have the same meaning as in Theorem 4.1).

Proof. This is a simple consequence of the Lebesgue dominated convergence theorem [Dunford and Schwartz, 1958]. Indeed, the integrand of $\Gamma(r_{12})$ is dominated in absolute value by the integrand of $\gamma(r_{12})$, because of inequality (4.5), and moreover one has

$$\lim_{\varepsilon\to 0} \varepsilon^{-2} B(r) = \frac{1}{2} (e^{-r}/r)^2$$

for all r except r = 0.

In other terms, one has:

$$\Gamma(r_{12}) \underset{\varepsilon\to 0}{\sim} \varepsilon^{\ell} \int_{\Lambda_\infty^k} \prod_{L\in\mathcal{L}\gamma} \frac{e^{-r_L}}{r_L} d\vec{r}_3 \dots d\vec{r}_{k+2} \qquad (4.7)$$

and $\Gamma(r_{12})$ is of order ε^{ℓ}, as well as $\gamma(r_{12})$. This shows that the contribution of $\Gamma(r_{12})$ to $w(r_{12})$ is of order $\varepsilon^{\ell-k}$ for small values of ε, according to Eq. (2.5). Therefore, the usual reordering of

2-graphs [Salpeter, 1958] according to the increasing values of ℓ-k (from 2 to infinity) applies here.

Now, we can complete the proof of assumption (ii). The latter can be expressed in the following precise form:

Theorem 4.5. Among the 2-graphs $\Gamma(r_{12})$ of order n* which have a finite associated 2-graph $\gamma(r_{12})$, those which are dominant at large distances are the chains made of n Debye-Hückel lines alternating with (n-1) Abe-Meeron lines, as illustrated in Figure 1.

Proof. Among the 2-graphs which have a given order ℓ-k = n, those which are dominant at large distances must have the largest possible number of isthmuses, according to Theorem 4.3.

In other terms, we have to maximize the linear function

$$i = k + n - \sum_{j=1}^{i+1} \ell_j \tag{4.8}$$

where ℓ_j is the number of lines of γ_j. Moreover, γ is an irreducible 2-prototype, and therefore k and the ℓ_j must satisfy the conditions:

$$2 \leq \ell_j \qquad j = 2, \ldots, i \tag{4.9}$$

$$0 \leq \ell_1 \quad , \quad 0 \leq \ell_{i+1} \tag{4.10}$$

$$k \leq 2(n - 1) \tag{4.11}$$

The last inequality is obtained by combining the relation ℓ = k+n to the inequality† $3k + 2 \leq 2\ell$.

The problem of maximizing i under the linear constraints (4.9) to (4.11) admits the trivial solution

$$k = 2(n - 1) \quad , \quad \ell_1 = \ell_{i+1} = 0 \quad , \text{ and } \quad \ell_j = 2$$

for j going from 2 to i, which corresponds to the 2-prototype of Figure 1. Moreover, this solution is clearly unique, because all the constraints are saturated. To end the proof, there remains only to remark that this 2-graph is finite and thus decays more

---

*A 2-graph $\Gamma(r_{12})$ is said to be of order n if its contribution to $w(r_{12})$ is of order ε^n. Note that one has then $\Gamma(r_{12}) \sim \varepsilon^{n+k}$ x Cste, according to Eq. (4.7).

†In a graph, the sum of the degrees of all the points is equal to twice the number of lines [Harary, 1969].

slowly than any other finite 2-graph of order n.

Finally, by resumming these 2-graphs, one finds [Cohen and Murphy, 1969; DelRio and DeWitt, 1969; Mitchell and Ninham, 1968] that the potential of mean force w(r) decays like exp $(-r/\Lambda_D)$, as indicated in the introduction.

V. DIVERGENT 2-PROTOTYPES

It is well-known that 2-graphs with k_{ij} lines e^{-r}/r joining pairs of points i and j are infinite if $k_{ij} \geq 3$. Such a problem does not occur any more in the Abe-Meeron theory, because the 2-graphs with $k_{ij} = 2,3, \ldots$ are resummed together and their infinite parts cancel each other.

It is less obvious that 2-graphs with $k_{ij} < 3$ lines joining points i and j can also be divergent, but this is nevertheless true. We will show this as an application of Theorem 3.2.

This explains why our proof of assumption (ii) is only formal (in the sense defined in the note following Eq. (1.3)). Indeed, if the 2-graph $\gamma(r_{12})$ associated to $\Gamma(r_{12})$ is infinite, lemma 4.4 fails to be valid, and so does the classification of the 2-graphs, because it is based on this lemma.

Even worse, we will show that there are 2-graphs belonging to the Abe-Meeron development of w(r) which are infinite. This shows that resummatins more sophisticated than the Abe-Meeron one would be necessary, to obtain a suitable theory of the one-component plasma.

We are now going to prove:

Theorem 5.1. A sufficient condition for a 2-graph with Debye-Hückel lines to be divergent is that

$$\ell > 3k + \lambda \qquad (5.1)$$

For any $k > 5$, there are simple irreducible 2-prototypes satisfying this condition.

Proof. The lower bound (3.16) cannot decay less rapidly at large distances than the upper bound (3.15). This implies that

$$\ell - 3k \leq \lambda \qquad (5.2)$$

If this condition is not satisfied, the inequality (3.15) can hold only if $\Gamma(r_0)$ is infinite. But r_0 is arbitrary, and then $\Gamma(r_{12})$ is infinite for any r_{12}. In other words, a sufficient condition for

$\Gamma(r_{12})$ to be divergent is that Eq. (5.1) holds true.

To construct such divergent 2-prototypes for any k, let us note that, according to Eq. (5.1), they must have a sufficiently large number of lines. We can for example consider a 2-graph with k field-points, and the maximum number of lines, that is:

$$\ell = \frac{1}{2}(k + 2)(k + 1) - 1 \tag{5.3}$$

This is clearly a simple irreducible 2-prototype because each field-point has a degree larger than two for $k \geq 2$. (And then, the associated 2-graph obtained by taking $f_L(r) = e^{-r}/r$ for any L actually occurs in the Abe-Meeron development of w(r), as can be seen from the definition of the $\beta_k(r)$). Moreover, we have also $\lambda = k$. By the relation (5.2), we then see that $\Gamma(r)$ is divergent as soon as $k > 5$, although all its lines are equal to e^{-r}/r and are thus integrable. We give some examples of divergent 2-prototypes in Figure 10.

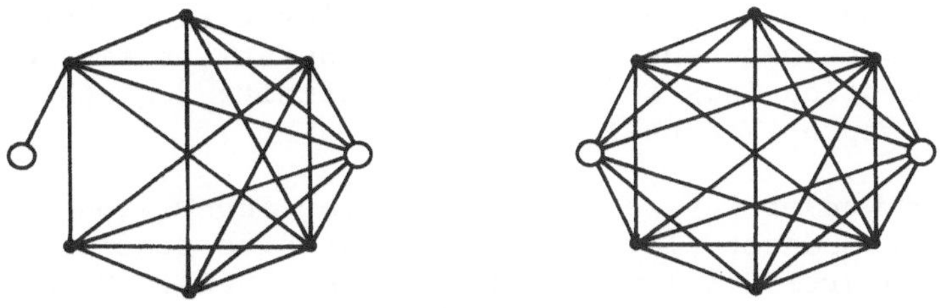

(a) (b)

Figure 10. Two examples of divergent 2-graphs. (a) $\ell = 20$, $k = 6$, $\lambda = 1$. (b) $\ell = 25$, $k = 6$, $\lambda = 6$.

Remark. The divergencies that we found were detected by looking at the behavior of 2-graphs at large distances, but it is clear that they come from the short distance behavior of the Debye-Hückel function.

VI. CONCLUSION AND COMMENTS

In this article, we have investigated the asymptotic behavior of the potential of mean force of a one-component plasma, for large values of the plasma parameter. We have shown that the exponential decay of a given 2-graph is determined by its local line-connectivity. We have also shown that 2-graphs dominant at large distances have at least one isthmus, and that their complementary decay is determined by their number of isthmuses. This has enabled us to justify the corrections to the Debye length for large values of the plasma parameter.

The results that we have obtained for 2-graphs (considered individually) are rigorous. On the contrary, there is still a very long way to go before one can prove (or disprove) rigorously the validity of the corrections to λ_D. Before reaching this goal, there are three main problems which remain to be solved.

The first one is common to all the models of ionized systems. It consists in proving rigorously the existence of λ_D. A first important step in this direction has been made recently [Brydges, 1977], for a Coulomb system on a lattice.

The second problem is still more general, because it is also encountered in the theory of neutral systems [Lavaud, to be published]. It consists in proving that the long distance behavior of w(r) is given by the sum of the 2-graphs which are dominant at large distances (this is what we called assumption (i)).

Finally, the third problem is particular to the Abe-Meeron theory of the one-component plasma. It consists in finding out suitable resummation laws such that the infinite 2-graphs that we exhibited in Chapter V cancel each other.

One could wonder why we made use of the Abe-Meeron theory to study the one-component plasma, because the divergencies that we found are of an artificial nature [Friedman, 1963]. Such divergencies would not occur if we made use of a one-component plasma model with a realistic short distance interaction [Morita, 1959; Ebeling, 1976; Deutsch, Furutani and Gombert, 1976]. In fact, one could even wonder why we made use of a one-component plasma model. Indeed, it has some well-known severe drawbacks. For example, the pressure becomes negative at sufficiently large densities [Lieb and Narnhofer, 1975]. So it is not clear whether this model is well-suited to the systems we are interested in, namely dense plasmas. Another drawback is that the value of the pressure P depends on the statistical ensemble that is chosen to define P [Lieb and Narnhofer, 1975]. So, one cannot speak of a thermodynamic function without specifying by which physical process it is defined.

Nevertheless, we have chosen to stick to the Abe-Meeron theory of the OCP because the problem that we solve here, namely the asymptotic behavior of 2-graphs with lines which decay exponentially*, can be treated in a very clean manner in this case. Moreover, our results can be generalized to two-component systems with a realistic short distance interparticle potential, where no 2-graphs are

*We thus assume the existence of the Debye length.

infinite and where the pressure is the same in all the statistical ensembles, and is everywhere positive[1]. This generalization can be made along the same lines that were used in Chapter IV. Finally, it could be interesting, for people working in quantum field theory, to study the mechanism of cancellations between the divergent 2-graphs of the Abe-Meeron theory, because one knows exactly, in this case, where the divergences come from and how they could be avoided.

ACKNOWLEDGEMENTS

I am very grateful to Prof. J. Chapelle for constant support and encouragements during this work, and to Professors M. Feix and G. Kalman for having given me the opportunity to include this article in the Proceedings of the Institute. I am also very grateful to Professors Ph. Choquard, C. Deutsch and H. E. DeWitt for very useful discussions.

Eq. (2.2)

Eq. (2.3)

$$\circ\!\!-\!\!\circ \quad \frac{1}{4\pi\varepsilon} \left[\text{diagram} + \text{diagram} + \text{diagram} \right]$$

$$+ \frac{1}{2!(4\pi\varepsilon)^2} \left[2\,\text{diagram} + 4\,\text{diagram} + \text{diagram} + \cdots \right] + \cdots$$

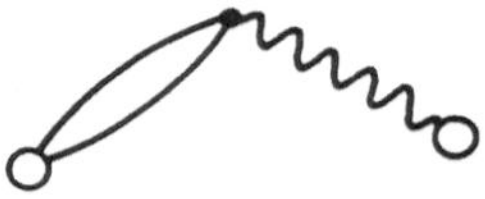

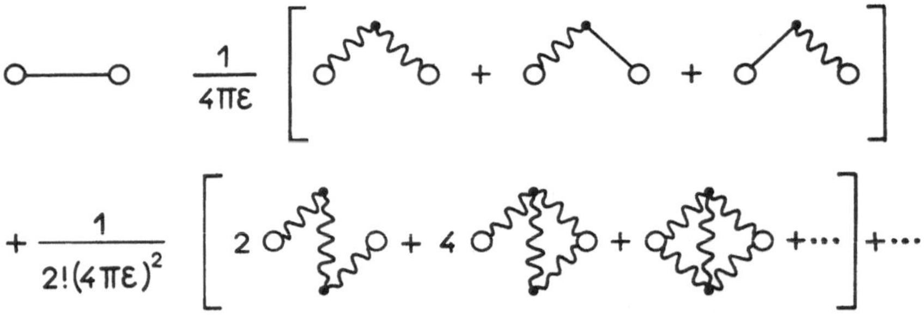

Eq. (2.6)

REFERENCES

Berge, C., 1973, Graphes et hypergraphes, Dunod.

Brydges, D., 1977, preprint, "A rigorous approach to Debye screening in dilute classical coulomb systems".

Cohen, E. G. D. and T. J. Murphy, 1969, Phys. Fluids 12, 1404.

DelRio, F. and H. E. DeWitt, 1969, Phys. Fluids 12, 791.

Deutsch, C. and Y. Furutani, 1975, J. Phys. A8, 183.

Deutsch, C., Y. Furutani and M. M. Gombert, 1976, Phys. Rev. A13, 2244.

Deutsch, C. and M. Lavaud, 1974, Phys. Rev. A9, 2598.

Dunford, N. and J. Schwartz, 1958, Linear Operators, Interscience, New York.

Ebeling, W., 1976, Theory of bound states and equilibrium ionization in plasmas and solids, Akademie Verlag, Berlin.

Edwards, S. F. and A. Lenard, 1962, J. Math. Phys. 3, 778.

Erdélyi, A., 1956, Asymptotic Expansions, Dover.

Ford, L. and D. Fulkerson, 1962, Flows in Networks, Princeton.

Friedman, H., 1963, Ionic Solution Theory, Interscience, New York.

Harary, F., 1969, Graph Theory, Addison Wesley, Reading, Mass.

Hardy, G. H., J. E. Littlewood and G. Polya, 1937, Inequalities, Cambridge University Press.

Hauge, E. H. and P. C. Hemmer, 1971, Physica Norvegica 5, 209.

Hirt, C. W., 1965, Phys. Fluids 8, 693.

Hsu, L., 1948, Duke Math. J. 15, 623.

Hsu, L., 1951, Am. J. Math. 73, 625.

Lavaud, M., 1977, Phys. Lett. 62A, 295.

Lavaud, M., "Estimates of general Mayer graphs. I. General theory and upper bounds with trees", submitted to J. Stat. Phys.

Lavaud, M., to be published, "Rigorous estimates of general Mayer graphs and applications. II: Long-range behavior of graphs with two root-points", Phys. Lett.

Lavaud, M., "Long range behavior of graphs with two root-points, occuring in the theory of ionized systems", submitted to J. Stat. Phys.

Lenard, A., 1961, J. Math. Phys. 2, 682.

Lieb, E. and Narnhofer, H., 1975, J. Stat. Phys. 12, 291.

Mitchell, D. J. and Ninham, B. W., 1968, Phys. Rev. 174, 280.

Morita, T., 1959, Progr. Theor. Phys. 22, 757.

Salpeter, E. E., 1958, Ann. Phys. 5, 183.

(1) The only rigorous result which is known is the existence of the thermodynamic functions. See Lieb, E. and Lebowitz, J., Adv. Math. 9, 316 (1972). The difficulty and the beauty of the proof are to be contrasted with the triviality of the result.

(2) These results were exposed in the course of a set of seminars, "Calcul des propriétés à l'équilibre des gas neutres et ionisés denses par la théorie de Mayer", given at the Laboratoire de physique et optique corpusculaire, Université Paris V., unpublished notes (1975-1976).

(3) Deutsch and Furutani [1975] and Deutsch, Furutani and Gombert [1976] find an asymptotic behavior in $Ar^{-1} \exp(-Br)$ for the 2-graphs of order 3 and 4. Their computed B's are non-integer real numbers. This is inexact, by our theorem 3.2. Their error comes from the fact that they Fourier transform $\Gamma(r)$, and make use of the equivalence $1 - a^2k^2 \sim (1 + a^2k^2)^{-1}$. This relation is exact for small values of k, but introduces poles in the complex plane which do not belong to the Fourier transform of $\Gamma(r)$. This gives thus a decay which is false. The same type of reasoning is used in the Ornstein-Zernike theory for the correlation function h(r), and is probably false too [Lavaud, to be published], for the same reason.

(4) They bound a given 2-graph by iterating the following algorithm. First, they choose a line, say (3,4). Then they split the domain of integration in two parts, one where $r_{34} < r_0$, and the other where $r_{34} \geq r_0$, where r_0 is defined in such a way that $f(r_0) = 1$. The integral in the first domain is bounded by the 2-graph obtained by deleting the line (3,4) because one has $f_{34} < 1$. Then, they evaluate the integral in the second domain by assuming that $f_{3j} \sim f_{4j}$ for any j, when $r_{34} < r_0$. This enables them to replace f_{3j} by f_{4j} in the integral (this is their pinching procedure). But this is <u>not</u>

possible, because the condition $r_{34} < r_0$ does not imply $f_{4j} \sim f_{3j}$ (one can have $f_{4j} = +\infty$ and $f_{3j} = 1$). This would be true only if one had moreover $r_{3j} >> r_0$. So, still other terms ought to be taken into consideration, the decay of which is not known.

(5) Deutsch et al call this quantity the degree of convection of the graph. In Deutsch, Furutani and Gombert [1976], they argue that $\kappa = \lambda$ for bridge graphs (i.e. simple irreducible graphs). This is nevertheless false, as can be seen in Figure 4c. In fact, these two quantities are generally different, in the same way that the number of points and lines in a graph are generally different (although one can clearly have graphs with the same number of points and lines, and similarly graphs with the same local point-connectivity κ and local line-connectivity λ). To see that their method cannot give upper bounds decaying like $e^{-\lambda r}$, it is sufficient to find a graph where <u>all</u> lines belong to <u>all</u> maximal sets of line-disjoint chains. So, the deletion of <u>any</u> line gives a graph with a local line-connectivity equal to λ-1. The graph (c) of Figure 4 has this property. We prove in this article that it decays exponentially like e^{-3r}, because $\lambda = 3$. But the pinching procedure of Deutsch et al gives an upper bound which is a sum of graphs, where at least one of them has one less line. So, even if their bounding procedure could be corrected, their bound could not decay faster than e^{-2r}, anyway.

ASYMPTOTIC BEHAVIOR OF THE CORRELATION FUNCTION

SOME EQUILIBRIUM PROPERTIES OF THE 2D STRONGLY COUPLED OCP

D. Merlini*

Centre de Recherche en Physique des Plasmas
Ecole Polytechnique Fédérale de Lausanne
21, Avenue des Bains, 1007 Lausanne, Switzerland

I. INTRODUCTION AND DEFINITION OF THE MODEL

Ever since the pioneering works of Onsager [1949] and Montgomery [1976] in two-dimensional (2D) hydrodynamic turbulence and on anomalous transport in plasmas pervaded by strong d.c. magnetic fields - modeled by Montgomery as a collection of charged filaments interacting through the 2D long range logarithmic potential, there has been a growing interest in 2D charged liquids.

Coulombian systems are known to play an important role in the nonrelativistic physics of matter and especially in the statistical mechanics of plasmas in the strongly coupled regime [Choquard, 1974]. In this lecture, we shall discuss some thermodynamic properties of the one-component plasma (OCP) in a strictly 2D world where the Coulombic potential is logarithmic. The model consists of N point charges (each having charge $-e$) in a domain Λ, immersed in a uniform neutralizing background of charge density

$$\rho_e = \rho e = \frac{Ne}{|\Lambda|}$$

The microscopic interaction potential between 2 point charges is given by

* Present address:
Department of Physics, College of William and Mary,
Williamsburg, Virginia 23185

$$V_{ij} = -e_i e_j \ \ell n \left(\frac{|\vec{x}_i - \vec{x}_j|}{\ell} \right) \qquad ;$$

we put $\ell = 1$ and in so doing, we adopt the usual point of view of statistical mechanics, that the zero point of the interaction potential does not depend on the shape of the system. The usual definition of the ν-d plasma parameter is

$$\gamma^{(\nu)} = \beta e^2 \ \rho^{(\nu-2)/\nu} \qquad ,$$

so that in the 2D case $\gamma = \beta e^2$ is independent of the density, reflecting the peculiar scaling property. We write the hamiltonian as

$$H_\Lambda = H_{bb} + H_{pp} + H_{pb}$$

where b and p denote, respectively, the background and the particles; the inverse Debye length is given by $k_D = (2\pi \rho \beta e^2)^{1/2}$.

II. H-STABILITY PROPERTY; 2D WIGNER LATTICES

The first question which arises is whether the 2D OCP H_Λ is bounded from below by an extensive quantity. This point is connected with the fact that, although it can rarely be proved, some many body potentials reach their minima at regular lattice structures.

Let Λ be a domain of reasonable shape and assume the system to be neutral, with ρ the fixed background density. Then there exists an extensive density-dependent lower bound on the hamiltonian of the 2D OCP expressed as

$$H_\Lambda(\vec{x}_1, \ldots, \vec{x}_N) \geq -Nb(\rho) = -N\left(\frac{3}{8} + \frac{1}{4} \ \ell n \ \pi\rho \right) e^2 \qquad , \qquad (2.1)$$

where

$$b = E + \frac{1}{4} e^2 \qquad ,$$

E being the self energy of a 2D ball of density ρ and total charge +e., i.e.

$$E = \left(\frac{1}{8} + \frac{1}{4} \ \ell n \ \pi\rho \right) e^2 \qquad ;$$

-b is simply the Coulomb energy of a neutral elementary system (consisting of just one particle and the background) in the configuration of lowest energy and may be understood in the light of the Onsager Lemma.

An heuristic derivation of the stability bounds in each dimension has been given [Sari, Merlini and Calinon, 1976] using Wigner Seitz dual cells. In three dimensions, the bounds coincide with the Wigner estimate of the lattice energy and the first rigorous proof was given by Lieb and Narnhofer [1975]; the corresponding 2D case is treated by Sari and Merlini [1976]. To make contact with other properties, the explicit bounds in d = 1,2,3 dimension read:

$$b^{(1)} = -\frac{e^2}{12\rho}$$

$$b^{(2)} = e^2\left(\frac{3}{8} + \frac{1}{4}\ln \pi\rho\right)$$

$$b^{(3)}(\rho) = \frac{9}{10}\cdot\left(\frac{4}{3}\pi\rho\right)^{1/3}$$

Bound (2.1) then provides a lower bound for the free and internal energies of the 2D OCP, i.e.

$$\beta f > -\rho(1 + \beta\rho b - \ln\rho) \quad , \qquad \beta = (kT)^{-1} \quad ,$$

$$\psi = \langle h\rangle \geq -\gamma\left(\frac{3}{8} + \frac{1}{4}\ln\pi\rho\right) \quad .$$

Remark: The above bounds hold for each domain Λ of reasonable shape. However, except for the N = 1 particle and Λ spherical where the ground state has the greatest symmetry, these bounds can never be reached. For N > 1 and large, the configurations of highest symmetry are then given by the Wigner lattices, i.e. the crystalline configurations of particles whose energies are and can be understood to be extremely close to the above bounds [Sari, Merlini and Calinon, 1976].

Then following the above remark, the method of dual cells, which takes exactly into account the actual shape of the polyhedron, yields accurate values of the energy in each dimension.

In zero order approximation we therefore have that [Sari, Merlini and Calinon, 1976]

$$U = -b + \begin{cases} 0.03170\ e^2 \\ 0.008628\ e^2 \\ 0.001588\ e^2 \end{cases} \text{for the} \begin{cases} \text{hexagonal} \\ \text{square} \\ \text{triangular} \end{cases} \text{lattice} \tag{2.2}$$

However, it can be shown that first- and second-order approximation calculations yield energies exactly up to 10^{-5}. The same method of cells then yields theoretical upper bounds for the free energy density of the 2D OCP: i.e., with $\Delta f = f - f_o$, where $f_o = \beta^{-1}\rho(\ell n\ \rho - 1)$.

$$-b \leq \frac{\beta\Delta f}{\rho} \leq -b + 0.253\gamma + 1 \qquad (2.3)$$

Relation (2.3) is valid for each value of γ. The method applies as well to the 3D case; with

$$\Gamma = \left(\frac{4}{3}\pi\rho\right)^{1/3} \beta e^2$$

[Sari and Merlini, 1976],

$$-0.9\Gamma \leq \frac{\beta\Delta f}{\rho} \leq -0.66\Gamma + 1 \qquad \forall\Gamma \quad .$$

The recent accurate power law of DeWitt in the solid phase of the 3D OCP turns out to lie between these two bounds.

Now the existence of well-defined lower and upper bounds (Eq. (2.3), although not sufficient, strongly suggests that the usual thermodynamic limit exists for the 2D OCP; this is summarized in the next chapter.

III. THERMODYNAMIC LIMIT. SCALING PROPERTY AND EQUATION OF STATE

As we have seen, the existence of two extensives, a lower and an upper bound (Eq. 2.3) strongly indicate that the usual thermodynamic limit for the free energy exists. In order to prove this, one adapts the "Cheese" Theorem of Lieb and Lebowitz [1972] to the OCP. A first point which must be kept in mind is that, in the OCP, the rigidity of the background (whose consequence is the appearance of non-translationally invariant potentials in a finite domain) does not allow one to put all particles in the standard sequences of balls; moreover, the 2D case presents some particular feature due to the long range nature of the logarithmic Coulomb potential. In any case, exploiting Newton's theorem of electrostatics, using Jensen's inequality and the knowledge of the H-stability bound (Eq. 2.1), it is shown following Lieb and Lebowitz [1972] that for the 2D OCP [Sari and Merlini, 1976]

$$g = -\beta f = \lim_{\substack{\Lambda\to\infty \\ N\to\infty \\ \frac{N}{|\Lambda|}=\rho}} \frac{1}{|\Lambda|} \ell n\ Z$$

exists for all domain Λ of reasonable shape. Since such a limit exists and is independent of the shape of Λ, the usual scaling property is obtained and the equation of state reads

$$\beta p = \beta\rho \frac{\nu - 2}{\nu} \langle h^{(\nu)} \rangle + \rho\left(1 - \frac{\gamma^{(\nu)}}{4} \delta\nu,2\right) \tag{3.1}$$

where $\langle h^{\nu} \rangle$ is the mean potential energy per particle in the ν - d OCP.

For the 2D OCP, the free energy takes the form

$$-\beta f(\rho,\gamma) = -\beta f^*(\gamma) - \rho \ln\rho(1 - \frac{\gamma}{4}) \tag{3.2}$$

where $f^*(\gamma=4)$ is the free energy density at the temperature

$$T_c = \frac{e^2}{4k\sqrt{\beta}} \quad ;$$

thus if one adopts the usual definition $\beta p = -\rho^2 \frac{\partial}{\partial\rho}\left(\frac{g(\rho,\beta)}{\rho}\right)$ for the 2D OCP then

$$\beta p = \rho(1 - \frac{\gamma}{4})$$

is the equation of state. Note that the scaling property leads immediately to the equation of state, independent of the knowledge of the existence of the thermodynamic limit as given by several authors [Hauge and Hemmer, 1971; Deutsch and Lavaud, 1974]. This situation analogously arises in the transport theory for the same 2D plasma model. It may be shown that a scaling property predicts an anomalous diffusion coefficient $D \sim 1/B$. However, even at equilibrium, such a scaling does not ensure the existence of D; in fact, D diverges [Taylor and McNamara, 1971]. In this context, it is not known if such a divergence is peculiar to the 2D case or if it is due to the approximations involved in the computation. It is expected that the investigation of such a question within the framework of a dynamics for damped non-markovian systems in the small region of γ will provide further information about the numerical value of D in the thermodynamic limit [Merlini et al, in preparation].

Notice further, that in the 3D case, the scaling property (Eq. 2.4), connected with the upper and lower bounds for the free energy, allows one to consider the two equations of state [Sari and Merlini, 1976]:

$$\beta p_1 = \rho(1 - \frac{3}{10} \Gamma)$$

$$\beta p_2 = \rho(1 - \frac{2}{9} \Gamma)$$

the first one being a rigorous lower bound for the pressure. In any case, the above simple expressions give very good agreement with the data of Hansen et al over a wide range of Γ values [Hansen and Pollock, 1973]. In particular, the values $\Gamma_1 = 2.5$ and $\Gamma_2 = 3.375$, which correspond to the maximum values of the p_1 and p_2 isotherms (in the p-ρ plane), bracket the critical Monte Carlo value $\Gamma \simeq 3$ which defines the onset of negative compressibility [Brush, Salin and Teller, 1966].

IV. APPROXIMATION SCHEMES FOR THE STRONGLY COUPLED 2D OCP

In this chapter we cite some new results for the static behavior of the 2D OCP obtained from analyses of two principal strongly coupled plasma approximation schemes which are non-perturbative in the plasma parameter [Calinon, Golden, Kalman and Merlini, to be published; Bakshi, Kalman and Silevitch, to be published]. We consider first the Totsuji-Ichimaru (TI) scheme [Totsiji and Ichimaru, 1973] formulated from the second BBGKY static equation connecting the equilibrium pair and triplet correlation function $g(r_{12})$ and $h(r_{12}, r_{23})$ with self-consistency guaranteed by supposing that the triplet can be decomposed into Mayer clusters of pair correlation functions. The resulting formulation

$$\rho g(k) = \frac{-k_D^2[1 + u(k)]}{k^2 + k_D^2[1 + u(k)]} , \tag{4.1a}$$

$$u(\vec{k}) = \frac{1}{(2\pi)^2} \int d^2\vec{p}\, \frac{\vec{k}\cdot\vec{p}}{p^2} [1 + \rho g(p)]\, g(|\vec{k} - \vec{p}|) , \tag{4.1b}$$

connects the Fourier transformed pair correlation function $g(k)$ to the 2D TI screening function $\vec{u}(k)$. Note that the combination of Eqs. (4.1a,b) results in a nonlinear integral equation for $g(k)$. Equation (4.1a) when written in terms of the structure factor $S(k) = 1 + \rho g(k)$ is

$$S(x) = \frac{x^2}{x^2 + 1 + u(x)} , \qquad x = k/k_D . \tag{4.2}$$

In the long wavelength limit ($x \to o$), we have from Eqs. (4.1b) and (4.2) that [Calinon, Golden, Kalman and Merlini, to be published]

$$u(x \to o) = -(\gamma/4)\, x^2 + [(\gamma/16) \int_o^\infty \frac{dy}{y} \left(\frac{dS(y)}{dy} \right)^2]\, x^4 + \ldots \tag{4.3}$$

so that

$$\lim_{x \to o} x^{-4}[x^2 - S(x)] = 1 - \gamma/4 \quad . \qquad (4.4)$$

Equation (4.4) exactly agrees with our recently established 2D OCP compressibility sum rule [Golden and Merlini, 1977]

$$\lim_{x \to o} x^2[\varepsilon(x,o) - 1] = (1 - \gamma/4)^{-1} \qquad (4.5)$$

and Eqs. (4.3) to (4.5) are valid in the strong coupling regime where γ can well exceed unity.

In the short wavelength limit ($x \to \infty$),

$$u(\vec{k}) \approx \frac{1}{(2\pi)^2} \int d^2\vec{p} \, \frac{\vec{k} \cdot \vec{p}}{p^2} \, g(|\vec{k} - \vec{p}|) \equiv u_{STLS}(\vec{k}) \quad ,$$

where the STLS subscript denotes the Singwi-Tosi-Land-Sjölander [1968] screening function. The asymptotic behavior of $g(k \to \infty)$ can therefore be described by a kind of "linear" (in u) STLS equation whose solution is [Calinon, Golden, Kalman and Merlini, to be published]

$$g(k) \sim \frac{C(\gamma)}{(k^2)^{1 + \gamma/2}} \quad , \qquad k \to \infty \quad ,$$

where $C(\gamma)$ is an appropriate γ-dependent constant.

One can assess the short range ($r \to o$) behavior of $g(r)$ by analyzing the configuration space version of the TI equations (4.1a,b) [Bakshi, Kalman and Silevitch, to be published]. From this configuration space version we found that for $\gamma < 2$ [Calinon, Golden, Kalman and Merlini, to be published],

$$g(r \to o) \sim -1 + ar^{\gamma}$$

in agreement with the binary approximation

$$g(r \to o) \sim -1 + e^{-\beta\phi(r)} \quad , \qquad \phi(r) = -e^2 \ln r/r_o$$

valid at short interaction distances in 2D Coulombian systems.

To gain more insight into the static behavior of the 2D OCP, we have also analyzed the 2D version of the STLS approximation scheme. This method amounts to dropping the non-equilibrium correlational part of the two-particle distribution function in the perturbed (by an external field) first BBGKY kinetic equation with self-consistency guaranteed by use of the linear static fluctuation-dissipation theorem. Here it is remarkable that the 2D nonlinear STLS equation, unlike its 3D counterpart, can be analytically solved: the ensuing exact solution involves a moving cut-off in k-space. Interesting analytical properties may be investigated leading to exact solutions in r-space featuring oscillations in g(r) [Calinon, Golden, Kalman and Merlini, to be published].

Concerning the numerical method for solving Eqs. (4.1), a correct method using generalized damping devices is involved and the solution up to relatively high values of γ has been obtained [Calinon, Golden, Kalman and Merlini, to be published; Calinon and Merlini, in preparation]; the numerical results agree with the theoretical predictions in the short- and long-range regions. Moreover, g(k) exhibits oscillations and a peak develops at about $\gamma = 2$.

Finally, connected with this remarkable change of the structure factor, it is shown that the fluctuation spectrum in the long-wavelength region is enhanced at $\gamma = 4$ [Calinon, Golden, Kalman and Merlini, to be published].

To conclude, the problem of the occurrence of inhomogeneous states and related singularities in g(k), in the strong coupling limit may be investigated and should provide new insight in favor or not of the existence of some kind of phase transition in such a continuous 2D model [Calinon, Golden, Kalman and Merlini, to be published].

REFERENCES

Bakshi, P., G. Kalman and M. B. Silevitch, to be published.

Brush, S. G., H. C. Sahlin, E. Teller, 1966, J. Chem. Phys. 45, 2102.

Calinon, R., K. I. Golden, G. Kalman and D. Merlini, to be published.

Calinon, R. and D. Merlini, in preparation.

Choquard, Ph., 1974, in Physical Reality and Mathematical Description, edited by Enz and Mehra, D. Reidel Publishing Company, Dordrecht-Holland, 516-532.

Deutsch, C. and M. Lavaud, 1974, Phys. Rev. A9, 2595.

Golden, K. I. and D. Merlini, 1977, Phys. Rev. 16A, 438.

Hansen, J. P. and E. Pollock, 1973, Phys. Rev. A8, 3110.

Hauge, E. H. and P. C. Hemmer, 1971, Phys. Norvegia 5, 209.

Lieb, E. H. and J. L. Lebowitz, 1972, Advances in Math 9, 316.

Lieb, E. H. and H. Narnhofer, 1975, J. Stat. Phys. 12, 291.

Merlini, D. et al, in preparation.

Montgomery, D., 1976, Physica 82C.

Onsager, L., 1949, Statistical Hydrodynamics, Nuovo Cimento 6 (Suppl.), 229.

Sari, R. R., D. Merlini, 1976, J. Stat. Phys. 14, 91.

Sari, R. R., D. Merlini and Calinon, R., 1976, J. Phys. A9, 1539.

Singwi, K. S., M. P. Tosi, R. H. Land and A. Sjölander, 1968, Phys. Rev. 176, 589.

Taylor, J. B. and B. McNamara, 1971, Phys. Fluids 14, 1492.

Totsuji, H. and S. Ichimaru, 1973, Progr. Theor. Phys. 50, 753.

GROUP TRANSFORMATION FOR PHASE SPACE FLUIDS

J. R. Burgan, J. Gutierrez*, A. Munier, E. Fijalkow**, M. R. Feix

CRPE/CNRS
Universite D'Orleans

I. INTRODUCTION

We want to study the time behavior of systems where long distance forces are predominant. Such is the case of plasmas, accelerator beam (where we are dealing with Coulomb forces plus electromagnetic confining external fields) and self-gravitating gas (galaxy, cluster of stars, etc.) where the Newtonian attraction competes against the thermal (ballistic) expansion. In many cases we can disregard the small irregularities due to the grain structure of the matter (with grain as big as a star in a galaxy!) and describe the interaction through a continuous field obtained by the solution of the Poisson equation. This is the well-known Vlasov Poisson system where the global description is obtained by considering the distribution function $f(\vec{x},\vec{v},t)$ in the six-dimensional phase space in contrast to a regular gas where we can usually deal with the first moments of f with respect to $\vec{v}$ (particle density, momentum, energy density, etc.). Consequently, we call such systems phase space fluids. A discussion of the relative properties of these a priori very different fluids is given by Feix [1975]. From the model maker's point of view adopted here they present great similarities.

The nonlinear solution of the Vlasov Poisson systems are of course very difficult to obtain analytically and most of the studies resort to numerical simulation. As soon as we introduce numerical algorithms the problem of the correctness of the long time behavior

---

*Supported by Instituto de Estudios Nucleares (J.E.N.) Spain
** UER Sciences

is raised. From our experience with differential equations this problem is best solved if we can find an asymptotic series based on a systematic study of the different terms. As a matter of fact, we begin to mix both analytical and numerical methods and find in this case the possibility of following some systems during thousands of periods.

Our group is developing ideas along this line and preliminary encouraging results are exposed in Bitoun, Nadeau, Guyard and Feix [1973] and Nadeau, Veyrier and Feix [1976].

Another "cornerstone" upon which this paper is built is the existence of singular solutions of the equations. As indicated by their names, singular solutions are to be opposed to the regular (usual) solutions. Singular solutions have a special ordinarily simple structure which is preserved during the motion. The problem is to know if these singular solutions are never representative of the others (which at the beginning do not possess the special structure) or on the contrary are very representative of the regular solutions which could, for example, go asymptotically to the singular solutions. These questions are of course practically unanswered. In plasma physics the BGK structures [Bernstein, Green and Kruskal, 1957] provide examples of such singular solutions. They are periodic steady states, eventually moving at a constant velocity and are self-supporting. Their stabilities are a very complex problem [Feix, 1975] and they can be "transient asymptotic" solutions as shown by numerical experiments [Morse and Nielson, 1969].

Steady state structures are usually not too difficult to obtain (although their number and their variety is a puzzle in plasma physics). Solutions involving the time are still more difficult. Recently the group technique has been used both in plasma physics (but on fluid models of nonlinear waves rather than on Vlasov-Poisson systems) and on models which represent one-dimensional equivalents of Navier Stokes fluid systems (Berger's, Kortweig De Vries equations, etc.). Although in these cases self similar group techniques solve the problem it is fair to mention that usually the problem has already been solved. Such is the case of the famous solitons structures [Shen and Ames, 1974; Zabusky and Kruskal, 1965]. Incidently, the method must be traced back to Boltzmann who used it to solve the heat diffusion equation.

Some confusion exists about the usefulness of the group method and the way the method should work. Usually we check the symmetry of the equations with respect to a continuous transformation group. We subsequently use this symmetry to reduce by one the number of variables. Numerical solutions are easier and sometimes analytical solutions are possible. But of course since we have eliminated from the very beginning one degree of freedom we cannot say how these solutions are affected by a modification of the initial (or boundary) conditions.

We will consider the self similar transformation of phase space and more precisely we will consider the transformation

$$\vec{\xi} = \vec{x}(t/T)^{-\alpha}$$

$$\vec{\eta} = \vec{V}(t/T)^{1-\alpha} \tag{1.1}$$

which will be justified later on. The meaning of the transformation is such that the $\vec{x},\vec{v},t$ dependence can be condensed into a $\vec{\xi},\vec{\eta}$ transformation where $\vec{\xi}$ and $\vec{\eta}$ are the rescaled coordinates of the phase space and describe such simple motion that the time behavior can consequently be taken into account by this rescaling. In Eq. (1.1) T is an arbitrary time and it convenient to consider that the initial time is not $t = 0$ but $t = T$. From the following consideration it is clear that we are simply going to obtain a generalization of the BGK (steady state) structures.

In fact, an interesting new point of view is to introduce both a rescaling like Eq. (1.1) and keep a time variation through the introduction of a new, also rescaled, time. In fact, we will introduce the following transformations

$$\theta(t)$$

$$\vec{\xi} = A(t)\ \vec{x}$$

$$\vec{\eta} = B(t)\ \vec{x} + C(t)\ \vec{V} \tag{1.2}$$

with a proper choice of $\theta(t)$, $A(t)$, $B(t)$ and $C(t)$. This transformation has an old story. Introduced in Courant and Snyder [1958] it has been rediscovered by R. H. Lewis [1968]. Our group has pointed out some of the interesting properties both from theoretical and practical points of view [Bitoun, Nadeau, Guyard and Feix, 1973; Guyard, Nadeau, Baumann and Feix, 1971]. Now it appears under a new aspect and we will learn how to use the freedom left, Eq. (1.2) being a continuous Lie group of transformations, to solve some problems.

Now we end this lengthy introduction with the title of the different chapters. In Chapter II we will review quickly some of the results of the classical self similar group theory applied to the Vlasov Poisson system (with the absorption of the time variation in the rescaling of the new phase space). In Chapter III we will show how the time rescaling can be worked out in the case of the heat equation (where only time and configuration space are involved). This example is especially interesting since it shows how the asymptotic solutions are, in this case, taken care of by the group transformations. In Chapter IV we will introduce the complete time phase space rescaling and will show the mathematical group structures. In Chapter V we will apply it to a very intriguing problem of cosmological theory, the structure of a self-gravitating system when the gravitational constant G varies with time, and we will show that the Dirac hypothesis with G varying inversely with the age of the universe corresponds to a very simple problem. In Chapter VI it will be shown that the problem of the motion of a charged particle in a uniform in space, time-varying magnetic field, is very similar to the preceding. In Chapter VII we will generalize the group transformation introduced in Chapter IV to the quantum case and in Chapter VIII we will come back to the Vlasov Poisson system for plasmas (we will study nonlinear oscillations for plane, cylindrical and spherical geometry), self-gravitating gas and the problem of the expansion of a beam under the space charge forces.

II. SELF SIMILAR GROUP

We consider a one-dimensional collisionless phase space fluid described by the Vlasov Poisson system. This fluid may be an electronic plasma (where for simplicity we suppose a motionless ion background of density N_o) or a one species population beam. The two cases are respectively labelled P and B. The gravitational case, labelled G, is identical to the B case with a change of sign in the Poisson law. Consequently, we have three independent variables $\vec{x},\vec{v},t$ and two functions $f(\vec{x},\vec{v},t)$ the phase space distribution and $E(\vec{x},t)$ the electric field connected by the following systems of equations

$$\frac{\partial f}{\partial t} + \vec{v}\,\frac{\partial f}{\partial \vec{x}} + \vec{E}\,\frac{\partial f}{\partial \vec{v}} = 0$$

$$\frac{\partial \vec{E}}{\partial \vec{x}} = \int_{-\infty}^{\infty} f(x,v,t)\, d\vec{v} - N_o \tag{2.1}$$

For simplicity we took $e = m = \varepsilon_o = 1$. As we mentioned already we introduce the transformation group

$$\bar{t} = a^{\beta} t; \quad \bar{x} = a^{\gamma} x; \quad \bar{v} = a^{\delta} v; \quad \bar{E} = a^{\varepsilon} E; \quad \bar{f} = a^{\omega} f$$

To get the formal invariance of the system we must consider the four following invariants.

$$\xi = x\left(\frac{t}{T}\right)^{-\alpha}; \quad \eta = v\left(\frac{t}{T}\right)^{1-\alpha}; \quad F = f\left(\frac{t}{T}\right)^{\alpha+1}; \quad \varepsilon = E\left(\frac{t}{T}\right)^{2-\alpha} \tag{2.2}$$

in Eq. (2.2) α is a real arbitrary number. T is a characteristic time and it is useful to consider the time origin at t = T. At this time

$$\xi = x; \quad \eta = v; \quad F = f; \quad \varepsilon = E \quad .$$

We obtain, after substitution for the Vlasov Poisson system

$$\eta \frac{\partial F}{\partial \xi} + \varepsilon \frac{\partial F}{\partial \eta} - \frac{1}{T} \{(\alpha - 1) \eta \frac{\partial F}{\partial \eta} + \alpha\xi \frac{\partial F}{\partial \xi} + (\alpha + 1) F\} = 0$$

$$\frac{\partial \varepsilon}{\partial \xi} = \int_{-\infty}^{\infty} F d\eta - N_o \left(\frac{t}{T} \right)^2 \tag{2.3}$$

From Eq. (2.3) we see that the transformation is possible in the B and G cases but not in the plasma case because the variable t is still present in Eq. (2.3). Nevertheless, a solution can be found in the following way: we translate the time origin at T writing $t = T = \tau$. Assuming $T \to \infty$ and keeping τ finite (otherwise as large as we like) we see that $N_o(t/T)^2 \to N_o$. Moreover, to avoid the trivial time-independent solution in Eq. (2.3) we must also let $\alpha \to \infty$ with $\alpha/T = \beta$. The transformation (2.2) becomes

$$\xi = x \exp -\beta\tau; \quad \eta = v \exp -\beta\tau; \quad F = f \exp \beta\tau; \quad \varepsilon = E \exp -\beta\tau. \tag{2.4}$$

and the Vlasov Poisson system becomes

$$\eta \frac{\partial F}{\partial \xi} + \varepsilon \frac{\partial F}{\partial \eta} = \beta(F + \xi \frac{\partial F}{\partial \xi} + \eta \frac{\partial F}{\partial \eta})$$

$$\frac{\partial \varepsilon}{\partial \xi} = \int_{-\infty}^{\infty} F d\eta - N_o \tag{2.5}$$

Now Eqs. (2.3) and (2.5) should be discussed and if possible solved in the B and P case. This has been partly done in Baranov [1976] and Burgan, Gutierrez, Fijalkow, Navet and Feix [to be published]. An interesting solution is obtained taking $\alpha = -1$ for the B case. Assuming that the self-consistent field has the form

$$\xi = 2\xi \; T^{-2} \tag{2.6}$$

we get for the solution of F

$$F = \psi(\eta \frac{T}{2} - \xi)$$

This is the "phase space stick" structure discussed in Burgan, Gutierrez, Fijalkow, Navet and Feix [to be published] and generalized for a plasma in Burgan, Gutierrez, Fijalkow, Navet and Feix [1977]. We see immediately that unphysical boundary conditions appear at $\xi = \pm \infty$ where particles with infinite velocity are allowed to appear.

It has been shown [Burgan, Gutierrez, Fijalkow, Navet and Feix, 1977] that these difficulties can be partly overcome through the concept of contamination which is based on the fact that particles on the extremities do not contribute to the field (as long as symmetry is conserved). In fact, this concept is going to be generalized and reintroduced in Chapter VIII and we will not discuss it any further for the moment.

III. TIME RENORMALIZATION

A. The Heat Diffusion Equation

To point out the important concept of time renormalization, which is the main concept introduced in this paper, and before treating the general case of phase space fluids we consider the one-dimensional heat equation

$$\frac{\partial\psi}{\partial t} = \chi \frac{\partial^2\psi}{\partial x^2} \tag{3.1}$$

In addition to the fact that it is on this equation that the self similar solution has been introduced for the first time by Boltzmann a nearly identical transformation will be used in our study of the Schrödinger equation (see Chapter VII).

We introduce a rescaling of x (space), t (time) and the function ψ with

$$x = \xi\, C(t)$$

$$\theta = \theta(t)$$

$$\psi = B(t)[\exp \Phi]\, \overline{\psi}(\xi,\theta) \tag{3.2}$$

We take $\Phi = k(t)\, x^2$. B(t), K(t) and C(t) are functions of time only. We introduce Eq. (3.2) in Eq. (3.1) and obtain for $\partial\psi/\partial t$ and $\partial^2\psi/\partial x^2$ respectively

$$\frac{\partial\psi}{\partial t} = B \frac{d\theta}{dt} \exp\Phi \frac{\partial\overline{\psi}}{\partial\theta} + B \exp\Phi \frac{\partial\overline{\psi}}{\partial\xi}\, x \frac{(-\dot{C})}{C^2} + \overline{\psi}\,\dot{B} \exp\Phi + \overline{\psi}\, Bx^2\dot{K} \exp\Phi \tag{3.3}$$

$$\frac{\partial^2\psi}{\partial x^2} = \frac{B}{C^2} \exp\Phi \frac{\partial^2\overline{\psi}}{\partial\xi^2} + 4K \exp\Phi \frac{\partial\overline{\psi}}{\partial\xi} \frac{B}{C}\, x$$

$$+ \overline{\psi}B(2K) \exp\Phi + \overline{\psi}4K^2B\, x^2 \exp\Phi \tag{3.4}$$

Let us try to leave the heat equation unchanged. We must

consequently equate the second, third and fourth terms of the right-hand-side of Eq. (3.3) with the corresponding terms of Eq. (3.4) (multiplied by χ). Moreover, $d\theta/dt$ must be equal to C^{-2} (equality of the coefficients of the first terms in the two right-hand-sides). Consequently, we have

$$\frac{d\theta}{dt} = C^{-2}$$

$$4\chi K = -\dot{C}/C$$

$$2B\chi K = \dot{B}$$

$$4\chi K^2 = \dot{K} \tag{3.5}$$

In Eqs. (3.3), (3.4) and (3.5), the dot indicates a time derivative. Equation (3.5) can be easily solved. Imposing the auxiliary condition $C = 1$ for $t = 0$ and introducing the arbitrary positive constant A we obtain

$$K = -\frac{1}{A + 4\chi t} \tag{3.6}$$

$$B = \frac{1}{(A + 4\chi t)^{1/2}} \tag{3.7}$$

$$C = 1 + \frac{4\chi t}{A} \tag{3.8}$$

$$\theta = \frac{A}{4\chi} \frac{t}{t + (A/4\chi)} \tag{3.9}$$

$\overline{\psi}(\xi,\theta)$ is the solution of

$$\frac{\partial\overline{\psi}}{\partial\theta} = \chi \frac{\partial^2\overline{\psi}}{\partial\xi^2}$$

Equation (3.9) is the important formula since $\theta \to A/4\chi$ when $t \to \infty$. If we know the solution during the finite time $0 - A/4\chi$ (using a numerical scheme, for example a fast Fourier transform) of the heat equation where we have simply changed the initial conditions, then analytical formula allow the prolongation of this solution for all time through the transformations (3.9), (3.10) and (3.11).

$$\psi(x,t) = \overline{\psi}(\xi,\theta) \frac{1}{(A + 4\chi t)^{1/2}} \exp - \frac{x^2}{A + 4\chi t} \tag{3.10}$$

$$x = \xi(1 + \frac{4\chi t}{A}) \tag{3.11}$$

The new initial conditions for the equation involving $\overline{\psi}$ are obtained noticing that for $t = \theta = 0$ we have $x = \xi$

$$\overline{\psi}(\xi,\theta=0) = A^{1/2} \exp \frac{\xi^2}{A} \psi(\xi=x,0) \tag{3.12}$$

The interesting characteristic of Eq. (3.10) is that we have divided the solution into two parts. One $\overline{\psi}(\xi,\theta)$ must be computed in the finite interval $[0,\theta_\ell = A/4\chi]$ (The point θ_ℓ does not possess any special property) while the second part exhibits, through the transformation, the asymptotic properties of the solution of the diffusion equation. If we want to solve Eq. (3.1) directly for large x and t we are faced with the problem of computing a very large number of Fourier components around $k = 0$. In fact, the inverse Fourier transform (to get $\psi(x,t)$) can be obtained by a saddle point integration method. The group transformation automatically takes care of this numerically difficult part of the integration.

Let us finally show the exact form of the asymptotic solution when x and $t \to \infty$ with $x^2/t \to$ finite limit, since we know that the heat diffusion involves an expansion in $\sqrt{t}$ of the initial heated zone. Equation (3.11) shows that $\xi \to 0$ and we must compute $\overline{\psi}(0,\theta_\ell)$. If we design by $g(x)$ the function $\psi(x,t = 0)$ we obtain through the usual Fourier transform formalism and, moreover, using Eq. (3.12)

$$\overline{\psi}(0,\theta_\ell) = \frac{1}{2\pi} \int_{-\infty}^{+\infty} dk \exp -\chi k^2 \theta_\ell \int_{-\infty}^{+\infty} g(x) A^{1/2} \exp \frac{x^2}{A} \exp ikx\, dx \tag{3.13}$$

In Eq. (3.13) we first perform the integration on k

$$\int_{-\infty}^{+\infty} \exp \left(ikx - \frac{k^2 A}{4}\right) dk = 2\sqrt{\frac{\pi}{A}} \exp - \frac{x^2}{A} \tag{3.14}$$

Introducing Eq. (3.14) in Eq. (3.13) we obtain after integration

$$\overline{\psi}(0,\theta_\ell) = \frac{1}{\sqrt{\pi}} \int g(x)\, dx$$

Neglecting A with respect to χt (since $t \to \infty$) we obtain

$$\psi = \sqrt{\frac{\pi}{\chi t}} \exp - \frac{x^2}{4\chi t} \left\{\frac{1}{2\pi} \int g(x)\, dx\right\} \tag{3.15}$$

Equation (3.15) is the well-known asymptotic formula obtained from

$$\psi(x,t) = \int_{-\infty}^{\infty} \exp -ikx\ g(k,0)\ \exp -\chi k^2 t\ dk$$

$$g(k,0) = \frac{1}{2\pi}\int_{-\infty}^{+\infty} \exp ikx\ g(x)\ dx \tag{3.16}$$

when the integration in Eq. (3.16) is performed by the saddle point method. Equation (3.10) is definitively an improvement since it is valid for any x and t but has the asymptotic form already built in.

B. Connection Between Fokker-Planck and the Heat Diffusion Equation

Let us consider the usual one-dimensional Fokker-Planck equation

$$\frac{\partial f}{\partial t} = \frac{\partial}{\partial v}\left\{A(t)\ vf + B(t)\ \frac{\partial f}{\partial v}\right\} \tag{3.17}$$

where the friction and diffusion coefficients A and B are functions of time. We want to obtain a simpler equation. Equation (3.17) involves the velocity but not the configuration space and consequently we just rescale v, θ and f with

$$\theta = \theta(t)$$

$$\eta = b(t)\ v$$

$$f = F(\eta,\theta)\ \Gamma(t) \tag{3.18}$$

Introducing Eq. (3.18) in Eq. (3.17) we obtain

$$AF + A\eta\frac{\partial F}{\partial \eta} + Bb^2\frac{\partial^2 F}{\partial \eta^2} = \frac{\dot{\Gamma}}{\Gamma}F + \frac{\partial F}{\partial \theta}\dot{\theta} + \frac{\partial F}{\partial \eta}\eta\frac{\dot{b}}{b} \tag{3.19}$$

as usual the dot indicates a time derivative.

The idea is to change Eq. (3.17) into the heat equation independent of time coefficients. We consequently identify in Eq. (3.19)

(i) the first term of the left-hand-side with the first term of the right-hand-side

(ii) the second term of the left-hand-side with the third term of the right-hand-side

(iii) the coefficient of $\partial^2F/\partial\eta^2$ (third term of the left-hand-side) with the coefficient of $\partial F/\partial\theta$ (second term of the right-hand-side). We get:

$$A = \frac{\dot{\Gamma}}{\Gamma} = \frac{\dot{b}}{b}$$

$$Bb^2 = \dot{\theta}$$

Introducing

$$\gamma(\mu) = \int_0^{\mu} A(\sigma)\, d\sigma$$

we obtain

$$b(t) = \exp \gamma(t)$$

$$\Gamma(t) = \exp \gamma(t)$$

$$\theta = \int_0^{t} B(\mu) \exp 2\gamma(\mu)\, d\mu$$

$$\partial^2 F/\partial\eta^2 = \partial F/\partial\theta \qquad (3.20)$$

The Fokker-Planck equation with time-dependent coefficients can consequently be reduced to the standard heat diffusion equation solved subsequently by any standard technique (including the group transformation mentioned previously).

IV. A USEFUL GROUP OF TRANSFORMATIONS

A. Derivation of the Transformation

We now treat the case of a phase space fluid. Let us consider the transformations given by Eq. (1.2) where $\vec{x}$ and $\vec{v}$ are the coordinate and velocity of a particle. We have

$$\vec{v} = \vec{\eta}/C(t) - B(t)/[A(t)\ C(t)] \cdot \vec{\xi}$$

We impose on the transformation two conditions

(1) The phase space element should be conserved, i.e.

$$d\vec{\xi}\ d\vec{\eta} = d\vec{x}\ d\vec{v}$$

implying that the Jacobian of the transformation must be unity; this imposes

$$A(t)\ C(t) = 1 \tag{4.1}$$

computing $d\vec{v}/dt$ and taking Eq. (3.1) into account.

$$\frac{d\vec{v}}{dt} = \frac{1}{C}\frac{d\theta}{dt}\frac{d\vec{\eta}}{d\theta} - \frac{\vec{\eta}}{C^2}\frac{dC}{dt} - B(t)\frac{d\theta}{dt}\frac{d\vec{\xi}}{d\theta} - \frac{dB}{dt}\vec{\xi} \tag{4.2}$$

(2) We want to keep the Hamiltonian formalism. Consequently, $d\vec{\xi}/d\theta = \vec{\eta}$ and the new force $\vec{\xi} = d\vec{\eta}/d\theta$ must be a function of $\vec{\xi}$ and θ only. Consequently, in the expression of $d\vec{v}/dt$ the friction term, i.e. the sum of the second and third terms in the left-hand-side of Eq. (3.2) must be zero.

$$\frac{dC}{dt}\cdot\frac{1}{C^2} + B(t)\frac{d\theta}{dt} = 0 \tag{4.3}$$

on the other hand, from a direct derivation of η in Eq. (1.2)

$$\frac{d\vec{\eta}}{d\theta} = \frac{dt}{d\theta}\left\{B\frac{d\vec{x}}{dt} + \vec{x}\frac{dB}{dt} + \frac{dC}{dt}\vec{v} + C\frac{d\vec{v}}{dt}\right\} \tag{4.4}$$

$\vec{x}$ and $\vec{v}$ being canonically conjugate we must again set equal to zero

$$\left(B + \frac{dC}{dt}\right)\vec{v}$$

and consequently $B = -dC/dt$. From this last relation and Eq. (4.3), we deduce

$$\frac{d\theta}{dt} = \frac{1}{C^2} \tag{4.5}$$

which, together with Eq. (4.1), defines completely the transformation characterized by the arbitrary function C(t). The "new field" $\vec{\varepsilon}$ acting on the particle in the "new phase" space, "new time" system is given by Eq. (4.4)

$$\vec{\varepsilon} = \frac{d\vec{\eta}}{d\theta} = C^3\frac{d\vec{v}}{dt} - C^3\frac{d^2C}{dt^2}\vec{\xi} = C^3\vec{E} - C^3\frac{d^2C}{dt^2}\vec{\xi} \tag{4.6}$$

Sometimes we will need the divergence of $\vec{\varepsilon}$ for Coulomb forces, i.e. forces for which

$$\frac{\partial\vec{E}}{\partial\vec{x}} = \sum_i q_i\ \delta(\vec{x} - \vec{x}_i) \tag{4.7}$$

From Eq. (4.6), $\vec{x} = \vec{\xi}\ C$ and the relations

$$\delta[C(\vec{\xi} - \vec{\xi}_i)] = \begin{cases} C^{-1}\delta(\vec{\xi} - \vec{\xi}_i) & \text{in one dimension} \\ C^{-2}\delta(\vec{\xi} - \vec{\xi}_i) & \text{in two dimensions} \\ C^{-3}\delta(\vec{\xi} - \vec{\xi}_i) & \text{in three dimensions} \end{cases}$$

we obtain

$$\frac{\partial \vec{\varepsilon}}{\partial \vec{\xi}} = \pm\, C^{4-d} \sum_i \delta(\vec{\xi} - \vec{\xi}_i) - d\, C^3 \frac{d^2C}{dt^2} \tag{4.8}$$

in the B and G cases with d the dimensionality of the system (d = 1, 2, 3 respectively for one-, two-, three-dimensional systems). The case of an electronic plasma with a continuous ion neutralizing background of density N_0 introduces a supplementary term

$$\frac{\partial \vec{\varepsilon}}{\partial \vec{\xi}} = C^{4-d} \sum_i \delta(\vec{\xi} - \vec{\xi}_i) - C^4 N_o - d\, C^3 \frac{d^2C}{dt^2} \tag{4.9}$$

We will use extensively Eqs. (4.8) and (4.9). In the Vlasov description

$$\sum_i \delta(\vec{\xi} - \vec{\xi}_i)$$

is replaced by

$$\int_{-\infty}^{\infty} F(\vec{\xi},\vec{\eta} \cdot \theta)\, d\vec{\eta} \quad .$$

The relation $d\vec{\xi}\, d\vec{\eta} = d\vec{x}\, d\vec{v}$ implies that $f(\vec{x},\vec{v},t) = F(\vec{\xi},\vec{\eta},\theta)$ and the Hamiltonian equations $\vec{\eta} = d\vec{\xi}/d\theta$; $\vec{\varepsilon} = d\vec{\eta}/d\theta$ mean that in the new phase space (with the new time) we keep invariant the Vlasov equation

$$\frac{\partial F}{\partial \theta} + \vec{\eta} \cdot \frac{\partial F}{\partial \vec{\xi}} + \vec{\varepsilon} \cdot \frac{\partial F}{\partial \vec{\eta}} = 0 \tag{4.10}$$

B. Group Structure of the Transformation

Each element of the transformation is characterized by the function C(t) which transforms $\vec{x},\vec{v},t$ in $\vec{\xi},\vec{\eta},\theta$. We can reiterate with another transformation characterized now by D(θ) which transforms $\vec{\varepsilon}\ \vec{\eta}\ \theta$ in $\vec{\lambda}\ \vec{\mu}\ \tau$. For a d-dimensional system, introducing I, the unit matrix of rank d, we have for the product of the two transformations, the new time τ given by

$$d\tau = D^{-2}\, d\theta = D^{-2}\, C^{-2}\, dt = (DC)^{-2}\, dt \tag{4.11}$$

For the new phase space the matrix transforming directly $\vec{x},\vec{v}$ in $\vec{\lambda},\vec{\mu}$ written.

$$\begin{pmatrix} D^{-1}I & 0 \\ -\frac{dD}{d\theta}I & DI \end{pmatrix}\begin{pmatrix} C^{-1}I & 0 \\ -\frac{dC}{dt}I & CI \end{pmatrix} = \begin{pmatrix} C^{-1}D^{-1}I & 0 \\ \left(-C^{-1}\frac{dD}{d\theta} - D\frac{dC}{dt}\right)I & CDI \end{pmatrix}$$

$$= \begin{pmatrix} C^{-1}D^{-1}I & 0 \\ -\frac{d}{dt}(CD)I & CDI \end{pmatrix} \qquad (4.12)$$

The "new field" $\vec{F}$ is given by

$$\vec{F} = D^3\vec{\varepsilon} - D^3\frac{d^2D}{d\theta^2}\vec{\lambda} = D^3\ C^3\ \vec{E} - D^3\ C^3\frac{d^2C}{dt^2}\vec{\xi} - D^3\frac{d^2D}{d\theta^2}\vec{\lambda} \qquad (4.13)$$

But $\vec{\xi} = D\vec{\lambda}$ and the two last terms of Eq. (3.13) can be written

$$-D^3\ C^3\left(D\frac{d^2C}{dt^2} + \frac{1}{C^3}\frac{d^2D}{d\theta^2}\right)\vec{\lambda} \qquad (4.14)$$

Taking into account

$$\frac{dD}{d\theta} = C^2\frac{dD}{dt} \qquad \text{and} \qquad \frac{d^2D}{d\theta^2} = C^4\frac{d^2D}{dt^2} + 2C^3\frac{dC}{dt}\cdot\frac{dD}{dt}$$

Eq. (4.14) can be written

$$-D^3\ C^3\left(D\frac{d^2C}{dt^2} + 2\frac{dC}{dt}\cdot\frac{dD}{dt} + C\frac{d^2D}{dt^2}\right) = -D^3\ C^3\frac{d^2}{dt^2}(DC) \qquad (4.15)$$

From Eqs. (4.11), (4.12), (4.13) and (4.15) we see that the successive applications characterized by C(t) and D(θ) are equivalent to the application characterized by CD. This demonstrates the abelian group structure of the transformation. Moreover, being characterized by an arbitrary function C(t) the group is continuous.

C. Time and Force Renormalization

In the new time, new phase space the motion equations are invariants and we can use all the maghematical models developed up to now; the only change will be in the computation of the force through Poisson law from Eq. (4.8) or (4.9). The new field is now the sum of three fields.

(i) The self-consistent field of the system given through Poisson law is computed in the same way except a multiplying factor C^{4-d} which varies with time.

(ii) In the P case the background ion field is multiplied by C^4.

(iii) Moreover, we must introduce a transformation field given by $-C^3(d^2C/dt^2)\,\vec{\xi}$.

On the other hand, the new time is given by

$$\theta = \int_0^t \frac{dt'}{C^2(t')}$$

and we see that the time is rescaled. Since we want to study the long time behavior of the system we can choose C(t) such that θ varies very slowly with t. In fact, we can select C(t) in such a way that θ goes to a limiting value θ_ℓ when $t \to \infty$. We consequently proceed to a renormalization of the time. This is in fact what has been done in Chapter III where $C(t) = 1 + (4\chi/A)\,t$. This implies $C(t) \to \infty$ with t and, of course, this time renormalization is obtained at the expense of a force increase. We show that in some physical problems we can select C(t) to renormalize, i.e. have a new θ going to a finite limit, or at least slow down considerably the time without introducing infinities in the forces.

V. GRAVITATIONAL SYSTEM WITH TIME-VARYING GRAVITATIONAL CONSTANT

We have seen that one consequence of the transformation was to introduce a function C(t) which, elevated at a proper power, scales the force. If a physical constant describing the interaction decreases with time C(t) can be selected as time increasing with a balance between the two effects while θ may still go to a finite limit. In fact, for the Dirac hypothesis, the problem of the dynamical evolution is very much simplified as if the Dirac law was an example invented to show the interest of the group method. The Dirac hypothesis [Dirac, 1973; 1938] states that the gravitational constant varies with time as

$$G(t) = G_o(T/t) \tag{5.1}$$

where t is the age of the universe. We prefer to shift the time and consider the present time as origin. Then replacing t by t + T with $T = \Omega^{-1}$ being the actual age of the universe

$$G(t) = G_o \frac{1}{1 + \Omega t} \tag{5.2}$$

From Eq. (4.8) considering a 3D system we notice that in the new phase space the field is affected by the factor G(t) C(t). Taking G(t) C(t) = G_o gives C(t) = 1 + Ωt and consequently the transformation field vanishes. But the new time θ is now

$$\theta = \frac{t}{1 + \Omega t} \tag{5.3}$$

and goes to the finite limit Ω^{-1} when $t \to \infty$. Consequently, the study for all time of an N-body gravitating problem in the Dirac hypothesis is identical to the study of an N-body gravitating system where G remains constant during a finite time interval equal to Ω^{-1}. As a consequence, if the initial conditions of the problem are, in a Vlasov description, a steady state in the $\vec{\xi}\ \vec{\eta}$ phase space, the group transformation describes completely the time evolution of the system. We have been able to renormalize the time without introducing infinities in the force due to the time decreases of G. The Dirac law is the only one which allows us to get rid of the variation of G without introducing a transformation field. We have two possibilities: either we stick to the $\vec{x},\vec{v},t$ space and introduce Eq. (5.2) or we consider the $\vec{\xi},\vec{\eta},\theta$ space and consider $\overline{G}$ as strictly constant.

We will, of course, refrain from claiming that one space is more physical than the other. We must simply point out that θ is the ephemerides time computed with the hypothesis of constant G. This time should be compared with the time given by an atomic clock, i.e. a time connected to the motion of electrons governed by electrostatic forces. The comparison of the two times was made by Van Flanders [1974]. Although it is very difficult to get rid of the different corrections, this last author claims a residual difference between the two times supporting Dirac's ideas with an $\Omega^{-1} = 10^{10}$ years.

Coming back to the more technical problems of computing orbits when G varies with time, we consider the motion of mass point in three dimensions attracted by the origin with a gravitational constant given by

$$G(t) = G_o \frac{1}{(1 + \Omega t)^{\alpha}} \tag{5.4}$$

We select C(t) to be able to deduce the asymptotic properties and more precisely

(i) To avoid increasing in the time transformation field and $\overline{G}(\theta)$

(ii) Within this last constraint to compress the time as much as possible with, if possible, a limiting θ_ℓ. Table I gives the choice of C(t), the expression in the new variables $\theta,\vec{\xi}$, of the transformation field, of the gravitational constant and, finally, the relation between θ and t.

TABLE I

α	C(t)	$\overline{G}(\theta)$	Trans. Field	θ(t)
$0<\alpha<\frac{1}{2}$	$(1+\Omega t)^{\alpha}$	G_o	$\frac{\alpha(1-\Omega)^2}{[1+(1-2\alpha)\Omega\theta]^2}\vec{\xi}$	$\Omega\theta=\frac{(1+\Omega t)^{1-2\alpha}-1}{1-2\alpha}$
$\alpha=\frac{1}{2}$	$(1+\Omega)^{1/2}$	G_o	$(\Omega^2/4)\vec{\xi}$	$\Omega\theta=\log(1+\Omega t)$
$\frac{1}{2}<\alpha<1$	$(1+\Omega t)^{1/2}$	$G_o\ \exp-(\alpha-\frac{1}{2})$	$(\Omega^2/4)$	$\Omega\theta=\log(1+\Omega t)$
$\alpha=1$	$1+\Omega t$	G_o	0	$\Omega\theta=\Omega t/(1+\Omega t)$
$\alpha>1$	$1+\Omega t$	$G_o(1-\Omega\theta)^{\alpha-1}$	0	If $t\to\infty$ $\Omega\theta\to 1$

For $\alpha \geq 1$ the transformation field is zero, $\overline{G}(\theta)$ is constant or goes to zero and $\theta \to \theta_\ell$. As a consequence, $\vec{\xi}$ goes to a limiting value ξ_ℓ and the asymptotic trajectory is $\vec{x} = \vec{\xi}_\ell(1 + \Omega t)$ with $\vec{x}$ increasing proportionally to the time.

In the case $1/2 < \alpha < 1$ the gravitational constant decreases exponentially with θ and the dominant field is $\Omega^2/4\vec{\xi}$. The asymptotic solution for $\vec{\xi}$ is consequently

$$\vec{\xi} = K \exp \frac{\Omega\theta}{2} \tag{5.5}$$

Combining with $\vec{x} = C\vec{\xi}$ the selected value of C(t) in this case and coming back to the time t, we find that again $\vec{x}$ increases as Ωt. In fact, situations where the dominant field is the transformation field correspond to ballistic motion. Consequently, for $\alpha > 1/2$ the asymptotic state corresponds to freely going particles with a uniform velocity. This is not surprising since $G(t) \to 0$ when $t \to \infty$, but this last condition is not sufficient. Indeed for $\alpha \leq 1/2$ we will see that another type of trajectory is possible. $\overline{G}(\theta)$ being a constant and the repulsive transformation field going to zero for $\theta \to \infty$, we can predict two types of trajectories. For the open one the gravitational field goes to zero as ξ^{-2}, while the transformation field goes to zero as ξ/θ^2, and in the case of ξ varying at least as θ the transformation field, although going to zero is still dominant.

The asymptotic time evolution is given by solving

$$\ddot{\vec{\xi}} = \frac{\alpha(1 - \alpha)}{(1 - 2\alpha)^2 \theta^2} \vec{\xi} \tag{5.6}$$

Seeking a solution of Eq. (5.6) of the form $\vec{\xi} = \vec{A}\theta^{\beta}$ we obtain $\beta = (1 - \alpha/1 - 2\alpha)$. Since for $\alpha < 1/2$, $\beta > 1$ the domination of the transformation field is enhanced. Now introducing $\vec{x} = C\vec{\xi} = \vec{A}C\theta^{\beta}$ and expressing θ as a function of t we obtain again an asymptotic solution in t for $\vec{x}$, i.e. the ballistic motion corresponding to the domination of the transformation field.

But, of course, the initial conditions of the problem can be such that in the $\vec{\xi}$ plane the trajectories are of closed type. The transformation field decreasing with θ, the asymptotic trajectory is an ellipse in ξ and although $G(t) \to 0$ the particle is still influenced by the center of attraction, turning around it with a distance going to infinity as $(1 + \Omega t)^{\alpha}$.

VI. MOTION OF A CHARGED PARTICLE IN A UNIFORM IN SPACE, TIME-VARYING MAGNETIC FIELD

If a time-varying G(t) is a simple cosmological hypothesis, the problem of a time-varying magnetic field is certainly important from a practical point of view. Let us assume a uniform in space, time-varying magnetic field, $\vec{B}(t) = B(t)\,\vec{e}$, where $\vec{e}$ is a unit vector. Now a time-varying $\vec{B}$ implies an electric field $\vec{E} = -\partial\vec{A}/\partial t$. Since the vector potential is

$$\vec{A} = \frac{1}{2} \vec{B} \times \vec{x}$$

we deduce

$$\vec{E} = \frac{\dot{B}(t)}{2} \vec{x} \times \vec{e} \tag{6.1}$$

Equation (6.1) must be handled with care. The electric field depends obviously on the point of origin and is zero on the line parallel to $\vec{e}$ and passing through that origin. A priori $\vec{B}(t)$ being uniform, it seems possible to take any point as the origin. The paradox is raised if we consider more carefully how the uniform magnetic field can be created.

Here we consider a cylindrical geometry, for example a solenoid of length L and radius R where both L and R go to infinity. This configuration imposes a cylindrical symmetry in the induced field and the axis of the solenoid is naturally the line where $\vec{E} = 0$. From

$$\vec{\xi} = C^{-1}\,\vec{x}$$

$$\vec{\eta} = C\vec{v} - (dC/dt)\,\vec{x}$$

$$\frac{d\theta}{dt} = C^{-2}$$

$$\vec{F} = \vec{E} + \vec{v} \times \vec{B} \tag{6.2}$$

we obtain

$$\frac{d\vec{\eta}}{d\theta} = C^3\,\vec{E} + \vec{\eta} \times C^2\,\vec{B} + C^3(dC/dt)\,\vec{\xi} \times \vec{B} - C^3(d^2C/dt^2)\,\vec{\xi} \tag{6.3}$$

Introducing Eq. (6.1) we see that

$$C^3\,\vec{E} + C^3(dC/dt)\,\vec{\xi} \times \vec{B} = C^3\left[\frac{1}{2}\frac{dB}{dt}\,C + B\,\frac{dC}{dt}\right]\vec{\xi} \times \vec{e} \tag{6.4}$$

If we take C(t) in such a way that $B(t)\,C^2(t) = B_o$, Eq. (6.4) cancels and we obtain

$$\frac{d\vec{\eta}}{d\theta} = \eta \times \vec{B}_o - C^3(d^2C/dt^2)\,\vec{\xi} \tag{6.5}$$

and we have to treat the motion of a particle in a constant magnetic field $\vec{B}_o$ in addition to the transformation field. Note that now C(t) is imposed by the relation

$$B(t)\,C^2(t) = B_o \tag{6.6}$$

We investigate more precisely the case where

$$B(t) = \frac{B_o}{(1 + \Omega t)^{2\alpha}} \tag{6.7}$$

and we list in Table II the value of the transformation field and the relation between θ and t. [In all cases the new magnetic field is B_o and $C(t) = (1 + \Omega t)^{\alpha}$].

In the new phase space for $\alpha < 1/2$ the transformation field goes to zero and the asymptotic trajectory is a circle (since $B = B_o$). Consequently, although the magnetic field goes to zero, the particle

TABLE II

α	Transformation Field	$\Omega\theta(t)$	$\Omega\theta_{\ell im}$	
$0<\alpha<\frac{1}{2}$	$\frac{\alpha(1-\alpha)\Omega^2}{[1+(1-2\alpha)\Omega\theta]^2}\vec{\xi}$	$\frac{(1+\Omega t)^{1-2\alpha}-1}{1-2\alpha}$	∞	
$\alpha = \frac{1}{2}$	$(\Omega^2/4)\ \vec{\xi}$	$\log(1+\Omega t)$	$\log \infty$	$\overline{B}(\theta) = B_o$
$\frac{1}{2}<\alpha<1$	$\frac{\alpha(1-\alpha)\Omega^2}{[1-(2\alpha-1)\Omega\theta]^2}\vec{\xi}$ repulsive force	$\frac{1}{2\alpha-1}\left\{1-\frac{1}{(1+\Omega t)^{2\alpha-1}}\right\}$	$\frac{1}{2\alpha-1}$	
$\alpha = 1$	0	$t/(1+\Omega t)$	1	
$1 < \alpha$	$\frac{-\alpha(\alpha-1)\Omega^2}{[1-(2\alpha-1)\Omega\theta]^2}\vec{\xi}$ attractive force	$\frac{1}{2\alpha-1}\left\{1-\frac{1}{(1+\Omega t)^{2\alpha-1}}\right\}$	$\frac{1}{2\alpha-1}$	

will always feel its presence, rotating and expanding around a line of force. The case $\alpha = 1/2$ is quite interesting, Eq. (6.5) is written

$$\frac{d^2\vec{\xi}}{d\theta^2} = \frac{d\vec{\xi}}{d\theta} \times \vec{B}_o + \frac{\Omega^2}{4}\vec{\xi} \tag{6.8}$$

Decoupling the motion in a first one perpendicular to the magnetic field (characterized by ξ_1 and ξ_2) and another parallel (this last one is a trivial displacement at uniform velocity) we get the eigenfrequencies of the perpendicular motion ω^2

$$\left(\omega^2 + \frac{\Omega^2}{4}\right)^2 - \omega^2 B_o^2 = 0 \tag{6.9}$$

Solutions of Eq. (6.9) are

$$2\omega = \pm\left[B_o \pm \sqrt{B_o^2 - \Omega^2}\right] \quad .$$

We must remember that due to the choice of units e = m = 1, B_o stands for the cyclotron frequency (e/m) B_o. We find two possibilities.

(1) $B_o > \Omega$ all the eigenvalues are real and the particle undergoes oscillations in the $\vec{\xi}$ plane. In the $\vec{x}$ plane x increases as $\sqrt{1 + \Omega t}$.

(2) $B_o < \Omega$ two of the eigenvalues are complex and one corresponds to a growing instability in $\vec{\xi}$. The particle has an asymptotic motion closer to the free motion. For $B_o = 0$ we recover, of course, the ballistic motion.

The asymptotic solutions for the case $1/2 < \alpha < 1$ are obtained noticing that the force is repulsive and that for $\theta \to \theta_\ell$ the motion is dominated by the transformation field

$$\frac{\alpha(1-\alpha)\,\Omega^2\,\theta_\ell^2}{(\theta_\ell - \theta)^2}\,\vec{\xi} = \frac{d^2\vec{\xi}}{d\theta^2} \tag{6.10}$$

We seek a solution of the form $\vec{\xi} = \vec{A}(\theta_\ell - \theta)^\beta$ and obtain $\beta = (\alpha - 1)/(2\alpha - 1)$. Taking account of the relation between θ and t, this gives an asymptotic solution for $\vec{x} = C\vec{\xi}$ of the form $x \propto \Omega t$ indicating that the final motion of the particle is a ballistic free motion (as expected since the transformation field dominates).

The last case $\alpha > 1$ implies an attractice force. If we suppose that the transformation field is still dominant (we will check a posteriori) we get the same equation (6.10) and the same value for β (but now β is positive and the particle falls to the origin). The transformation field decreases like

$$(\theta_\ell - \theta)^{(\alpha-1)/(2\alpha-1)-2} = (\theta_\ell - \theta)^{(1-3\alpha)/(2\alpha-1)}$$

while the magnetic force decreases as the velocity, i.e.

$$(\theta_\ell - \theta)^{(\alpha-1)/(2\alpha-1)-1} \quad .$$

The ratio of the two forces (transformation/magnetic) varies as $(\theta_\ell - \theta)^{-1}$ and indeed for $\theta \to \theta_\ell$ the transformation field is the dominant one, $\xi \to 0$ as

$$(\theta_\ell - \theta)^{(\alpha-1)/(2\alpha-1)}$$

and $x = C\xi$ varies as Ωt, implying in the regular $\vec{x}$ space a ballistic motion for large t.

Figures 1 to 5 illustrate this different concept. Figures 1, 2 and 3 show what happens when $\alpha = .45$ (i.e. a little bit below the critical point $\alpha = 0.5$). In all cases $\Omega = 1$. For $B_o = 10$ the particles are nicely trapped in the $\vec{\xi}$ plane indicating an expanding spiral motion in x. Notice that for $B_o = 0.1$ (Figure 2) the particle gets on its circular orbit around $\theta = 140$. Such a value corresponds to $t \simeq 10^{12}$. This is an astonishing long time (and incidently supposes that the field stays uniform on the very large corresponding distances).

Figures 4 and 5 illustrate the case $\alpha = .55$, above the critical value $\alpha = .5$. Here the case $B_o = 10$ is the interesting one. We see 14 revolutions in the $\vec{\xi}$ plane followed by the rather sudden escape of the particle. Again the corresponding time is very large $\theta = 8$, $t = 357$ and practical space and time limitations will limit the possibility of seeing these transitions. Nevertheless, it is interesting to see that the $\vec{\xi}\ \vec{\eta}\ \theta$ space with its high degree of compression in space and time is indeed the best suited for description of the motion.

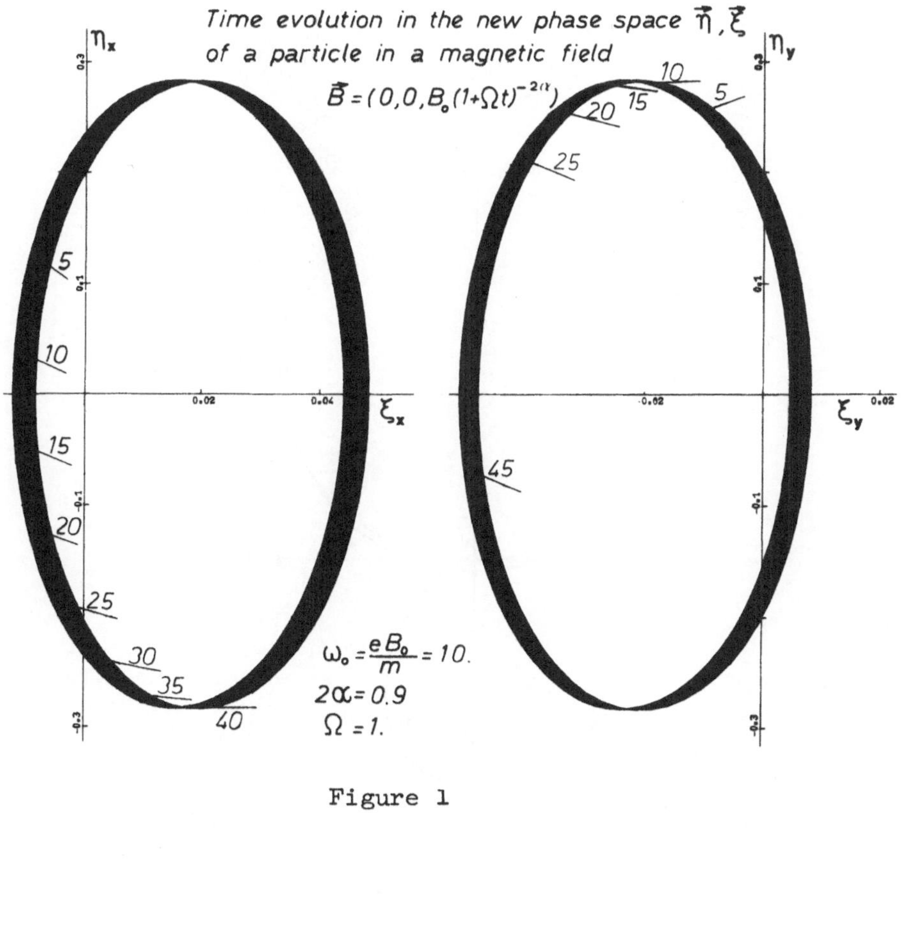

Figure 1

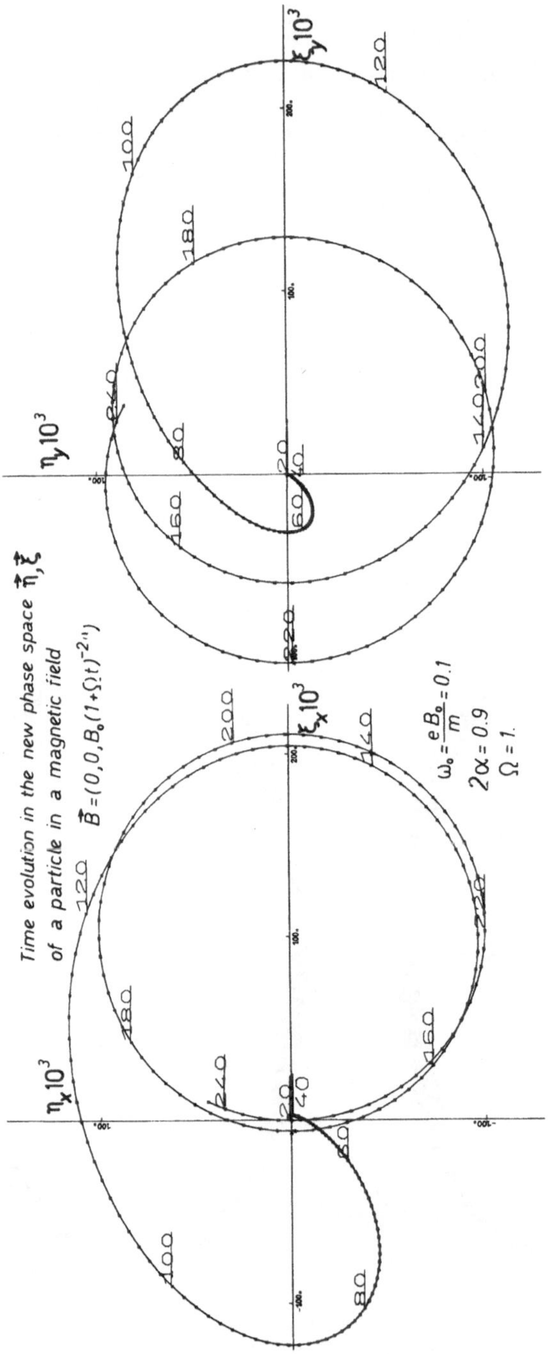
Time evolution in the new phase space $\vec{\eta}, \vec{\xi}$
of a particle in a magnetic field
$\vec{B} = (0, 0, B_o(1+\Omega t)^{-2\alpha})$
$\omega_o = \frac{e B_o}{m} = 0.1$
$2\alpha = 0.9$
$\Omega = 1.$
$\eta_x 10^3$
$\xi_x 10^3$
$\eta_y 10^3$
$\xi_y 10^3$

Figure 2

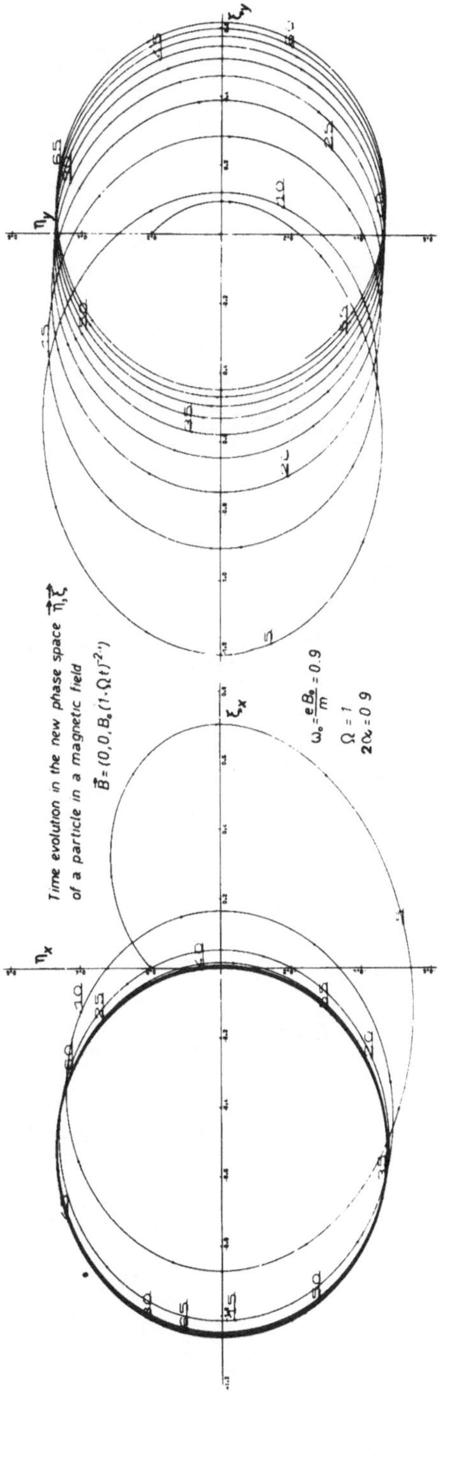
Time evolution in the new phase space
of a particle in a magnetic field
ω₀ = eB₀/m = 0.9
Ω = 1
2α = 0.9

Figure 3

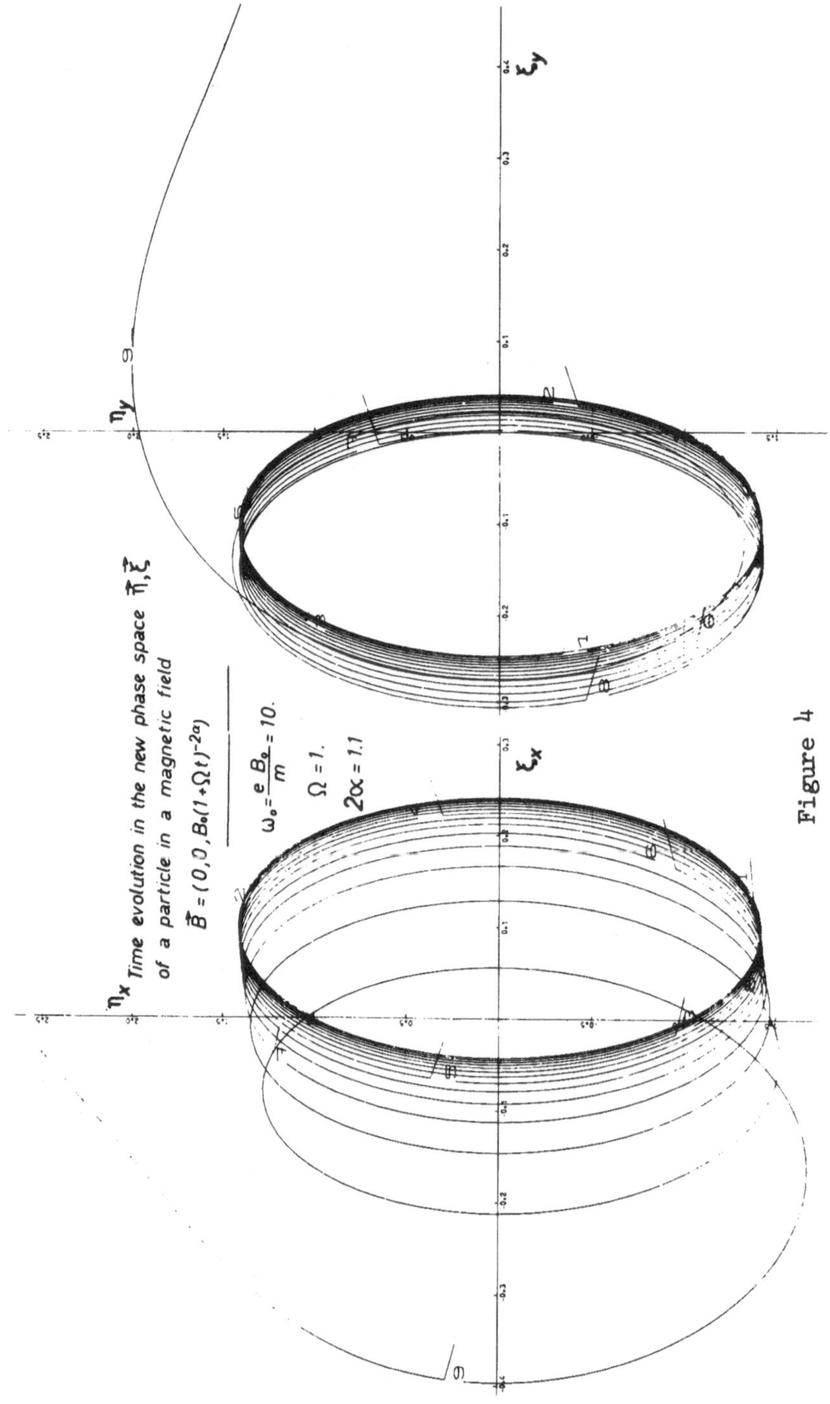
Time evolution in the new phase space η⃗, ξ⃗
of a particle in a magnetic field
B⃗ = (0,0,B₀(1+Ωt)^-2α)
ω₀ = eB₀/m = 10.
Ω = 1.
2α = 1.1
ηx
ηy
ξx
ξy

Figure 4

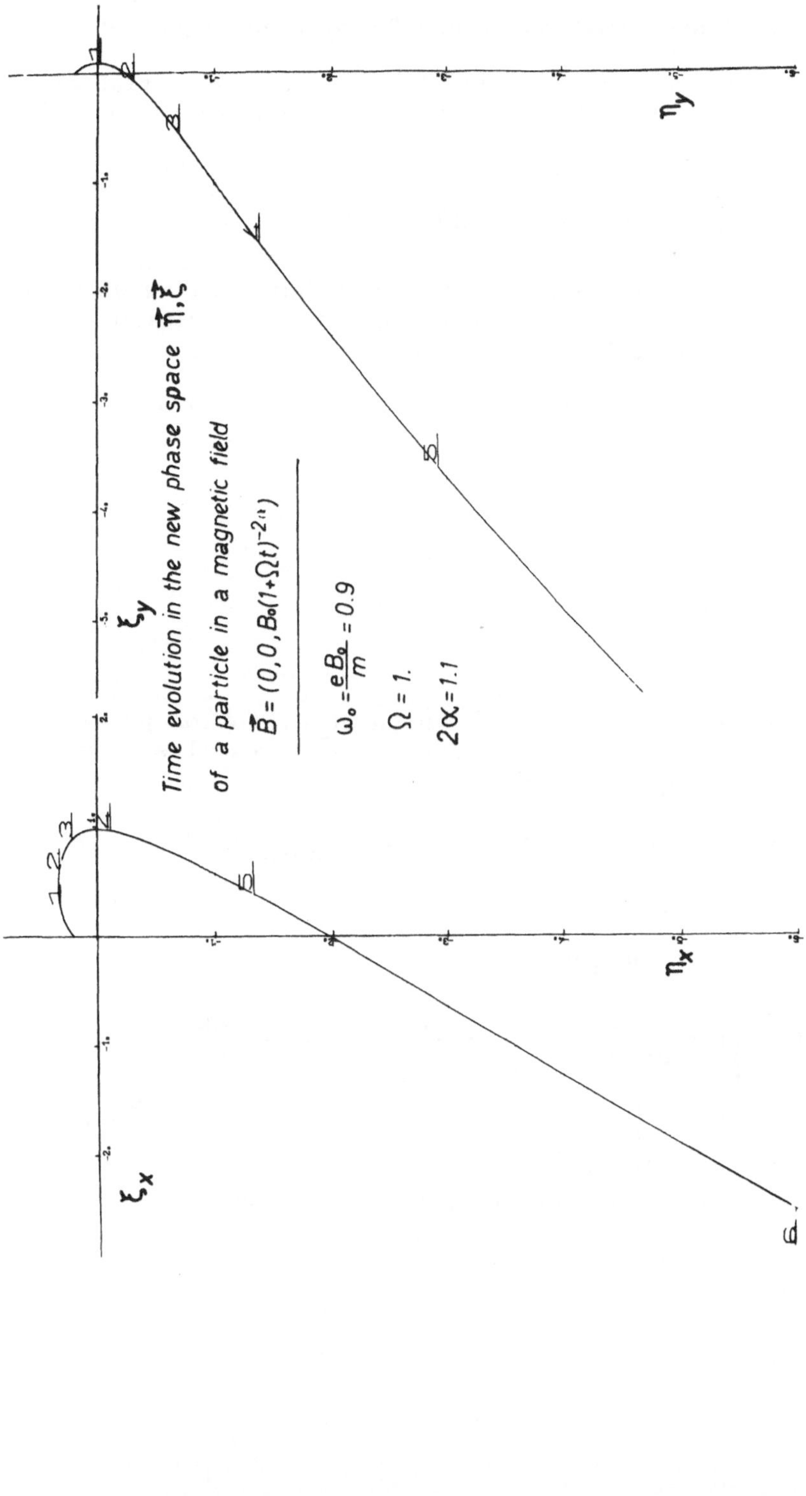
Time evolution in the new phase space $\vec{\eta}, \vec{\xi}$
of a particle in a magnetic field
$\vec{B} = (0, 0, B_0(1+\Omega t)^{-2\alpha})$
$\omega_0 = \frac{eB_0}{m} = 0.9$
$\Omega = 1.$
$2\alpha = 1.1$
ξ_y
η_y
ξ_x
η_x

Figure 5

VII. THE GROUP OF TRANSFORMATION IN QUANTUM MECHANICS

A. Transformation of the Schrödinger Equation

Let us suppose one-dimensional (to simplify) motion of a particle in a potential V(x). The description is through the Schrödinger equation

$$i \frac{\partial \psi}{\partial t} = - \frac{1}{2} \frac{\partial^2 \psi}{\partial x^2} + V(x)\ \psi = H\psi \tag{7.1}$$

we introduce the transformation $\xi = xC^{-1}$ but dealing with the Schrödinger formalism we cannot yet consider a relation between η and x and v (we will see later on).

Now we introduce the following tranformation on ψ

$$\psi(x,t) = B(t) \exp i\ \phi\ \overline{\psi}(\xi,\theta) \tag{7.2}$$

with

$$\phi = x^2\ K(t)$$
$$\theta = \theta(t) \tag{7.3}$$

Notice that Eqs. (7.2) and (7.3) are, except for the factor i, identical to relations (3.2) in the heat diffusion problem. Computing $\partial\psi/\partial t$, $\partial^2\psi/\partial x^2$ and introducing these results we obtain for the Schrödinger equation (7.1):

$$H\psi = \left\{ - \frac{1}{2} \frac{B}{C^2} \frac{\partial^2 \overline{\psi}}{\partial \xi^2} - \frac{B}{C} \frac{\partial \overline{\psi}}{\partial \xi} (2i\ x\ K) + B\overline{\psi}(2x^2\ K^2) - i\ KB\overline{\psi} + VB\overline{\psi} \right\} \exp i\ \phi \tag{7.4}$$

$$i \frac{\partial \psi}{\partial t} = \left\{ i\ B \frac{d\theta}{dt} \frac{\partial \overline{\psi}}{\partial \theta} - i \frac{B}{C^2}\ x \frac{\partial \overline{\psi}}{\partial \xi} \frac{dC}{dt} - B\overline{\psi}\ x^2 \frac{dK}{dt} + i \frac{dB}{dt} \overline{\psi} \right\} \exp i\ \phi \tag{7.5}$$

We want to leave invariant the Schrödinger equation, i.e. we want to write

$$- \frac{1}{2} \frac{\partial^2 \overline{\psi}}{\partial \xi^2} + \overline{V}\ \overline{\psi} = i \frac{\partial \overline{\psi}}{\partial \theta} \tag{7.6}$$

consequently in Eq. (7.4) = Eq. (7.5) we must suppress the term in $\partial\overline{\psi}/\partial\xi$

$$\frac{2K}{C} = \frac{dC}{dt}\frac{1}{C^2} \tag{7.7}$$

Moreover, equating the first terms in the right-hand-side of Eqs. (4.15) and (7.5)

$$\frac{d\theta}{dt} = \frac{1}{C^2} \tag{7.8}$$

Finally getting rid of the terms in $i\overline{\psi}$

$$-K(t) = \frac{1}{B}\frac{dB}{dt} \tag{7.9}$$

Combining Eqs. (7.7), (7.8) and (7.9), we obtain

$$B^2C = 1 \tag{7.10}$$

$$K = (1/2C)\ dC/dt \tag{7.11}$$

while the potential $\overline{V}$ in Eq. (7.6) becomes

$$\overline{V} = VC^2 + \left(\frac{dK}{dt} + 2K^2\right) x^2\ C^2 = VC^2 + \frac{1}{2}\,C^3\,\frac{d^2C}{dt^2}\,\xi^2 \tag{7.12}$$

We see that as in the classical case the new potential $\overline{V}$ is the sum of the rescaled physical potential V plus a transformation potential.

If we introduce the fields $\varepsilon = -\partial\overline{V}/\partial\xi$ and $E = -\partial V/\partial x$ we get

$$\varepsilon = C^3\ E - C^3\,\frac{d^2C}{dt^2}\,\xi \tag{7.13}$$

Equation (7.13) is strictly identical to Eq. (4.6).

B. Introduction of the Wigner Distribution Function

How can we recover the second of Eq. (6.2), i.e., the relation between η and v and x? The question is interesting and is answered by considering the Wigner distribution function (3.8) f(x,v). It must be pointed out that the physical meaning of this function is not clear since this function has not all the good properties characterizing such a distribution (the biggest difficulty being that f can be negative). From a Schrödinger ψ function Wigner defines a distribution in phase space through the relation

$$f(x,v) = \frac{1}{2\pi} \int \psi(x + \frac{\Delta}{2}) \; \psi^*(x - \frac{\Delta}{2}) \; \exp -i \; v\Delta \; d\Delta \tag{7.14}$$

from Eq. (7.2)

$$\begin{aligned} \psi(x + \frac{\Delta}{2}) &= \overline{\psi}(\xi + C^{-1} \frac{\Delta}{2}) \exp \{iK(t)(C\xi + \frac{\Delta}{2})^2\} \; B(t) \\ \psi^*(x - \frac{\Delta}{2}) &= \overline{\psi}^*(\xi - C^{-1} \frac{\Delta}{2}) \exp \{-iK(t)(C\xi - \frac{\Delta}{2})^2\} \; B(t) \end{aligned} \tag{7.15}$$

We introduce Eq. (7.15) in Eq. (7.14) and use Eq. (7.11)

$$f(x,v) = \frac{1}{2\pi} \int B^2 \; \overline{\psi}(\xi + C^{-1} \frac{\Delta}{2}) \; \overline{\psi}^*(\xi - C^{-1} \frac{\Delta}{2}) \exp -i\Delta(V - C\xi) d\Delta$$

we introduce $\Delta' = C^{-1}\Delta$ and notice Eq. (7.10) that $B^2C = 1$. We obtain

$$f(x,v) = \frac{1}{2\pi} \int \overline{\psi}(\xi + \frac{\Delta'}{2}) \; \overline{\psi}^*(\xi - \frac{\Delta'}{2}) \exp -i\Delta'(CV - \dot{C}\xi) \; d\Delta' \tag{7.16}$$

Defining as in the classical case $\eta = CV - \dot{C}x = CV - \frac{dC}{dt} x$; $\xi = C^{-1}x$

$$f(x,v) = F(\xi,\eta) \tag{7.17}$$

and the phase space distribution is invariant as in the classical case. This is indeed the great advantage of the Wigner distribution.

The classical concepts can be used but, of course, the Wigner distribution does not obey a classical Vlasov equation.

C. Application to the Problem of Quantum Harmonic Oscillator with Varying Frequency

To end this section on quantum transformations let us consider a potential for a one-dimensional oscillator given by

$$V(x) = \frac{1}{2} \Omega^2(t) \; x^2 \tag{7.18}$$

the classical problem involves the solution of the equation

$$\frac{d^2x}{dt^2} + \Omega^2(t) \; x = 0 \tag{7.19}$$

Although the following result has already been somewhat established by Lewis and Riesenfeld [1968] let us show that if we know a solution of Eq. (7.19) we can solve the quantum case. We transform the equation through a given C(t) and from Eqs. (7.18) and (7.12) we get

$$\overline{V} = \frac{1}{2}\,\Omega^2\, C^4\, \xi^2 + \frac{1}{2}\, C^3\, \frac{d^2C}{dt^2}\, \xi^2 = \frac{1}{2}\, C^3\, \xi^2 \left[\frac{d^2C}{dt^2} + \Omega^2(t)\, C\right]$$

If we choose $\ddot{C} + \Omega^2 C = 0$, $\overline{V} = 0$ and the problem is reduced to the problem of free particles. But a difficulty occurs if $C(T) = 0$ for $t = T$. Then θ goes to infinity and we cannot pass the time $t = T$. Since another, easy to solve, problem is the quantum oscillator with a fixed frequency $\Omega^2(t)$ and ω^2 we can select C as a solution of

$$\frac{d^2C}{dt^2} + \Omega^2\, C = \frac{\omega^2}{C^3} \tag{7.20}$$

and we have

$$\overline{V} = \frac{1}{2}\,\omega^2\,\xi^2 \quad ,$$

a problem with a known solution. We must simply solve Eq. (7.20). It can be shown that if $C(0) > 0$ the solution of Eq. (7.20) never goes to zero and consequently we can obtain the solution for an arbitrarily long time.

VIII. VLASOV POISSON SYSTEM

A. One-Dimensional Problem-Beam

We go back to the Vlasov Poisson system. We begin by considering a one-dimensional beam problem B with an initial condition as indicated in Figure 6. It must be pointed out that we have put a limit V_o on the modulus of the possible velocities. Although physically such an absolute cut-off does not exist, the number of particles with velocities greater than, say, $5V_T$ is very small and can be supposed equal to zero. Moreover, the beam is limited in space from $x = -L$ to $x = L$ and is homogeneous in this interval. To compute its evolution we select a C(t) such that $C(0) = 1$ and $(dC/dt) = 0$ at $(t = 0)$. Consequently, at initial time $t = \theta = 0$ the two phase spaces x,v and ξ,η coincide. The Poisson equation (4.9) can be written

$$\frac{\partial\varepsilon}{\partial\xi} = C^3 \int F(\xi,\eta)\, d\eta - C^3\, \frac{d^2C}{dt^2} \tag{8.1}$$

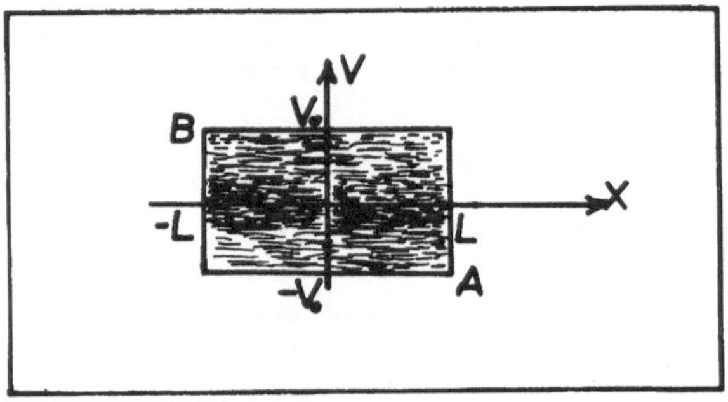

Figure 6

In the ξ plane at $\theta = 0$, $F(\xi,\eta) = F(\eta)$ for $-L < \xi < L$ and the density $\int F \, d\eta = N$.

We select C such that:

$$C^3 \left[N - \frac{d^2C}{dt^2} \right] = 0$$

with $C(0) = 1$, $dC/dt(0) = 0$. We get, of course,

$$C = 1 + \Omega^2 t^2 \quad \text{with } \Omega^2 = N/2 \quad . \tag{8.2}$$

If we remember that we took $e = m = \varepsilon_o = 1$ we see that Ω^2 is simply the plasma frequency associated with a density which is half of the initial density of the beam. Now the possibility of selecting such C(t) has two important consequences.

The new field ε is zero and since the beam is uniform it will stay zero. Being homogeneous and with no field acting on it, the plasma is on a steady state with no time evolution.

In fact, things are more complex. The steady state character of the beam implies an infinite length. If boundaries exist, as they do, we must use the concept of contamination already introduced by Burgan, Gutierrez, Fijalkow, Navet and Feix [1977]. We briefly remind the reader.

Due to the one-dimensional character of the problem, supposing moreover a space symmetry, particles outside a limited region do not create any field inside, this property remaining true in cylindrically and spherically symmetrical structures. As long as these particles do not penetrate physically into this zone they will not influence the behavior of the particles in this zone. When particles located initially in the inner and outer zones, respectively, cross a "contamination process" is generated. Obviously the evolution of the contaminated zones is given by the evolution of the two boundaries and more precisely by the symmetric trajectories of point A and B of Figure 6. Now the trajectory of point A, for example, is a simple uniform motion (in the ξ,η,θ space) with

$$\xi_A = L - V_o \theta \tag{8.3}$$

and the region $|\xi| < \xi_A$ is the uncontaminated zone. The interesting point is the following: when $t \to \infty$, $\theta \to$ limit θ_ℓ and the contamination will stop if $L > V_o \theta_\ell$ leaving the central zone $|\xi| < L - V_o \theta_\ell$ always uncontaminated. Introducing Eq. (8.2) into Eq. (4.5) we obtain the relation between θ and t

$$\Omega\theta = \frac{1}{2} \left\{ \frac{\Omega t}{1 + \Omega^2 t^2} + \text{Arc tan } \Omega t \right\} \tag{8.4}$$

and when $t \to \infty$, $\Omega\theta \to \pi/4$. The percentage of uncontaminated particles in Figure 6 is consequently $1 - \pi V_o/4\Omega L$.

Figures 7 and 8 show an illustration of this concept. It can be noticed that in the usual phase space we have a simple expansion. Figure 8 indicates that indeed η is an invariant. The computer experiments are of Lagrangian type with plane superparticles, the motion of which is computed through Newton and Coulomb laws.

B. One-Dimensional Problem-Plasma

For a plasma the corresponding Poisson equation (4.9) is written

$$\frac{\partial \varepsilon}{\partial \xi} = C^3 \int F(\xi,\eta)\, d\eta - C^4 - C^3 \frac{d^2C}{dt^2} \tag{8.5}$$

Starting from a uniform but bounded electron plasma P as in Figure 6 with an initial density n_o for the electrons and N_o for the motionless ions we select C(t) satisfying

$$C(0) = 1 \quad ; \qquad \frac{dC}{dt}(0) = 0$$

and

$$n_o = C\, N_o + \frac{d^2C}{dt^2} \tag{8.6}$$

introducing the plasma frequency $\Omega_p^2 = N_o\, e^2/m\, \varepsilon_o$ the solution of Eq. (8.6) is

$$C(t) = \frac{n_o}{N_o} + \left(1 - \frac{n_o}{N_o}\right) \cos \Omega_p\, t \tag{8.7}$$

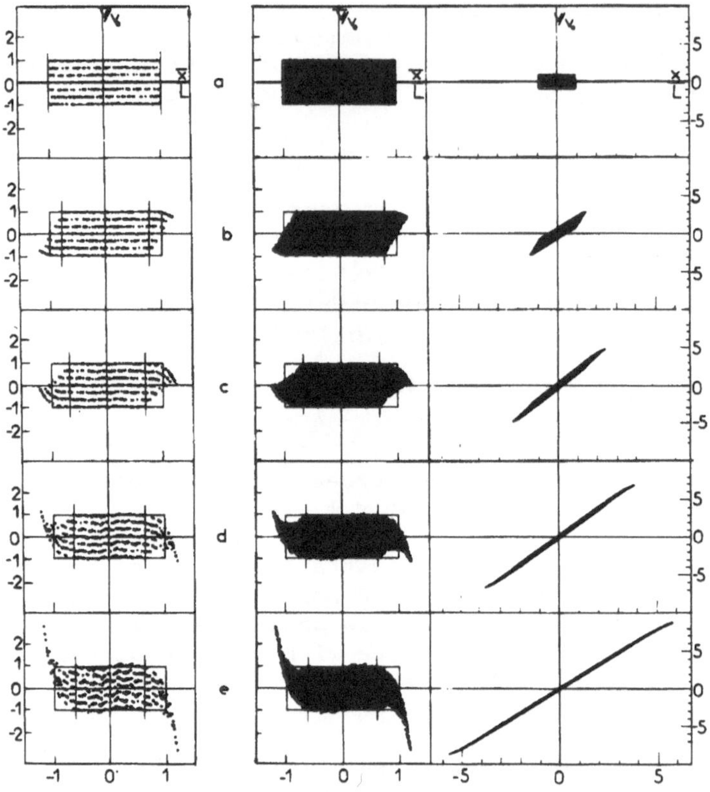

Figure 8 Figure 7

Figure 7. Evolution in the phase spaces xv (right) and $\xi\ \eta$ (left) of a horizontal water bag rod. At $t = 0$ the cut-off velocities are $\pm V_o$ and the rod length is 2L. We define Ω as the plasma frequency corresponding to the density $AV_o = 1/2$ of the density at $t = 0$. (A being the value of the phase space density inside the bag). The problem is entirely characterized by the value of the parameter $L\Omega/V_o$ (here equal to 2). Figures are respectively given for Ωt = a) 0; b) 0.5; c) 1.0; d) 1.5; e) 2.0. The uncontaminated zone is indicated only on the $\xi\eta$ phase space and is a rectangle delineated by the straight lines $\pm V_o$ and the two vertical lines whose abscisses are given by $\pm \xi_c/L = 1 - V_o/2\Omega L\ (\tau/(1 + \tau^2) + \text{Arc bg}\ \tau)$. $\tau = \Omega t$. Here for $\tau \rightarrow \infty$ the percentage of uncontaminated particles goes to $1 - \rho\ V_o/4\Omega L = 0.607\ 3$.

Figure 8. Same diagram $\xi\eta$ as in Figure 7 with indication of the evolution of seven groups of particles with initial velocities $\pm V_o$, $\pm 2V_o/3$ and 0. It can be seen that in the uncontaminated zone η is an invariant.

We see two things:

If $n_o/N_o < 1/2$, there will be a value $t = t_o$ for which C(t) will take the value 0 and for this value, θ will be infinite. We will consequently not be able to compute the behavior beyond t_o.

If $n_o/N_o > 1/2$, C(t) is always positive but does not increase with time; when $t \to \infty$, $\theta \to \infty$ and the contamination is always total, irrespective of the size of the system. The time of total contamination is L/V_o in the θ scale. For the uncontaminated zone the density varies as $n = n_o C^{-1}(t)$.

$$n(t) = \frac{n_o}{(n_o/N_o) + (1 - n_o/N_o) \cos \Omega_p t} \tag{8.8}$$

The results are very similar to the cold plasma case (3.10) with density in the center oscillating at frequency ω_p. We notice that since $C(t) \to 0$ implies $\theta \to \infty$ we always have total contamination before n(t) belows up in the case $n_o/N_o < 1/2$.

C. One-Dimensional Problem-Gravitating Gas

The case of gravitating gas with an initial situation described as in Figure 6 is similar to the B case except for a sign in front of $N = \int F d\eta$ and a solution for C(t) given by

$$C = 1 - \Omega^2 t^2 \qquad \text{with } \Omega^2 = N/2 \tag{8.9}$$

Ω^2 is now the Jean's frequency associated with a density equal to half the initial density of the system.

As in the B case the system is stationary in the ξ, η phase space with

$$\begin{aligned} \xi &= x(1 - \Omega^2 t^2)^{-1} \\ \eta &= v(1 - \Omega^2 t^2) + 2\Omega^2 t x \end{aligned} \tag{8.10}$$

η being an invariant in the uncontaminated zone. Of course the contamination becomes total in a time always smaller than Ω^{-1} since the relation between θ and t is now

$$\Omega\theta = \frac{1}{2}\left(\frac{\Omega t}{1 - \Omega^2 t^2} + \text{Arc th } \Omega t\right) \tag{8.11}$$

If 2L is the length of the system the time of total contamination is given by the solution of the equation $L/V_o = \theta$. At time t the density in the uncontaminated zone is $N(1 - \Omega^2 t^2)^{-1}$ and for a sufficiently large L a very large increase of the density is consequently possible. Figure 9 shows the situation for $L\Omega/V_o = 1$. The variance of η and the shrinking of the uncontaminated zone are well illustrated.

D. Homogeneous Systems in Cylindrical and Spherical Geometry (Beam and Gravitational Gas)

An advantage of our group transformation analysis is that the results can be generalized to higher dimensions although it must be pointed out that in order to apply the contamination concept we must deal with systems where the particles from outside do not create a field for the particle inside which implies a cylindrical and a spherical geometry in respectively two-dimensional and three-dimensional problems. N will be the initial density in the B and G cases with $N/2 = \Omega^2$ and n_o and $N_o = \Omega_p^2$ the electrons and ion density in the P case. We must select $C(t)$ satisfying, in two dimensions (cylindrical):

$$\Omega_p^2 C^2 + 2C \frac{d^2C}{dt^2} = n_o \qquad \text{P}$$

$$2C \frac{d^2C}{dt^2} = \Omega^2 \qquad \text{B}$$

$$2C \frac{d^2C}{dt^2} = -\Omega^2 \qquad \text{G} \quad , \qquad (8.12)$$

in three-dimensions (spherical):

$$\Omega_p^2 C^3 + 3C^2 \frac{d^2C}{dt^2} = n_o \qquad \text{P}$$

$$3C^2 \frac{d^2C}{dt^2} = \Omega^2 \qquad \text{B}$$

$$3C^2 \frac{d^2C}{dt^2} = -\Omega^2 \qquad \text{G} \qquad (8.13)$$

In all cases $C(t)$ and dC/dt for $t = 0$ are respectively 1 and 0 to have identity of the $\vec{x},\vec{v}$ and $\vec{\xi},\vec{\eta}$ phase spaces at $t = \theta = 0$.

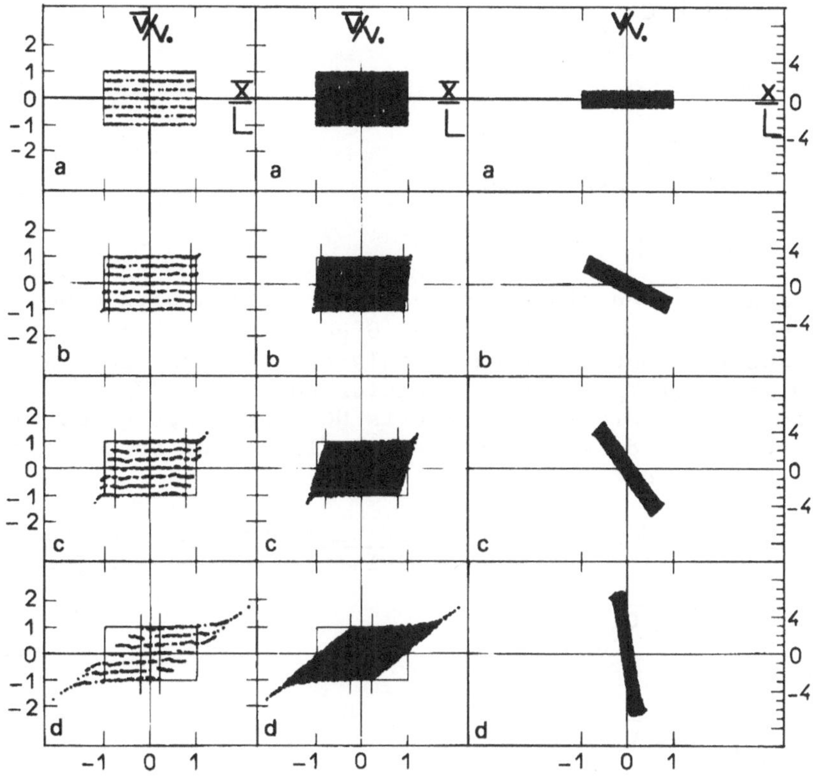

Figure 9. Evolution in the phase spaces x,v (right) and ξ η (center and left) for a limited gravitational horizontal "rod" solution. At initial time the boundaries are ($\overline{x} = \xi$; $\overline{v} = \eta$) $\overline{x} = x = \pm L$, $\overline{v} = v = \pm V_o$. Here the parameter characterizing the problem is $L\Omega/V_o = 4$. Figures, obtained using 3,000 particle codes are given respectively for $\Omega\tau$ = a) 0; b) 0.3; c) 0.6; d) 0.9. In this case, the collapse and the crossing of all particles in the neighborhood of the center bring a total contamination for $\Omega\tau = 0.92$. It can be seen (left) that, as long as the particles remain uncontaminated, η is an invariant. We indicate the evolution of seven groups of particles with initial velocities $\pm V_o$, $\pm 2/3\ V_o$, $\pm 1/3\ V_o$, and 0.

(1) The two-dimensional beam problem. In the two-dimensional B problem the relations between C, t and θ are given by the following parametric solutions:

$$\Omega t = 2 \int_0^{\sqrt{\log C}} \exp Z^2 \, dZ \tag{8.14}$$

$$\Omega\theta = 2 \int_0^{\sqrt{\log C}} \exp -Z^2 \, dZ \tag{8.15}$$

$C(t) > 1$ and $C(t)$, an always increasing function with time, goes to infinity when $t \to \infty$. Consequently, θ goes to a finite limit $\Omega\theta_\ell = \sqrt{\pi}$ and the contamination is only partial provided, of course, we can put a limit V_o on the velocities and that $R/V_o > \theta_\ell$. The density in the uncontaminated zone is NC^{-2} and this gives the law of decrease of the density under the influence of space charge and ballistic effect for a cylindrical beam.

(2) The two-dimensional gravitating problem. In the two-dimensional G problem, relations (8.14) and (8.15) are just inversed.

$$\Omega\theta = 2 \int_0^{\sqrt{-\log C}} \exp Z^2 \, dZ \tag{8.16}$$

$$\Omega t = 2 \int_0^{\sqrt{-\log C}} \exp -Z^2 \, dZ \tag{8.17}$$

$C(t) \leq 1$ and $C(t)$, an always decreasing function with time, equal zero for $\Omega t_c = \sqrt{\pi}$. Before that time, total contamination has taken place and in the uncontaminated zone we can follow the density increase in the center C^{-2} through Eq. (8.17) which gives C as a function of t.

(3) The three-dimensional beam problem. In the three-dimensional B problem, Eq. (8.13) B can be easily integrated giving

$$\Omega t = \sqrt{\frac{3}{2}} \{\sqrt{C(C-1)} + \text{Arc ch} \sqrt{C}\} \tag{8.18}$$

$$\Omega\theta = \sqrt{6} \sqrt{\frac{C-1}{C}} \tag{8.19}$$

$C(t) \geq 1$ is always increasing. θ goes to the finite limit $\sqrt{6}\,\Omega^{-1}$ and provided a sufficiently large system the contamination is partial. NC^{-3} gives the evolution of the density in the center of the spherical system where C is given as a function of time through Eq. (8.18).

(4) The three-dimensional gravitational problem. In the three-dimensional G problem, the integration is quite similar to the preceding one. We get

$$\Omega t = \sqrt{\frac{3}{2}}\,\{\text{Arc cos}\,\sqrt{C} + \sqrt{C(1 - C)}\} \tag{8.20}$$

$$\Omega\theta = \sqrt{6}\sqrt{\frac{1 - C}{C}} \tag{8.21}$$

$C(t) \leq 1$ is always decreasing. Contamination is total in a time always smaller than $\Omega t_c = \pi\sqrt{3/8}$.

E. Plasma Case: Two and Three Dimensions

The plasma case is a little bit more difficult. We must solve Eqs. (8.12) and (8.13) P. The density is given by

$$n(t) = n_o\, C^{-d} \tag{8.22}$$

d being the dimension (d = 2 or 3). Deriving Eq. (8.22) we get dn/dt and d^2n/dt^2

$$\frac{d^2n}{dt^2} = \frac{d + 1}{dn_o}\, C^d\, \left(\frac{dn}{dt} \right)^2 - dn_o\, C^{-d-1}\, \frac{d^2C}{dt^2} \tag{8.23}$$

Introducing d^2C/dt^2 taken from Eqs. (8.12) and (8.13) in Eq. (8.23),

$$\frac{d^2n}{dt^2} = \frac{d + 1}{d}\,\frac{1}{n}\,\left(\frac{dn}{dt} \right)^2 - n^2 + \Omega_p^2\, n \tag{8.24}$$

we write $P = dn/dt$

$$\frac{d^2n}{dt^2} = P\,\frac{dP}{dn} = \frac{d}{dn}\left(\frac{P^2}{2}\right) \tag{8.25}$$

Introducing Eq. (8.25) in (8.24) we obtain the first degree equation (P^2 function, n variable)

$$\frac{dP^2}{dn} = 2\,\frac{d+1}{d}\,\frac{1}{n}\,P^2 - 2n^2 + 2\Omega_p^2\,n \tag{8.26}$$

the solution of which is

$$P^2 = K_o\,n^{2(d+1)/d} - d\,\Omega_p^2\,n^2 - 2n^{2(d+1)/d}\int n^{-2/d}\,dn \tag{8.27}$$

We must distinguish between d = 2 and d = 3.

(1) Plasma case 3D. d = 3 gives no problem in the last integration of Eq. (8.27). k_o is obtained noticing that for t = 0 $n = n_o$ and $dn/dt = P = 0$.

$$K_o = 3\Omega_p^2\,n_o^{-2/3} + 6n_o^{1/3} \tag{8.28}$$

We finally obtain the relation between t and n

$$t = \int_{n_o}^{n} x^{-1}\left\{3\Omega_p^2\left[\left(\frac{x}{n_o}\right)^{2/3} - 1\right] + 6n_o\left[\left(\frac{x}{n_o}\right)^{2/3} - \frac{x}{n_o}\right]\right\}^{-1/2} dx \tag{8.29}$$

The denominator of Eq. (8.29) cancels for $n = n_o$. It is easy to find the other values for which it is zero. One is negative and without interest, the other is (taking $\Omega_p^2 = N_o = 1$),

$$n_1 = n_o\left[\frac{\sqrt{1+8n_o}+1}{4n_o}\right]^3 \tag{8.30}$$

if $n_o < 1$, $n_1 > 1$ and if $n_o > 1$, $n_1 < 1$. Consequently the density is periodic and oscillates between these two values. The period is given by twice the integral (8.29) and taken between the values n_o and n_1 (or n_1 and n_o). In contradistinction to the plane case, n_o can be as small as we like without appearance of a maximum n_1 going to infinity. (In the plane case $n_1 = n_o/(2n_o - 1)$.)

Equation (8.29) can be transformed in order to get for the period T a formula numerically more tractable. If n_o is the electron density (N_o being equal to 1) we have

$$T = 4\sqrt{\frac{3}{b+2}}\int_0^{\pi/2}\frac{\sqrt{b^2 + tg^2\,\phi}}{\sqrt{\frac{2b+d}{b+2} + tg^2\,\phi}}\,d\phi \quad ; \quad 2b = \sqrt{1+8n_o} - 1 \tag{8.31}$$

Different limiting cases are recovered: $n_o = 1$ gives $T = 2\pi$; $n_o = 0$ (or $n_o = \infty$) gives $T = \pi\sqrt{3}$.

(2) Plasma case 2D (and 4D). In Eq. (8.27) for d = 2 the integral $\int n^{-2/d}\, dn = \qquad n$, we obtain K_o through $n = n_o$, $P = 0$, for $t = 0$

$$K_o = 2\Omega_p^2\, n_o^{-1} + 2 \log n_o \tag{8.32}$$

t is obtained by integration between n_o and n

$$t = \int_{n_o}^{n} 2^{-1/2} \left\{ \Omega_p^2\, x^2 \left(\frac{x}{n_o} - 1 \right) - x^3 \log \frac{x}{n_o} \right\}^{-1/2} dx \tag{8.33}$$

As in the preceding case, the expression between brackets cancels for $x = n_o$ and $x = n_1$ (one greater, the other smaller, than N_o) the density oscillates between these two values and is periodic with the half period given by the integral (8.33) taken from n_o to n_1 (or n_1 to n_o). As in the three-dimensional cases, no restrictions are put on the value of n_o, and n remains finite. Equations (8.29) and (8.33) indicate that in cylindrical and spherical geometry the frequency of oscillation is a function of n_o in contradistinction to the result obtained for plane geometry as given by Eq. (8.8) where Ω_p is the oscillation frequency and does not depend on n_o.

Nevertheless, in the linearized limit we recover in the three cases the plasma frequency Ω_p. The equation for C can be written

$$n_o = \Omega_p^2\, C^d + d\, C^{d-1} \frac{d^2C}{dt^2} \tag{8.34}$$

We suppose that $n_o = N_o\,(1 + \varepsilon) = \Omega_p^2\,(1 + \varepsilon)$ and ε is small since n_o is very close to N_o. Introducing $\Omega_p t = \tau$ we rewrite Eq. (8.34)

$$1 + \varepsilon = C^d + d\, C^{d-1} \frac{d^2C}{d\tau^2} \tag{8.35}$$

We write $C = 1 + \psi$ and linearize neglecting terms in $\psi^2\ \psi^3$. Equation (8.34) becomes

$$\varepsilon/d = \psi + \frac{d^2\psi}{d\tau^2} \tag{8.36}$$

The solution of Eq. (8.36) with $\psi = 0$, $d\psi/d\tau = 0$ for $\tau = 0$ is

$$\psi = \varepsilon/d\,(1 - \cos\tau) \tag{8.37}$$

Introducing Eq. (8137) in $n = N_o\ (1 + \varepsilon)(1 + \psi)^{-d}$ and retaining only the first terms in ε we get

$$n = N_o\ (1 + \varepsilon \cos \tau) \tag{8.38}$$

Equation (8.38) describes the usual linearized plasma oscillations phenomena.

Finally we would like to point out a curiosity of a four-dimensional plasma (whatever it could be !). As in the preceding case, we cancel the field ε by a proper selection of the function C.

$$n_o = \Omega_p^2\ C^4 + 4C^3 \frac{d^2C}{dt^2} \tag{8.39}$$

But Eq. (8.39) can be exactly solved. Introducing the usual initial conditions C(0) = 1, dC/dt (0) = 0, we obtain

$$C = \sqrt{\frac{1 + \cos \Omega_p t}{2} + \frac{n_o}{N_o}\,\frac{1 - \cos \Omega_p t}{2}} \tag{8.40}$$

describing oscillation of electron density between the values $n = n_o$ for $t = 0$ and $n = n_o\ C^{-4}$ ($\Omega_p t = \pi$), i.e., between n_o and N_o^2/n_o. Moreover, from Eq. (8.40) it is obvious that the frequency is independent of n_o and $T = 2\pi$ for all n_o (as in the one-dimensional case) with no restriction put on the values of n_o.

It can also be noticed that the equation for C can be written

$$C + d\,\frac{d^2C}{dt^2} = \frac{n_o}{C^{d-1}} \tag{8.41}$$

If $n_o \to 0$ we neglect the right-hand-side of Eq. (8.41) and obtain a solution $C = \cos\,(t/\sqrt{d})$, of course for $t/\sqrt{d} = \pi/2$, $C \to 0$, and n_o/C^{d-1} cannot be neglected and in fact the curve becomes symmetric with respect to the axis $t = \pi\sqrt{d}/2$. As a consequence we can write for the period of plasma for $d > 1$

$$T_d(n_o \Rightarrow 0) = \pi\sqrt{d} \tag{8.42}$$

Equation (8.42) is in agreement with Eq. (8.31) in the three-dimensional case and the above-mentioned relation $T_4\ (n_o) = 2\pi$.

Again we must point out that these results are valid for homogeneous warm system in regions where contamination has not yet taken place. They generalize and give the solution of the equations given by Dawson [1959] in the cold plasma case.

IX. CONCLUSION

Several concepts have emerged from the different parts of this paper. The more important is certainly the possibility of time renormalization for partial derivative equations both in configuration and phase spaces. This time renormalization allows for certain types of equations (heat, gravitational systems in the Dirac's hypothesis, for example) an analytical expression of the asymptotic properties which are taken care of through time rescaling of the variables and functions as in more orthodox group techniques (for example, self similar transformation) while the general time evolution involves the solution of the same equation on a finite time interval, with no special treatment needed for the end point.

The second interesting concept is the force rescaling. We should really say that this force rescaling is the second aspect of the transformation, complementary to the time renormalization. In fact, we like to renormalize the time while keeping the force small. We especially want to avoid infinities near $\theta = \theta_\ell$ to keep the advantage of an easy and fast numerical treatment in the $\xi\ \eta$ space. If a physical constant ($\vec{B}(t)$ or $G(t)$) decreases quickly enough, this is possible (cases $\alpha \geq 1$ in Tables I and II).

On the other hand, $G(t)$ (for example) can have a too slow decrease to allow a time renormalization which does not introduce infinities on self-consistent or external forces. Nevertheless a time compression is quite useful from a numerical point of view. Moreover, through the introduction of the two competing fields (transformation and physical) it allows predictions on type of trajectories. In that respect the results obtained both for the magnetic and gravitational field variations for $\alpha < 1/2$ are quite interesting since we show that a rescaling of the forces giving G and B independent of time, and consequently able to support a periodic motion in the new space, is counterbalanced by a time decreasing transformed field. This shows the possibility of "adiabatic" motions for some initial conditions in the G case and no restriction on these conditions on the magnetic field cases for variation respectively slower than $t^{-1/2}$ and t^{-1}. This last result should be compared with the result obtained by Lewis [1968a] for $B(t) = At^{-n/n+1}$ (Eq. 3.15) for which the non-adiabatic effects cancel.

If the time renormalization is useful in a direct problem, its use can be crucial in an inverse problem. For example, the problem of computing an initial heat flux $\psi(x,0)$ knowing $\psi(x,T)$ is a very unstable numerical problem when T is large. It would be quite interesting to study it with the techniques introduced in Chapter III. The freedom of transferring information from different wavelength and choosing the value of θ must be of great help for the obtention of a stable numerical scheme.

The third interesting concept is the concept of contamination. This concept is useful in the theory of bounded homogeneous systems (plane or cylindrically or spherically symmetric). The homogeneous character of the system is not a completely self-preserving property since boundary inhomogeneities propagate inside the system. The determination of the uncontaminated zones in which the theory applies can be handled by the transformation. Contamination can be partial (in a homogeneous beam of dimension larger than the Debye length) or total (either in a time proportional to the length of the system (plasma) or in a finite time irrespective of the size of the system (gravitational)).

Also, the transformation solves in a rather elegant way the problem of time-dependent quantum harmonic oscillator which is directly obtained when the classical solution is known.

Finally, let us point out the general philosophy of group techniques as applied in Eqs. (2.5), (3.1) and (3.11) compared to the generalized form adopted here. In both cases time transformation of variables and functions are introduced which take care of the entire time variation in the "classical" (self similar) point of view. The price to pay is that only a subset of the general initial conditions can be treated and we generally do not know if initial conditions close to these will give solutions converging to or diverging from these singular solutions.

The great advantage is that there is no longer an explicit time-dependence (one variable less). In our treatment we keep explicit time-dependence (through the introduction of an explicit time variable). Then arbitrary initial conditions can be treated; the number of independent variables and functions is not changed but the numerical scheme is greatly simplified. If time renormalization can be introduced the asymptotic properties will be given by the transformation.

Very likely the group transformation to be used depends on the problem we want to solve and this invites more precise study of systems of equations like (2.3) or (2.5) in self-similar technique and other (nonlinear) transformations between ξ η and x v in our system.

Two final remarks are in order:

· The group technique is a methodology looking for problems

(and finding quite a few of them). The present paper is a good illustration of this philosophy.

- This methodology is both analytical and numerical with analytical studies used to simplify and make faster and more efficient the numerical work. We come back to the general preoccupations of our group as indicated in Bitoun, Nadeau, Guyard and Feix [1973] and Nadeau, Veyrier, and Feix [1976].

ACKNOWLEDGEMENTS

For fruitful discussions and cooperation on the numerical simulation work the authors wish to thank Dr. M. Navet.

We would like also to thank Miss F. Fournier for the arduous and meticulous typing of the manuscript.

REFERENCES

Baranov, V. B., 1976, Sov. Phys. Tech. Phys. 21, 720.

Bernstein, I. B., J. M. Greene and M. D. Kruskal, 1957, Phys. Rev. 108, 546.

Bitoun, J., A. Nadeau, J. Guyard and M. R. Feix, 1973, J. of Compl Phys. 12, 315-330.

Burgan, J. R., J. Gutierrez, E. Fijalkow, M. Navet, M. R. Feix, 1977, J. de Physique Lettres 38, L161.

Burgan, J. R., J. Gutierrez, E. Fijalkow, M. Navet, M. R. Feix, to be published, "Self similar solutions for Vlasov and water bag models".

Courant, E. D. and H. D. Snyder, 1958, Annals of Phys. 3, 1.

Dawson, J. M., 1959, Phys. Rev. 113, 383.

Dirac, P. A. M., 1938, Proc. Roy. Soc. A155, 199.

Dirac, P. A. M., 1973, Proc. Roy. Soc. A333, 403.

Feix, M. R., 1975, "Some problems and methods in computational plasma physics", Proc. Culham S. R. C. Symp. on Turbulence and Nonlinear Effects in Plasma, edited by B. E. Keen and E. W. Laing, pp. 139-183.

Feix, M. R., 1975, "Crossfertilization between Plasma, Stellar Dynamics and Hydrodynamics", Dynamics of Stellar Systems, edited by A. Hayli, Reidel, pp. 179-194.

Guyard, J., A. Nadeau, G. Baumann and M. R. Feix, 1971, J. Math. Phys. 12, 488.

Hsuan, H. C. S., K. E. Lonngren and W. F. Ames, 1974, J. of Eng. Math. 8, 303.

Kalman, G., 1960, Ann. Phys. 10, 1.

Lewis, H. R., 1968, J. of Math. Phys. 9, 1976.

Lewis, H. R. and W. G. Riesenfeld, 1968, J. Math. Phys. 10, 1458.

Morse, R. L. and C. W. Nielson, 1969, Phys. of Fluids 12, 2418.

Moyal, J. E., 1949, Proc. Cambridge Phil. Soc. 45, 99.

Nadeau, A., J. P. Veyrier, and M. R. Feix, April 1976, "Numerical and analytical alternative to the Krylov Bogoliubov method. Applications to slightly nonlinear automonous and mono-autonomous systems", 2nd European Conf. on Comp. Phys., edited by D. Biskampf, Munich.

Shen, H. and W. F. Ames, 1974, Phys. Lett. 49A, 313.

Van Flandern, T. C., 1974, Bull. Ann. Astr. Soc. 6, 206.

Zabusky, N. J. and M. D. Kruskal, 1965, Phys. Rev. Lett. 15, 240.

EQUATION OF STATE OF REACTING STRONGLY COUPLED PLASMAS

F. J. Rogers

University of California, Lawrence Livermore Laboratory

P. O. Box 808, Livermore, California 94550

I. INTRODUCTION

The equation of state of complex reacting mixtures of partially ionized gases is almost always obtained from a free energy model. Typically this is a pseudo ideal gas free energy minimization calculation in which a particular aspect of the interaction of an atom (or ion) with its surroundings is invoked to cut off the divergence of the atomic partition function [Mchesney, 1964]. This results in some uncertainty in the equation of state even in the zero coupling limit. The Debye Hückel electrostatic free energy is often added to the ideal gas free energy to account for ionic coupling. In more sophisticated calculations additional additive free energy terms to account for the finite size of the atoms (and ions) and other effects are also included [Graboske, Harwood and DeWitt, 1971]. These model approaches grew out of a need to obtain the equation of state of astrophysical mixtures at a time when a consistent fundamental theory was not available. Considerable progress on the theory has been made in recent years. But even now most of the work is concerned with hydrogen plasmas. The recent monograph of Ebeling, Kraeft and Kremp [1977] gives an excellent review of this literature. In the present paper we give a brief review of the theory for plasmas having $Z > 1$. A more detailed description of most of the material can be found in Rogers and DeWitt [1973] and Rogers [1974].

II. PLASMA ACTIVITY EXPANSION

A. Renormalization to Account for the Formation of Composites

The general starting place of this work is an activity expansion of the grand partition function. Several important features of the problem can be demonstrated by way of a simple example. Consider the most basic ionization problem of electrons (e) and nuclei (α) in equilibrium with one-electron composites (c), i.e.

$$e + \alpha \rightleftarrows c \quad . \tag{2.1}$$

Truncation of the activity series at two-body terms gives

$$P/kT = z_e + z_\alpha + z_e^2\, b_{ee} + 2z_e z_\alpha b_{e\alpha} + z_\alpha^2\, b_{\alpha\alpha} \quad , \tag{2.2}$$

subject to the conditions

$$\rho_i = N_i/V = z_i \frac{\partial(P/kT)}{\partial z_i} \quad , \qquad i=(e,\alpha) \tag{2.3}$$

where,

$$z_i = (2s_i + 1)\; e^{\mu_i/kT} \tag{2.4}$$

is the activity,

$$\lambda_i = (2\pi\hbar^2/M_i kT)^{1/2} \tag{2.5}$$

is the deBroglie wavelength, and the b_{ij} are second cluster coefficients. Classical approximations to b_{ee} and $b_{\alpha\alpha}$ are sufficient for the present discussion, i.e.

$$b_{ii} = 2\pi \int_0^\infty dr\; r^2 \left(e^{-z_i^2 e^2 r/kT} - 1 \right) \tag{2.6}$$

which diverges in the first three orders of perturbation theory as $r \rightarrow \infty$. The strong attraction of the electron-ion term always requires a quantum mechanical treatment. A useful expression for $b_{e\alpha}$ was obtained by Beth and Uhlenbeck [Rogers and DeWitt, 1973]:

$$b_{e\alpha} = b_{e\alpha}^b + b_{e\alpha}^f \tag{2.7}$$

where,

$$b_{e\alpha}^b = \sqrt{2}\,\lambda_{e\alpha}^3 \sum_{n\ell} (2\ell + 1)\; e^{-E_{n\ell}/kT} \tag{2.8}$$

is the bound state contribution and

$$b_{e\alpha}^f = \frac{\sqrt{2}\,\lambda_{e\alpha}^3}{\pi} \int_0^\infty dp \sum_\ell (2\ell + 1) \frac{d\delta_\ell}{dp}\; e^{-p^2/2\mu_{e\alpha}} \tag{2.9}$$

is the scattering state contribution. In Eqs. (2.8)-(2.9) $\lambda_{e\alpha}$ is the deBroglie wavelength for particles of reduced mass $\mu_{e\alpha}$, δ_ℓ is the phase shift and p is the relative momentum.

$b_{e\alpha}$ has divergences for large r similar to those of b_{ee} and $b_{\alpha\alpha}$. However it can be shown that the divergences in the bound state sum are completely compensated by the continuum state terms and have nothing to do with the large r divergence [Larkin, 1960; Kopyshev, 1969; Ebeling, 1969; Ebeling, 1974; Rogers, 1977]. After compensation the result for large r is

$$b^b_{e\alpha} = \sqrt{2}\,\lambda^3_{e\alpha} \sum_n^{n_{max}} \sum^{\ell_{max}} (2\ell + 1)(e^{-\beta E_{n\ell}} - 1 + \beta E_{n\ell})$$

$$b^f_{e\alpha} = \pi(\beta Z e^2)\, r^2 + \pi(\beta Z e^2)^2\, r - \frac{\pi(\beta Z e^2)}{3} \log\left(\frac{\lambda_{e\alpha}}{r}\right) + 0(\lambda^3_{e\alpha}\, \beta E_{1s}) \tag{2.10}$$

where $E_{n\ell}$, n_{max}, and ρ_{max} are functions of r for states whose Dehr radius is of order r, but are hydrogenic over the range of ($n\ell$) that contributes to $b^b_{e\alpha}$. The r^2 divergence in the cluster coefficients cancels out due to electrical neutrality. The second order term is the lowest order ring diagram. The classical sum over these diagrams gives the Debye-Hückel result. A quantum statistical mechanical evaluation of the ring diagrams has been given by DeWitt [1962]. Summation of the latter diagrams starting at third order gives an expression that resembles the third cluster coefficient of the dynamic screened Coulomb potential [Nakayama and DeWitt, 1964]. In the limit $\lambda/\lambda_D \to 0$ this potential goes over to the Debye-Hückel form and the latter sum gives

$$s_{ij} = b_{ij}(\lambda_D) - b^1_{ij}(\lambda_D) - b^2_{ij}(\lambda_D) \quad . \tag{2.11}$$

where $b_{ij}(\lambda_D)$ is the second cluster coefficient for a Debye potential, $b^1_{ij}(\lambda_D)$ and $b^2_{ij}(\lambda_D)$ are the first and second order perturbation terms. In this limit the many body plasma part of the problem remains classical while the few body part displays the necessary uncertainty principle effects at short distances. Alternatively we can obtain the same result by replacing the Boltzmann factors of the classical theory by Slater sums. This is the method used in the formal development.

Due to the rearrangement just mentioned, which summed certain types of diagrams from all cluster coefficients. Eq. (2.2) is replaced by a properly behaved activity series. It is

$$\frac{P}{kT} = z_e + z_\alpha + S_R + z_e^2 s_{ee} + 2z_e z_\alpha s_{e\alpha} + z_p^2 s_{\alpha\alpha} \tag{2.12}$$

where

$$S_R = 1/12\pi \lambda_D^3, \quad \lambda_D = [kT/4\pi e^2(z_e + Z^2 z_p^2)]^{1/2} \quad ,$$

is the Debye-Hückel correction obtained by a sum over the ring diagrams. Equation (2.12) includes the possibility of the formation of composites. To see how this comes about assume zero coupling to the ideal gas, so that Eq. (2.12) reduces to

$$\begin{aligned} P/kT &= z_e + z_\alpha + 2z_e z_\alpha s^b_{e\alpha} \\ &= \rho_e{}^* + \rho_\alpha{}^* + \rho_{e\alpha} \quad , \end{aligned} \tag{2.13}$$

where $s^b_{e\alpha}$ is the non-compensating part of the bound state sum similar to Eq. (2.10), but, now involving the Debye energy levels. $\rho_e{}^*$, $\rho_p{}^*$ and $\rho_{e\alpha}$ are the equilibrium numbers of free electrons, free protons, and one electron composites, respectively. Due to electrical neutrality

$$\rho_e{}^* = (Z - 1)\, \rho_\alpha + \rho_\alpha{}^*, \quad \rho_{e\alpha} = \rho_\alpha - \rho_\alpha{}^* \quad , \tag{2.14}$$

For the special case Z = 1 symmetry requires that $z_e = z_\alpha$ and it follows from Eqs. (2.3) and (2.12) that

$$z_e = z_p = \left(-1 + \sqrt{1 + 8\rho_e s^b_{e\alpha}}\right)/4s^b_{e\alpha} \tag{2.15}$$

and

$$\begin{aligned} P/kT &= 2\left[-1 + \sqrt{1 + 8\rho_e s^b_{e\alpha}}\right]/4s^b_{e\alpha} \\ &+ \left[-1 + \sqrt{1 + 8\rho_e s^b_{e\alpha}}\right]^2/8s^b_{e\alpha} \quad . \end{aligned} \tag{2.16}$$

At high temperature $s^b_{e\alpha} \to 0$ so that $PV/N_e kT \to 2$ indicating complete ionization. As $T \to 0$ $s^b_{e\alpha} \to \infty$ and $PV/N_e kT \to 1$ indicating the formation of one electron composites. The important point is that the product $2z_e z_\alpha s^b_{e\alpha}$ plays the role of an activity, $z_{e\alpha}$, for one electron composites, so that Eq. (2.12) takes the form

$$\begin{aligned} P/kT &= z_e + z_\alpha + z_{e\alpha} + S_R(z_e + Z^2 z_\alpha) \\ &+ z_e^2 s_{ee} + 2z_e z_\alpha s^f_e + z_\alpha^2 s_{\alpha\alpha} \end{aligned} \tag{2.17}$$

Since there are no two body terms involving electrons on nuclei scattering from composite ions, it is apparent that Eq. (2.17) is short of terms. These terms come from the third cluster coefficients which are composed of three conceptually different parts corresponding to: (1) the formation of two electron composites; (2) scattering of electrons and nuclei from one electron composites; (3) scattering between three unbound particles. These parts enter the activity expansion at first, second and third order in powers of the activity, respectively. The entire activity series must be renormalized on this basis. The first order terms in the revised series correspond to the Saha equation. Scattering states enter only in higher order terms. Because of this natural separation of bound and scattering state terms it is essential that effective compensations be taken into account before the separation is made. Otherwise the high temperature term will not be properly ordered, i.e. the Saha term will predict too many composites.

B. Renormalization to Account for the Plasma Coupling of Composite Particles

Equation (2.12) has another obvious shortcoming. The ring term and the Debye length that appears in the s_{ij} only involve the activities of electrons and nuclei, whereas, it is apparent that one electron composites must somehow be included. The resolution of this problem is complicated and involves finding Taylor series expansions of functions of $\lambda_D[z_e + Z^2 z_\alpha + (Z^{-1})^2 z_{e\alpha}]$ in the complete expansion Eq. (2.17). As a result of this second type of renormalization the energy levels of the Debye potential that enter $s^b_{e\alpha}$ are shifted, to first order, back to their isolated atom (ion) values, i.e.

$$E_{n\ell}(\lambda_D) = E_{n\ell}(\lambda_D) - Ze^2/\lambda_D \qquad (2.18)$$

This is consistent with a result of Jackson and Klein [1974]. The result of this renormalization which includes all terms through 5/2 powers in the activity, allowing the possibility of the formation of many particle composites, is given in Rogers [1974]. Composite particles enter this expansion similar to fundamental particles although, due to the fact that composite particle activities are coupled to the many body system through their λ_D dependence, there are some differences. If terms of type $b^b_{e\alpha}$ of Eq. (2.10) are not included in the resummation required to eliminate the scattering state divergences, an expansion in which composite activities enter exactly like fundamental particles is obtained. This will be discussed in detail elsewhere.

C. Renormalization to Account for Strong Ion Coupling

When Z >> 1 Eq. (2.12) has an additional shortcoming. This is easily seen by truncating P/kT at the Debye Hückel term and assuming there are no composites. The activities are given by [Rogers and DeWitt, 1973]

$$z_e = \rho_e\, e^{-\partial S/\partial \rho_e} = \rho_e\, e^{-\Lambda/2} \tag{2.19}$$

$$z_\alpha = \rho_\alpha\, e^{-\partial S/\partial \rho_e} = \rho_\alpha\, e^{-Z^2\Lambda/2} \tag{2.20}$$

where S is the Mayer S function and

$$\Lambda = \beta e^2/kT\ \lambda_D\ (\rho_e + z^2\ \rho_\alpha) \tag{2.21}$$

corresponds to the approximation $S = S_R(\rho_e + Z^2\ \rho_\alpha)$.

In general, Eqs. (2.19)-(2.20) cannot be used to obtain z_e and z_α since, as already discussed, when composites are formed $S_R = S_R[\rho_e{*} + Z^2\rho_2{*} + (Z-1)^2\ \rho_{e\alpha}]$ where $\rho_e{*}$, $\rho_\alpha{*}$, and $\rho_{e\alpha}$ depend on (V,T). Equations (2.19)-(2.20) show that for sufficiently large values of Z, $z_\alpha/\rho_\alpha \ll 1$ when $\Lambda \ll 1$. The expansion in powers of the activity given by Eq. (2.12) is only valid when $z_\alpha/\rho_\alpha \gtrsim 1/2$ and is not applicable to this situation. In fact it predicts that the nuclei fall out of the interaction terms altogether, whereas the density expansion predicts the electrons fall out of the problem, i.e. the Debye-Hückel pressure correction in the canonical formulism is given by

$$P_{DH} = \frac{-1}{24\pi\ \lambda_D^3} = \frac{-1}{24\pi}\ [4\pi\beta e^2(\rho_e + Z^2\rho_\alpha)]^{3/2}$$

$$= \frac{-1}{24\pi}\ (4\pi\beta e^2 Z^2\rho_\alpha)^{3/2}\ \left(1 + \frac{3}{2}\ \frac{\rho_e}{Z^2\rho_\alpha}\ ---\right) \tag{2.22}$$

Neither result is entirely correct. The difficulty is that the Debye-theory only applies to weakly correlated motion, whereas, due to the high Z the nuclear motion is strongly correlated even at very low density. By adding higher S_n corrections [Rogers, 1974] in the density expansion it is possible to show that the electron part decreases in importance as Z is increased and goes over to the result for ion mixtures in a neutralizing electron background. As a first approximation to the interaction correction at high temperature we can use the multi-component fitting formula for the Monte-Carlo results worked out by H. E. DeWitt [DeWitt and Rogers, 1976]. In order to account for the formulation of composites it is convenient to express this formula in terms of activities. To accomplish this we equate the canonical expression for the pressure to that given by the grand canonical theory,

$$\frac{P}{kT} = \rho_e{}^* + \rho_I + S(Z_{ef}^2\ \Lambda) - \rho_I\ \partial S(Z_{ef}^2\ \Lambda)/\partial\rho_I \tag{2.23}$$

$$= z_e + z_I\ g(Z_{ef}^2\ \Lambda_z) \tag{2.24}$$

where the electrons are uncoupled from the ions which are treated by a one fluid model, i.e.

$$\rho_I = \rho_\alpha{}^* + \rho_{e\alpha} + \rho_{ee\alpha} + --- \tag{2.25}$$

$$z_I = z_\alpha + z_{e\alpha} + z_{ee\alpha} + --- \tag{2.26}$$

$$Z_{ef}^2 = \langle Z\rangle^{1/3} \langle Z^{5/3}\rangle \tag{2.27}$$

$$\langle Z^n\rangle = (z_\alpha\ Z^n + z_{e\alpha}\ (Z - 1)^n + ---)/z_I \tag{2.28}$$

$$\Lambda_Z = \beta_e^2/\lambda_D\ [z_e + Z^2\ z_\alpha + (Z - 1)^2\ z_{e\alpha} + ---] \tag{2.29}$$

By using the relation

$$z_I = \rho_I\ e^{-\partial S/\partial\rho_I} \tag{2.30}$$

to eliminate z_I from Eq. (2.24) and since the electrons are uncoupled $z_e = \rho_e{}^*$, the function g can be tabulated from

$$g(Z_{ef}^2\ \Lambda_z) = -1 + \frac{\rho_I + S - \rho_I \frac{\partial S}{\partial\rho_I}}{\rho_I\ e^{-\partial S/\partial\rho_I}} \tag{2.31}$$

Equation (2.24) can now be solved in the usual way through the relation

$$\rho_e = z_e \frac{\partial(\rho/kT)}{\partial z_e}\ , \quad \rho_\alpha = z_\alpha \frac{\partial(\rho/kT)}{\partial z_\alpha} \tag{2.32}$$

The composite activities are also involved in Eq. (2.32) since they are built up from products of z_e and z_α. Since the ions are now strongly coupled to the ideal gas they have a significant effect on the ionization equilibrium and always increase the state of ionization from that obtained in a Saha calculation.

Going beyond Eq. (2.24) to obtain an expression that includes electron corrections as the temperature is reduced for a given Z, or for intermediate values of Z, is complicated. The addition of any electron coupling invalidates the relation $z_e = \rho_e^*$ and would require the tabulation of g as a function of many variables. The difficulty can be removed by a third renormalization in which the most important part of each term in the expansion of P/kT in terms of the Mayer S function, i.e.

$$\frac{P}{kT} = z_e + z_\alpha + z_{e\alpha} + \ldots + S(z_e, z_\alpha, z_{e\alpha} \ldots)$$

$$+ \frac{1}{2} \sum_{i=(e,\alpha,e\alpha,\ldots)} z_i \left(\frac{\partial S}{\partial z_i}\right)^2 + \ldots \tag{2.33}$$

are first summed together, the second most important terms summed, etc. This results in an expansion of the following form,

$$\frac{P}{kT} = z_e + z_\alpha + z_{e\alpha} + \ldots + P_1 + P_2 + \ldots \tag{2.34}$$

where

$$P_1 = S + \sum_i z_i \left(e^{(\partial S/\partial z_i)} - 1 - \frac{\partial S}{\partial z_i}\right)$$

$$P_2 = \frac{1}{2} \sum_{ij} z_i z_j \frac{\partial^2 S}{\partial z_i \partial z_j} \left(e^{(\partial S/\partial z_i)} - 1\right)\left(e^{(\partial S/\partial z_j)} - 1\right) \tag{2.35}$$

Equation (2.34) is valid over the entire range of Z and T and for all densities in the fluid phase provided phase transition or phase separation is not taking place. A full description of the results of this section including calculation for silicon at all stages of ionization will be given elsewhere.

III. CONCLUDING REMARKS

This paper has given a brief review of the current state of theoretical procedures for calculating the equation of state of complex mixtures of reacting plasmas under various conditions. No mention was made of electron degeneracy and exchange effects, diffraction corrections for λ/λ_D finite, or relativistic effects for high Z. These effects can be very important, but only quantitatively effect the discussion here. A more detailed description of much of the material can be found in the cited literature. The procedures described all start from a rigorous basis, but require approximations along the way whose effect can be evaluated. The resultant calculations are much more involved than those in the models mentioned in the beginning, but, at the least, they can be used to improve these models. This work is by no means complete and much remains to be done. A very important test of the theory, for instance, can be made bv attempting to explain recent conductivity measurements of P.P. Kulik and his collaborators [Yermokin, Kallavkin, Kovaliov, Koslov, Kulik and Pallo, 1973]. A study of the possibility of phase separation in high Z plasmas is another interesting problem.

REFERENCES

Beth, E. and G. E. Uhlenbeck, 1937, Physica 4, 915.

DeWitt, H. E., this Volume.

DeWitt, H. E., 1962, J. Math. Phys. 3, 1216.

DeWitt, H. E. and F. J. Rogers, 1976, Lawrence Livermore Laboratory Report UCRL-50028.

Ebeling, W., 1969, Ann. Physik 22, 33, 383, 392.

Ebeling, W., 1974, Physica 73, 573, 593.

Ebeling, W., W. D. Kraeft, D. Kremp, 1977, Theory of Bound States and Ionization Equilibrium in Plasmas and Solids, Akademie-Verlag, Berlin.

Graboske, H. C., D. J. Harwood and H. E. DeWitt, 1971, Phys. Rev. A3, 1419.

Jackson, J. L. and L. S. Klein, 1969, Phys. Rev. 177, 352.

Kopyshev, V. P., 1969, Sov. Phys. JETP 28, 684.

Larkin. A. I., 1960, Sov. Phys. JETP 11, 1363.

Mchesney, M., 1964, Can. J. of Phys. 42, 2473.

Nakayama, T. and H. E. DeWitt, 1964, J. Quant. Spectrosc. Radiat. Transfer 4, 623.

Rogers, F. J., 1974, Phys. Rev. A10, 2441.

Rogers, F. J., 1977, Phys. Lett. 61A, 358.

Rogers, F. J. and H. E. DeWitt, 1973, Phys. Rev. A8, 1061.

Yermokin, N. Y., V. M. Kallavkin, B. M. Kovaliov, A. M. Koslov, P. P. Kulik and A. V. Pallo, 1973, 11th Int'l Conf. on Phenomena in Ionized Gases, Prague, p. 422.

INDEX